Lecture Notes in Bioinformatics 4414

Subseries of Lecture Notes in Computer Science

T0181585

Sepp Hochreiter   Roland Wagner (Eds.)

Bioinformatics
Research
and Development

First International Conference, BIRD 2007
Berlin, Germany, March 12-14, 2007
Proceedings

 Springer

Volume Editors

Sepp Hochreiter
University of Linz, Institute of Bioinformatics
Altenbergerstr. 69, 4040 Linz, Austria
E-mail: hochreit@bioinf.jku.at

Roland Wagner
University of Linz, Institute of FAW
Altenbergerstrasse 69, 4040 Linz, Austria
E-mail: rrwagner@faw.uni-linz.ac.at

Library of Congress Control Number: 2007921873

CR Subject Classification (1998): H.2.8, I.5, J.3, I.2, H.3, F.1-2

LNCS Sublibrary: SL 8 – Bioinformatics

ISSN   0302-9743
ISBN-10   3-540-71232-1 Springer Berlin Heidelberg New York
ISBN-13   978-3-540-71232-9 Springer Berlin Heidelberg New York

Typesetting: Camera-ready by author, data conversion by Scientific Publishing Services, Chennai, India
Printed on acid-free paper   SPIN: 12029112   06/3142   5 4 3 2 1 0

Preface

This volume contains the papers which were selected for oral presentation at the first Bioinformatics Research and Development (BIRD) conference held in Berlin, Germany during March 12-14, 2007. BIRD covers a wide range of topics related to bioinformatics like microarray data, genomics, single nucleotide polymorphism, sequence analysis, systems biology, medical applications, proteomics, information systems.

The conference was very competitive. From about 140 submissions only 36 were selected by the Program Committee for oral presentation at BIRD and for publication in these proceedings. The acceptance rate was 1/4. The decisions of the Program Committee were guided by the recommendations of several reviewers for each paper. It should be mentioned that these proceedings have companion proceedings published by the Austrian Computer Society where selected poster presentations of the BIRD conference are included.

The invited talk titled "From Flies to Human Disease" by Josef Penninger, one of the leading researcher in genetic experiments for investigating disease pathogenesis, was very inspiring and gave new insights into future bioinformatics challenges.

Many contributions utilized microarray data for medical problems or for modeling biological networks. In the field of systems biology new approaches were introduced, especially for flux distribution, for regulatory modules where the inverse problem theory is applied, and for inferring regulatory or functional networks. Several contributions are related to medical problems: therapy outcome prognosis, SNP data, genome structure, genetic causes of addiction, HIV virus, MEDLINE documents, analyzing TP53 mutation data, investigation satellite-DNA data, or testing bioactive molecules. Another main topic was sequence analysis, where contributions propose new coding schemes, new profiles, and new kernels and where Markovian approaches for genome analysis and protein structure were developed. A major research field was proteomics, which can be sub-divided into measurement- and structure-related issues. Another research field focused on databases, text retrieval, and Web services.

The first BIRD conference was a great success to which many persons contributed. The editors want to thank first of all the authors, whose work made this volume possible. Then we want to thank once again the invited speaker, Josef Penninger. We thank especially the Program Committee members, who ensured the high quality of the conference and the publications in this volume, provided their experience and expert knowledge, and invested time to select the best submission. The conference was only possible by the excellent support of the local Organizing Committee Camilla Bruns and Klaus Obermayer (Technical University of Berlin), whom we want to thank. Many thanks also to the President of the Technical University of Berlin, Kurt Kutzler, for hosting this conference. We thank the Web administrators (A. Anjomshoaa, A. Dreiling, C. Teuschel) who took care of the online submission process and registration. Thanks to the "Arbeitskreis Bioinformatik" of the Austrian Computer Society for the support of the BIRD conference. Many thanks also to Noura Chelbat, Djork-Arne Clevert, and Martin Heusel for their support in setting up this first BIRD conference.

Most of all we wish to express our gratitude to Gabriela Wagner (DEXA Society), who managed the conference, which includes organizing the submission and the review process, setting up and coordinating the decisions of the Program Committee, being responsible for the final paper versions, planning and supervising the technical program schedule including the banquet, looking after the editorial and printing issues, and many more. Last but not least our thanks to Rudolf Ardelt, Rector of the Johannes Kepler University of Linz, Austria, who enabled this conference by establishing Bioinformatics at the JKU Linz.

January 2007   Sepp Hochreiter
                                                               Roland Wagner

Organization

Program Committee

General Chair:
Roland Wagner, FAW, University of Linz, Austria

Program Chair:
Sepp Hochreiter, University of Linz, Austria

Local Organizing Committee:
Klaus Obermayer, TU Berlin, Germany
Camilla Bruns, TU Berlin, Germany

Program Committee (Steering):
Amos Bairoch, Swiss Institute of Bioinformatics, Switzerland
Pierre Baldi, University of California, Irvine, USA
Philip E. Bourne, UCSD, USA
Bruno Buchberger, RISC, Linz
Cornelius Frömmel, Georg-August-Universität Göttingen, Germany
David Gilbert, University of Glasgow, UK
Amarnath Gupta, SDSC of the University of California San Diego, USA
Knut Reinert, FU Berlin, Germany
Erik Sonnhammer, Stockholm University, Sweden
Hershel Safer, Weizmann Institute of Science, Israel

Poster Session Chair:
Josef Küng, FAW, Austria
Jürgen Palkoska, FAW, Austria

Program Committee:
Jörg Ackermann, Univesity of Applied Sciences, Berlin, Germany
Tatsuya Akutsu, Kyoto University, Japan
Mar Alba, IMIM Barcelona, Spain
Patrick Aloy, University of Barcelona, Spain
Francois Artiguenave, INRA, France
Brian Aufderheide, Keck Graduate Institute of Applied Life Sciences, USA
Joel S. Bader, Johns Hopkins University, USA
Amos Bairoch, Swiss Institute of Bioinformatics, Switzerland
Pierre Baldi, University of California, Irvine, USA
Simon Barkow, ETH Zurich, Switzerland
Leonard Barolli, Fukuoka Institute of Technology, Japan
Christian Bartels, Novartis Pharma AG, Switzerland
Khalid Benabdeslem, LORIA & IBCP, France

Paul Bertone, EMBL-EBI, UK
Howard Bilofsky, University of Pennsylvania, USA
Christian Blaschke, Bioalma Madrid, Spain
Jacek Blazewicz, Poznan University of Technology, Poland
Andreas M. Boehm, Rudolf Virchow Center for Experimental Biomedicine,
    Germany
Uta Bohnebeck, TTZ Bremerhaven, Germany
Philip E. Bourne, UCSD, USA
Eric Bremer, Children's Memorial Hospital Chicago, USA
Bruno Buchberger, University of Linz, Austria
Philipp Bucher, Swiss Institute of Bioinformatics, Switzerland
Gregory Butler, Concordia University, Canada
Rita Casadio, University of Bologna, Italy
Michele Caselle, INFN , Italy
Sònia Casillas Viladerrams, Universitat Autònoma de Barcelona, Spain
Alistair Chalk, RIKEN Yokohama Institute, Genome Sciences Center, Japan
Noura Chelbat, University of Linz, Austria
Phoebe Chen, Deakin University, Australia
Francis Y.L. Chin, The University of Hong Kong, Hong Kong
Jacques Colinge , University of Applied Sciences Hagenberg, Austria
James M. Cregg, Keck Graduate Institute of Applied Life Sciences, USA
Charlotte Deane, Oxford University, UK
Coral del Val Muñoz, University of Granada, Spain
Sabine Dietmann, GSF - National Research Center for Environment and Health,
    Germany
Zhihong Ding, University of California, Davis, USA
Anna Divoli, SIMS UC Berkeley, USA
Arjan Durresi, Louisiana State University, USA
Jacob Engelbrecht, KVL - The Royal Veterinary and Agricultural University,
    Denmark
Nick Fankhauser, University of Bern, Switzerland
Domenec Farre, Center for Genomic Regulation, Spain
Pedro Fernandes, Inst.Gulbenkian de Ciência, Portugal
Milana Frenkel-Morgenstern, Weizmann Institute of Science, Israel
Rudolf Freund, European Molecular Computing Consortium, Austria
Christoph M. Friedrich, Fraunhofer SCAI, Germany
Holger Froehlich, University of Tübingen, Germany
Cornelius Frömmel, Georg-August-Universität Göttingen, Germany
Georg Fuellen, Ernst-Moritz-Arndt-University Greifswald, Germany
Toni Gabaldon, CIPF, Bioinformatics Department, Spain
Michal J. Gajda, InternationalInstitute of Molecular nd Cell Bology, Poland
David J. Galas, Battelle Memorial Institute, USA
Barbara Gawronska, School of Humanities and Informatics, Skövde, Sweden
Dietlind Gerloff, The University of Edinburgh, UK
David Gilbert, University of Glasgow, UK
Mark Girolami, University of Glasgow, UK
Aaron Golden, National University of Ireland, Ireland

Joaquin Goni, University of Navarra, Spain
Pawel Gorecki, Warsaw University, Poland
Gordon Gremme, Zentrum für Bioinformatik, Universität Hamburg, Germany
Roderic Guigo, IMIM Barcelona, Spain
Amarnath Gupta, SDSC of the University of California San Diego, USA
Hendrik Hache, Max Planck Institute for Molecular Genetics, Germany
John Hancock, MRC Mammalian Genetics Unit, UK
Artemis Hatzigeorgiou, Penn Center forBioinformatcs, USA
Amy Hauth, Université de Montréal, Canada
Marti Hearst, SIMS, UC Berkeley , USA
Jaap Heringa, Free University, Amsterdam, The Netherlands
Javier Herrero, EMBL-EBI, UK
Hanspeter Herzel, Humboldt University Berlin , Germany
Winston Hide, University of the Western Cape, South Africa
Des Higgins, University College Dublin, Ireland
Rolf Hilgenfeld, University of Lübeck, Germany
Ralf Hofestaedt, University of Bielefeld, Germany
Robert Hoffmann, Memorial Sloan-Kettering Cancer Center, USA
Hermann-Georg Holzhütter, Humboldt-Universität Berlin, Charité, Germany
Chun-Hsi Huang, University of Conneticut, USA
Tao-Wie Huang, National Taiwan University, Taiwan
Dan Jacobson, National Bioinformatics Network, South Africa
Markus Jaritz, Novartis Institutes for Biomedical Research GmbH & Co KG,
    Austria
Rafael C. Jimenez, University of the Western Cape, South Africa
Jaap Kaandorp, University of Amsterdam, The Netherlands
Lars Kaderali, Deutsches Krebsforschungszentrum, Germany
Alastair Kerr, University of Edinburgh, UK
Ju Han Kim, Seoul National University College of Medicine, Korea
Sung Wing Kin, National University of Singapore, Singapore
Juergen Kleffe, Charite, Germany
Erich Peter Klement, University of Linz, Austria
Ina Koch, Technical University of Applied Sciences Berlin, Germany
Martin Krallinger, National Center of Cancer Research (CNIO), Spain
Roland Krause, Max Planck Institute for Infection Biology, Germany
Judit Kumuthini, Cranfield University, UK
Josef Küng, University of Linz, Austria
Tony Kusalik, University of Saskatchewan, Canada
Michael Lappe, Max Planck Institute for Molecular Genetics, Germany
Pedro Larranaga, UPV/EHU, Spain
Gorka Lasso-Cabrera, University of Wales, Swansea, UK
Jorge Amigo Lechuga, Centro Nacional de Genotipado, Spain
Doheon Lee, KAIST, Korea
Sang Yup Lee, KAIST, Korea
Marc F. Lensink, SCMBB, Belgium
Christina Leslie, Columbia University, USA
Dmitriy Leyfer, Gene Network Sciences, Inc., USA

Guohui Lin, University of Alberta, Canada
Michal Linial, The Hebrew University of Jerusalem, Israel
Stefano Lise, University College London, UK
Mark Lohmann, University of Cologne, Germany
Francois Major, Université de Montréal, Canada
Paul Matthews, The European Bioinformatics Institute, Hinxton, UK
Patrick May, Konrad-Zuse-Zentrum Berlin, Germany
Shannon McWeeney, Oregon Health & Science University, USA
Engelbert Mephu Nguifo, Université d'Artois, France
Henning Mersch, Forschungszentrum Jülich, Germany
Satoru Miyano, University of Tokyo, Japan
Francisco Montero, Universidad Complutense Madrid, Spain
Yves Moreau, Katholieke Universiteit Leuven, Belgium
Robert F. Murphy, Carnegie Mellon University, USA
Sach Mukherjee, University of California, Berkeley, USA
Tim Nattkemper, University of Bielefeld, Germany
Jean-Christophe Nebel, Kingston University, UK
Steffen Neumann, Leibniz-Institut für Pflanzenbiochemie, Germany
See Kiong Ng, Institute for Infocomm Research, Singapore
Pierre Nicodeme, CNRS-LIX, France
Mario Nicodemi, Universitá di Napoli "Federico II", Italy
Grégory Nuel, CNRS, France
Klaus Obermayer, Technical University Berlin, Germany
Baldo Oliva, IMIM Barcelona, Spain
Bjorn Olsson, University of Skovde, Sweden
Philipp Pagel, GSF - National Research Center for Environment and Health,
     Germany
Jürgen Palkoska, University of Linz, Austria
Utz J. Pape, Max Planck Institute for Molecular Genetics, Germany
Jose M. Peña, Linköping University , Sweden
Graziano Pesole, University of Bari and Istituto Tecnologie Biomediche del C.N.R.,
     Italy
Francisco Pinto, ITQB, University of Lisbon, Portugal
Thomas Ploetz, University of Dortmund, Germany
Adam Podhorski, CEIT, Spain
Heike Pospisil, University of Hamburg, Germany
Jagath Rajapakse, Nanyang Technological University, Singapore
Ari J Rantanen, University of Helsinki, Finland
Axel Rasche, Max Planck Institute for Molecular Genetics, Germany
Thomas Rattei, Technical University of Munich, Germany
Animesh Ray, Keck Graduate Institute, USA
Dietrich Rebholz, European Bioinformatics Institute, UK
Knut Reinert, FU Berlin, Germany
Stefan Rensing, University of Fribourg, Germany
Mark Robinson, University of Hertfordshire, UK
Peter Robinson, Humboldt-Universität, Germany
Hugues Roest Crollius, Dyogen, CNRS, France

Paolo Romano, National Cancer Research Institute (IST), Italy
Elsbeth Roskam, BIOBASE GmbH, Germany
Cristina Rubio-Escudero, University of Granada, Spain
Manuel Rueda, IRBB, Parc Cientific de Barcelona, Spain
Rob Russell, EMBL, Germany
Kim Rutherford, FlyMine (An Integrated DB for Drosophila and Anopheles
    Genomics), UK
Hershel Safer, Weizmann Institute of Science, Israel
Yasubumi Sakakibara, Keio University, Japan
Guido Sanguinetti, University of Sheffield, UK
David Sankoff, University of Ottawa, Canada
Clare Sansom, Birkbeck College, UK
Ferran Sanz Carreras, Biomedical Research Park of Barcelona, Spain
Reinhard Schneider, EMBL , Germany
Michael Schroeder, Technical University of Dresden, Germany
Joerg Schultz, University of Würzburg, Germany
Angel Sevilla, Universidad de Murcia, Spain
Ian Shadforth, Bioinformatics - Cranfield Health, UK
Amandeep S. Sidhu, University of Technology, Sydney, Australia
Florian Sohler, University of Munich, Germany
Erik Sonnhammer, Stockholm University, Sweden
Rainer Spang, Max Planck Institute for Molecular Genetics, Germany
William Stafford Noble, University of Washington, USA
Fran Supek, University of Zagreb, Croatia
Ashish V Tendulkar, Kanwal Rekhi School of Information Technology, India
Todt Tilman, HAN University, The Netherlands
A Min Tjoa, University of Technology Vienna, Austria
Andrew Torda, University of Hamburg, Germany
Todd Treangen, Universitat Politecnica de Catalunya, Spain
Oswaldo Trelles, University of Malaga, Spain
Michael Tress, CNIO, Spain
Elena Tsiporkova, R&D Group, Flemish Radio & Television, Belgium
Koji Tsuda, Max Planck Institutes Tübingen, Germany
Dave Ussery, The Technical University of Denmark, Denmark
P. Van der Spek, Erasmus MC, The Netherlands
Jean-Philippe Vert, Center for Computational Biology, Ecole des Mines de Paris,
    France
Allegra Via, University of Rome "Tor Vergata", Italy
Jordi Villà i Freixa, Research Unit on Biomedical Informatics (GRIB) of IMIM/UPF,
    Spain
Pablo Villoslada, University of Navarra, Spain
Susana Vinga, INESC-ID, Portugal
Dirk Walther, Max Planck Institute for Molecular Plant Physiology, Germany
Lusheng Wang, City University of Hong Kong, Hong Kong
Georg Weiller, Australian National University, Australia
Bertram Weiss, Schering AG, Germany
David Wild, Keck Graduate Institute, USA

Table of Contents

Session 4: Medical, SNPs, Genomics II

Session 5: Systems Biology

Session 6: Sequence Analysis I (Coding)

Session 7: Sequence Analysis II

Session 8: Proteomics I

Session 9: Proteomics II (Measurements)

Session 10: Proteomics III (Structure)

Session 11: Databases, Web and Text Analysis

Bayesian Inference of Gene Regulatory Networks Using Gene Expression Time Series Data

Nicole Radde[1] and Lars Kaderali[2]

[1] Center for Applied Computer Science (ZAIK), University of Cologne
Weyertal 80, 50931 Köln, Germany
[2] German Cancer Research Center (dkfz)
Im Neuenheimer Feld 580, 69120 Heidelberg, Germany
`radde@zpr.uni-koeln.de`, `l.kaderali@dkfz-heidelberg.de`

Abstract. Differential equations have been established to model the dynamic behavior of gene regulatory networks in the last few years. They provide a detailed insight into regulatory processes at a molecular level. However, in a top down approach aiming at the inference of the underlying regulatory network from gene expression data, the corresponding optimization problem is usually severely underdetermined, since the number of unknowns far exceeds the number of timepoints available. Thus one has to restrict the search space in a biologically meaningful way.

We use differential equations to model gene regulatory networks and introduce a Bayesian regularized inference method that is particularly suited to deal with sparse and noisy datasets. Network inference is carried out by embedding our model into a probabilistic framework and maximizing the posterior probability. A specifically designed hierarchical prior distribution over interaction strenghts favours sparse networks, enabling the method to efficiently deal with small datasets.

Results on a simulated dataset show that our method correctly learns network structure and model parameters even for short time series. Furthermore, we are able to learn main regulatory interactions in the yeast cell cycle.

Keywords: gene regulatory network, ordinary differential equations, network inference, Bayesian regularization, *Saccharomyces cerevisiae*, cell cycle.

One of the major challenges in systems biology is the development of quantitative models to infer interactions between various cell components. The aim is to understand the temporal response of a system to perturbations or varying external conditions and to provide a detailed description of the system's dynamic behavior. For this task, ordinary differential equations (ODEs) have been established in recent years to model gene regulatory – or, more general, biochemical – networks, since they are particularly suited to describe temporal processes in detail (see, e.g. [24]). ODEs can show a rich variety of certain dynamic behaviors like convergence to a steady state after perturbation, multistationarity, switch-like behavior, hysteresis or oscillations. All these phenomena can be observed in

S. Hochreiter and R. Wagner (Eds.): BIRD 2007, LNBI 4414, pp. 1–15, 2007.

biological systems as well. Furthermore, when based on chemical reaction kinetics, model parameters correspond directly to reaction rates, for example in binding or degradation processes. Thus model parameters have a biological meaning and might be veryfied in separate experiments for model evaluation. When analyzing an ODE model, one can take advantage of the profound theory in the research field of differential equations (see, e.g. [11,17]) and use available software tools (for example [7]) for an analysis of the system. Thus, ODEs are well suited to analyze a dynamical system from a theoretical point of view. However, there are several complications in practice.

In a top down approach, the network inference problem is usually formulated as an optimization problem of an objective function depending on model parameters and experimental data. A conventional objective function is the sum of squared errors between measurements and model predictions, which is minimized with respect to model parameters. However, this often turns out to be an ill-posed problem, since the number of parameters to be estimated far exceeds the number of time points available for this purpose. In addition, gene expression data, which are most often used for this task, are very noisy due to the variability within and between cells, and due to the complicated measurement process. Typically, classical regression methods fail on such data, since they overfit the data. One has to restrict the solution space in order to get biologically meaningful results.

The easiest way to do this is to reduce the dimension of the search space by including only a few components into the learning process or choosing a simple model class to keep the number of unknowns small. This is not always possible, because many regulatory subsystems involve a lot of different components which are important and cannot be neglected in the model. Also the predictive power of simple models is often too limited. Boolean networks for instance are useful to give a qualitative insight into the processes, but cannot make quantitative statements. In the field of differential equations, simple linear relations between regulator and regulated components are sometimes used, such that every interaction is described with one single parameter. These models can provide a good approximation when the system under consideration works near an operating point, but they are not suited to give a comprehensive view of the whole system under various conditions [12]. Obviously, nonlinear phenomena cannot be captured, but transcriptional regulations as well as protein-protein interactions and most other biological interactions are known to be highly nonlinear.

If we do not want to reduce the number of parameters drastically in advance, we have to narrow the domain in the search space. That can be done by modifying the objective function or by introducing constraints to the optimization problem. All these constraints should be biologically motivated and reflect prior knowledge about the system. A reasonable way to integrate prior information into the network inference process is a Bayesian learning approach, which has been demonstrated by a couple of authors recently [3,4,19].

Beal *et al.* [3] inferred regulatory interactions from expression data by maximization of the marginal likelihood with a modification of the EM algorithm.

They use a linear state space model, a class of dynamic Bayesian networks, and include hidden factors into the model, e.g. genes of which the expression values are not measured. A linear Bayesian network was also used by Rogers and Girolami [19], who inferred gene regulatory networks with knockout experiments and a hierarchical prior distribution over edge weights which limits the number of regulators for each gene. So far, they have only evaluated their method on simulated data. Bernard and Hartemink [4] infer dynamic Bayesian networks and designed a prior distribution using information about transcription factor binding sites. They showed that the number of timepoints needed for network inference can drastically be reduced with their prior distribution.

In this paper, we introduce a method to infer gene regulatory networks with a Bayesian regularized ODE model, which is based on chemical reaction kinetics, and thus contains nonlinear regulation functions. We discretize the model and embed it into a probabilistic framework. The structure of the network and the model parameters are learned by maximization of the posterior probability, using suitable prior distributions over network parameters. In particular, we introduce a prior distribution over the network structure that favors sparse networks with only a few significant interactions between network components. This is achieved by a hierarchical prior distribution with independent normal distributions with mean zero over edge weights. The corresponding variances follow in turn a gamma distribution.

The paper is organized as follows: In Section 1, we introduce our ODE model and the Bayesian learning framework. Results for a simulated and a real dataset on the yeast cell cycle are shown in Section 2, which are discussed in Section 3. The paper concludes with a summary and an outlook.

1 Methods

1.1 Model

A gene regulatory network can be represented as a directed graph $G = (V, E)$ with sign labeled edges. The set $V = \{v_1, \ldots, v_n\}$ of nodes correspond to genes and an edge from node j to node i indicates that the product of gene j positively or negatively regulates the expression of gene i via binding to the promoter of gene i. A regulation function $r f_{ij}(x_j)$ is assigned to every edge e_{ij} in the network and describes the effect on the concentration of gene product i, denoted with x_i, in dependence of the regulator's concentration x_j. This effect can be activating or inhibiting. The total effect on i is assumed to be the sum of the single effects, implying the assumption that all regulators act independently. This is a rough, but frequently made simplification. We write the temporal change of x_i as

$$\dot{x}_i(t) = s_i - \gamma_i x_i(t) + \sum_{i=1}^{n} r f_{ij}(x_j) =: f_i(x(t)). \tag{1}$$

Here, s_i and γ_i denote basic synthesis and degradation rates that determine the dynamic behavior of component i when the concentrations of all regulators are zero. The regulation functions are parameterized using sigmoidal functions of the form

$$rf_{ij}(x_j) = k_{ij} \frac{x^{m_{ij}}}{x_j^{m_{ij}} + \theta_{ij}^{m_{ij}}} \tag{2}$$

with edge weights k_{ij} which indicate the maximal influence of component j on i, and Hill-coefficient m_{ij}. The parameter θ_{ij} is a threshold that determines a minimum concentration for the regulator to have a significant influence on the regulated component. Equation (2) can be derived from chemical reaction kinetics considering the binding process of a transcription factor to a specific binding site as a reversible chemical reaction in equilibrium [13,25]. Such a regulation function is of course a simplification, but in contrast to a simple linear relation it provides stability of the model due to it's boundedness. See also [8,9,18] for more details on the model used.

1.2   Bayesian Regularization

We apply a Bayesian approach to infer model parameters from time series expression data. To simplify notation, let ω be a parameter vector containing all parameters of our model. These are synthesis and degradation rates for every node, as well as weights k_{ij}, Hill coefficients m_{ij} and threshold values θ_{ij} for each gene pair (i, j). For numerical integration of the ODE system, we use a simple Euler discretization, where the time derivative on the left hand side is approximated with a difference quotient, and add normally distributed noise with mean 0 and variance σ_ξ to the output. This noise term reflects our assumption that the data are not biased and that the noise results from several independent sources. According to this, the concentration vector at time $t + \Delta t$ is a normally distributed random variable with variance σ_ξ and a mean that is determined by the ODE model:

$$x_i(t + \Delta t) = \underbrace{x_i(t) + \Delta t \cdot f_i(\omega, x(t))}_{h_i(\omega, x(t))} + \eta \tag{3}$$

The hypothesis $h_i(\omega, x(t))$ is the most probable value for $x_i(t + \Delta t)$, which is disturbed by the noise term $\eta \sim \mathcal{N}(0, \sigma_\xi)$.

Assuming that noise terms for different time points and different network components are not correlated, the likelihood function $\mathcal{L}$ is a product of normal distributions:

$$\mathcal{L}(\omega, x(t)) = p(D \mid \omega) \tag{4}$$

$$= \prod_{i=1}^{n} \prod_{t=1}^{T-1} \frac{1}{\sqrt{2\pi\sigma_\xi^2}} \exp\left[-\frac{1}{2\sigma_\xi} \left(h_i(\omega, x(t)) - x_{i,exp}(t + \Delta t) \right)^2 \right] \tag{5}$$

The likelihood function can be used as an objective function for parameter estimation, but is inappropriate for large networks, since it tends to overfit the data. Thus we use Bayes' Theorem to refine the objective function. A point estimate for the parameter vector ω, denoted with $\hat{\omega}$, is then obtained via maximizing the a posterior probability (MAP), that is the probability for a certain vector ω, given the data D:

$$\hat{\omega}_{MAP} = arg \max_{\omega} p(\omega \mid D) = arg \max_{\omega} \frac{p(D \mid \omega)p(\omega)}{p(D)}. \tag{6}$$

The posterior distribution is proportional to the product of the likelihood function and the prior distribution $p(\omega)$ over network parameters. Note that the denominator $p(D)$ is independent of ω and can be neglected in a maximization. The maximum a posterior estimator is computationally much less expensive than an estimate that takes the whole shape of the posterior distribution into account, which can be achieved using Monte Carlo methods (see e.g. [15]).

The prior distribution $p(\omega)$ reflects knowledge we may have about the system modeled, such as knowledge from biological experts or general insights about the structure of gene regulatory networks. The aim is to reduce the search space drastically in comparison with the maximum likelihood estimation (MLE). We use gamma distributions

$$g(x) = \lambda^{\gamma} x^{\gamma-1} \exp(-\lambda x) \Gamma^{-1}(\gamma) \tag{7}$$

for synthesis- and degradation rates, which make sure that the corresponding values neither become negative nor too large. For the weights k_{ij} we choose a hierarchical distribution which assures that only a few edges have significantly high values, whereas most of the edges have weights close to zero. This is achieved with independent normal distributions for k_{ij} with mean 0 and standard deviations which are drawn from a gamma distribution:

$$p(k) = \prod_{i=1}^{n} \int_{0}^{\infty} \mathcal{N}(0, \sigma_{ij}) g(\sigma_{ij}) d\sigma_{ij} \tag{8}$$

Figure 1 shows a scheme of the resulting distribution for k_{ij} in a two dimensional example. As can be seen, solutions where only one of the edge weights is distinct from zero are more likely than solutions where both are nonzero. This reflects our expectation of sparse networks. A similar prior distribution has recently been used on regression parameters in a Bayesian approach to predict survival times of cancer patients from gene expression data [14]. As in the network inference problem, the aim here is to drive the predictor to sparse solutions, since only few genes should exhibit a significant influence on survival.

We use a conjugate gradient descent to minimize the negative logarithm of the posterior distribution (see [14] for details).

2 Results

2.1 Simulated Data

Data Simulation: We simulated a network of seven genes with interaction graph $G = (V, E)$ as shown in Figure 2. This network is designed according to

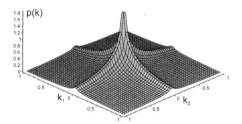

Fig. 1. The two dimensional hierarchical prior distribution with parameters $\lambda = 1$ and $\gamma = 1.0001$ for edge weights k_{ij}

the regulatory network in yeast and contains many negative and positive feedback loops and selfregulations, which can cause oscillating behavior or multistationarity [10,22]. Parameters used in the simulation were $s_i = 1, \gamma_i = 0.1, k_{ij} = \pm 2, \theta_{ij} = 5$ and $m_{ij} = 2$. Data simulation was carried out with time steps $\Delta t = 1$. Initial concentrations were drawn from a uniform distribution over the interval $[0, 5]$, and three time points per initial concentration vector were simulated according to the discretized model (3). We varied the number of timepoints (tp) used for learning and the noise level σ_ξ.

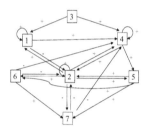

Fig. 2. Structure of the simulated network

Parameter Estimation: We carried out a conjugate gradient descent to solve the optimization problem (6). The thresholds θ_{ij} and the Hill coefficients m_{ij} were fixed to values $m_{ij} = 2$ and $\theta_{ij} = 5$, respectively, since they are numerically hard to learn. To test how strongly results depend on these parameters, we compared several different values, with no significant differences in the results. Further details of the optimization process are listed in the Appendix.

Table 1 shows mean squared errors per estimated model parameter, i.e synthesis and degradation rates s_i and γ_i and edge weights k_{ij}, for MLE in comparison with the MAP estimation for different noise levels and varying number of timepoints. It can be observed that the MAP estimate outperforms MLE especially for higher noise levels.

Network Structure: Since our method aims at the inference of the network structure and the estimation of parameters simultaneously, we performed a receiver operator characteristic (ROC) analysis to score the inferred regulations.

Table 1. Mean squared errors for model parameters. The number of timepoints used for learning is denoted with tp, σ_ξ is the noise level. From left to right: a. Errors for synthesis rates in % b. Errors for degradation rates in % c. Errors for edge weights, absolute values.

$tp\backslash\sigma_\xi$	0.5	2	3
40 (MLE)	4	64	144
40 (MAP)	15	36	67
70 (MLE)	3	50	114
70 (MAP)	6	19	72

$tp\backslash\sigma_\xi$	0.5	2	3
40 (MLE)	77	124	278
40 (MAP)	9	6	13
70 (MLE)	12	198	447
70 (MAP)	9	9	8

$tp\backslash\sigma_\xi$	0.5	2	3
40 (MLE)	1.2	19.3	43.5
40 (MAP)	1.3	1.4	1.4
70 (MLE)	1.5	23.4	52.7
70 (MAP)	1.2	1.2	1.7

Therefore, we set a threshold value t for absolute values of the edge weights k_{ij} and assume a regulation from j to i to be present if $\| k_{ij} \| \geq t$. ROC curves are achieved by varying t from 0 to $\max\{k_{ij} \mid i,j = 1,\ldots,n\}$ and calculating sensitivity and specificity for the corresponding network structures.

Figure 3 shows ROC curves for noise levels $\sigma_\xi = \{2,3\}$ and 40 and 70 timepoints, respectively. As a high sensitivity and specificity are desireable, a good classifier's ROC curve would be positioned in the upper left corner of the graph. Guessing would on average lead to a diagonal where sensitivity equals 1-specificity.

Figure 3 indicates that MLE fails in case of 40 timepoints, as the corresponding ROC curves are not better than choosing edges randomly, whereas MAP is able to infer parts of the network structure. With 70 timepoints, both approaches perform better compared to the results with 40 timepoints, and also here MAP outperforms MLE.

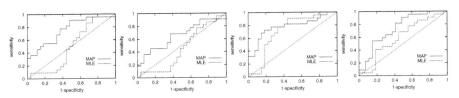

Fig. 3. ROC curves for the Bayesian and the ML approach. From *left* to *right*: a. $\sigma_\xi = 2, tp = 40$ b. $\sigma_\xi = 3, tp = 40$ c. $\sigma_\xi = 2, tp = 70$ d. $\sigma_\xi = 3, tp = 70$.

Figure 4 shows inferred networks for the ML (*left*) and MAP (*right*) approach for a noise level $\sigma_\xi = 2$ and 70 timepoints. The 17 edges with highest weights are marked in bold. Solid lines indicate true positives, dashed lines false positives, and thin lines false negatives. Both approaches reveal many regulations, 12 of 17 edges are true positives for MLE, 14 are found in the MAP approach.

The quality of a ROC curve can be represented by the area under the curve (AUC). This value is independent of the threshold t. The AUC lies between 0.5 and 1 and increases with increasing performance of the classifier. Random guessing whether an edge is present or not would on average yield an AUC value of 0.5. AUC values for MLE and MAP can be seen in Figure 5 for different noise

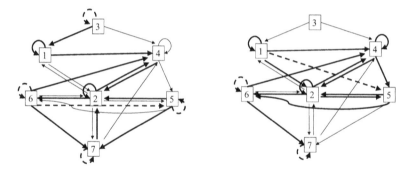

Fig. 4. 17 identified regulations with highest edge weights for the MLE (*left*) and the MAP (*right*) approach. True posititives are marked in *bold*, false positives are marked with *bold dashed lines*, false negatives correspond to *thin lines*. 70 timepoints were used for learning, noise level $\sigma_\xi = 2$.

levels and 20, 40 and 70 timepoints. Shown is the area under the ROC curve under various conditions. The left plot shows how performance deteriorates with increasing noise levels, for a fixed dataset size of 40 timepoints. The right plot shows how performance increases with increasing number of timepoints, for a fixed noise level of $\sigma_\xi = 2$.

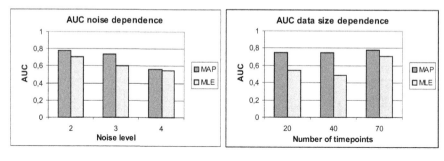

Fig. 5. AUC values for the ML and MAP approach, with respect to varying noise levels (*left*, with 70 timepoints) and varying number of timepoints (*right*, with noise $\sigma_\xi = 2$)

Although ROC analysis is a very coarse-grained method to evaluate the performance of our approach, since it does not consider exact parameter values and even neglects values for several model parameters as synthesis or degradation rates, the AUC value turns out to be a good measure of the method's overal performance. Performance in ROC analysis correlates very well with mean squared errors of learned network parameters and also with predictive power of the inferred model.

We conclude from our analysis of the simulated dataset that a Bayesian approach with appropriate prior distributions improves network inference compared to a ML approach, especially in case of noisy datasets with only a few timepoints, typical for microarray measurements. Moreover, as seen in Figure 5, it reduces the minimal number of timepoints needed to draw meaningful conclusions. In the following, we will evaluate the performance of the approach presented on a real biological dataset.

2.2   The Yeast Cell Cycle

The cell cycle is one of the best known regulatory mechanisms. A proper function of the cell cycle machinery is essential for the organisms to survive. Dysfunctions often lead to programmed cell death or to phenotypes that are not able to survive a long time and show significant changes in the cell cycle. As the cell cycle is highly conserved among eucaryotes, many key regulatory processes in yeast are also found in higher organisms and it is often possible to draw conclusions from yeast experiments to higher eucaryotes. Surveys of control mechanisms of the yeast cell cycle can be found in [2,6].

We applied our approach to a dataset of Spellman *et al.* [21], who measured gene expression values of the whole yeast genome. They conducted a cluster analysis and reported approximately 800 genes to be cell cycle regulated. The dataset consists of log ratios between synchronized cells and control experiments and contains 69 timepoints in total. These are divided into four different time series, which arise due to four cell synchronzation methods used.

We analyzed measurements of eleven genes including cln1, cln2, cln3, clb5, clb6, cdc20, cdc14, clb1, clb2, mcm1 and swi5, which are known to be involved in the cell cycle. The reference network in Figure 6 was used for evaluation and is a reduction of the regulatory network specified in Li *et al.* [16]. Details about the interactions can be found in Table 2 in the Appendix. Nodes that contain more than one gene were represented by their means, missing values were replaced by means of concentrations of consecutive and subsequent time points. Figure 7 shows ROC curves for MLE and MAP estimation. The corresponding AUC values are 0.61 and 0.68, respectively. The parameters used for the optimization are listed in the Appendix. This plot shows that both, MLE and MAP, are better than guessing and reveal some of the main regulatory interactions. Inferred

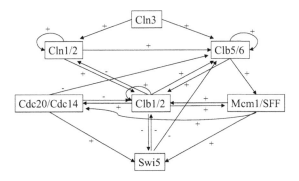

Fig. 6. Regulatory network of the yeast cell cycle which was used as a reference to evaluate our results. A descriptions for each interaction is given in Table 2.

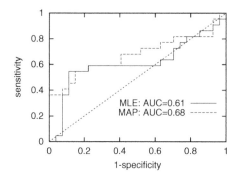

Fig. 7. Receiver operator characteristics for the regulatory network of the yeast cell cycle

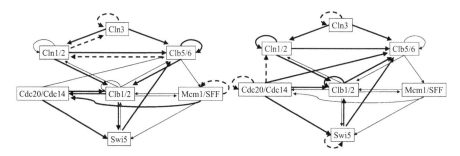

Fig. 8. 16 identified regulations with highest edge weights for the MLE (*left*) and the MAP (*right*) approach. True posititives are marked in *bold*, false positives are marked with *bold dashed lines*, false negatives correspond to *thin lines*.

networks for MLE (*left*) and MAP (*right*) can be seen in Figure 8. The 16 edges with highest inferred weights are shown in bold. True positives are indicated by continuous bold lines, dashed bold lines correspond to false positives, thin lines are referenced in literature, but were not revealed in our approach. Many true regulations are revealed in both approaches. The MAP estimate identifies more true regulations than MLE between different genes. Interestingly, it reports a couple of artificial selfregulations, the reason for this requires further investigation.

3   Discussion

We have evaluated our method on a simulated and a real-world dataset. Results on the simulated data show that the method is able to infer the underlying regulatory network from time series data, even in the typical case of only few time points and a high level of noise. We are able to estimate model parameters such as synthesis and degradation rates or interaction strenghts, which have direct biological relevance and may be validated experimentally. The Bayesian method is appropriate for sparse datasets, when additional information about the system can be exploited. This is typically the case for regulatory systems within a cell. In general, all kinds of information can be modeled as a prior probability distribution over network parameters, such that it is possible to integrate multiple data sources into the learning process.

Results on the yeast cell cycle show that analysis of a real-world dataset is far more complex than simulated data. The reasons for this are manifold. For example, our model considers only regulations at the transcriptional level, where the regulation affects the amount of RNA, which is assumed to be a measure for the corresponding protein. This is a necessary restriction when inferring regulations from microarray data, as these data only show transcriptional regulations. However, the yeast cell cycle involves several kinds of different interactions like for example protein modifications like phosphorylation, labeling for degradation or complex formation of different proteins. Many of these regulations are not visible in microarray data, and one would have to include other data sources like information about binding sites or protein-protein interactions to discover them.

Both, the ODE model and the Bayesian regularization can be extended to capture also posttranscriptional regulations, but this leads automatically to some other problems. Phosphorylation of a protein for example is much faster than transcriptional regulation, such that an extended model would include differential equations on different time scales, leading to a stiff system. For such systems, explicit numerical solvers like the Euler method often fail and more involved implicit methods are required, which need a lot more computing time.

Nonlinear relations are much harder to learn than linear ones. For the results presented here, we have fixed the two nonlinear parameters θ_{ij} and m_{ij} and only estimated the linear parameters s_i, γ_i and k_{ij}. It is possible to include both parameters into the learning process as well, and numerical problems are avoided when choosing very strict prior distributions for them. This is possible for both parameters. Most Hill-coefficients are known to be in a range between 2 and 4 [20]. When working with normalized data, it is also reasonable to fix θ_{ij} to a value somewhere around the mean, reflecting the assumption that it is not much higher than the actual concentrations of network components.

A technical issue is the choice of proper parameters for the distributions over model parameters. Although the hierarchical prior distribution already simplifies choosing the variance parameter of the normal distribution on k_{ij}, since instead of a fixed value just a general distribution has to be specified, this is still not straightforward. There is a clear trade-off between bias and variance in the choice of these parameters, and careful engineering is required at this point.

4 Conclusions and Outlook

We have presented a Bayesian approach to infer gene regulatory networks from time series expression data. Our method is appropriate for sparse and noisy data, as overfitting is prevented by suitably chosen prior distributions over network parameters. A specifically designed prior distribution over edge weights favours sparse networks with only a few edges having significant weights. The method shows good performance on simulated data and outperforms maximum likelihood estimation in case of few timepoints and noisy data, the typical setting with gene expression datasets. We are able to estimate parameters of our quantitative ODE model and to infer the structure of the underlying regulatory network. We were furthermore able to reconstruct main regulations of the yeast cell cycle network.

As quantitative models of gene networks deepen the understanding of cellular processes at the genomic level, a lot of efforts are currently under way to estimate model parameters from experimental data. Most of the time, the quantity and quality of these data do not ad hoc allow for a detailed quantitative modeling approach, making a restriction of the solution space unavoidable. A Bayesian learning approach provides an appropriate framework for this problem. It can include various sources of information about the system under consideration into the inference process. Furthermore, by modeling knowledge or expectations as probability distributions, we can also express how sure we are about this prior information. This makes the Bayesian approach a very attractive method to get a comprehensive survey of complicated regulation processes within a cell, that can only be gained involving several data sources which provide information about different regulation levels.

Future work will encompass the inclusion of information on posttranscriptional regulation into the learning program. Similarly, transcription factor binding sites can be evaluated to modify prior distributions for network interaction parameters. Such information is easily modeled into the prior distribution on the network parameters, and can be expected to yield significantly improved results. Similarly, topological features of regulatory networks required to enable them to show certain behaviors such as oscillations or multistationarity observed in real networks can be enforced through appropriate prior distributions. The Bayesian approach presented here will be extended and adapted accordingly in the future.

Acknowledgments. We would like to thank the BMBF for funding this work through the Cologne University Bioinformatics Center (CUBIC).

References

1. Alberts,B., Johnson,J., Lewis,J., Raff,M., Roberts,K., Walker,P.: Molecular Biology of the Cell. Garland Publishing (2002)
2. Baehler,J.: Cell-cycle control of gene expression in budding and fission yeast. Annu.Rev.Genet.39 (2005) 69–94
3. Beal,M.J., Falciani,F., Ghahramani,Z., Rangel,C. and Wild,D.L.: A bayesian approach to reconstructing genetic regulatory networks with hidden factors. Bioinformatics 21(3) (2005) 349–356
4. Bernard,A. and Hartemink,A.J.: Informative structure priors: joint learning of dynamic regulatory networks from multiple types of data. Pacific Symposium on Biocomputing (2005) 459–470
5. Chen,L., Aihara,K.: A Model of Periodic Oscillation for Genetic Regulatory Systems. IEEE Transactions on Circuits and Systems 49(10) (2002) 1429–1436

6. Chen,C.C., Calzone,L., Csikasz-Nagy,A., Cross,F.R., Novak,B., Tyson,J.J.: Integrative Analysis of Cell Cycle Control in Budding Yeast. Mol.Biol.Cell 15(8) (2004) 3841–3862

7. Ermentrout,B.: Simulating, Analyzing and Animating Dynamical Systems: A Guide to XPPAUT for Researchers and Students. Soc. for Industrial & Applied Math. edn.1 (2002)

8. Gebert,J., Radde,N., Weber,G.-W.: Modeling gene regulatory networks with piecewise linear differential equations. To appear in EJOR, Chall.of Cont.Opt. in Theory and Applications (2006)

9. Gebert,J., Radde,N.: Modeling procaryotic biochemical networks with differential equations. in AIP Conference Proc.839 (2006) 526–533

10. Gouze,J.L.: Positive and negative circuits in dynamical systems. J.Biol.Sys.6(21) (1998) 11–15

11. Guckenheimer,J., Holmes,P.: Nonlinear Oscillations, Dynamical Systems, and Bifurcations of Vector Fields. New York: Springer Series (1983)

12. Gustafsson, M. et al.: Constructing and analyzing a large-scale gene-to-gene regulatory network - Lasso constrained inference and biological validation. IEEE/ACM Trans. Comp. Biol. Bioinf.2 (2005) 254–261

13. Jacob,F., Monod,J.: Genetic regulatory mechanisms in the synthesis of proteins. J.Mol.Biol.3 (1961) 318–356

14. Kaderali,L., Zander,T., Faigle,U., Wolf,J., Schultze,J.L., Schrader,R.: CASPAR: A Hierarchical Bayesian Approach to predict Survival Times in Cancer from Gene Expression Data. Bioinformatics 22 (2006) 1495–1502

15. Kaderali,L.: A hierarchical Bayesian approach to regression and its application to predicting survival times in cancer. Shaker Verlag (2006).

16. Li,F., Long,T., Lu,Y., Ouyang,Q. and Tang,C.: The yeast cell-cycle is robustly designed. PNAS 101(14) (2004) 4781–4786

17. Luenberger,D.G.: Introduction to Dynamic Systems. John Wiley & Sons (1979)

18. Radde,N., Gebert,J. and Forst,C.V.: Systematic component selection for gene network refinement. Bioinformatics 22(21) (2006) 2674-2680.

19. Rogers,S. and Girolami,M.: A bayesian regression approach to the inference of regulatory networks from gene expression data. Bioinformatics 21(14) (2005) 3131–3137

20. Savageau,M. and Alves,R.: Tutorial about "Mathematical Representation and Controlled Comparison of Biochemical Systems". ICMSB2006, Muenchen (2006)

21. Spellman,P.T., Sherlock,G. *et al.*: Comprehenssive Identification of Cell Cycle-regulated Genes of the Yeast *Saccharomyces cerevisiae* by Microarray Hybridization. Mol.Biol.Cell 9 (1998) 3273–3297

22. Thomas, R.: On the relation between the logical structure of systems and their ability to generate mutliple steady states or sustained oscillations. Springer Series in Synergetics 9 (1981) 180–193

23. Tyson,J.J., Csikasz-Nagy,A., Novak,B.: The dynamics of cell cycle regulation. BioEssays 24 (2002) 1095–1109

24. Voit,E.: Computational Analysis of Biochemical Systems. Cambridge University Press, Cambridge

25. Yagil,G., Yagil,E.: On the relation between effector concentration and the rate of induced enzyme synthesis. Biophys.J.11(1) (1971) 11–27

Appendix

Details of the Optimization Procedure

Gradient descent was started with $s_i = \gamma_i = 0.1$. All edge weights k_{ij} were initially set to 0. For the simulated network, hyperparameters for gamma distributions over synthesis and degradation rates were set to $\gamma_s = 2$, $\lambda_s = 1$, $\gamma_\gamma = 1.0001$ and $\lambda_\gamma = 2$. Parameters for the gamma distribution over standard deviations σ_{ij} in equation (8) were set to $\gamma_\sigma = 1.2$ and $\lambda_\sigma = 1.5$. The corresponding parameters for the yeast cell cycle network were set to $\gamma_s = \gamma_\gamma = 0.01$, $\lambda_s = \lambda_g = 0.1$, $\gamma_\sigma = 1.7$ and $\lambda_\sigma = 5$. Thresholds and Hill coefficients were set to $\theta_{ij} = 1$ and $m_{ij} = 2$ for $i, j = 1, \ldots, n$.

Regulations of the Yeast Cell Cycle

Table 2. Regulations during the yeast cell cycle. Most descriptions and corresponding references are listed in the supplement of Li *et al.* [16].

Regulation	Description
Cln3→Cln1/2	The Cln3/Cdc28-complex activates the transcription factor complex SBF by phosphorylation. SBF acts as a transcription factor for Clb1,2.
Cln3→Clb5/6	The transcription factor complex MBF is activated by Cln3 and facilitates transcription of Clb5,6.
Cln1/2→Cln1/2	Cln1/2 activates its own transcription factor complex SBF.
Cln1/2→Clb5/6	Cln1/2-Cdc28-complex triggers degradation of Sic1, which in turn inactivates the Clb5/6-Cdc28 complex.
Cln1/2→Clb1/2	Twofold regulation: Cln1/2-Cdc28-complex inhibits Cdh1, which accelerates degradation of Clb1/2. Moreover, Cln1/2-Cdc28 phosphorylates Sic1 for degradation, which in turn inactivates Clb1/2-Cdc28 by binding.
Clb1/2→Cln1/2	Clb1/2 inactivates the transcription factor complex SBF, which triggers transcription of Cln1/2
Clb1/2→Clb1/2	Twofold regulation: Clb1/2 phosphorylate Sic1, which in turn inactivates Clb1/2. Additionally, Clb1/2 inactivates Cdh1 by phosphorylation, which in turn controls degradation of Clb1/2.
Clb1/2→Clb5/6	Clb1/2 phosphorylates Sic1, which inactivates the Clb5/6-Cdc28-complex. Maybe there is a further negative regulation from Clb1/2 to MBF, the transcription factor of Clb5/6.
Clb1/2→Mcm1/SFF	Clb1/2 phosphorylates the complex.
Clb1/2→Swi5	Clb1/2 phosphorylates Swi5 such that it cannot enter the nucleus.
Clb1/2→Cdc20	Ccdc20 is a subunit of the anaphase promoting complex (APC), which is activated by Clb1/2 via phosphorylation.
Clb5/6→Clb1/2	Twofold regulation: Clb5/6 phosphorylate Sic1 and Cdh1, which both inhibit Clb1/2.
Clb5/6→Clb5/6	Clb5/6 phosphorylates Sic1, Sic1 in turn inhibits the Clb5/6-Cdc28-complex.
Clb5/6→Mcm1/SFF	Clb5/6 initiate DNA replication. In this phase G2/M, Mcm1/SFF is activated via binding of Ndd1.
Cdc20&Cdc14→Clb1/2	Direct interaction: Cdc20/APC degrades Clb1/2. Twofold indirect interaction: Cdc14 can dephosphorylate and thus activate Sic1 and Cdh1, which inhibit Clb1/2.
Cdc20&14→Clb5/6	Cdc20 presents Clb5/6 to the APC for ubiquination. Moreover, Cdc14 dephosphorylates Sic1, which inhibits Clb5/6.
Cdc14→Swi5	Cdc14 dephosphorylates and thus activates Swi5.
Mcm1/SFF→Clb1/2	Mcm1/SFF is the transcription factor of Clb1/2.
Mcm1/SFF→Cdc20	Transcription control is assumed to depend on Mcm1/SFF.
Mcm1/SFF→Swi5	Mcm1/SFF is the transcription factor of Swi5.
Swi5→Clb1/2	Swi5 is the transcription factor of Sic1, which inhibits Clb1/2.
Swi5→Clb5/6	Swi5 is the transcription factor of Sic1, which inhibits Clb5/6.

Biological Network Inference Using Redundancy Analysis

Patrick E. Meyer, Kevin Kontos, and Gianluca Bontempi

ULB Machine Learning Group
Computer Science Department
Université Libre de Bruxelles
1050 Brussels – Belgium
{pmeyer,kkontos,gbonte}@ulb.ac.be
http://www.ulb.ac.be/di/mlg/

Abstract. The paper presents MRNet, an original method for inferring genetic networks from microarray data. This method is based on
Maximum Relevance – Minimum Redundancy (MRMR), an effective
information-theoretic technique for feature selection.

MRNet is compared experimentally to Relevance Networks (RelNet)
and ARACNE, two state-of-the-art information-theoretic network inference methods, on several artificial microarray datasets. The results
show that MRNet is competitive with the reference information-theoretic
methods on all datasets. In particular, when the assessment criterion attributes a higher weight to precision than to recall, MRNet outperforms
the state-of-the-art methods.

1 Introduction

The data flood phenomenon that biology is experiencing has propelled scientists
toward the view that biological systems are fundamentally composed of two
types of information: genes, encoding the molecular machines that execute the
functions of life, and transcriptional regulatory networks (TRNs), specifying how
and when genes are expressed [1].

Two of the most important challenges in computational biology are the extent to which it is possible to model these transcriptional interactions by large
networks of interacting elements and the way that these interactions can be effectively learnt from measured expression data [2]. The reverse engineering of
TRNs from expression data alone is far from trivial because of the combinatorial nature of the problem and the poor information content of the data [2].
However, progress has been made over the last few years and effective methods
have been developed. Well-known state-of-the-art methods used for TRN inference are *Boolean Network models* [3,4,5], *Bayesian Network models* [6,7,8] and
Association Network models [9,10,11,12].

This paper will focus on information theoretic approaches which typically rely
on the estimation of mutual information [9,12] from expression data in order to
measure the statistical dependence between genes.

S. Hochreiter and R. Wagner (Eds.): BIRD 2007, LNBI 4414, pp. 16–27, 2007.
© Springer-Verlag Berlin Heidelberg 2007

Information-theoretic network inference methods have recently held the attention of the bioinformatics community also for very large networks [9,12]. This paper introduces an original information-theoretic method, MRNet, inspired by a recently proposed feature selection algorithm, the Maximum Relevance – Minimum Redundancy (MRMR) algorithm [13,14]. This algorithm has been used with success in supervised classification problems to select a set of non redundant genes which have explanatory power for the targeted phenotype [14,15]. The MRMR selection strategy consists in selecting a set of variables that both have high mutual information with the target variable (maximum relevance) and are mutually maximally dissimilar (minimum redundancy). The advantage of this approach is that redundancy among selected variables is avoided and that the trade-off between relevance and redundancy is properly taken into account.

The proposed MRNet strategy consists in (i) formulating the network inference problem as a series of input/output supervised gene selection procedures where each gene in turn plays the role of the target output, and (ii) adopting the MRMR principle to perform the gene selection for each supervised gene selection procedure. The rationale is that the MRMR selection relies on a square matrix of bivariate mutual information that can be computed once for all, making the network inference computationally affordable for large numbers of genes.

The paper benchmarks MRNet against two state-of-the-art information-theoretic network inference methods, namely RelNet and ARACNE. The comparison relies on six different artificial microarray datasets obtained with two different generators.

The outline of the paper is as follows. Section 2 reviews the two state-of-the-art network inference techniques based on mutual information. Section 3 introduces our original approach based on MRMR. The experimental framework and the results obtained on artificially generated datasets are presented in Sects. 4 and 5, respectively. Section 6 concludes the paper.

2   Information-Theoretic Network Inference

Let $X_i \in \mathcal{X}$, $i = 1, \ldots, n$, be a discrete random variable denoting the expression level of a gene, where $\mathcal{X}$ is the set of expression levels of all genes and n is the number of genes. Network inference methods based on information theory [9,12] rely on the availability of the mutual information matrix (MIM) whose generic element

$$a_{ij} = I(X_i; X_j) = \sum_{x_i \in \mathcal{X}} \sum_{x_j \in \mathcal{X}} p(x_i, x_j) \log \left(\frac{p(x_i, x_j)}{p(x_i)p(x_j)} \right) \tag{1}$$

is the mutual information between the genes X_i and X_j. This measure has the important advantage of making no assumptions about the form of dependence (e.g. linear) between variables.

2.1 Relevance Network (RelNet) Models

The Relevance Network [9] approach consists in inferring a genetic network where each pair of genes $\{X_i, X_j\}$ is linked by an edge whose weight is the mutual information $I(X_i; X_j)$. A threshold value I_0 is used to infer the network by eliminating all edges whose weights are beyond that threshold. As a result, the complexity of this method is $O(n^2)$ since all pairwise interactions are considered. This method was successfully applied in [16] to infer relationships between RNA expression and chemotherapeutic susceptibility.

2.2 ARACNE

The Algorithm for the Reconstruction of Accurate Cellular Networks (ARACNE) method [12] is based on the Data Processing Inequality [17]. This inequality states that, if gene X_1 interacts with gene X_3 through gene X_2, then

$$I(X_1; X_3) \leq \min\left[I(X_1; X_2), I(X_2; X_3)\right] \ . \tag{2}$$

Like RelNet, the ARACNE inference procedure begins by assigning to each pair of nodes a weight equal to the mutual information. It then scans all triplets of nodes and removes the weakest edge of each triplet, which is interpreted as a redundant indirect relationship. Eventually, a threshold value is used to eliminate the weakest edges. Note that the method is proved to recover the underlying network (from the exact MIM) provided that the network is a tree and has only pairwise interactions. ARACNE's complexity is $O(n^3)$ since the algorithm considers all triplets of genes. Note that in [12], the method has been able to recover components of the TRN in mammalian cells. Furthermore, ARACNE has shown higher experimental performances than Bayesian Networks and Relevance Networks [12].

3 Our Proposal: Minimum Redundancy Network (MRNet) Models

We propose to infer a network using the Maximum Relevance – Minimum Redundancy (MRMR) feature selection method. The idea consists in performing a series of supervised MRMR gene selection procedures where each gene in turn plays the role of the target output.

The MRMR method has been introduced in [13,14] together with a best-first search strategy for performing filter selection in supervised learning problems. Consider an input/output learning task where the output is denoted by Y and V is the set of input variables. The method aims at selecting a set $S \subset V$ of inputs that have high mutual information with the target Y (maximum relevance) and low mutual information between them (minimum redundancy). The method is initialised by setting $S = X_0$ where $X_0 = \arg\max_{X_i \in V} I(X_i; Y)$. Then for a given set S of selected variables, the criterion updates S by choosing the variable

$X_j \in V \setminus S$ that maximises the score $s_j = u_j - z_j$, where u_j is a relevance term and z_j is a redundancy term. More precisely,

$$u_j = I(X_j; Y) \tag{3}$$

is the mutual information of X_j with the target variable Y, and

$$z_j = \frac{1}{|S|} \sum_{X_k \in S} I(X_j; X_k) \tag{4}$$

expresses the average redundancy of X_j to the already selected variables $X_k \in S$. At each step of the algorithm, the variable

$$X_j^{\mathrm{MRMR}} = \arg \max_{X_j \in V \setminus S} (u_j - z_j) \tag{5}$$

is selected. This variable is expected to allow an efficient trade-off between relevance and redundancy. It has been shown in [18] that the MRMR criterion is an optimal "pairwise" approximation of $I(X_i; X_j|S)$.

The MRNet approach consists in repeating the selection procedure for each target gene by putting $Y = X_i$ and $V = \mathcal{X} \setminus \{X_i\}$, $i = 1, \dots, n$, where $\mathcal{X}$ is the set of expression levels of all genes. For each pair $\{X_i, X_j\}$, the MRMR method returns a score according to (5) where $Y = X_i$. A specific network can then be inferred by deleting all edges whose score lies below a given threshold I_0.

Note that this approach infers directed networks. For comparison purposes, however, the direction of the edges is ignored.

An effective implementation of MRMR can be performed by using the approach proposed in [19]. This implementation demands a $O(f \times n)$ complexity for selecting f features using a best first search strategy. It follows that MRNet requires a $O(f \times n^2)$ complexity since the feature selection step is repeated for each of the n genes. This means that the method can infer edges in a network with a complexity ranging between $O(n^2)$ and $O(n^3)$ according to the value of f. Note that the lower this value, the lower the number of incoming edges per node to infer and consequently the lower the resulting complexity.

4 Experiments

The experimental framework consists of four steps (see Fig. 1): the artificial network and data generation, the computation of the mutual information matrix, the inference of the network and finally the validation of the results. This section details each step of the approach.

4.1 Network and Data Generation

In order to assess the results returned by our algorithm and compare it to other methods, we benchmarked it on artificially generated microarray datasets. This

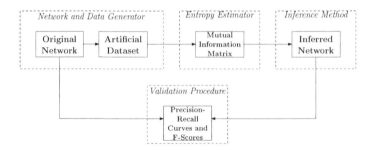

Fig. 1. An artificial microarray dataset is generated from an original network. The inferred network can then be compared to this *true* network.

approach allows us to compare an inferred network to the *true* underlying network, i.e. the one used to (artificially) generate the microarray dataset (see Fig. 1).

We used two different generators of artificial gene expression data, specifically the data generator described in [20], hereafter referred to as the sRogers generator, and the SynTReN generator [21]. The two generators, whose implementations are freely available on the World Wide Web, are sketched in the following paragraphs.

sRogers generator. The sRogers generator produces the topology of the genetic network according to an approximate power-law distribution on the number of regulatory connections out of each gene. The normal steady-state of the system is evaluated by integrating a system of differential equations. The generator offers the possibility to obtain $2k$ different measures (k wild-type and k knock-out experiments). These measures can be replicated R times, yielding a total of $N = 2kR$ samples. After the optional addition of noise, a dataset containing normalised and scaled microarray measurements is returned.

SynTReN generator. The SynTReN generator generates a network topology by selecting subnetworks from *E. coli* and *S. cerevisiae* source networks. Then, transition functions and their parameters are assigned to the edges in the network. Eventually, mRNA expression levels for the genes in the network are obtained by simulating equations based on Michaelis-Menten and Hill kinetics under different conditions. As for the previous generator, after the optional addition of noise, a dataset containing normalised and scaled microarray measurements is returned.

Generation. We generated six datasets, varying the number of genes and samples, using both generators (see Table 1 for details).

4.2 Mutual Information Matrix and Inference Methods

Each mutual information measure is computed using the Miller-Madow entropy estimator [22]. In order to use that estimator, the data are first quantised with

Table 1. The six artificial datasets generated, where n is the number of genes and N is the number of samples

Dataset	Generator	Topology	n	N
dR1	sRogers	power-law tail	2000	1000
dR2	sRogers	power-law tail	1000	750
dR3	sRogers	power-law tail	600	600
dS1	SynTReN	E. coli	500	500
dS2	SynTRen	E. coli	300	300
dS3	SynTReN	E. coli	50	500

a well known unsupervised method: the equal frequency intervals [23]. The complexity of a mutual information estimation is $O(N)$, where N is the number of samples. As a result, the MIM computation is of $O(N \times n^2)$ complexity, where n is the number of genes. In order to benchmark MRNet versus RelNet and ARACNE, the same Miller-Madow MIM is used for the three inference approaches. Note however that other entropy estimators exist [22]. For example [12] adopted Gaussian Kernel Estimators [24].

4.3 Validation

A network inference problem can be seen as a binary decision problem where the inference algorithm plays the role of a classifier: for each pair of nodes, the algorithm either adds an edge or not. Each pair of nodes is thus assigned a positive label (an edge) or a negative label (no edge).

A positive label (an edge) predicted by the algorithm is considered as a true positive (TP) or as a false positive (FP) depending on the presence or not of the corresponding edge in the underlying true network, respectively. Analogously, a negative label is considered as a true negative (TN) or a false negative (FN) depending on whether the corresponding edge is present or not in the underlying true network, respectively.

The decision made by the algorithm can be summarised by a confusion matrix (see Table 2).

Table 2. Confusion matrix

edge	actual positive	actual negative
inferred positive	TP	FP
inferred negative	FN	TN

Provost et al. [25] recommends the use of receiver operator characteristic (ROC) curves when evaluating binary decision problems in order to avoid effects related to the chosen threshold. However, ROC curves can present an overly optimistic view of an algorithm's performance if there is a large skew in the class distribution, as typically encountered in network inference.

To tackle this problem, precision-recall (PR) curves have been cited as an alternative to ROC curves [26]. Let the precision quantity

$$p = \frac{TP}{TP + FP} \; , \tag{6}$$

measure the fraction of real edges among the ones classified as positive and the recall quantity

$$r = \frac{TP}{TP + FN} \; , \tag{7}$$

also known as true positive rate, denote the fraction of real edges that are correctly inferred. These quantities depend on the threshold chosen to return a binary decision. The PR curve is a diagram which plots the precision (specifically, we used the interpolated precision as defined in [27]) versus recall for different values of the threshold on a two-dimensional coordinate system.

The F_1-measure [28] is defined as the harmonic mean of precision and recall. It combines these quantities with an equal weight and is computed as

$$F_1(p, r) = \frac{2pr}{p + r} \; . \tag{8}$$

The weighted version of the F_1-measure is obtained by computing the weighted harmonic mean of precision and recall, the F_α-measure:

$$F_\alpha(p, r) = \frac{(1 + \alpha)pr}{\alpha p + r} \; , \tag{9}$$

where $\alpha \in (0, +\infty)$ is the weight of recall in the weighted harmonic mean; the weight of precision being 1. Consequently, by taking values of α smaller than 1, one attributes relatively more importance to precision than to recall.

As precision and recall depend on the classification threshold, the F-measures vary with this threshold.

The following section presents the results by means of PR curves and F_α-measures for varying values of α.

5   Results

PR-curves for the six artificial datasets generated (see Table 1) are shown in Figs. 2 and 3. For each algorithm, the best F-measure for each dataset is shown for varying values of α in Tables 3 and 4.

We observe that ARACNE often outperforms RelNet, in particular when precision has a higher weight than recall. MRNet appears competitive with ARACNE when precision is as important than recall (see Table 3). However, as precision is getting more important than recall, MRNet outperforms the two other methods (see Table 4). High precision is an interesting property in biological network inference since testing the existence of an affinity between a target gene and a regulator gene is expensive in time and materials.

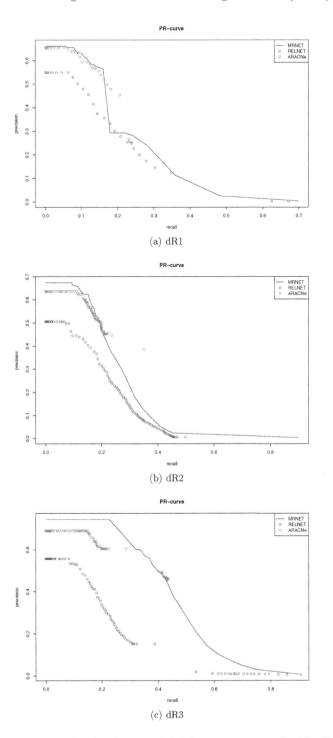

Fig. 2. PR-curves for the three artificial datasets generated with sRogers

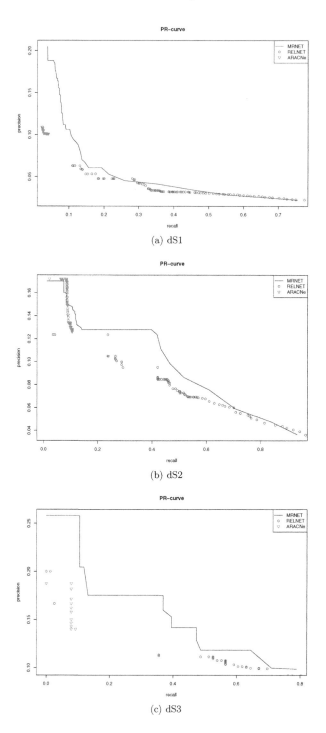

(a) dS1

(b) dS2

(c) dS3

Fig. 3. PR-curves for the three artificial datasets generated with SynTReN

Table 3. Best F_α-measures with $\alpha = 1$ (precision as important as recall). For each dataset, the best measure is shown in boldface.

Dataset	RelNet	ARACNE	MRNet
dR1	0.24	**0.28**	0.26
dR2	0.25	**0.36**	0.29
dR3	0.25	**0.45**	**0.45**
dS1	0.09	0.06	**0.10**
dS2	0.16	0.12	**0.19**
dS3	0.18	0.11	**0.24**

Table 4. Best F_α-measures with $\alpha = \frac{1}{4}$ (precision more important than recall). For each dataset, the best measure is shown in boldface.

Dataset	RelNet	ARACNE	MRNet
dR1	0.29	0.37	**0.38**
dR2	0.31	0.38	**0.39**
dR3	0.32	0.49	**0.52**
dS1	0.07	0.08	**0.13**
dS2	0.13	0.14	**0.15**
dS3	0.13	0.15	**0.20**

6 Conclusion and Future Works

A new network inference method, MRNet, has been proposed. This method relies on an effective method of information-theoretic feature selection called MRMR. Similarly to other network inference methods, MRNet relies on pairwise interactions between genes.

MRNet was compared experimentally to two state-of-the-art information-theoretic network inference methods, namely relevance networks and ARACNE. Six different artificial microarray datasets were generated with two different generators. Our preliminary results appear promising. Indeed, MRNet is competitive with the other information-theoretic methods proposed in the literature on all datasets. In particular, when precision has a higher weight than recall, MRNet outperforms the other methods.

Generation of a large number of artificial datasets, including datasets with low sample to feature ratios, for the significance assessment of the results is being conducted. Experiments on real microarray datasets to infer the regulatory networks of *E. coli* and *S. cerevisiae* are also in progress. The sensitivity of the MRNet algorithm to the mutual information estimator and to the noise component will be the topic of future investigations.

Acknowledgements

This work was supported by the Communauté Française de Belgique under ARC grant no. 04/09-307.

References

1. Ideker, T., Galitski, T., Hood, L.: A new approach to decoding life: Systems biology. Annual Review of Genomics and Human Genetics **2** (2001) 343–372
2. van Someren, E.P., Wessels, L.F.A., Backer, E., Reinders, M.J.T.: Genetic network modeling. Pharmacogenomics **3** (2002) 507–525
3. Liang, S., Fuhrman, S., Somogyi, R.: REVEAL, a general reverse engineering algorithm for inference of genetic network architectures. In Altman, R.B., Dunker, A.K., Hunter, L., Klein, T.E., eds.: Pacific Symposium on Biocomputing. Volume 3., Singapore, World Scientific Publishing (1998) 18–29
4. Akutsu, T., Miyano, S., Kuhara, S.: Inferring qualitative relations in genetic networks and metabolic pathways. Bioinformatics **16** (2000) 727–734
5. Shmulevich, I., Dougherty, E., Kim, S., Zhang, W.: Probabilistic Boolean networks: a rule-based uncertainty model for gene regulatory networks. Bioinformatics **18** (2002) 261–274
6. Friedman, N., Linial, M., Nachman, I., Pe'er, D.: Using Bayesian Networks to Analyze Expression Data. Journal of Computational Biology **7** (2000) 601–620
7. Pe'er, D., Regev, A., Elidan, G., Friedman, N.: Inferring subnetworks from perturbed expression profiles. Bioinformatics **17** (2001) S215–S224
8. van Berlo, R.J.P., van Someren, E.P., Reinders, M.J.T.: Studying the conditions for learning dynamic bayesian networks to discover genetic regulatory networks. SIMULATION **79** (2003) 689–702
9. Butte, A.J., Kohane, I.S.: Mutual information relevance networks: Functional genomic clustering using pairwise entropy measurments. Pacific Symposium on Biocomputing **5** (2000) 415–426
10. de la Fuente, A., Bing, N., Hoeschele, I., Mendes, P.: Discovery of meaningful associations in genomic data using partial correlation coefficients. Bioinformatics **20** (2004) 3565–3574
11. Schäfer, J., Strimmer, K.: An empirical Bayes approach to inferring large-scale gene association networks. Bioinformatics **21** (2005) 754–764
12. Margolin, A., Nemenman, I., Basso, K., Wiggins, C., Stolovitzky, G., Dalla Favera, R., Califano, A.: ARACNE: An algorithm for the reconstruction of gene regulatory networks in a mammalian cellular context. BMC Bioinformatics **7** (2006) S7
13. Tourassi, G., Frederick, E., Markey, M., Floyd Jr, C.: Application of the mutual information criterion for feature selection in computer-aided diagnosis. Medical Physics **28** (2001) 2394
14. Ding, C., Peng, H.: Minimum redundancy feature selection from microarray gene expression data. Journal of Bioinformatics and Computational Biology **3** (2005) 185–205
15. Meyer, P.E., Bontempi, G.: On the use of variable complementarity for feature selection in cancer classification. Lecture Notes in Computer Science **3907** (2006) 91–102

16. Butte, A., Tamayo, P., Slonim, D., Golub, T., Kohane, I.: Discovering functional relationships between RNA expression and chemotherapeutic susceptibility using relevance networks. Proceedings of the National Academy of Sciences **97** (2000) 12182–12186
17. Cover, T.M., Thomas, J.A.: Elements of Information Theory. John Wiley, New York (1990)
18. Peng, H., Long, F., Ding, C.: Feature selection based on mutual information: Criteria of max-dependency, max-relevance, and min-redundancy. IEEE Transactions on Pattern Analysis and Machine Intelligence **27** (2005) 1226–1238
19. Merz, P., Freisleben, B.: Greedy and local search heuristics for unconstrained binary quadratic programming. Journal of Heuristics **8** (2002) 1381–1231
20. Rogers, S., Girolami, M.: A Bayesian regression approach to the inference of regulatory networks from gene expression data. Bioinformatics **21** (2005) 3131–3137
21. Van den Bulcke, T., Van Leemput, K., Naudts, B., van Remortel, P., Ma, H., Verschoren, A., De Moor, B., Marchal, K.: SynTReN: a generator of synthetic gene expression data for design and analysis of structure learning algorithms. BMC Bioinformatics **7** (2006) 43
22. Paninski, L.: Estimation of entropy and mutual information. Neural Computation **15** (2003) 1191–1253
23. Dougherty, J., Kohavi, R., Sahami, M.: Supervised and unsupervised discretization of continuous features. Proceedings of the Twelfth International Conference on Machine Learning **202** (1995) 194–202
24. Beirlant, J., Dudewica, E., Gyofi, L., van der Meulen, E.: Nonparametric entropy estimation: An overview. Journal of Statistics (97)
25. Provost, F., Fawcett, T., Kohavi, R.: The case against accuracy estimation for comparing induction algorithms. In: Proceedings of the Fifteenth International Conference on Machine Learning, Morgan Kaufmann, San Francisco, CA (1998) 445–453
26. Bockhorst, J., Craven, M.: Markov networks for detecting overlapping elements in sequence data. In Saul, L.K., Weiss, Y., Bottou, L., eds.: Advances in Neural Information Processing Systems 17. MIT Press, Cambridge, MA (2005) 193–200
27. Chakrabarti, S.: Mining the Web: Discovering Knowledge from Hypertext Data. Morgan Kaufmann (2003)
28. van Rijsbergen, C.J.: Information Retrieval. Buttersworth, London (1979)

A Novel Graph Optimisation Algorithm for the Extraction of Gene Regulatory Networks from Temporal Microarray Data

Judit Kumuthini[1], Lionel Jouffe[2], and Conrad Bessant[1]

[1] Cranfield University, Barton Road, Silsoe, Beds, UK
[2] Bayesia – 6, rue Léonard de Vinci – BP0102 – 53001 Laval Cedex - France
s.j.kumuthini.s02@cranfield.ac.uk

Abstract. A gene regulatory network (GRN) extracted from microarray data has the potential to give us a concise and precise way of understanding how genes interact under the experimental conditions studied [1, 2]. Learning such networks, and unravelling the knowledge hidden within them is important for drug targets and to understand the basis of disease. In this paper, we analyse microarray gene expression data from *Saccharomyces cerevisiae,* to extract Bayesian belief networks (BBNs) which mirror the cell cycle GRN. This is achieved through the use of a novel structure learning algorithm of Taboo search and a novel knowledge extraction technique, target node (TN) analysis. We also show how quantitative and qualitative information captured within the BBN can be used to simulate the nature of interaction between genes. The GRN extracted was validated against literature and genomic databases, and found to be in excellent agreement.

Keywords: Gene regulatory networks, Bayesian belief networks, Taboo search and knowledge extraction.

1 Introduction

The entire network system of mechanisms that governs gene expression is a very complex process, regulated at several stages of protein synthesis [3]. However, the activity of genes is a result of protein and metabolite regulation, and proteins themselves are gene products. Thus, how genes are regulated in a biological system, i.e. which genes are expressed, when and where in the organism, and to what extent, is crucial for determination of the regulatory network.

A GRN is not only the representation of the biochemical activity at the system level, it also provides a large-scale, coarse-grained view of the physiological state of an organism at the mRNA level. It not only provides an overall view of the system at the genetic level, it also provides the precise way of gene interaction under the experimental conditions studied and the dynamic properties of those underlying interactions [1, 2]. The knowledge within a gene regulatory network might provide valuable clues and lead to new ideas for treating complex diseases. Such knowledge can aid pharmaceutical research to fulfil nonexistent drug and gene therapy, providing

S. Hochreiter and R. Wagner (Eds.): BIRD 2007, LNBI 4414, pp. 28–37, 2007.

possibilities to tailor target drugs for individual patients, to unravel the causal factor of some serious genetic diseases, and to better understand side effects.

In recent years, several groups have developed methodologies for the extraction of regulatory networks from gene expression data. Spellman *et al* [4] used DNA microarrays to analyse mRNA levels in yeast cell cultures that had been subjected to synchronised cell cycle arrest using three independent methods. This protocol allowed the inclusion of data from previously published studies such as Cho *et al.*, 1998 [5], and identified 800 genes that met an objective minimum criterion for cell cycle regulation. In contrast, traditional methods identified 104 genes as cell cycle regulated, while the study by Cho *et al.*, 1998, using a manual decision process identified 421 genes. In fact, the Spellman study included 304 of the genes identified by Cho *et al*, perhaps indicating that the technical differences in the way the two studies were carried out may have contributed to the differences between the results. Spellman *et al* in particular, employed a statistical approach for identification of genes associated in cell cycle process, using a larger and diverse number of experiments.

In this paper we describe the use of the novel Bayesian network methodology of Taboo search to extract GRNs from temporal microarray data of the yeast cell cycle. From a statistical viewpoint, a Bayesian network (BN) efficiently encodes the joint probability distribution of the variables describing an application domain. BNs are represented in a graphical annotated form that allows us to quickly understand the qualitative part of the encoded knowledge. The nodes of a BN correspond to random variables describing the domain and the arcs that connect the nodes correspond to direct probabilistic relations between these variables.

We go on to interrogate the networks learnt using TN analysis. This shows how all the genes in the GRN are associated with a specific gene of interest, and allow us to interactively evaluate how changes in the expression of selected genes impact on the behaviour of other genes.

2   Methods

2.1   Gene Expression Data

The data was obtained from a cell cycle analysis project, published by [4], whose aim was to identify all genes regulated by the cell cycle. This time series data has been used by [1, 6, 7], Yihui *et al.*, 2002, Anshul *et al.*, and others for genome wide gene expression and cell cycle regulation studies. The data is freely available to the public from the yeast cell cycle project site, http://genome-www.stanford.edu/cellcycle/. The Spellman dataset contained temporal gene expression measurements of the mRNA levels of 6177 *S. cerevisiae* ORFs. This project considers the cell cycle regulation of yeast, which traditional laboratory analysis indicates consists of six separable phases. These are M/G1, late G1 (SCB regulated), late G1 (MCB regulated), S, S/G2, and G2/M. In our analysis, late G1 (SCB regulated) is denoted as G1, and late G1 (MCB regulated) is denoted as Late G1. At the time of writing, the KEGG database for the yeast cell cycle pathway only contains 88 genes, and these are grouped according to cell cycle phases, G1, S, G2 and M.

2.2 Data Acquisition and Pre-processing

Four data sets of yeast cell cycle synchronization experiments, alpha, *Cdc28*, elutriation [4] and *Cdc15* [5] were selected for this study. This normalised microarray log ratio data had some missing values, accounting for 5.29% of the total data in the case of the 161 selected genes (yeast cell cycle pathway from KEGG plus 104 genes known to be cell cycle associated from [4]) which in general were distributed randomly throughout the datasets. However, we found 16 genes had many missing values (up to 34% of total) across the *Cdc15*-based synchronization experiments. The *Structural EM Algorithm*, [8] implemented in BayesiaLab (Bayesia, Paris) was used to infer the missing data. We found that this resulted in very small variations in the marginal probabilities for gene expression level of a gene when compared across extracted GRNs of varying size, for example GRNS containing 31, 104, and 161 genes.

The number of time points for each synchronization are 18 (alpha), 24 (*Cdc15*), 17 (*Cdc28*) and 14 (elutriation), giving a total of 73 measurement samples for each gene. The 104 genes that were identified as cell cycle regulated, using traditional methods prior to 1998 [4] and yeast cell cycle pathway of KEGG (comprising 88 genes) were chosen for this gene regulatory network study. The 31 genes that were common in both were used for validating extracted networks and for comparison with published literature and KEGG, SGD and MIPS databases.

2.2.1 Data Discretization
The log ratio expression levels were discretized into three classes, a common practice in Bayesian structure learning: over expressed (designated as 'over') where the expression level is greater than 0, 'under' where the expression level is below 0 and 'normal' when the expression level is 0. Because the selected genes were all cell cycle regulated, very few fell into the 'normal' category within the datasets used, even if a tolerance of ±0.5 was employed. When the variables are limited to two states, the conditional probability table (CPT) will specify 2^K distributions, where K is the number of nodes in the network. This is a general representation (known as *multinomial models*), which can describe any discrete conditional distribution. In this study, we discretized the dataset, as using continuous variables would have meant assigning a pre-defined distribution which is dependent on the assumed parameter for the dataset such as its mean, for example, linear Gaussian.

However, the major drawback with the above general representation is that the number of free parameters is exponential in the number of parents. Another potential problem is, when discretizing data for application of the model, we risk losing information. For example, when discretizing the gene expression levels into the three categories, we need to decide on the control expression thresholds against which the values are compared. In such cases, we must test for the *robustness* of results in response to various modifications of the thresholds.

2.3 Learning Gene Regulatory Networks Using Bayesian Analysis

Using BNs to analyse gene expression data consists of considering probability distributions over all possible states observed in the data for the system in question, namely a cell or an organism and its environment. In this study, the nodes of the BN denote the expression level of individual genes. However, other attributes that affect the system, such as experimental conditions, the time/stage that the sample is taken from, techniques involved, etc. can be captured by including additional nodes to denote these attributes [7].

For *n,* the number of genes, the variables are $X_1.....X_n$ and we use the data that describes the experiments to find the probabilistic dependencies that hold between these variables. In specifying the conditional probability distributions (CPDs), one can choose from several representations. Suppose the parents of a variable X (i.e. the variables that have a direct probabilistic relation with X) are $(Y_1,.....,Y_n)$. The choice of representation depends on the type of variables we are dealing with. As described in section 2.2.1, we chose here to descretize the expression levels of the genes. This data pre-processing allows us to work with discrete random variables. As X and $Y_1.......,Y_n$ take discrete values from a finite set, the CPDs can be represented with conditional probability tables (CPTs), where $P(X | Y_1......,Y_n)$ specifies the probability of values for X given each joint assignments of $Y_1......,Y_n.$

There are two main approaches to learn the structure, i.e. finding the arcs of Bayesian networks from data. The first one, the constraint-based approach, consists in constructing the network according to the conditional independence relations found in data. The other approach is to define an evaluation function (or score) that accounts for the candidate network quality with respect to data, and to use a search algorithm to find a network optimizing that score. Score-based algorithms are less sensitive to the quality of the available data than the constraint-based algorithms, which have to use independence tests that suffer from a lack of reliability for conditional parts of high-dimension.

The learning algorithm used in this study to extract GRNs is based on the Minimum Description Length score [9]. This score, based on Information Theory, is two fold. One part is used to score how well the network fits data; the second part represents the complexity of the network. This structural part prevents over-fitting and can be viewed as a probabilistic significance threshold that is set automatically (see [8] for details). The search strategy employed to find a network that optimizes that score is the so called Taboo search [10]. Briefly, this algorithm works like the greedy search algorithm, but, whereas the greedy strategy consists in always choosing the operation that leads to the best score improvement (here adding, deleting and inverting arcs), the Taboo search always selects the best operation, even if this operation degrades the score. By accepting such score deterioration, we expect to escape from local optima. The name Taboo comes from the list of the states that has to be maintained in order to prevent exploring already visited states.

Each gene of the GRN was tagged according to their cell cycle phase. The relation between two genes (the arc connecting their two nodes) is identified as positive or negative based on the analysis of the CPT. If the conditional probability of the parent and child nodes being of the same state (Over, Under of Normal) is over 50%, the relationship is designated as positive, otherwise it is designated as negative.

2.4   Knowledge Extraction from Bayesian Networks

Target Node Analysis
TN analysis measures the influence of conditional dependencies of the network variables on the knowledge of the TN and it also allows the visualization of the quantity of information gain contributed by each node to the knowledge of the TN. The brightness of the squares appear at the centre of the nodes is proportional to this quantity of information. For the specified outcome, (modality) of the TN, specifically focusing the analysis algorithms on that node. The information gain of two random variables, X relative to Y, which represents conceptually the average amount of information about X that can be gained by observing Y.

3   Results and Discussion

Figure 1 shows the BN derived from the microarray data using the Taboo search algorithm, for the 31 genes that are common in both cell cycle pathway of KEGG and the 104 genes traditionally identified as cell cycle regulated.[1] To evaluate and validate the results, it is important to appreciate the nature of gene regulation within the cell cycle. The eukaryotic cell cycle involves both continuous events and periodic events

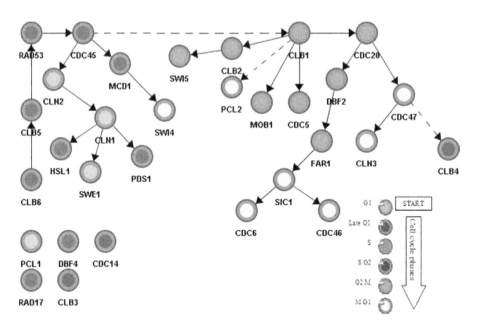

Fig. 1. GRN of common genes found in KEGG and yeast cell cycle project. The BN derived from microarray data, for the 31 genes that are common in both cell cycle pathway of KEGG and 104 genes traditionally identified as cell cycle regulated using a Taboo search algorithm. The cell cycle phases are colour coded according to cell cycle phase as determine by the Stanford cell cycle project.

[1] Reference: http://genomewww.stanford.edu/cellcycle/data/rawdata/KnownGenes.doc.

such as cell growth, DNA synthesis and mitosis. The cell cycle starts with cell size dependent START checkpoint, at this stage the level of cyclins such as *Cln* and *Clb* dramatically increases. It has been proved that cyclins are periodically expressed and at least 11 of them are known to be involved in the control of G1, G2 and DNA synthesis [11].

Over expression of G1 cyclin *Cln2* inhibits the assembly of pre-replicative complexes (pre-RCs) [12]. *Cln3*, a particular G1 cyclin, is a putative sensor of cell size, which acts by modulating the levels of other cyclins. In addition to cyclin accumulation, the activity of a cylin-dependent kinase (*Cdc*) which is an effector of START, is induced; this is the gene product of *Cdc28*. *Cdc28* couples with G1 cyclins that activate its kinase potential, [13] and [14]. Cyclin dependent kinase in yeast is denoted by *Cdc* (cell division control).

APC (anaphase promoting complex) proved to be activated by *Cdc20* or *Cdh1* and to cyclical accumulation of the *Sic1* inhibitor. Also *Clb* high/inhibitors, (*Sic1*/APC-*Cdh1*) low Clb low/inhibitors high, [15].

It is found that any one of the *Clns* can do the essential jobs of the other two, if the cell is large enough. The double mutant *Clb3 Clb4* is normal [16], so their roles can be played by other *Clbs*. Because a *Clb5 Clb6* mutant cell carries out DNA synthesis whereas a cell with all six *Clb* genes deleted (*clb1– 6*) does not, *Clb1–4* can trigger DNA synthesis in the absence of *Clb5–6* [17]. Only the *Clb1–2* pair is special in the sense that at least one of them is necessary for completing mitosis [18].

3.1 Validation of the Extracted Genetic Associations

All of the relationships found were successfully validated against the literature. Here, two examples are provided as case studies.

Far1 and Sic1: Both *Far 1* and *Sic1* are CDK inhibitors. Substrates of the *Cdc28-Cln1* CDK are *Far1p, Sic1p,* and the so far not identified proteins p41 and p94 [19]. *Sic1* is a P40 inhibitor of *Cdc28p-Clb5* protein kinase complex. Upon its destruction by programmed proteolysis, cyclin-Cdk1 activity is induced. This degradation then triggers the G1-S transition [20] *Far1*, a pheromone-activated inhibitor of *Cln-Cdc28* kinases, is dispensable for arrest of *Cln⁻ Sic1* cells by pheromone, implying the existence of an alternate *Far1*-independent arrest pathway. These observations define a pheromone-sensitive activity able to catalyze "Start" only in the absence of p40[Sic1][21].

Cln2 and Far1: The *Far1* gene is required for cell cycle arrest but not for a number of other aspects of the pheromone response. Genetic evidence suggests that *Far1* is required specifically for inactivation of the G1 cyclin *Cln2* [22].

3.2 Target Node Analysis

The TN analysis report presented below in Table 1. This shows the *relative* significance of the association of all other genes to *Sic1*. The top ten significant genes are shown in the first column, and only the rest of the selected genes in Figure 2 and not in the top ten are shown here in row 2 for illustration. The TN Analysis Report shows relative significance of all the other genes in the network (160 genes in this case), including those that are independent (zero significance), with respect to the selected (target) node.

The derived gene regulatory network captures many known relationships between known cell cycle regulated genes. For example, the immediate cluster of Sic1 i.e. direct child node of Sic1, in Figure 2, contains Ash1, Tec1, Chs1, Egt2, Pcl9, Cts1, Rme1, Cdc6, and Far1. This can be compared to: "Sic1" cluster comprises 27 genes, including Egt2, Pcl9, Tec1, Ash1, Sic1, Chs1, and Cts1. These genes are strongly cell cycle regulated and peak in late M or at the M/G1 boundary" [4]. The strength of the association of other genes relative to gene Mcd1 (Mitotic Chromosome Determinant) is shown in Figure 2. Mcd1 is an essential gene required for sister chromatid cohesion in mitosis and meiosis [23].

Mcd1 (Scc1) along with Irr1 (Scc3), Smc1, and Smc3 forms the yeast cohesion complex. The shade of the squares in nodes Pol30, Msh2 and Msh6 represent close association with Mcd1. This is backed by biological literature: [7] found that Mcd1 and Msh6, both of which bind to DNA during mitosis, are in the top 3 scoring Markov relations (functionally related). "Msh6 forms a complex with Msh2p to repair both single-base & insertion-deletion mispairs [24]. Pol30 and Mcd1, are found to be co-regulated at the end of S phase [25].

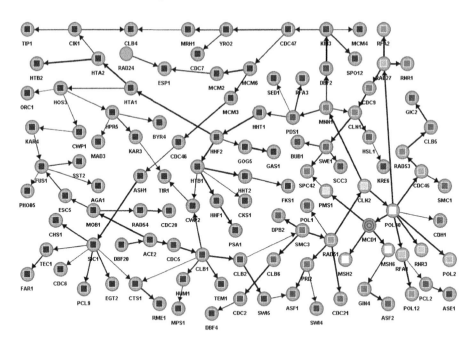

Fig. 2. Unsupervised learning, extracting information contribution of the arcs and the nodes to the overall BN. The *Mcd1* is set as the target node (TN) represented by the wavy circular node. The shade of the squares in the centre of the node represents the amount of information brought to the TN. The lighter the shade of the square the more the information contributed to the (TN). The thickness of the arc represents the information contributed to the global structure. The strength of the arc is proportional to the strength of the probabilistic relation.

Table 1. Node relative significance with respect to the information gain brought by the node to the knowledge of *Sic1*

Gene	Relative Significance Top10	Node Relative significance with respect to *Sic1*	
Ash1	1	Gene	Relative significance of other genes
Pcl9	0.8085		
Chs1	0.7522		
Egt2	0.7458	*Clb5*	0.001
Tec1	0.5002	*Clb1*	0.0005
Cdc6	0.3652	*Clb2*	0.0004
Cts1	0.3067	*Swe1*	0.0003
Far1	0.2134	*Swi5*	0.0002
Rme1	0.1425	*Swi4*	0.0001
Cln2	0.0434	*Cdc20*	0.0001

[7] identified Mcd1 as one of the dominant genes of the 800 genes identified as cell cycle regulated by [4], in terms of how they tend to dominate the order relations (i.e. they appear before other genes) in the extracted Bayesian network. As pointed out by [7], that dominant genes in the order relations may be indicative of potential causal sources of the cell-cycle process. Figure 3 includes 6 of the 13 genes (out of 800 genes) listed by [7] as 10 dominant genes: Mcd1 (2nd most dominant out of the 800), Cln2 (3rd) – a G1 cyclin involved in initiation of cell cycle, Rfa2 (5th) – DNA replication factor involved in nucleotide excision repair, Pol30 (9th) - required for DNA synthesis and DNA repair, Cln1 (11th) - G1 cyclins involved in initiation of cell cycle, and Msh6 (13th) - protein required for mismatch repair in mitosis and meiosis. Our results back the findings of [7], with all 6 of the genes dominating the order.

4   Conclusions

We employed a Bayesian probabilistic framework using Taboo search to identify the relations and dependencies between the genes known to be cell cycle regulated in yeast. Our methodology allows for high confidence results, which are learned entirely statistically from datasets containing gene expression levels, rather than known associations in the published literature and genome databases. However, the complexity of the network pathways and the discrepancies of updated information across these databases and the difficulty in integrating GRNs extracted using alternative methods are still remaining problems.

From the results obtained, we are confident our approach developed in this piece of work can be easily and efficiently applied to learning a larger scale network even when partial genetic regulation is unknown. We believe this is a sensible approach, as at this stage, a number of genes in yeast that are cell cycle regulated are yet to be conclusively identified. For example, [6] identified 822 genes whose posterior probabilities of being Periodically expressed are greater than 0.95. Among these 822 genes, only 540 are in the list of 800 genes detected by [4].

The TN analysis is a novel approach for extracting knowledge hidden within a GRN. In particular it gives the opportunity to visualise within a GRN how the gene of interest influences the network given the gene expression data. This feature could be used for monitoring gene expression levels in GRN by modifying the observation of genes to any level.

References

1. Dewey, T.G., *From microarrays to networks: mining expression time series.* Drug Discovery Today, 2002. **7**(20): p. s170-s175.
2. Brazhnik, P., A. de la Fuente, and P. Mendes, *Gene networks: how to put the function in genomics.* Trends in Biotechnology, 2002. **20**(11): p. 467-472.
3. Lewin, B., *Genes VII.* 1999, Oxford: Oxford University Press.
4. Spellman, P.T., et al., *Comprehensive identification of cell cycle-regulated genes of the yeast Saccharomyces cerevisiae by microarray hybridization.* Molecular Biology Of The Cell, 1998. **9**(12): p. 3273-3297.
5. Cho, R.J., et al., *A Genome-Wide Transcriptional Analysis of the Mitotic Cell Cycle.* Molecular Cell, 1998. **2**(1): p. 65-73.
6. Lu, X., et al., *Statistical resynchronization and Bayesian detection of periodically expressed genes.* Nucleic Acids Research, 2004. **32**(2): p. 447-455.
7. Friedman, N., et al., *Using Bayesian networks to analyse expression data.* Journal of Computational Biology, 2000. **7**(3/4): p. 601-620.
8. Friedman, N., *Learning Bayesian Networks with Local Structure.* Proceedings of the Twelfth International Conference on Uncertainty in Artificial Intelligence, 1996.
9. Grunwald, P., *Model Selection Based on Minimum Description Length.* Journal of Mathamatical Psychology, 2000. **1**(44): p. 133-152.
10. Fred Glover, M.L., *Tabu Search.* 1997: Kluwer Academic Publishers. 408.
11. Donaldson, A.D. and J.J. Blow, *The regulation of replication origin activation.* Current Opinion in Genetics & Development, 1999. **9**(1): p. 62-68.
12. Tanaka, S. and J.F.X. Diffley, *Deregulated G1-cyclin expression induces genomic instability by preventing efficient pre-RC formation.* Genes & Development, 2002. **16**(20): p. 2639-2649.
13. Biggins, S. and A.W. Murray, *Sister chromatid cohesion in mitosis.* Current Opinion In Genetics & Development, 1999. **9**(2): p. 230-236.
14. Mata, J. and P. Nurse, *Discovering the poles in yeast.* Trends In Cell Biology, 1998. **8**(4): p. 163-167.
15. Cross, F.R., *Two redundant oscillatory mechanisms in the yeast cell cycle.* Developmental Cell, 2003. **4**(5): p. 741-752.
16. Schwob, E. and K. Nasmyth, *CLB5 and CLB6, a new pair of B cyclins involved in DNA replication in Saccharomyces cerevisiae.* Genes & Development, 1993. **7**(7A): p. 1160-1175.
17. Schwob, E., et al., *The B-type cyclin kinase inhibitor p40SIC1 controls the G1 to S transition in S. cerevisiae.* Cell, 1994. **79**(2): p. 233-244.
18. Surana, U., et al., *The role of CDC28 and cyclins during mitosis in the budding yeast S. cerevisiae.* Cell, 1991. **65**(1): p. 145-161.

19. Andrews, B. and V. Measday, *The cyclin family of budding yeast: abundant use of a good idea.* Trends In Genetics: TIG, 1998. **14**(2): p. 66-72.
20. Lew DJ, e.a., *Cell cycle control in Saccharomyces cerevisiae,* in *Molecular and Cellular Biology of the Yeast Saccharomyces: Cell Cycle and Cell Biology,* B.J.a.J.E. Pringle JR, Editor. 1997, Cold Spring Harbor Laboratory Press: New York. p. 607-695.
21. Tyers, M., *The cyclin-dependent kinase inhibitor p40SIC1 imposes the requirement for Cln G1 cyclin function at Start.* Proceedings Of The National Academy Of Sciences Of The United States Of America, 1996. **93**(15): p. 7772-7776.
22. Valdivieso, M.H., et al., *FAR1 is required for posttranscriptional regulation of CLN2 gene expression in response to mating pheromone.* Molecular And Cellular Biology, 1993. **13**(2): p. 1013-1022.
23. Guacci, V., D. Koshland, and A. Strunnikov, *A Direct Link between Sister Chromatid Cohesion and Chromosome Condensation Revealed through the Analysis of MCD1 in S. cerevisiae.* Cell, 1997. **91**(1): p. 47-57.
24. Habraken, Y., et al., *ATP-dependent assembly of a ternary complex consisting of a DNA mismatch and the yeast MSH2-MSH6 and MLH1-PMS1 protein complexes.* The Journal Of Biological Chemistry, 1998. **273**(16): p. 9837-9841.
25. Werner-Washburne, M., et al., *Comparative analysis of multiple genome-scale data sets.* Genome Research, 2002. **12**(10): p. 1564-1573.

Analysing Periodic Phenomena
by Circular PCA

Matthias Scholz

Competence Centre for Functional Genomics (CC-FG),
Institute for Microbiology, Ernst-Moritz-Arndt-University Greifswald, Germany
matthias.scholz@functional-genomics.de
http://www.functional-genomics.de

Abstract. Experimental time courses often reveal a nonlinear behaviour. Analysing these nonlinearities is even more challenging when the observed phenomenon is cyclic or oscillatory. This means, in general, that the data describe a circular trajectory which is caused by periodic gene regulation.

Nonlinear PCA (NLPCA) is used to approximate this trajectory by a curve referred to as nonlinear component. Which, in order to analyse cyclic phenomena, must be a closed curve hence a circular component. Here, a neural network with circular units is used to generate circular components.

This circular PCA is applied to gene expression data of a time course of the intraerythrocytic developmental cycle (IDC) of the malaria parasite *Plasmodium falciparum*. As a result, circular PCA provides a model which describes continuously the transcriptional variation throughout the IDC. Such a computational model can then be used to comprehensively analyse the molecular behaviour over time including the identification of relevant genes at any chosen time point.

Keywords: gene expression, nonlinear PCA, neural networks, nonlinear dimensionality reduction, *Plasmodium falciparum*.

1 Introduction

Many phenomena in biology proceed in a cycle. These include circadian rhythms, the cell cycle, and other regulatory or developmental processes such as the cycle of repetitive infection and persistence of malaria parasites in red blood cells which is considered here.

Due to an individual behaviour of molecules over time, the resulting data structure becomes nonlinear as shown, for example, in [1] for a cold stress adaptation of the model plant *Arabidopsis thaliana*. In this context, nonlinearity means that the trajectory of the data describes a curve over time. For periodic processes, this curve is closed and hence cannot be well described by a standard (open) nonlinear component.

S. Hochreiter and R. Wagner (Eds.): BIRD 2007, LNBI 4414, pp. 38–47, 2007.

Therefore, the objective is to visualise and analyse the potential circular structure of molecular data by a nonlinear principal component analysis which is able to generate circular components.

Nonlinear principal component analysis (NLPCA) is generally seen as a nonlinear generalisation of standard linear *principal component analysis* (PCA) [2,3]. The principal components are generalised from straight lines to curves. Here, we focus on a neural network based nonlinear PCA, the *auto-associative neural network* [4,5,6,7].

To generate circular components, Kirby and Miranda constrained network units to work in a circular manner [8]. In the fields of atmospheric and oceanic sciences, this circular PCA is applied to oscillatory geophysical phenomena [9]. Other applications are in the field of robotics to analyse and control periodic movements [10]. Here, we demonstrate the potential of circular PCA to biomedical applications. The biological process, analysed here, is the intraerythrocytic developmental cycle (IDC) of *Plasmodium falciparum*.

P. falciparum is the most pathogenic species of the *Plasmodium* parasite, which causes malaria. The three major stages of *Plasmodium* development take place in the mosquito and upon infection of humans in liver and red blood cells. The infection of red blood cells (erythrocytes) recurs with periodicity of around 48 hours. This intraerythrocytic developmental cycle (IDC) of *P. falciparum* is responsible for the clinical symptoms of the malaria disease. A better understanding of the IDC may provide opportunities to identify potential molecular targets for anti-malarial drug and vaccine development.

2   NLPCA – Nonlinear PCA

The nonlinear PCA (NLPCA), proposed by Kramer in [4], is based on a multi-layer perceptron (MLP) with an auto-associative topology, also known as an autoencoder, replicator network, bottleneck or sandglass type network. Comprehensive introductions to multi-layer perceptrons can be found in [11] and [12].

The auto-associative network performs the identity mapping. The output $\hat{x}$ is enforced to equal the input x with high accuracy. This is achieved by minimising the square error $\| x - \hat{x} \|^2$.

This is no trivial task, as there is a 'bottleneck' in the middle, a layer of fewer nodes than at the input or output, where the data have to be projected or compressed into a lower dimensional space Z.

The network can be considered as two parts: the first part represents the extraction function $\Phi_{extr} : \mathcal{X} \to \mathcal{Z}$, whereas the second part represents the inverse function, the generation or reconstruction function $\Phi_{gen} : \mathcal{Z} \to \hat{\mathcal{X}}$. A hidden layer in each part enables the network to perform nonlinear mapping functions.

In the following we describe the applied network topology by the notation $[l_1\text{-}l_2\text{-}\ldots\text{-}l_S]$ where l_s is the number of units in layer s: the input, hidden, component, or output layer. For example, [3-4-1-4-3] specifies a network with three units in the input and output layer, four units in both hidden layers, and one unit in the component layer, as illustrated in Figure 1.

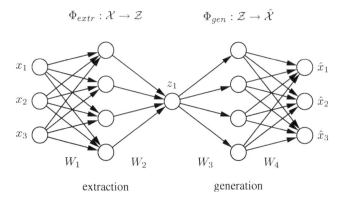

$$\Phi_{extr} : \mathcal{X} \rightarrow \mathcal{Z} \qquad\qquad \Phi_{gen} : \mathcal{Z} \rightarrow \hat{\mathcal{X}}$$

extraction   generation

Fig. 1. Standard auto-associative neural network. The network output $\hat{x}$ is required to be equal to the input x. Illustrated is a [3-4-1-4-3] network architecture. Biases have been omitted for clarity. Three-dimensional samples x are compressed (projected) to one component value z in the middle by the extraction part. The inverse generation part reconstructs $\hat{x}$ from z. The sample $\hat{x}$ is usually a noise-reduced representation of x. The second and fourth hidden layer, with four nonlinear units each, enable the network to perform nonlinear mappings. The network can be extended to extract more than one component by using additional units in the component layer in the middle.

2.1   Circular PCA

Kirby and Miranda [8] introduced a circular unit at the component layer that describes a potential circular data structure by a closed curve. As illustrated in Figure 2, a circular unit is a pair of networks units p and q whose output values z_p and z_q are constrained to lie on a unit circle

$$z_p^2 + z_q^2 = 1 \tag{1}$$

Thus, the values of both units can be described by a single angular variable θ.

$$z_p = \cos(\theta) \qquad \text{and} \qquad z_q = \sin(\theta) \tag{2}$$

The *forward propagation* through the network is as follows: First, equivalent to standard units, both units are weighted sums of their inputs z_m given by the values of all units m in the previous layer.

$$a_p = \sum_m w_{pm} z_m \qquad \text{and} \qquad a_q = \sum_m w_{qm} z_m \tag{3}$$

The weights w_{pm} and w_{qm} are of matrix W_2. Biases are not explicitly considered, however, they can be included by introducing an extra input with activation set to one.

The sums a_p and a_q are then corrected by the radial value

$$r = \sqrt{a_p^2 + a_q^2} \tag{4}$$

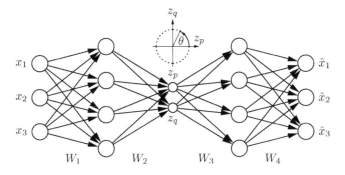

Fig. 2. Circular PCA network. To obtain circular components, the auto-associative neural network contains a circular unit pair (p, q) in the component layer. The output values z_p and z_q are constrained to lie on a unit circle and hence can be represented by a single angular variable θ.

to obtain circularly constraint unit outputs z_p and z_q

$$z_p = \frac{a_p}{r} \quad \text{and} \quad z_q = \frac{a_q}{r} \tag{5}$$

For *backward propagation*, we need the derivatives of the error function

$$E = \frac{1}{2} \sum_n^N \sum_i^d [x_i^n - \hat{x}_i^n]^2 \tag{6}$$

with respect to all network weights w. The dimensionality d of the data is given by the number of observed variables, N is the number of samples.

To simplify matters, we first consider the error e of a single sample x, $e = \frac{1}{2} \sum_i^d [x_i - \hat{x}_i]^2$ with $x = (x_1, \ldots, x_d)^T$. The resulting derivatives can then be extended with respect to the total error E given by the sum over all n samples, $E = \sum_n e^n$.

While the derivatives of weights of matrices W_1, W_3, and W_4 are obtained by standard back-propagation, the derivatives of the weights w_{pm} and w_{qm} of matrix W_2 which connect units m of the second layer with the units p and q are obtained as follows: We first need the partial derivatives of e with respect to z_p and z_q:

$$\tilde{\sigma}_p = \frac{\partial e}{\partial z_p} = \sum_j w_{jp} \sigma_j \quad \text{and} \quad \tilde{\sigma}_q = \frac{\partial e}{\partial z_q} = \sum_j w_{jq} \sigma_j \tag{7}$$

where σ_j are the partial derivatives $\frac{\partial e}{\partial a_j}$ of units j in the fourth layer.

The required partial derivatives of e in respect to a_p and a_q of the circular unit pair are

$$\sigma_p = \frac{\partial e}{\partial a_p} = (\tilde{\sigma}_p z_q - \tilde{\sigma}_q z_p) \frac{z_q}{r^3} \quad \text{and} \quad \sigma_q = \frac{\partial e}{\partial a_q} = (\tilde{\sigma}_q z_p - \tilde{\sigma}_p z_q) \frac{z_p}{r^3} \tag{8}$$

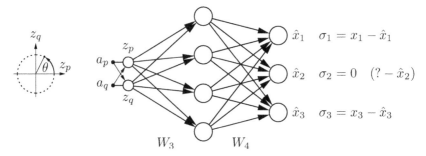

Fig. 3. Inverse circular PCA network. As inverse model, only the second part of the auto-associative neural network (Figure 2) is used. Now, the values a_p and a_q are unknown inputs and have to be estimated together with all weights w of matrices W_3 and W_4. This is done by propagating the partial errors σ_i back to the input (component) layer. Beside a higher efficiency, the main advantage is that the inverse model can be applied to incomplete data. If one value x_i of a sample vector x is missing, the corresponding partial error σ_i is set to zero, thereby ignoring the missing value but still back-propagating all others.

The final back-propagation formulas for all n samples are

$$\frac{\partial E}{\partial w_{pm}} = \sum_n \sigma_p^n z_m^n \qquad \text{and} \qquad \frac{\partial E}{\partial w_{qm}} = \sum_n \sigma_q^n z_m^n \qquad (9)$$

2.2 Inverse NLPCA Model

In this work, NLPCA is applied as an inverse model [1]. Only the second part, the generation or reconstruction part, of the auto-associative neural network is modelled, see Figure 3. The major advantage is that NLPCA can be applied to incomplete data. Another advantage is a higher efficiency since only half of the network weights have to be estimated.

Optimising the second part as inverse model means that the component layer becomes the input layer. Thus, in circular PCA as inverse model we have to find suitable values for all network weights as well as for a_p and a_q as input. Hence, the error function E depends on both the weights w and the component layer inputs a_p and a_q

$$E(w, a_p, a_q) = \frac{1}{2} \sum_n^N \sum_i^d [x_i^n - \hat{x}_i^n(w, a_p, a_q)]^2 \qquad (10)$$

The required partial derivatives of E with respect to the weights w of matrix W_3 and W_4 can be obtained by standard back-propagation, see [1], while the derivatives with respect to a_p and a_q are given by equation (8).

2.3 Artificial Data

The performance of NLPCA is illustrated in Figure 4 for the three described variants: the standard auto-associative network (NLPCA), the inverse model

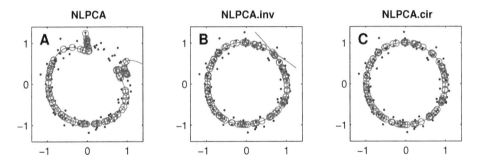

Fig. 4. Nonlinear PCA (NLPCA). Shown are results of three variants of NLPCA applied to a two-dimensional artificial data set of a noisy circle. **(A)** The standard NLPCA cannot describe a circular structure completely. There is always a gap. **(B)** The inverse NLPCA can provide self-intersecting components and hence approximates the circular data structure already quite well but the circular PCA **(C)** is most suitable since it is able to approximates the data structure continuously by a closed curve.

with standard units (NLPCA.inv) and with circular units (NLPCA.cir). NLPCA is applied to data lying on a unit circle and disturbed by Gaussian noise with standard deviation 0.1. The standard auto-associative network cannot describe a circular structure completely by a nonlinear component due to the problem to map at least one point on the circle onto two different component values. This problem does not occur in inverse NLPCA since it is only a mapping from component values to the data. However, the resulting component is an intersecting circular loop with open ends. Thus, a closed curve solution as provided by circular PCA would be more appropriate to describe the circular structure of the data.

2.4 Experimental Data

Circular PCA is used to analyse the transcriptome of the intraerythrocytic developmental cycle (IDC) of the malaria parasite *Plasmodium falciparum* [15], available at http://malaria.ucsf.edu/. The 48-hour IDC is observed by a sampling time of one hour thereby providing a series of time points 1, 2, ..., 48. Since two time points, 23 and 29, are missing, the total number of expression profiles (samples) is 46. Each gene is represented by one or more oligonucleotides on the microarray. The samples of individual time points (Cy5) were hybridised against a reference pool (Cy3). The $\log_2(\text{Cy5/Cy3})$ ratio is used in our analysis. Due to the sometimes large number of missing data in the total set of 7,091 oligonucleotides, we removed all oligonucleotides of more than 1/3 missing time observations (more than 15 missing time points). The considered reduced data set contains the $\log_2$ ratios of hybridisations of 5,800 oligonucleotides at 46 time points.

Identifying the optimal curve (the circular component) in the very high-dimensional data space of 5,800 variables is difficult or even intractable with a number of 46 data points. Therefore, the 5,800 variables are linearly reduced to 12

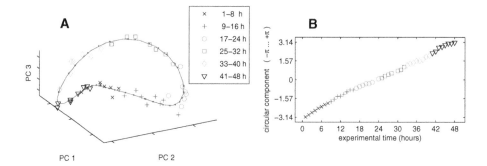

Fig. 5. Circular PCA. (A) The data describe a circular structure which is approximated by a closed curve (the circular component). The circular component is one single curve in the 5,800 dimensional data space. Visualised is the reduced three dimensional subspace given by the first three components of standard (linear) PCA. **(B)** The circular component (corrected by an angular shift) is plotted against the original experimental time. It shows that the main curvature, given by the circular component, explains the trajectory of the IDC over 48 hours.

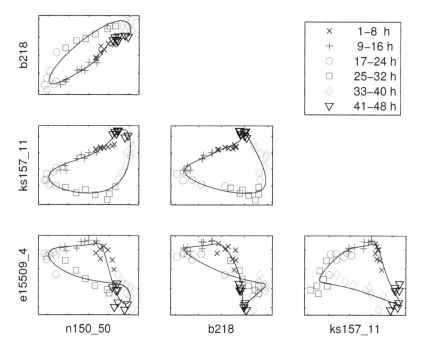

Fig. 6. Pair-wise scatter plot of four selected oligonucleotides of importance at 12, 24, 36, and 48 hours respectively, see also Table 1. The curve represents the circular component which approximates the trajectory of the 48 hour IDC.

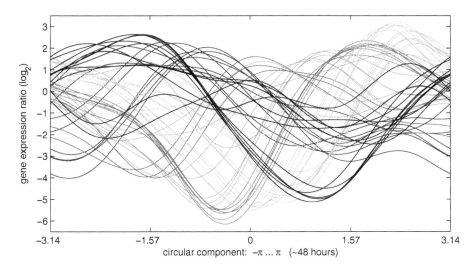

Fig. 7. Gene expression curves. Plotted are shapes of oligonucleotide response curves over the 48-hour IDC time course. Shown are the top 50 oligonucleotides of genes of highest response at any time. Nearly all of them show a period of 48 hours: one up- and down-regulation within the 48-hour time course. However, the time of activation is differently for individual genes.

principal components, each of which is a linear combination of all oligonucleotides. To handle missing data, a PCA algorithm, based on a linear neural network working in inverse mode [1], is used. Alternatively, *probabilistic PCA* (PPCA) (http://lear.inrialpes.fr/~verbeek/software) [16], based on [17], can be used as PCA algorithm for missing data.

Circular PCA is applied to the reduced data set of 12 linear components. It describes a closed curve explaining the circular structure of the data, as shown in Figure 5 and 6. To achieve circular PCA, a network of a [2-5-12] architecture is used, where the two units in the first layer are the circularly constrained unit pair (p, q). Using the inverse 12 eigenvectors the curve can be mapped back into the 5,800-dimensional original data space. The circular component represents the 48 hour time course of the IDC observation, as shown in Figure 5B.

Thus, circular PCA provides a model of the IDC, which gives us to any chosen time point, including interpolated time points, the corresponding gene expression values. The neural network model is given by a function $\hat{x} = \Phi_{gen}(\theta)$ which maps any time point, represented by a angular value θ onto a 5,800-dimensional vector $\hat{x} = (\hat{x}_1, ..., \hat{x}_{5800})^T$ representing the response of the original variables. Thus, circular PCA provides approximated response curves of all oligonucleotides, as shown in Figure 7 for the top 50 oligonucleotides of genes of highest response (highest relative change at their expression level).

In standard PCA we can present the variables that are most important to a specific component by a rank order given by the absolute values of the corresponding eigenvector, sometimes termed loadings or weights. As the components are curves in nonlinear (circular) PCA, no global ranking is possible. The rank

Table 1. Candidate genes. At specific times, exemplarily shown for 12 and 36 hours, the most important genes can be provided. Listed are the identified oligonucleotides and, if available, the corresponding PlasmoDB gene identifier of the *Plasmodium* genome resource *PlasmoDB.org* [13,14]. Note that a single gene may be represented by more than one oligonucleotide.

	12 hours			36 hours	
$\hat{v}_i$	Oligo ID	PlasmoDB ID	$\hat{v}_i$	Oligo ID	PlasmoDB ID
-0.06	i6851_1	—	0.06	ks157_11	PF11_0509
-0.06	n150_50	PF14_0102	0.06	a10325_32	PFA0110w
-0.05	e24991_1	PFE0080c	0.06	a10325_30j	—
-0.05	opfg0013	—	0.06	b70	PFB0120w
-0.05	c76	PFC0120w	0.06	a10325_30	PFA0110w
-0.05	n140_2	PF14_0495	0.06	i14975_1	PF07_0128
-0.05	opff72487	—	0.06	f739_1	PF07_0128
-0.05	kn5587_2	MAL7P1.119	0.06	opfl0045	PFL1945c
0.05	f24156_1	PFI1785w	0.05	a10325_29	PFA0110w
0.05	d44388_1	PF10_0009	0.05	ks75_18	PF11_0038
...	...	...	...	...	...

order is different for different positions on the curved component, meaning that the rank order depends on time. The rank order for each individual time point is given by the values of the tangent vector $v = \frac{d\hat{x}}{d\theta}$ on the curve at a specific time θ. To compare different times, we use l_2-normalised tangents $\hat{v}_i = v_i/\sqrt{\sum_i |v_i|^2}$ such that $\sum_i (\hat{v}_i)^2 = 1$. Large values $\hat{v}_i$ point to genes of high changes on their expression ratios and hence may have an importance at the considered time point. A list of 10 most important genes at 12 hours and 36 hours is exemplarily given in Table 1.

3   Conclusions

Circular PCA as special case of nonlinear PCA (NLPCA) was applied to gene expression data of the intraerythrocytic developmental cycle (IDC) of the malaria parasite *Plasmodium falciparum*. The data describe a circular structure which was found to be caused by the cyclic (nonlinear) behaviour of gene regulation.

The extracted circular component represents the trajectory of the IDC. Thus, circular PCA provides a noise reduced model of gene responses continuously over the full time course. This computational model can then be used for analysing the molecular behaviour over time in order to get a better understanding of the IDC.

With the increasing number of time experiments, nonlinear PCA may become more and more important in the field of molecular biology. This includes the analysis of both: non-periodic phenomena by standard NLPCA and periodic phenomena by the circular variant.

Acknowledgement

This research is funded by the German Federal Ministry of Education and Research (BMBF) within the programme of *Centres for Innovation Competence* of the BMBF initiative *Entrepreneurial Regions* (Project No. ZIK 011).

References

1. M. Scholz, F. Kaplan, C.L. Guy, J. Kopka, and J. Selbig. Non-linear PCA: a missing data approach. *Bioinformatics*, 21(20):3887–3895, 2005.
2. I. T. Jolliffe. *Principal Component Analysis*. Springer-Verlag, New York, 1986.
3. K.I. Diamantaras and S.Y. Kung. *Principal Component Neural Networks*. Wiley, New York, 1996.
4. M. A. Kramer. Nonlinear principal component analysis using auto-associative neural networks. *AIChE Journal*, 37(2):233–243, 1991.
5. D. DeMers and G. W. Cottrell. Nonlinear dimensionality reduction. In D. Hanson, J. Cowan, and L. Giles, editors, *Advances in Neural Information Processing Systems 5*, pages 580–587, San Mateo, CA, 1993. Morgan Kaufmann.
6. R. Hecht-Nielsen. Replicator neural networks for universal optimal source coding. *Science*, 269:1860–1863, 1995.
7. M. Scholz and R. Vigário. Nonlinear PCA: a new hierarchical approach. In M. Verleysen, editor, *Proceedings ESANN*, pages 439–444, 2002.
8. M. J. Kirby and R. Miranda. Circular nodes in neural networks. *Neural Computation*, 8(2):390–402, 1996.
9. W. W. Hsieh. Nonlinear multivariate and time series analysis by neural network methods. *Reviews of Geophysics*, 42(1):RG1003.1–RG1003.25, 2004.
10. K.F. MacDorman, R. Chalodhorn, and M. Asada. Periodic nonlinear principal component neural networks for humanoid motion segmentation, generalization, and generation. In *Proceedings of the Seventeenth International Conference on Pattern Recognition (ICPR), Cambridge, UK*, pages 537–540, 2004.
11. C.M. Bishop. *Neural Networks for Pattern Recognition*. Oxford University Press, 1995.
12. S. Haykin. *Neural Networks - A Comprehensive Foundation*. Prentice Hall, 2nd edition, 1998.
13. J.C. Kissinger, B.P. Brunk, J. Crabtree, M.J. Fraunholz, and B. Gajria, et al. The plasmodium genome database. *Nature*, 419(6906):490–492, 2002.
14. M.J. Gardner, N. Hall, and E. Fung, et al. Genome sequence of the human malaria parasite *plasmodium falciparum*. *Nature*, 419(6906):498–511, 2002.
15. Z. Bozdech, M. Llinas, B.L. Pulliam, E.D. Wong, J. Zhu, and J.L. DeRisi. The Transcriptome of the Intraerythrocytic Developmental Cycle of *plasmodium falciparum*. *PLoS Biology*, 1(1):E5, 2003.
16. J.J. Verbeek, N. Vlassis, and B. Kröse. Procrustes analysis to coordinate mixtures of probabilistic principal component analyzers. Technical report, Computer Science Institute, University of Amsterdam, The Netherlands, 2002.
17. S. T. Roweis, L. K. Saul, and G. E. Hinton. Global coordination of locally linear models. In T. G. Dietterich, S. Becker, and Z. Ghahramani, editors, *Advances in Neural Information Processing Systems 14*, pages 889–896, Cambridge, MA, 2002. MIT Press.

Identification of Cold-Induced Genes in Cereal Crops and *Arabidopsis* Through Comparative Analysis of Multiple EST Sets

Angelica Lindlöf[1], Marcus Bräutigam[2], Aakash Chawade[2], Björn Olsson[1], and Olof Olsson[2]

[1] School of Humanities and Informatics, University College of Skövde, Box 408, 541 28 Skövde, Sweden
[2] Dept. of Cell and Molecular Biology, Göteborg University, Göteborg, Box 462, 403 20 Göteborg, Sweden
{angelica.lindlof, bjorn.olsson}@his.se, {marcus.brautigam, aakash.chawade, olof.olsson}@molbio.gu.se

Abstract. Freezing tolerance in plants is obtained during a period of low non-freezing temperatures before the winter sets on, through a biological process known as cold acclimation. Cold is one of the major stress factors that limits the growth, productivity and distribution of plants, and understanding the mechanism of cold tolerance is therefore important for crop improvement. Expressed sequence tags (EST) analysis is a powerful, economical and time-efficient way of assembling information on the transcriptome. To date, several EST sets have been generated from cold-induced cDNA libraries from several different plant species. In this study we utilize the variation in the frequency of ESTs sampled from different cold-stressed plant libraries, in order to identify genes preferentially expressed in cold in comparison to a number of control sets. The species included in the comparative study are oat (*Avena sativa*), barley (*Hordeum vulgare*), wheat (*Triticum aestivum*), rice (*Oryza sativa*) and *Arabidopsis thaliana*. However, in order to get comparable gene expression estimates across multiple species and data sets, we choose to compare the expression of tentative ortholog groups (TOGs) instead of single genes, as in the normal procedure. We consider TOGs as preferentially expressed if they are detected as differentially expressed by a test statistic and up-regulated in comparison to all control sets, and/or uniquely expressed during cold stress, i.e., not present in any of the control sets. The result of this analysis revealed a diverse representation of genes in the different species. In addition, the derived TOGs mainly represent genes that are long-term highly or moderately expressed in response to cold and/or other stresses.

1 Introduction

Cold acclimation is a biological process that increases freezing tolerance of an organism, which prior to exposure to sub-zero temperatures is subjected to cold stress at temperatures between 0-10°C [1-6]. Many species have this ability to acclimate to cold and thereby survive harsh winter conditions. In nature, the tolerance is obtained

S. Hochreiter and R. Wagner (Eds.): BIRD 2007, LNBI 4414, pp. 48–65, 2007.

during a period of low non-freezing temperatures before the winter sets on. Plants that cold acclimate are either chilling tolerant and can resist cool, but non-freezing, temperatures (below 10°C) or freezing-tolerant with resistance also to temperatures below 0°C.

The cold acclimation process is biologically complex and involves numerous physiological, cellular and molecular changes [4, 5], such as alterations in the phytohormone homeostasis, increase of osmolytes (e.g. sugar, proline and betaine), membrane modifications and increased levels of antioxidants. These changes result from numerous changes in gene expression, where hundreds of genes are either up- or down-regulated as a response to the cold stress signal. Many of these genes are also induced by drought and high salt concentrations. In addition to physical injuries that may be caused by freezing, the low temperatures limit the growth, productivity and distribution of the plants.

In this study, we identify several genes involved in the cold-response, by analyzing ESTs derived from cold stressed plant cDNA libraries. The analysis is conducted for the crop species oat, wheat, barley and rice. In addition, *Arabidopsis thaliana*, the major model organism for plants, is included for comparison. The crops included in the study are all agriculturally important species, where yields are limited by low temperatures, and where the development of cold-hardy cultivars that are planted in autumn and will survive harsh winter conditions will be of major importance for increasing yield levels. The mentioned plants also represent different levels of cold tolerance: rice is chilling-sensitive and in most cases unable to cold-acclimate, Arabidopsis is chilling-tolerant and can withstand temperatures just above zero, oat is slightly more chilling-tolerant and can withstand temperatures just below zero, while barley and especially wheat are freezing-tolerant and can survive sub-zero temperatures.

In the genomic era that we are now facing, sequencing of whole genomes is becoming more common, providing us with the full repertoire of genes present in an organism [7]. One of the main goals is to identify and elucidate the function of each gene, and which biological processes the gene product participates in. However, although complete genome sequencing has become possible, there are many organisms for which this is not an option. For example, plant genomes show large variations in genome size, with *Arabidopsis thaliana* having a genome of 125 megabase pairs (Mbp), oats 13 000 Mbp and rice 460 Mbp. In addition, most plants have genomes much larger than the mammalian ones, whereas Arabidopsis and rice have reasonably small genomes. Plant genomes also contain many repeated regions and transposable elements. Altogether, this makes whole genome sequencing both expensive and complicated, and in some cases impossible with present technologies. Consequently there are relatively few plant nuclear gene sequence entries in the public databases. An alternative to whole genome sequencing is to randomly sequence expressed genes from cDNA libraries, so called "expressed sequence tags" (ESTs) [8-10]. This is due to EST sequencing being a relatively economical way of surveying expressed genes. The trade-off, however, is that it will not result in a full collection of all genes. On the other hand, the assembled transcripts also reflect one of the major benefits of EST sequencing, i.e. that it generates information showing which genes are induced during the biological process under study.

The analysis of gene expression patterns from different conditions is a valuable tool in the discovery of genes participating in a specific biological process. Gene expression levels in unbiased cDNA libraries can be estimated by using the cognate frequencies of gene transcripts. The variation in the frequency of ESTs sampled from different cDNA libraries has been used, for example, for detecting genes appearing to be differentially expressed in a biological sample and for identifying genes with similar expression patterns in multiple cDNA libraries [11-13]. During the last years, several test statistics have been proposed for detecting differentially expressed genes in multiple EST sets and the approach has frequently been used for identifying genes with expression patterns that differ between different tissues [11, 14-19].

The aim of this study is to identify genes preferentially expressed in EST sets originating from cold-stressed plant cDNA libraries in comparison to several control sets. These genes are presumably induced by cold and it is therefore likely that they participate in the cold acclimation process. The intention is twofold: to identify the overexpression of already established cold-regulated genes, but also to identify novel genes that could provide additional information on the acclimation process. In addition, the results will indicate if there are any major differences between the sets of identified genes for the different species, or if there is a limited number of cold-responsive genes that are common to all these species.

In order to identify preferentially expressed genes we apply some of the proposed test statistics to the EST sets. Additionally, in order to facilitate the identification of expressed genes and their corresponding expression values, as well as the comparison of expressed genes in different libraries, we derive the expression of orthologous genes instead of single genes. This is done by mapping the ESTs to tentative ortholog groups (TOGs), as inferred from a clustering of tentative contigs (TCs) collected from the EGO database [20] using the OrthoMCL algorithm [21]. We consider genes as preferentially expressed if they are detected as differentially expressed and up-regulated in comparison to all control sets. Additionally, since the test statistics are highly conservative and sometimes detect relatively few genes compared to the size of the EST set, we also include genes that are uniquely expressed during cold stress.

2   EST Sets and Analysis

Seven cold stress EST sets and eleven control sets were included in the study (Table 1). The sets were downloaded from dbEST[1], except for the *A. sativa* cold stress set where we used our in-house derived library [22]. ESTs with a significant match to mtDNA, cpDNA or rRNA sequences were identified and removed before clustering and assembly, as we are mainly interested in nuclear DNA. The identification was based on a BLASTn homology search against the Arabidopsis mitochondrial genome, the *T. aestivum* chloroplast genome and the vector database provided by NCBI, respectively, using an E value threshold of 10^{-8} (Figure 1a). The sequences in each

---

[1] NCBI Expressed Sequence Tags Database, http://www.ncbi.nlm.nih.gov/dbEST

EST set were separately clustered and assembled with CLOBB [23] and Cap3 [24], respectively.

One of the sets from *Arabidopsis* and the set from rice originates from subtracted libraries, whereas the remaining are unbiased EST sets. The subtracted libraries are only included for the purpose of comparison of methods for identifying induced genes. The selection of data sets was based on stress type, but also on tissue origin. The unbiased cold stress EST sets should be comparable to one or several control experiments, which in this study consist of drought stress, etiolated and/or unstressed libraries. These control libraries mainly originate from the same tissue in order to avoid detecting tissue-specific genes. This criterion restricted the number of selectable EST sets in dbEST, and the sets used here were considered to be the most suitable for the comparative study. The comparative study was performed species by species, i.e. each cold stress set was compared to control sets from the same species.

3   Identification of Expressed Ortholog Groups

The Institute of Genome Research (TIGR) have collected ESTs from various research projects and clustered the sequences into tentative consensus (TC) sequences for each species [25]. The TIGR database contains TCs from a large number of plant species, which makes it attractive for similarity searches, since the availability of gene sequences is limited for some of the species included in this study. TIGR have also clustered the sequences into ortholog groups, which are stored in the EGO (Eukaryotic Gene Orthologs) database [20], and where each cluster consists of TCs representing orthologous genes. Utilizing ortholog groups facilitates the identification of expressed genes in EST sets from different species, since the ESTs can be mapped to a group according to the best similarity match among the TCs, disregarding species origin. In addition, it makes it possible to compare the expression of genes across multiple species and sets, since an expression value can be inferred by counting the number of ESTs per ortholog group.

However, the clusters in EGO are redundant, meaning that a TC is commonly present in more than one ortholog group. This complicates the derivation of expression values, since many of the ESTs will be mapped to several groups, thus yielding an overestimated expression value. Instead, in order to still take advantage of using ortholog groups but to avoid the redundancy in EGO, we downloaded TCs from a number of selected plant species (*Arabidopsis thaliana, Hordeum vulgare, Triticum aestivum, Sorghum bicolor, Zea mays*, and *Oryza sativa*) included in the EGO database (Table 2) and clustered them into a set of non-redundant ortholog groups using the algorithm OrthoMCL [21] (Fig. 1b). The OrthoMCL algorithm was developed to cluster proteins from whole genomes and identifies orthologs as reciprocal best hits between at least two species. In addition, since it is difficult to separate "recent" paralogs (i.e. gene duplications that occurred after speciation) many clusters contain both orthologs and recent paralogs.

Table 1. EST sets included in the comparative study. Species: name of the species. Stress: the type of stress the plants were exposed to. ESTs: number of ESTs in the set after removal of non-nuclear DNA sequences. Tissue: tissue of origin. Description: experimental details in brief. UniGene: NCBI UniGene library number.

Species	Set	Stress	ESTs	Tissue	Description	UniGene
Arabidopsis thaliana	AtCI I	Cold	1240	Whole plant	Subtracted, samples collected after 1h, 2h, 5h, 10h, 24h in 4°C.	10443
	AtCI II	Cold	18736	Rosette	Samples collected after 1h, 2h, 5h, 10h, 24h in 4°C.	10441
	AtCI III	Cold	2038	Rosette	Samples collected after 1h, 2h, 5h, 10h, 24h in 4°C.	10438
	AtDI I	Drought	4215	Rosette	Control	10442
	AtDI II	Drought	3368	Rosette	Control	10439
Avena sativa	AsCI	Cold	8216	Seedling shoot	Samples collected after 4h, 16h, 32h and 48h in 4°C. Dark incubated.	-
	AsR	None	2510	Root	Control	-
	AsL	None	2173	Leaf	Control	-
	AsEL	Etiolated	2308	Leaf	Control	-
Hordeum vulgare	HvCI	Cold	4406	Seedling shoot	Samples collected after 2d in 5°C.	7260
	HvER	Etiolated	5241	Seedling root	Control	7120
	HvEL	Etiolated	2275	Seedling shoot	Control	7259
	HvDI	Drought	4815	Seedling shoot	Control	7261
Triticum aestivum	TaCI	Cold	1175	Seedling shoot	Samples collected after 2d in 5°C.	5509
	TaEL	Etiolated	2432	Seedling shoot	Control	5467
	TaER	Etiolated	4571	Seedling root	Control	5428
Oryza sativa	OsCI	Cold	3059	Seed and seedling shoot	Subtracted, samples collected after 4-6d, 7-9d, 10-14d in 10/13 °C.	11160

The TCs were clustered into 16,824 tentative ortholog groups (TOGs), where each group has >1 sequence. The remaining sequences, i.e. singletons, are not included in the output from OrthoMCL. Instead, they were manually added and given ortholog group numbers sequentially. In total, 25,093 TOGs were inferred from the 91,753 downloaded TCs.

Each contig and singleton from each EST set was mapped to the TOG containing the most similar TC. The most similar TC was identified through a BLASTn similarity search of the contigs and singletons against the TCs (Fig. 1a). We used an E

Table 2. Number of TCs for each species

Species	TCs
Hordeum vulgare	13 874
Triticum aestivum	17 131
Sorghum bicolor	12 596
Zea mays	15 661
Oryza sativa	17 323
Arabidopsis thaliana	15 168
Total	91 753

value threshold of 10^{-80} as an attempt to assure that the sequences were matched to a true ortholog. An expression value for each TOG was thereafter inferred by counting the number of ESTs in the contigs and singletons that were mapped to the TOG (Fig. 1c). Expression values were calculated for each set, hence giving an expression profile for each TOG.

The number of contigs and singletons that had a significant match varied among the EST sets, where rice and *Arabidopsis* had the highest percentage of matches and oat the lowest (Fig. 2). The lower number of significant matches for oat is probably due to there not being any oat sequences among the TCs downloaded from TIGR

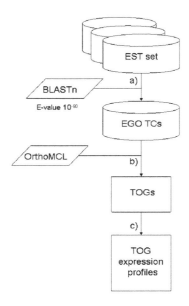

Fig. 1. Overview of the derivation of TOG expression profiles. a) The best EGO TC similarity match for each contig and singleton in each set was identified through a BLASTn search. b) The downloaded TCs from the EGO database were clustered into a set of non-redundant tentative ortholog groups (TOGs) using the OrthoMCL algorithm. c) Each contig and singleton was mapped to the TOG containing the most similar TC, and an expression profile was inferred for each TOG by summing the number of ESTs included in the TCs mapped to it.

that can match oat-specific genes. In total 8,971 TOGs received an expression value in at least one of the 18 EST sets, of which 7,460 TOGs had an expression value in at least one of the cold stress sets.

4 Cold Acclimation Genes in Arabidopsis Identified Through Microarray Analysis

Arabidopsis thaliana has for a century been the model organism for the majority of processes in plants and it is the most studied species regarding cold acclimation in plants [26]. Several hundred cold-responsive genes in Arabidopsis have been identified through high-throughput gene expression profiling experiments [16, 27-31]. We therefore constructed a database, AtCG (Arabidopsis thaliana Cold-Induced Genes), of genes induced by cold in *Arabidopsis*, as reported from five previous microarray studies [28-32], and studied their TOG expression values in the different cold stress sets. For a gene to be included in the database, it had to be up-regulated in at least one time step in at least two of the five independent studies [28-32]. These criteria applied to 1,271 genes, for which coding sequences were downloaded from the TAIR web site[2]. The downloaded genes were searched by sequence similarity against the downloaded TCs from TIGR using BLASTn (*E*=0.0), where the stringent *E* value was used to assure that genes were mapped to the correct corresponding TC. The distribution of the number of significant similarity matches per AtCG gene in the different cold stress EST sets reveals that a majority of them did not receive a hit (Fig. 3). In addition, of those that did get a hit, the majority are weakly expressed with only one matching EST member. This means that a large proportion of the genes will not be detected as preferentially expressed in the cold sets by the statistical tests. It also indicates that microarray and EST sequencing studies seem to identify different

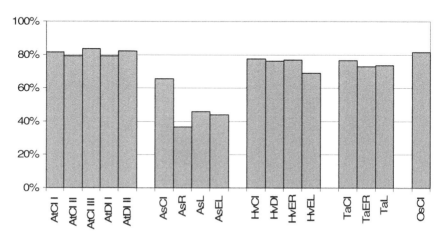

Fig. 2. Percentage significant similarity matches for each species and EST set

---

[2] http://www.arabidopsis.org/tools/bulk/sequences/index.jsp

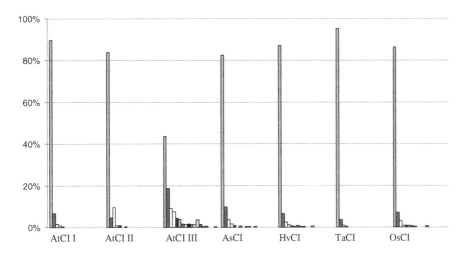

Fig. 3. Distribution of number of significant similarity matches for the AtCG genes in the different cold stress EST sets. The AtCI III set had the least uneven distribution, with ~45% of AtCG genes having 0 matching ESTs. For all other EST sets, more than 80% of the AtCG genes had 0 matches. The intervals in the histogram were set (from left to right) to: 0, 1, 2, 3, 4, 5, 6, 7, 8, 9, 10, 20, 30, 40, 60, 80, 100 or more ESTs.

sets of cold regulated genes. The subsequent question was therefore: which cold regulated genes have been picked up by the EST sequencing experiments?

5 Preferentially Expressed TOGs

In a study by Romualdi *et al.* (2001), different test statistics for detecting differentially expressed genes in multiple EST sets were compared. The conclusion was that the general χ^2 test was the most efficient in multiple libraries, especially for weakly expressed genes, and that the Audic and Claverie (1997) test was the most efficient for pairwise comparisons. Thus, we applied the test statistics developed by Audic and Claverie (AC), the χ^2 2x2 test (Chi), the R statistic [18] and the general χ^2 test over multiple data sets (MultChi) to the unbiased sets in order to detect differentially expressed TOGs in the cold sets. Fisher's exact test and Greller and Tobin's test (1999) were excluded, as they appear to be too conservative [12]. The web tool developed by Romualdi *et al.* (2003) was used to infer differentially expressed TOGs from each test.

In the initial testing of the statistics we applied the R statistic both pairwise and across all sets, even though it is intended for comparison over multiple sets. The *p*-value threshold was set to 0.05 and TOGs were detected with and without Bonferroni correction. In this initial testing the larger unbiased EST set from *Arabidopsis*, AtCI II, in comparison to the drought stress sets, AtDI I and AtDI II, were used as control sets. The number of detected TOGs varies among the tests (see Table 3), revealing that the test statistics do not entirely detect the same ones. Additionally, which TOGs are detected depends on the control set used, which can be deduced from the small intersection between sets of TOGs derived from the two control sets (Table 3, column 'Intersection'). Comparing the results gained with and without Bonferroni correction reveals that the tests detect a substantially larger number of TOGs when Bonferroni correction is not used. Additionally, a larger number of TOGs are detected when the R statistic is applied pairwise, since the union of the pairwise-detected TOGs is larger than the number inferred from the multi-comparison.

When using EST gene expression studies the focus is on genes that are preferentially expressed in a cDNA library compared to another. However, since the tests only consider if there is a significant differential expression in the libraries, without preference to any particular library, there is no report on which library the gene is preferentially expressed in. Consequently, a proportion of the reported TOGs might be preferentially expressed in the control libraries. However, these genes are less interesting, as we want to extract those that are expressed during cold stress. Preferential expression of genes in a control set does not automatically indicate that these are down-regulated in cold. For example, housekeeping genes, which are expressed at a relatively constant level disregarding cold stress or not, may show a lower proportion of the total number of transcripts during over-expression of cold-responsive genes.

In order to derive preferentially expressed TOGs, we calculated the relative EST frequency for each detected TOG in each set and marked them as up-regulated if the frequency was higher in the cold stress set than in the control set. TOGs were considered as preferentially expressed if they were: 1) detected as differentially expressed by any of the test statistics used; and 2) up-regulated in comparison to all control sets. This applied to 55 TOGs, which corresponds to 1.4% of the total number of TOGs expressed in AtCI II.

The most important aspect of the initial study is that only a minor portion of the TOGs were detected as preferentially expressed. However, many of the TOGs are uniquely expressed in cold, i.e. have an expression value in the cold stress set but not in any of the control sets. Of the 2253 (58% of totally derived from AtCI II) TOGs that are uniquely expressed in this set, merely five are detected by any of the test statistics. There is a possibility that these TOGs represent genes that are cold induced, since they do not have a corresponding expression in any of the control sets. On the other hand, many of them are weakly expressed with only one EST member. Considering the larger size of the cold stress EST set compared to the control sets, these TOGs might contain false positives that have not been discarded in the comparison to the control sets (the weakly expressed genes have probably not been sampled in the controls). As an attempt to decrease the numbers of false positives but still pick up presumably cold induced genes among these uniquely expressed, we only included TOGs with ≥3 EST members, even though there is a risk of missing some cold induced genes. Setting a threshold reduced the number of uniquely expressed TOGs to 627 (16% of totally derived from AtCI II), but in total increased the number of derived TOGs to 653 (17%).

Table 3. The number of differentially expressed TOGs in the AtCI II set in comparison to the control sets, as derived by the different test statistics. BF indicates if Bonferroni correction was used (Y) or not (N). AtCI II vs. DI I/II represents the number of detected TOGs in the AtCI II in comparison to AtDI I or AtDI II, respectively. Intersection is the number of TOGs in the intersection of TOGs detected in comparison to AtDI I and AtDI II, respectively; Union is the number of TOGs in the union of TOGs detected in comparison to AtDI I and AtDI II, respectively; Cold vs. All represents the number of TOGs detected by the multi-comparison tests applied to all sets.

Test	BF	AtCI II vs. DI I	AtCI II vs. DI II	Intersection	Union	Cold vs. All
AC	Y	17	19	7	29	-
Chi	Y	24	8	2	30	-
R	Y	17	6	2	21	23
MultChi	Y	24	8	2	30	20
AC	N	226	135	45	316	-
Chi	N	360	214	54	520	-
R	N	269	243	66	446	302
MultChi	N	360	214	54	520	314

Table 4. Results from the identification of preferentially expressed TOGs. TOGs: total number of TOGs expressed in the specified set. ≥ESTs: the threshold set for the minimum number of ESTs in the TOGs. Test statistics: number of TOGs derived using the specified test statistics. Uniq. expr.: number of TOGs uniquely expressed in the cold stress set. Pref. expr.: total number of preferentially expressed TOGs. AtCG: number of TOGs corresponding to a gene in the AtCG database.

EST set	TOGs	≥ESTs	Test statistics	Uniq. expr.	Pref. expr.	AtCG
Unbiased						
AtCI II	3875	3	55 (1.4%)	627 (16.2%)	653 (16.6%)	90 (13.8%)
AtCI III	913	2	410 (44.8%)	338 (37.0%)	410 (44.9%)	43 (10.5%)
AsCI	1762	2	59 (3.3%)	458 (26.0%)	473 (26.8%)	36 (7.6%)
HvCI	1985	2	24 (1.2%)	117 (5.9%)	136 (6.9%)	11 (8.1%)
TaCI	598	1	121 (20.2%)	285 (47.7%)	308 (51.5%)	16 (5.2%)
Subtracted						
AtCI I	668	1	-	-	668 (100%)	88 (13.2%)
OsCI	1431	2	-	-	540 (37.7%)	59 (10.9%)

The combined method of deriving preferentially expressed TOGs was thereafter applied to the remaining unbiased EST sets (see Table 4). We used different thresholds for the minimum number of ESTs when deriving uniquely expressed TOGs, since the size of the cold stress sets differ markedly (Table 4, column '≥ESTs'). For the subtracted EST sets no comparison to any control set was performed as this has already been done experimentally.

6 Comparison of Derived TOGs

The number of derived preferentially expressed TOGs varies among the sets (Table 4), with the largest number being derived for AtCI I and the smallest for barley.

Comparing the results with the previously identified cold-induced genes from the *Arabidopsis* microarray experiments (the AtCG set), several genes can be found among the derived TOGs. Although most of the AtCG genes have not been sampled in any of the cold experiments included here or are weakly expressed, about one third of the expressed genes have been identified by the method. In the AtCI II set a total of 491 TOGs correspond to an AtCG gene, where 251 are represented by more than two ESTs (the specified threshold for this set), and of those, 35.9% (90 TOGs) were identified by the method. For AtCI III a total of 119 TOGs with more than one EST member correspond to an AtCG gene and of those 36.1% (43 TOGs) were derived by the method. For the oat, barley and wheat sets 24.3%, 44.3% and 14.4%, respectively, were derived. In the subtracted sets these numbers correspond to 38.4% and 43.7% in the AtCI I and rice set, respectively.

However, there are many other TOGs that are preferentially expressed, but are not included among the AtCG genes, which is shown by the low percentages of AtCG genes among the total number of derived TOGs (Table 4). In the unbiased sets 5.2-13.8% of the TOGs correspond to an AtCG gene, which resembles the number of AtCG genes identified in the sets from the subtracted libraries, i.e.13.2% and 10.9% in AtCI I and rice, respectively. There is also a clear difference in the number of detected AtCG genes in *Arabidopsis* and rice, in comparison to the remaining crop species, where substantially larger numbers of genes can be found among the first two species. This possibly indicates a difference in the cold response among the closely related species oat, barley and wheat, compared to the less cold tolerant species *Arabidopsis* and the very cold sensitive rice.

Further analysis of the results reveals that the derived TOGs differ considerably among the sets, as the percentage in the intersection of TOGs among pairs of sets is generally low (1.0-11.5%) (Table 5). However, again there is a slight trend that the *Arabidopsis* and rice sets are more alike, as they have more TOGs in common, and that oat resembles *Arabidopsis* and rice more than barley or wheat. At first sight, these results could indicate that the method has problems in deriving cold induced genes, but when examining the annotation of the TOGs, we found that many of them have been shown to be related to cold and/or other stresses (a listing of annotations of the most highly expressed TOGs is available from the authors on request).

For example, there is a dehydrin, At1g54410, which is induced by water stress, as well as different heat shock proteins that respond to stress [33]. There are also the cold acclimation proteins WCOR410b and WCOR615, and a low temperature responsive RNA binding protein. The cold-induced COR410 (Wcor410) is a also dehydrin [34], which is expressed during both water-deficiency and cold stress. Additionally, the hits include alcohol dehydrogenase (ADH), which has been shown to increase in seedlings and leaves of *Arabidopsis*, rice and maize plants exposed to low temperature [35, 36]. Jeong *et al.* [37] isolated a glyceraldehyde-3-phosphate dehydrogenase, from the mushroom *Pleurotus sajor-caju*, which showed an increase in expression when subjected to cold, but also in response to other stresses such as salt, heat and drought. When this gene was continuously expressed in transgenic potato plants an improved salt tolerance was achieved [37]. Hurry *et al.* (2000) established that low phosphate levels is important for triggering cold acclimatization of leaves in *Arabidopsis* and that, amongst others, fructose-1,6-bisphosphate aldolase, transketolase and glyceraldehyde-3-phosphate dehydrogenase increased in expression in wild-type plants subjected to cold stress and these can also be found among the derived TOGs that are most highly expressed.

Further, two glutathione-S-transferase genes are represented among the most expressed TOGs in barley. A low temperature regulated glutathione-S-transferase has been isolated in a freezing tolerant potato species, which did not accumulate in a freezing sensitive potato species [38]. Berberich *et al.* (1995) isolated a translation elongation factor 1 in maize, which increased in expression at low temperature in leaves, but, however, decreased in roots. This is also among the most expressed TOGs in maize.

Among the most expressed TOGs for oat and barley, four and three, respectively, chlorophyll a/b binding proteins are represented. The oat plants were incubated in the dark during the cold treatment, and for the barley experiments dark incubation preceded the cold treatment. This possibly triggered the photoacclimation process, which is additionally indicated by the presence of photosystem I reaction center subunit XI (PSI-L) among the most highly expressed TOGs in the two sets. Light harvesting chlorophyll a/b binding proteins (LHCII) are parts of the light-harvesting complex (LHC), which transfers excitation energy between photosystem I and II, and under changing light conditions the LHCIIs balance this energy between the two photosystems [39]. PSI-L has also been shown to increase in expression during light [40]. On the other hand, several chlorophyll a/b binding proteins have been shown to respond to cold in a wheat microarray experiment [41], which can be explained as a response to the imbalance in the energy absorbed through photochemistry and the energy utilized through metabolism caused by the shift from warm to low temperatures. Light has also previously been shown to activate cold-stress signaling, in an *Arabidopsis* mutant study conducted by Kim *et al.* (2002), and it was suggested that cross-talk occurs between cold- and light-signaling pathways.

There is also a large representation of ribosomal proteins among the derived TOGs. This possibly indicates an increased metabolic activity in the cell as a response to cold, which requires an increased synthesis of proteins. On the other hand, although ribosomal proteins are primarily part of the ribosome, the function of which is to catalyze the mRNA-directed protein synthesis, many of them have a function outside the ribosome. For example, a mutation in the 50S ribosomal protein L6 affected the membrane stability of cells, which at low temperature caused an influx of compounds that are normally unable to penetrate [42]. Hence, this protein is possibly important for the membrane stability during cold stress. Additionally, the 60S ribosomal protein L10 is also represented among the microarray-identified cold-induced genes in *Arabidopsis* (the AtCG set).

According to the GO annotation of the TCs, a dozen transcription factors (TFs) can be found among the derived TOGs in each of the different sets, except for barley where only two TFs were represented. However, this is most likely not a complete set of expressed TFs, since many of the TOGs have an unknown function, and a number of TOGs are presumably included among these unknowns. The identified TF TOGs follow the trend with very little overlap among the sets. Then again, when considering TF family membership, some families are represented in almost all sets. For example, the AP2/EREBP domain-containing TFs are represented in oat, rice and the *Arabidopsis* sets, bZIPs are represented in rice, wheat, the AtCI II and AtCI III sets, and MYBs are represented in all sets except rice. Particularly interesting is the fact that the ICE1 TF [43] has been detected in the AtCI II set and that there is a WRKY TF represented in oat [44]. The WRKYs constitute a large family of plant specific TFs, of which some are involved in cold and drought responses.

In addition, studying the GO term annotation of the TCs in the derived TOGs reveals that for a large proportion the molecular function and/or biological process is unknown. An average of 38.2% of the TOGs in the unbiased sets and 31.4% in the subtracted sets are unknowns according to the GO annotation, and on average 51.1% and 38.6%, respectively, have an unknown biological process.

Many of the derived preferentially expressed TOGs correspond to genes previously shown to be induced during a longer period of stress. RNA extracted from pooled tissues picked from plant individuals at multiple time points will inevitably foremost represent transcripts of such genes. Highly or moderately long-term expressed genes will be sampled at several time points and therefore have a larger representation in the pooled sample than those that are only transiently expressed. Pooled samples were used in the construction of cDNA libraries for the oat, *Arabidopsis* and rice cold stress EST sets. For the barley and wheat cold stress sets, the RNA was extracted from plants exposed to cold during two days. Here, the transcripts will represent a combination of genes that are either long-term expressed or transiently expressed after some time of cold treatment. Consequently, the derived TOGs mainly represent genes that are highly or moderately long-term expressed during cold. One example is alcohol dehydrogenase (ADH), the expression level of which in *Arabidopsis*, rice and maize plants was elevated during several days of cold treatment [35, 36]. In a microarray study by Gulick *et al.* (2000) where wheat plants were exposed to cold, the cold acclimation gene WCOR14a was up-regulated at all three of the sampled time points, i.e. 1, 6 and 36 days. Similar patterns of expression have been shown for Heat shock protein family 70 in spinach and tomato [33], a perennial ryegrass glycine-rich RNA binding protein (LpGRP1) [45], and several others among the derived TOGs.

Table 5. Percentage of preferentially expressed TOGs shared by pair of cold stress EST sets

	AtCI I	AtCI II	AtCI III	AsCI	HvCI	TaCI	OsCI
AtCI I	-	5.1	6.5	4.4	2.3	1.0	5.7
AtCI II		-	8.1	1.3	11.5	1.0	2.4
AtCI III			-	2.8	0.6	1.6	3.5
AsCI				-	2.4	2.8	4.0
HvCI					-	1.6	1.7
TaCI						-	1.3

The over-representation of long-term expressed genes is further illustrated by the absence of the important CBF TFs among the derived TOGs. These TFs are transiently expressed very early in the cold acclimation process [46]. Coding sequences for 40 CBF genes were downloaded [47], of which 28 had a significant similarity match against a TC ($E \leq 10^{-80}$). These genes were mapped to seven different TOGs, where the three CBF genes in *Arabidopsis*, for example, were clustered together in a separate TOG. Of the seven TOGs, only one is among the derived preferentially expressed and is included among those derived from the oat cold stress set.

7 Discussion and Conclusions

Cold acclimation is an essential process for plants in temperate climates to survive harsh winter conditions. Low temperature is a major factor limiting the growth, productivity and distribution of the plants. In addition, physical injuries may result from freezing at subzero temperatures. The low temperature response is highly complex and the challenge is to distinguish genes responsible for the regulation of the acclimation from those related to the metabolic adjustments to the temperature shift. In order to do this, gene expression studies, such as cDNA microarrays and EST sequencing, are valuable for identification of genes contributing to the acclimation process. Consequently, such techniques have been used to study the cold acclimation process in various species. Several EST sets have been generated from cold stress plant cDNA libraries, which represent a rich source of information on genes participating in the process.

   In this study we analyzed cold stress EST sets from a number of crop species as well as *Arabidopsis thaliana,* with the aim of identifying genes preferentially expressed in cold. Instead of deriving expression profiles for individual genes in the different species, we inferred expression values for orthologous genes, in order to get comparable expression estimates across multiple species and sets. This proved to be a successful approach, since expression values were easily derived and many TOGs with a function related to cold stress could be identified.

   On the other hand, the approach is not entirely faultless and has limitations. For some of the TOGs the OrthoMCL algorithm clustered too many TCs together. For example, the largest TOG consists of 188 TC with rather diverse annotations, and should preferably be split up into smaller groups. Unfortunately, the consequence of the erroneous clustering is that it will generate an overestimated expression value for some of the TOGs. This flaw partly depends on the algorithm not separating orthologs and paralogs. Hence, ESTs that correspond to different orthologs will be mapped to the same TOG. It also depends on the fact that TCs originate from ESTs, which are not full-length gene sequences and therefore many of the TCs are partial sequences that are difficult to cluster into correct, separate groups.

   Several test statistics were applied and assumed to correctly detect differentially expressed genes. This assumption was based on a comparative study by Roumaldi *et al.* (2001) where it was concluded that the statistics are capable of detecting all differentially expressed genes when dealing with highly expressed genes that also have a large extent of differential expression. On the other hand, when the expression levels and/or the extent of differential expression decrease, the number of false negatives increase. In addition, Roumaldi *et al.* did not investigate how many false positives were detected by the statistics, and therefore an extended study should be conducted in order to get an objective view of the performance of the statistics. In this study the number of false positives is unknown, although our analysis of the annotation of the TOGs indicates that the number is limited.

   The results show an unexpectedly diverse representation of genes in the different cold sets, adding valuable information on potential mechanisms behind the acclimation process. However, it can not be excluded that a portion of these genes are identified as a result of limitations in the method. For example, the experimental conditions when constructing the cDNA library and sequencing of ESTs affects which genes are sampled, especially considering the time points of sampling and the temperatures, but also other variables such as the humidity and growth medium. This is supported by a microarray study by Vogel *et al.* (2004), aimed at identifying a set of genes that were cold-responsive both in plants grown in soil in pots and plants grown on plates containing solid culture medium. In the plate experiments both root and shoot tissue was used in the expression study, while only shoot tissue was used in the soil experiment. The results showed that only about 30% of the genes were differentially expressed in both growth media, and hence, there was a diverse representation of identified cold-regulated genes. Vogel *et al.* concluded that this likely reflected the differences in the culture conditions as well as differences in harvested tissue. The EST sets used in this study also differ in both growth media and tissue, and consequently, the diverse representation is also here likely to reflect some of these differences.

Additionally, the control sets used also affect both how many and which specific TOGs are derived. For example, the AtCI sets were compared to drought stress sets, which most likely excluded cold induced genes that are also drought induced, whereas for oat and wheat etiolated and/or unstressed sets were used as controls. Another factor is that induced genes most probably differ among the species, which is further supported by a comparative microarray study between winter and spring wheat in response to cold [41], where the results showed a clear difference between genes induced by cold in the two cultivars.

The method identified a large number of genes that were not identified in the microarray studies for *Arabidopsis*. Thus, this study indicates that the two different types of expression studies identify different sets of cold-induced genes. The EST sequencing technique has few parameters that affect the results. In addition, it works even if probes are missing and it focuses on expressed genes. However, it is less sensitive than microarrays, since the sets commonly originate from a pooled cDNA library including several time points. Microarray studies give a higher sensitivity and are commonly designed to gain information on several time points. On the other hand, there are many parameters in the analysis of the microarray experiments that can give very diverse results. Consequently, the methods should be used as complements to each other, instead of being seen as mutually exclusive.

In conclusion, the method used here to identify preferentially expressed TOGs among multiple EST sets, despite its limitations, was able to identify many genes with a function related to cold and/or other stresses. Although the results showed a highly diverse representation of genes, the TOGs mainly represented genes that are long-term expressed. Thus, the results generated here will provide useful information on genes that participate and are required over a longer time in the cold acclimation process. The method also identified a large proportion of genes with unknown function and the next step is to elucidate their role in the cold acclimation process.

References

1. Browse, J. and Z. Xin, *Temperature sensing and cold acclimation. Current Opinion in Plant Biology,*, 2001. **4**: p. 241-246.
2. Kim, H.J., et al., *Light signalling mediated by phytochrome plays an important role in cold-induced gene expression through the C-repeat/dehydration responsive element (C/DRE) in Arabidopsis thaliana.* Plant J, 2002. **29**(6): p. 693-704.

3. Sharma, P., N. Sharma, and R. Deswal, *The molecular biology of the low-temperature response in plants.* BioEssays, 2005. **27**: p. 1048-1059.
4. Smallwood, M. and D.J. Bowles, *Plants in cold climate.* The Royal Society, 2002. **357**: p. 831-847.
5. Stitt, M. and V. Hurry, *A plant for all seasons: alterations in photosynthetic carbon metabolism during cold acclimation in Arabidopsis.* Current Opinion in Plant Biology, 2002. **5**: p. 199-206.
6. Thomashow, M.F., *Plant Cold Acclimation: Freezing Tolerance Genes and Regulatory Mechanisms. .* Annual Reviews in Plant Physiology and Plant Molecular Biology., 1999. **50**: p. 571-599.
7. Venter, J.C., et al., *Massive parallelism, randomness and genomic advances.* Nature Genetics, 2003. **33**: p. 219-227.
8. Lindlof, A., *Gene identification through large-scale EST sequencing processing.* Applied Bioinformatics, 2003. **2**: p. 123-129.
9. Mayer, K. and H.W. Mewes, *How can we deliver the large plant genomes? Strategies and perspectives.* Current Opinion in Plant Biology, 2002. **5**: p. 173-177.
10. Rudd, S., *Expressed sequence tags: alternative or complement to whole genome sequences?* Trends in Plant Science, 2003. **8**: p. 321-329.
11. Fei, Z., et al., *Comprehensive EST analysis of tomato and comparative genomics of fruit ripening.* The Plant Journal, 2004. **40**: p. 47-59.
12. Romualdi, C., S. Bortoluzzi, and G.A. Danieli, *Detecting differentially expressed genes in multiple tag sampling experiments: comparative evaluation of statistical tests.* Human Molecular Genetics, 2001. **10**: p. 2133-2141.
13. Wu, X.-L., et al., *Census of orthologous genes and self-organizing maps of biologically relevant transcriptional patterns in chickens (Gallus gallus).* Gene, 2004. **340**: p. 213-225.
14. Audic, S. and J.-M. Claverie, *The Significance of Digital Gene Expression Profiles.* Genome Research, 1997. **7**: p. 986-995.
15. Claverie, J.-M., *Computational methods for the identification of differential and coordinated gene expression.* Human Molecular Genetics, 1999. **8**: p. 1821-1832.
16. Jung, S.H., J.Y. Lee, and D.H. Lee, *Use of sage technology to reveal changes in gene expression in arabidopsis leaves undergoing cold stress. .* Plant Molecular Biology, 2003. **52**: p. 553-567.
17. Schmitt, A.O., et al., *Exhaustive mining of EST libraries for genes differentially expressed in normal and tumour tissue.* Nucleic Acids Research, 1999. **27**: p. 4251-4260.
18. Stekel, D.J., Y. Git, and F. Falciani, *The Comparison of Gene Expression from Multiple cDNA libraries.* Genome Research, 2000. **10**: p. 2055-2061.
19. Strausberg, R.L., et al., *In Silico analysis of cancer through the Cancer Genome Anatomy Project.* TRENDS in Cell Biology, 2001. **11**: p. S66-S70.
20. Lee, Y., et al., *Cross-referencing Eukaryotic Genomes: TIGR Orthologous Gene Alignments (TOGA).* Genome Research, 2002. **12**: p. 493-502.
21. Li, L., J.C.J. Stoeckert, and D.S. Roos, *OrthoMCL: Identification of Ortholog Groups for Eukaryotic Genomes.* Genome Research, 2003. **13**: p. 2178-2189.
22. Bräutigam, M., et al., *Generation and analysis of 9792 EST sequences from cold acclimated oat, Avena sativa.* BMC Plant Biology, 2005. **5**: p. 18.
23. Parkinson, J., D.B. Guiliano, and M. Blaxter, *Making sense of EST sequences by CLOBBing them.* BMC Bioinformatics, 2002. **3**: p. 31.
24. Huang, X. and A. Madan, *CAP3: A DNA sequence assembly program. .* Genome Research, 1999. **9**: p. 868-877.

25. Quackenbush, J., et al., *The TIGR Gene Indices: reconstruction and representation of expressed genes.* Nucleic Acids Research, 2000. **28**: p. 141-145.
26. Zhang, J.Z., R.A. Creelman, and J.-K. Zhu, *From Laboratory to Field. Using information from Arabidopsis to Engineer Salt, Cold and Drought Tolerance in Crops.* Plant Physiology, 2004. **135**: p. 615-624.
27. Chen, W., et al., *Expression profile matrix of arabidopsis transcription factor genes suggests their putative functions in response to environmental stresses.* Plant Cell, 2002. **14**: p. 559-574.
28. Fowler, S. and M.F. Thomashow, *Arabidopsis transcriptome profiling indicates that multiple regulatory pathways are activated during cold acclimation in addition to the cbf cold response pathway.* . Plant Cell, 2002. **14**: p. 1675-1690.
29. Kreps, J.A., et al., *Transcriptome changes for arabidopsis in response to salt, osmotic, and cold stress.* Plant Physiology, 2002. **130**: p. 2129-2141.
30. Seki, M., et al., *Monitoring the expression profiles of 7000 arabidopsis genes under drought, cold and high-salinity stresses using a full-length cdna microarray.* Plant Journal, 2002. **31**: p. 279-292.
31. Vogel, J.T., et al., *Roles of the cbf2 and zat12 transcription factors in configuring the low temperature transcriptome of arabidopsis.* Plant Journal, 2005. **41**: p. 195-211.
32. Hannah, M.A., A.G. Heyer, and D.K. Hincha, *A Global Survey of Gene Regulation during Cold Acclimation in Arabidopsis thaliana.* PLoS Genetics, 2005. **1**: p. e26.
33. Li, Q.-B., D.W. Haskell, and C.L. Guy, *Coordinate and non-coordinate expression of the stress 70 family and other molecular chaperones at high and low temperature in spinach and tomato.* Plant Molecular Biology, 1999. **39**: p. 21-34.
34. Danyluk, J., et al., *Differential expression of a gene encoding an acidic dehydrin in chilling sensitive and freezing tolerant gramineae species.* FEBS Letter, 1994. **344**: p. 20-24.
35. Christie, P.J., M. Hahn, and V. Walbot, *Low-temperature accumulation of alcohol dehydrogenase-1 mRNA and protein activity in maize and rice seedlings.* Plant Physiology, 1991. **95**: p. 699-706.
36. Jarillo, J.A., et al., *Low temperature Induces the Accumulation of Alcohol Dehydrogenase mRNA in Arabidosis thaliana, a Chilling-Tolerant Plant.* Plant Physiology, 1993. **101**: p. 833-837.
37. Jeong, M.J., S.C. Park, and M.O. Byun, *Improvement of salt tolerance in transgenic potato plants by glyceraldehyde-3 phosphate dehydrogenase gene transfer.* Mol Cells, 2001. **12**(2): p. 185-9.
38. Seppanen, M.M., et al., *Characterization and expression of cold-induced glutathione S-transferase in freezing tolerant Solanum commersonii, sensitive S. tuberosum and their interspecific somatic hybrids.* Plant Science, 2000. **153**(2): p. 125-133.
39. Yang, D.H., H. Paulsen, and B. Andersson, *The N-terminal domain of the light-harvesting chlorophyll a/b-binding protein complex (LHCII) is essential for its acclimative proteolysis.* FEBS Lett, 2000. **466**(2-3): p. 385-8.
40. Toyama, T., H. Teramoto, and G. Takeba, *The level of mRNA transcribed from psaL, which encodes a subunit of photosystem I, is increased by cytokinin in darkness in etiolated cotyledons of cucumber.* Plant Cell Physiol, 1996. **37**(7): p. 1038-41.
41. Gulick, P.J., et al., *Transcriptome comparison of winter and spring wheat responding to low temperature.* Genome, 2005. **48**(5): p. 913-23.
42. Bosl, A. and A. Bock, *Ribosomal mutation in Escherichia coli affecting membrane stability.* Mol Gen Genet, 1981. **182**(2): p. 358-60.
43. Chinnusamy, V., et al., *Ice1: A regulator of cold-induced transcriptome and freezing tolerance in arabidopsis.* . Genes & Development, 2003. **17**: p. 1043-1054.

44. Wu, K.L., et al., *The WRKY family of transcription factors in rice and Arabidopsis and their origins.* DNA Res, 2005. **12**(1): p. 9-26.
45. Shinozuka, H., et al., *Gene expression and genetic mapping analyses of a perennial ryegrass glycine-rich RNA-binding protein gene suggest a role in cold adaptation.* Mol Genet Genomics, 2006: p. 1-10.
46. Thomashow, M.F., *So What's new in the field of plant cold acclimation? Lots!* Plant Physiology, 2001. **125**: p. 89-93.
47. Skinner, J.S., et al., *Structural, functional and phylogenetic characterization of a large CBF gene family in barley.* Plant Molecular Biology, 2005. **59**: p. 533-551.

Mining Spatial Gene Expression Data for Association Rules

Jano van Hemert[1] and Richard Baldock[2]

[1] National e-Science Institute, University of Edinburgh, UK
`jvhemert@nesc.ac.uk`
[2] MRC Human Genetics Unit, Edinburgh, UK
`Richard.Baldock@hgu.mrc.ac.uk`

Abstract. We analyse data from the Edinburgh Mouse Atlas Gene-Expression Database (EMAGE) which is a high quality data source for spatio-temporal gene expression patterns. Using a novel process whereby generated patterns are used to probe spatially-mapped gene expression domains, we are able to get unbiased results as opposed to using annotations based predefined anatomy regions. We describe two processes to form association rules based on spatial configurations, one that associates spatial regions, the other associates genes.

Keywords: association rules, gene expression patterns, in situ hybridization, spatio-temporal atlases.

1   Introduction

Association rules are popular in the context of data mining. They are used in a large variation of application domains. In the context of gene expression and images, we can classify previous studies into three categories:

1. Association rules over gene expression from micro-array experiments.
2. Association rules over features present in an image.
3. Association rules over annotated images.

The first category aims to find rules that show associations between genes, and perhaps other things, such as the type of treatment used in the experiment [1,2]. Typical examples of such rules are: if gene a expresses then there is a good chance that gene b also expresses. The second type leads to rules that say something about the relationships between features in images. We found only a few studies in this direction. One which first uses a vocabulary to annotate items found in tiled images and then creates rules that describe the relationships within tiles [3], another which aims to discriminate between different textures by using associations rules [4,5]. Last, the third category extracts rules that show how annotations of images are associated, which is useful to find further images of interest based on an initial set found from a search query [6]. This is the typical concept of "customers who bought this also bought that".

S. Hochreiter and R. Wagner (Eds.): BIRD 2007, LNBI 4414, pp. 66–76, 2007.
© Springer-Verlag Berlin Heidelberg 2007

We introduce a novel application of mining for association rules in results of *in situ* gene expression studies. In earlier studies in which association rules were applied to gene expression results, these results originated from microarray experiments, where the aim is to find associations between genes [1,2] in the context of broad tissue types. Here in contrast, we will operate on accurate spatial regions with patterns derived from *in situ* experiments. This type of accurate data enables us to extract two types of interesting association rules, first we can extract the same type of relationships between genes. However, we can also extract rules expressed in the form of spatial regions, thereby providing knowledge on how areas in an embryo are linked spatially. The only other study in the direction of spatial association rules the authors are aware of, is solely based on synthetic data [7].

In the next section we describe the Edinburgh Mouse Atlas Project, a spatio-temporal framework for capturing anatomy and gene expression patterns in developing stages of the mouse. Then, in Section 3 we explain the concepts and process of extracting association rules. The spatial framework and association rules are combined in Section 4, which forms the basis for our experiments and results in Section 5. A discussion is given in Section 6.

2   Edinburgh Mouse Atlas Project

EMAGE (`http://genex.hgu.mrc.ac.uk/`) is a freely available, curated database of gene expression patterns generated by *in situ* techniques in the developing mouse embryo [8]. It is unique in that it contains standardized spatial representations of the regions of gene expression for each gene, denoted against a set of virtual reference embryo models. As such, the data can be interrogated in a novel and abstract manner by using space to define a query. Accompanying the spatial representations of gene expression patterns are text descriptions of the sites of expression, which also allows searching of the data by more conventional text-based methods terms.

Data is entered into the database by curators that determine the level of expression in each *in situ* hybridization experiment considered and then map those levels on to a standard embryo model. An example of such a mapping is given in Figure 1. The strength of gene expression patterns are classified either as no expression, weak expression, moderate expression, strong expression, or possible detection.

In this study we restrict to a subset of the data contained in the database. This subset of data originates from one study [9] and contains 1618 images of *in situ* gene expression patterns in a wholemount developing mouse embryo model of Theiler Stages 16, 17, and 18 [10]. The study includes 1030 genes; a subset of genes were screened two or three times. By mapping the strong and moderate expression patterns of these images on to the two-dimensional model for Theiler Stage 17 shown in Figure 1(b), we can work with all these patterns at the same time.

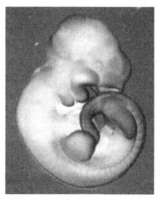 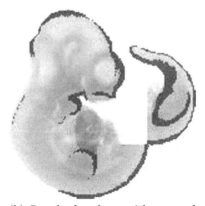

(a) Original image shows expression of Hmgb1 in a mouse
embryo at Theiler Stage 17

(b) Standard embryo with mapped
levels of expression (red=strong, yel-
low=moderate, blue=not detected)

Fig. 1. An example of curating data; the gene expression in the original image on the left is mapped on to the standard embryo model of equal developmental stage on the right (entry EMAGE:3052 in the online database).

3   Association Rules

Various algorithms exist to extract association rules, of which the Apriori [11] algorithm is the most commonly used and we too shall use it in this study. It entails a two-step process, defined below, which consists of first generating the set of frequent itemsets, from which association rules are extracted that are above a certain confidence level.

Definition 1

1. *Given a set of items I, the input consists of a set of* transactions D, *where each transaction T is a non-empty subset of items taken from the* itemset I, *so $T \subseteq I$.*
2. *Given an itemset $T \subseteq I$ and a set of transactions D, we define the* support *of T as $support_D(T)$ equals the proportion of transactions that contain T to all transactions $|D|$.*
3. *By setting a* minimum support level α, *where $0 \leq \alpha \leq 1$, we define* frequent itemsets *to be itemsets where $support_D(T) \geq \alpha$.*

Definition 2

1. *An* association rule *is a pair of disjoint itemsets, the antecedent $A \subseteq I$ and the consequent $C \subseteq I$, where $A \Rightarrow C$ and $A \cap C = \emptyset$.*
2. *The concept of support of an association rule carries over from frequent itemsets as $support_D(A \Rightarrow C) = support_D(A \cup C)$.*

3. We define the confidence *of an association rule* $A \Rightarrow C$ *as:*

$$confidence_D(A \Rightarrow C) = \frac{support_D(A \cup C)}{support_D(A)}$$

In other words, frequent itemsets are items that frequently occur together in transactions with respect to some user defined parameter, i.e., the minimum support. In an analog way, the confidence of association rules shows how much we can trust a rule, i.e., a high confidence means that if A occurs, there is a high chance of C occurring with respect to the set of transactions.

The prototypical example to illustrate association rules uses the domain of the supermarket. Here a transaction is someone buying several items at the same time. An itemset would then be something as $\{jam, butter, bread\}$. If this itemset is also a frequent itemset, i.e., it meets the minimum support level, then a possible association rule would be $\{jam, butter\} \Rightarrow \{bread\}$.

We provide the definition of *lift* [12], which is a popular measure of interestingness for association rules. Lift values larger than 1.0 indicate that transactions $(A \Rightarrow C)$ containing the antecedent (A) tend to contain the consequent (C) more often than transactions that do not contain the antecedent (A). Lift is defined as,

Definition 3. $lift(A \Rightarrow C) = confidence(A \Rightarrow C)/support(C)$

4   Applying Association Rules on the Spatial Patterns

We create a set of "probe patterns" by laying a grid over the standard embryo model. Each point in the grid is a square of size 5×5 pixels. The whole image is 268×259 pixels.

To create a relationship between the gene expression patterns in the images, we first build a matrix of similarities between those patterns and the probe patterns. Each element in the matrix is a measure of similarity between the corresponding probe pattern and gene-expression region. We calculate the similarity as a fraction of overlap between the two and the total of both areas. This measurement is intuitive, and commonly referred to as the Jaccard index [13]:

$$similarity(d_1, d_2) = \frac{S(d_1 \wedge d_2)}{S(d_1 \vee d_2)},$$

where $S(x)$ calculates the size of the pattern x. The pattern $d_1 \wedge d_2$ is the the disjunction or intersection of the image d_1 and the probe pattern d_2, while the pattern $d_1 \vee d_2$ is the union of these regions. The similarity is higher when the overlapping area is large and the non-overlapping areas are small.

In Figure 2, two gene expression pattern images are shown, together with their patterns in the standard embryo model. Also, three examples of probe patterns are shown. We pair probe patterns with gene expression patterns and then calculate the Jaccard Index, which gives us a measure of overlap. A high number means much of the two patterns overlap, where 1.0 would mean the

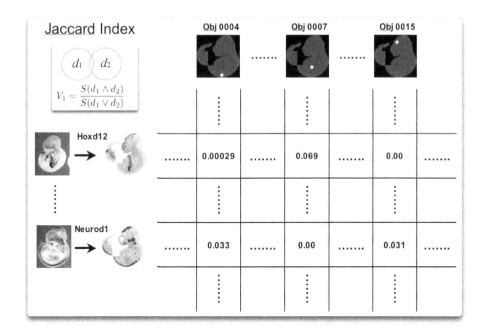

Fig. 2. Similarity matrix. Each original gene-expression assay shown on the left-hand side is mapped on to the standard model embryo, a Theiler Stage 17 wholemount. The probe patterns, shown at the top are then compared with the mapped gene-expression patterns using the Jaccard Index as a similarity measure. These are depicted in the table. The actual simularity matrix has 1618 rows and 1675 columns.

unlikely event of total overlap. A very small number, such as 0.00029, means only a little area of the probe pattern overlaps with a large gene expression pattern, and 0.0 would mean no overlap occurs at all. This is important to note as later we will use a threshold to filter these latter two occurrences.

From the similarity matrix, or *interaction matrix*, two different sets of transactions are constructed, which in turn lead to two different types of association rules.

1. The items I are genes from the data set, where a transaction $T \subseteq I$ consists of genes that all have an expression pattern intersecting with the same probe pattern.
2. The items I are the probe patterns, where a transaction $T \subseteq I$ consists of probe patterns all intersecting with the expression patterns in the same image.

To create the first type of transactions, we take for each probe pattern r, every gene g from which its associated gene expression pattern g_e satisfies the minimum similarity β, i.e., similarity$(r, g_e) > \beta$, to form the itemset.

The second type of transactions is created in a similar way. For each gene expression pattern g in the database we create an itemset that consists of a set of probe patterns that intersect with the gene expression pattern g_e. Each probe pattern r must satisfy the minimum similarity β, i.e.., similarity$(r, g_e) > \beta$, to get included in the itemset.

When the transactions are formed, the standard Apriori algorithm is used to create frequent itemsets, where the minimum support is set to different values for the different types. The common procedure is to start with a high minimum support, e.g., 90%, which often does not yield any results, and then reduce the threshold as far as the algorithm will support it. At some point either the amount of memory required or the amount of time will render the algorithm useless, at which time we stop and take the results from previous tried minimum support level. From the frequent itemsets we build association rules. Generally we want rules we can be confident about, hence we set the minimum confidence level to 0.97.

5 Experiments and Results

We generated a 1675 square probe patterns of size 5×5 to cover the whole standard embryo model. These parameters were chosen first to match the number of images used in this study. Also, the 5×5 probe patterns allow sufficiently large transactions and a sufficiently number of transactions. When forming transactions, we used a minimum similarity of $\beta = 0.005$. This latter parameter setting was chosen after first performing the whole process with $\beta = 0.00$, which resulted in frequent itemsets and consequently, association rules, with a very high support (above 80%). This generally happens when items are over-represented, and then dominate the analysis. This is caused by gene expression patterns that cover an extremely large area of the embryo. As such patterns will intersect always they do not make an interesting result for finding associations. Also, such association rules are obvious ones.

For example, when using no minimum similarity, e.g., $\beta = 0.00$, and searching for associations by genes, the highest supported association rule is Hmgb2 $\Rightarrow$ Hmgb1 with a support of 0.881, a confidence of 0.993, and a lift of 1.056. These genes are known to be highly associated and are common to many processes [14], hence they express over much of the embryo.

By using the similarity measure to our advantage we can filter out these over-represented genes as in these cases the similarity measure will be small due to the large non-overlapping areas of expression. We found a threshold of $\beta = 0.005$ is sufficient to exclude these patterns. Smaller values rapidly decrease the number of genes left after filtering, and this will make the analysis useless. The runtime of the Apriori algorithm on these data sets is a few seconds on a 2.33Ghz Intel Core Duo chip.

5.1 Associations by Genes

Table 1 shows the association rules found when transactions are genes that have regions of expression intersecting with the same probe pattern. Here a minimum support level of 0.06 is used, and a minimum confidence of 0.97. The lift values

Table 1. Association rules based on itemsets of genes and where a transaction is a set of genes exhibiting expression all intersecting with the same probe pattern, created with a minimum support of 0.06 and a minimum confidence of 0.97.

Rule	Antecedent	Consequent	Support	Confidence	Lift
1	Lhx4 Otx2	Dmbx1	0.065	0.971	10.871
2	Lhx4 Dmbx1	Otx2	0.065	0.990	9.226
3	Brap Zfp354b	9830124H08Rik	0.060	0.979	10.225
4	9830124H08Rik Trim45	Brap	0.061	0.979	11.813
5	9130211I03Rik Zfp354b	9830124H08Rik	0.062	0.980	10.230

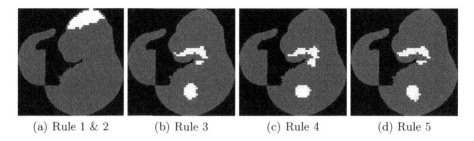

(a) Rule 1 & 2   (b) Rule 3   (c) Rule 4   (d) Rule 5

Fig. 3. Spatial regions corresponding to the conjunction of all transactions, i.e., probe patterns, that lead to each association rule.

of the resulting rules are high; rule 2 for instance has the smallest lift of 9.2. This tells us that for all these rules, if the antecedent occurs, it is far more likely that the consequent occurs than that the consequent does not occur. In other words, these rules are trustworthy under assumption of the provided data.

We can provide more background information about each rule by showing the transactions they were inferred from. As in this case the transactions are the probe patterns, we can display them all at once in the standard embryo model. Figure 3 shows the spatial regions within the embryo model of each rule. Note that rule 1 and rule 2 have the same set of genes and therefore reference the same spatial region. Both these rules associate the genes Lhx4, Otx2, and Dmbx1 in the midbrain. Rules 3, 4, and 5, all associate genes in the forebrain, 1st branchial arch and forelimb. However, rule 4 is quite different to rules 3 and 5, as it has two continuous regions instead of three. Important to note is the accuracy of the expression patterns in the latter three rules. All three rules share common genes, but the differences in genes are cause for subtle changes in the patterns seen in Figure 3.

5.2 Associations by Spatial Regions

When calculating the association rules based on spatial regions representing items in transactions, which are in turn representing probe patterns intersecting with patterns of the same gene, we end up with millions of rules. The reason for this is that many rules will state obvious knowledge. This is a common effect in data mining exercises, and it allows one to verify the quality of knowledge extracted. In the context of spatial transactions, most of the rules we extracted state: if a probe pattern exhibits expression of gene g, then it is very likely a neighbouring probe pattern will also exhibit expression of gene g. Although from a data miner's perspective, such rules aid in improving trust in the correctness of the system, they do not provide new information to biologists, as gene expression patterns are inherently local.

The next step, after extracting the association rules, is to select the rules that are newsworthy. As local rules are reasonably obvious, many of these rules state that if a gene exhibits expression in a probe area than it is likely the adjacent probe area also exhibits the same gene. We therefore sort all the rules by the relative distances of the probe patterns in each rule. This way, rules that show relationships between probe patterns further away from each other can be distinguished from association rules operating on local patterns. This sorting uses the sum of the absolute Euclidean distance of every pair of probe patterns in each rule divided by the number of pairs. The higher this sum, the larger the relative distances between pairs of probe patterns. We will show the two rules with the largest average distance between the probe patterns.

Figure 4 shows an association rule based on probe patterns as items. The three square patterns in Figure 4(a) form the antecedent of the rule, while the square pattern in Figure 4(b) forms the consequent of the rule. It has a support of 0.130, which is quite high. The confidence of 1.00 means that if the patterns in the antecedent are expressing a particular gene then under the data provided, the regions in the consequent *must* also exhibit expression of the same gene. The lift is 3.53; this tells us that rules which contain the antecedent will be more likely to also contain the pattern in the consequent than not contain the pattern in the consequent. This confirms the rule provides valuable information about the data.

This rule is extracted from gene expression patterns involving the following list of genes: 9130211I03Rik 9830124H08Rik Abcd1 Ascl3 BC012278 BC038178

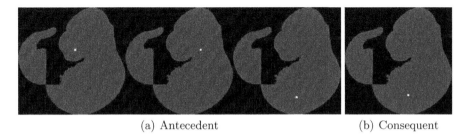

(a) Antecedent   (b) Consequent

Fig. 4. Spatial region based association rule with a support of 0.130, a confidence of 1.00, and a lift of 3.53. This should be interpreted as "if a gene exhibits expression in all areas shown in (a), then it shall also be exhibiting expressing in (b)". The average Euclidean distance between the probe patterns is 85.0 pixels.

Bhlhb2 Brap Cebpa Chek2 Cited4 Cops6 Creb3l4 Cxxc1 Elf2 F830020C16Rik Fgd5 Gmeb1 Jmjd2a Jrk Mefv Mid2 Mtf1 Nab2 Neurog3 Phtf1 Piwil1 Pole3 Prdm5 Rfx1 Rnf20 Rxra Snf8 Tcf12 Tcfcp2 Thap7 Trim11 Trim34 Wbscr14 Zbtb17 Zfp108 Zfp261 Zfp286 Zfp354b Zfp95.

Figure 5 shows a similar association rule. It has a support of 0.101, a confidence of 1.00, and a lift of 3.53. More of these rules were extracted and show us there is a strong relationship between the areas which project onto the developing eye/forebrain, hindbrain and limb/kidney. It involves the following list of genes: 9130211I03Rik Abcd1 Ascl3 BC012278 BC038178 Bhlhb2 Brap Brca1 Cebpa Chek2 Cited4 Cxxc1 Elf2 F830020C16Rik Fgd5 Gmeb1 Jmjd2a Jrk Mid2 Mtf1 Nab2 Neurog3 Phtf1 Piwil1 Pole3 Prdm5 Rfx1 Rnf20 Snai2 Snf8 Tcf12 Tcfcp2 Thap7 Trim11 Trim34 Trim45 Wbscr14 Zbtb17 Zfp108 Zfp113 Zfp261 Zfp354b Zfp454 Zfp597 Zfp95.

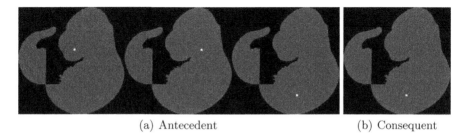

(a) Antecedent   (b) Consequent

Fig. 5. Spatial region based association rule with a support of 0.101, a confidence of 1.00, and a lift of 3.53. This should be interpreted as "if a gene exhibits expression in all areas shown in (a), then it shall also be exhibiting expressing in (b)". The average Euclidean distance between the probe patterns is 84.5 pixels.

6 Discussion

We show a novel application of association rules extracted from unique data; accurate spatial localization of gene expression patterns derived from *in situ* experiments. Approaching the data from two different angles allows us to derive two different types of associations. The first type shows relationships between genes, which form a popular field of study for data mining applications. The second type extracts rules that link spatial regions of expression patterns within the embryo. So far, just a few studies consider such rules and they operate on synthetic data. In this study we have shown that this type of analysis can yield potentially interesting and novel associations. The next step will include both automatic (e.g. GO term enrichment and pathway analysis) and manual analysis of the biological meaning of these rules. The identification of spatial associations within the embryo using unbiased sampling is, so far as we can determine, completely novel.

Using our approach of generated probe patterns to represent items in conjunction with a similarity measure between a gene expression pattern and a particular probe pattern provides two major advantages. It allows to filter out expression patterns that exhibit an area so large they threaten to dominate the frequent itemsets. These expression patterns generally do not add much knowledge as they show ubiquitous gene-expression over almost the whole embryo.

A further advantage is a significant decrease in the influence of the human bias that is present in the annotations of *in situ* images. Similar experiments as reported in this study, but performed on the annotations proved difficult due to a lack of richness, often resulting in support levels too low to consider useful, for instance less than 3%. There are a number of reasons why human annotations lack richness compared to the spatial annotations present in the curated database. Most important is the interest of the annotator as this will result in a focus on particular features present in the image. Then there is the knowledge of the annotator, which will significantly influence what features will be picked up for annotation, and may also result in error. Last we mention time constraints, which always form a major bottleneck in interactive annotations. The result is that manual text annotations are at best low-resolution descriptions of the data but are typically partial and biased observations.

Last, we want to mention the results we showed from tracing back the transactions that are the cause for a certain association rule. In other analyses, such as supermarket data, these transactions are rather meaningless. However, here in the spatial context and in the biological context of genes, these transactions provide supporting evidence to the biologist to understand and verify the consequence of the association rules.

Our near future goal is to provide an interface freely accessible via any web browser to steer the process of the extraction of both types of association rules. By providing a feedback mechanism we then allow biologists to evaluate the usefulness of the rules.

Longer term goals are to include the temporal aspects present in the database, thereby providing rules that state relationships between different stages of development. Also, providing comparative analysis of association rules across species, specifically between the mouse and human model, to highlight similarities and differences in gene and spatial interactions.

References

1. Creighton, C., Hanash, S.: Mining gene expression databases for association rules. Bioinformatics **19**(1) (2003) 79–86
2. Becquet, C., Blachon, S., Jeudy, B., Boulicaut, J., Gandrillon, O.: Strong-association-rule mining for large-scale gene-expression data analysis: a case study on human sage data. Genome Biology **3**(12) (2002)
3. Tešić, J., Newsam, S., Manjunath, B.: Mining image datasets using perceptual association rules. In: Proceedings of SIAM Sixth Workshop on Mining Scientific and Engineering Datasets in conjunction with the Third SIAM International Conference. (2003)
4. Rushing, J., Ranganath, H., Hinke, T., Graves, S.: Using association rules as texture features. IEEE Transactions on Pattern Analyses and Machine Intelligence (2001)

5. Bevk, M., Kononenko, I.: Towards symbolic mining of images with association rules: Preliminary results on textures. Intelligent Data Analysis **10**(4) (2006) 379–393

6. Malik, H., Kender, J.: Clustering web images using association rules, interestingness measures, and hypergraph partitions. In: ICWE '06: Proceedings of the 6th international conference on Web engineering, New York, NY, USA, ACM Press (2006) 48–55

7. Ordonez, C., Omiecinski, E.: Discovering association rules based on image content. In: Proceedings of the IEEE Advances in Digital Libraries Conference (ADL'99), Baltimore, Maryland (1999)

8. Christiansen, J., Yang, Y., Venkataraman, S., Richardson, L., Stevenson, P., Burton, N., Baldock, R., Davidson, D.: Emage: a spatial database of gene expression patterns during mouse embryo development. Nucleic Acids Research **34** (2006) 637–641

9. Gray, P., et al.: Mouse brain organization revealed through direct genome-scale tf expression analysis. Science **306**(5705) (2004) 2255–2257

10. Theiler, K.: The House Mouse Atlas of Embryonic Development. Springer Verlag, New York (1989)

11. Agrawal, R., Imielinski, T., Swami, A.: Mining association rules between sets of items in large databases. In Buneman, P., Jajodia, S., eds.: Proceedings of the 1993 ACM SIGMOD International Conference on Management of Data, Washington, D.C. (1993) 207–216

12. Brin, S., Motwani, R., Ullman, J.D., Tsur, S.: Dynamic itemset counting and implication rules for market basket data. In Peckham, J., ed.: Proceedings ACM SIGMOD International Conference on Management of Data. (1997) 255–264

13. Jaccard, P.: The distribution of flora in the alpine zone. The New Phytologist **11**(2) (1912) 37–50

14. Pallier, C., Scaffidi, P., Chopineau-Proust, S., Agresti, A., Nordmann, P., Bianchi, M., Marechal, V.: Association of chromatin proteins high mobility group box (hmgb) 1 and hmgb2 with mitotic chromosomes. Mol. Biol. Cell **14**(8) (2003) 3414–3426

Individualized Predictions of Survival Time Distributions from Gene Expression Data Using a Bayesian MCMC Approach

Lars Kaderali

German Cancer Research Center (dkfz), Theoretical Bioinformatics,
Im Neuenheimer Feld 580, 69120 Heidelberg, Germany
l.kaderali@dkfz-heidelberg.de

Abstract. It has previously been demonstrated that gene expression data correlate with event-free and overall survival in several cancers. A number of methods exist that assign patients to different risk classes based on expression profiles of their tumor. However, predictions of actual survival times in years for the individual patient, together with confidence intervals on the predictions made, would provide a far more detailed view, and could aid the clinician considerably in evaluating different treatment options. Similarly, a method able to make such predictions could be analyzed to infer knowledge about the relevant disease genes, hinting at potential disease pathways and pointing to relevant targets for drug design. Here too, confidences on the relevance values for the individual genes would be useful to have.

Our algorithm to tackle these questions builds on a hierarchical Bayesian approach, combining a Cox regression model with a hierarchical prior distribution on the regression parameters for feature selection. This prior enables the method to efficiently deal with the low sample number, high dimensionality setting characteristic of microarray datasets. We then sample from the posterior distribution over a patients survival time, given gene expression measurements and training data. This enables us to make statements such as *"with probability 0.6, the patient will survive between 3 and 4 years"*. A similar approach is used to compute relevance values with confidence intervals for the individual genes measured.

The method is evaluated on a simulated dataset, showing feasibility of the approach. We then apply the algorithm to a publicly available dataset on diffuse large B-cell lymphoma, a cancer of the lymphocytes, and demonstrate that it successfully predicts survival times and survival time distributions for the individual patient.

1 Introduction

Gene expression profiles have been shown to correlate with survival in several cancers [1,13]. One important goal is the development of new tools to diagnose cancer [5] and to predict treatment response and survival based on expression profiles [14,12]. A classification of patients in different risk classes has been

S. Hochreiter and R. Wagner (Eds.): BIRD 2007, LNBI 4414, pp. 77–89, 2007.

demonstrated to be a first step [12], and can aid the clinician in evaluating different treatment options. However, predicting actual survival times together with
confidence intervals on the predictions would be a valuable progress. Furthermore, an analysis of such a predictor could hint at relevant genes that correlate
with survival, which may provide new information on pathogenesis and ethiol-
ogy, and may further aid in the elucidation of disease pathways and the search
for novel targets for drug design.

One of the main statistical problems in gene expression data analysis stems
from the large number of genes measured simultaneously, typically in the range of
several thousands to tens of thousands, while usually only a few dozen to several
hundred patients are available, even in the larger studies. This causes problems
with multiple testing in hypothesis testing approaches, and yields regression
models with thousands of unknowns but only a small number of data points
to estimate model parameters, and is a severe limitation [11]. For this reason,
dimension reduction methods such as clustering or principal components analysis
are typically carried out prior to further analysis. Unfortunately, criteria used in
dimension reduction are often unrelated to the prediction model used to analyse
the data later and to survival, and may remove relevant genes.

By combining dimension reduction and parameter estimation for regression
parameters into one single step, one can avoid the problems stemming from a
separate dimension reduction. Feature selection can be combined into the regression using a Bayesian approach, penalizing overly complex models using a
special feature selection prior on the regression parameters. On can then maximize the posterior, and use the resulting regression parameters for predictions
on new patients [7].

However, from a theoretical point of view, instead of using the maximum
a-posteriori regression parameters for predictions, one should average over *all*
possible regression parameter values, weighting each with its probability when
making predictions. Let $\mathcal{D}$ be the training data given, let $\mathbf{x}$ be the vector of gene
expression values measured for a new patient, and let t be the patient's survival
time. Then, instead of maximizing the posterior $p(\theta|\mathcal{D})$ and using the resulting
vector θ to predict, for example, the mean of a distribution $p(t|\mathbf{x}, \theta)$, predictions
ought to be based on the distribution

$$p(t|\mathbf{x}, \mathcal{D}) = \int_{\Omega} p(t|\mathbf{x}, \theta) p(\theta|\mathcal{D}) \, d\theta, \tag{1}$$

where the integration is over the full parameter space Ω. Computationally, this
requires an integration over the full (and potentially very high dimensional)
parameter space, and is far more involved than a simple maximization of the
posterior $p(\theta|\mathcal{D})$.

In this report, we evaluate equation (1) using a Markov Chain Monte Carlo
(MCMC) approach. This allows us to sample survival times directly from the
distribution $p(t|\mathbf{x}, \mathcal{D})$, which are used to compute means and variances over sur-
vival times for the individual patient. The distribution over predicted survival
times can be visualized, providing far more information than just a mere predicted risk class or a single survival time in years. In addition, we demonstrate

sampling from the posterior distribution $p(\theta|\mathcal{D})$, which provides relevance values with confidence intervals for the distinct gene expression values measured, revealing relevant survival related genes.

After a presentation of the method, we show its application to a simulated dataset first, and then present results on a publicly available clinical dataset on diffuse large B-cell lymphoma, a cancer of the white blood cells.

2 Approach

2.1 The Cox Regression Model

Let $\mathbf{x} \in \mathbb{R}^n$ be the vector of gene expression measurements for a given patient of interest, where n is the number of genes measured, and let T be a random variable corresponding to the survival time of the patient. The Cox regression model [3] postulates that the risk of failure in the next small time interval Δt, given survival until t, is given by

$$\lambda(t|\mathbf{x}, \theta) = \lim_{\Delta t \downarrow 0} \frac{P\{T < t + \Delta t | t \leq T\}}{\Delta t} = \lambda_0(t)e^{<\theta,\mathbf{x}>}, \qquad (2)$$

where $\theta \in \mathbb{R}^n$ is a vector of regression parameters, and $\lambda_0(t)$ is an arbitrary baseline hazard function describing the hazard when $\mathbf{x} = 0$. Given $\mathbf{x}$ and θ, the hazard completely specifies the survivor function $F(t) = P\{T \geq t\}$ and the probability density function $f(t) = -dF(t)/dt$.

Survival data in clinical studies is usually right-censored, i.e., for some patients, only a lower bound on the survival time is known. Under the conditions of random censoring over time, independence of the censoring times for distinct patients from one another, and independence of the censoring process from $\mathbf{x}$ and θ, the maximum likelihood solution to the regression problem is the parameter vector θ maximizing

$$p(\mathcal{D}|\theta) \propto \prod_{j=1}^{m} f(t^{(j)}|\mathbf{x}^{(j)}, \theta)^{1-\delta^{(j)}} F(t^{(j)}|\mathbf{x}^{(j)}, \theta)^{\delta^{(j)}}, \qquad (3)$$

where m is the number of patients in the clinical study and $\mathbf{x}^{(j)}$ and $t^{(j)}$ are the vector of the gene expression measurements for patient j and the patients survival time or censoring time, respectively. The indicator variable $\delta^{(j)}$ is 1 if $t^{(j)}$ is censored, and 0 otherwise.

2.2 Bayesian Extension of the Cox Model

In the setting where one measures far more genes than one has patients in the study, $n >> m$, typical of microarray measurements, the optimization problem of maximizing equation (3) with respect to θ is underdetermined. A predictor based on a vector θ obtained from such a maximization will usually overfit the data. This problem can be controlled using Bayesian regularization.

The posterior distribution over the regression parameters θ given the training data $\mathcal{D}$ is given by

$$p(\theta|\mathcal{D}) = \frac{p(\mathcal{D}|\theta)p(\theta)}{p(\mathcal{D})} \propto p(\mathcal{D}|\theta)p(\theta). \qquad (4)$$

Here, $p(\mathcal{D}|\theta)$ is given by the likelihood (3), according to the Cox regression model. $p(\theta)$ is a prior distribution over the regression parameters θ, reflecting our subjective prior beliefs about the parameters, and $p(\mathcal{D}) = \int p(\mathcal{D}|\theta)p(\theta)\, d\theta$.

The central assumption we make at this point is that most of the gene expression measurements x_i for given genes i are not related to survival. This is modeled using independent mean-zero normal distributions on the components θ_i of the parameter vector θ. Furthermore, we assume that most of the normal distributions are strongly peaked around zero, by using independent gamma distributions on the standard deviations of the normal distributions on the regression parmameters. Thus, the prior distribution $p(\theta)$ on the regression parameters is given by

$$p(\theta) = \prod_{i=1}^{n} \int_{0}^{\infty} \frac{1}{\sqrt{2\pi\sigma_i^2}} \exp\left[-\frac{1}{2}\frac{\theta_i^2}{\sigma_i^2} \right] \frac{a^r \sigma_i^{r-1}}{\Gamma(r)} e^{-a\sigma_i}\, d\sigma_i, \qquad (5)$$

where a and r are scale and shape parameters of the gamma distribution, and $\Gamma(\cdot)$ is the gamma function. Figure 1 shows the resulting prior for the two-dimensional case. The plot clearly shows how the prior favours solutions with most of the weights θ_i in the proximity of zero, thus penalizing models with a large number of relevant genes.

One can now maximize the posterior (4) and use the resulting weights θ to make predictions for new cases [7]. However, from a theoretical point of view, one should not focus on just one single underlying vector of regression parameters θ. In fact, the distribution $p(\theta|\mathcal{D})$ may have multiple modes. To determine the relevance of individual genes, the full distribution $p(\theta|\mathcal{D})$ should be considered. Similarly, predictions of the survival time t for a new patient with gene expression values $\mathbf{x}$, given the training data $\mathcal{D}$, should be based on the full distribution

$$p(t|\mathbf{x}, \mathcal{D}) = \int_{\Omega} p(t|\mathbf{x}, \theta, \mathcal{D})p(\theta|\mathcal{D})\, d\theta. \qquad (6)$$

Mean and variance of the distributions $p(\theta|\mathcal{D})$ and $p(t|\mathbf{x}, \mathcal{D})$ can then be used to asses the relevance of individual gene expression measurements and to make survival time predictions for newly diagnosed patients, respectively.

2.3   Hybrid Monte Carlo Algorithm

Unfortunately, the integrals required to calculate means and variances of equations (4) and (6) are not analytically tractable. In addition, the integration in (6) over θ has to be carried out in a very high dimensional space, depending

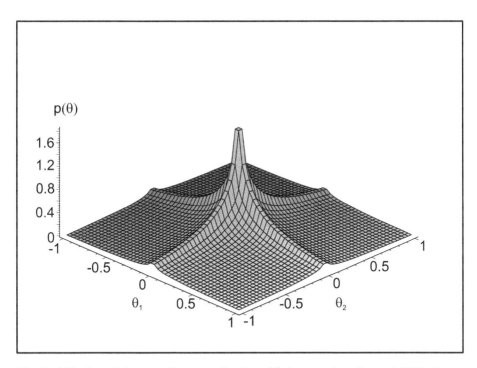

Fig. 1. 3-D plot of the two-dimensional prior $p(\theta)$ for $a = 1$ and $r = 1.0001$. It can clearly be seen how the prior favors sparse regression parameter vecors with many of the θ_i near zero.

on the number of gene expression values measured. Computationally, this is a challenging problem.

We propose to use the Hybrid Monte Carlo (HMC) algorithm for this purpose. The algorithm was developed by [4] for problems arising in quantum chromodynamics, good reviews can be found in [10] and [2]. HMC samples points from a given multi-dimensional distribution $\pi(\nu)$, expressed in terms of a potential energy function $E(\nu) = -\ln \pi(\nu)$, by introducing momentum variables $\rho = (\rho_1, ..., \rho_n)$ with associated kinetic energy $K(\rho)$, and iteratively sampling for the momentum variables from $K(\rho)$ and following the Hamiltonian dynamics of the system in time. Doing so, the algorithm generates a sequence of points distributed according to $\pi(\nu)$. Moments of the distribution $\pi(\nu)$ can then be estimated from the samples drawn. In our application, estimates derived from $p(\theta|\mathcal{D})$ will then provide some information on the relevance of individual genes.

The optimum survival time prediction y for a given patient depends on the error function $\mathrm{err}(t, t^{true})$ used to measure the error of predicting a survival time t when the true survival time is t^{true}. When the squared error function $\mathrm{err}(t, t^{true}) = (t^{true} - t)^2$ is used, it is optimal to predict the mean

$$\int_0^\infty t p(t|\mathbf{x}, \mathcal{D})\, dt = \int_0^\infty \int_\Omega t p(t|\mathbf{x}, \theta) \frac{p(\mathcal{D}|\theta)p(\theta)}{p(\mathcal{D})}\, d\theta\, dt. \qquad (7)$$

This expression can be approximated in two steps:

1. Draw $\theta^{(k)}$ from $p(\mathcal{D}|\theta)p(\theta)/p(\mathcal{D})$, using the Hybrid Monte Carlo algorithm.
2. Given $\theta^{(k)}$, draw $t^{(k)}$ from $p(t|\mathbf{x}, \theta^{(k)})$. This is easily done for the exponential distribution used in the Cox regression model using rejection sampling.

These two steps are repeated a large number of times, say for $k = 1, ..., N$, and (7) is then approximated as

$$\int_0^\infty t p(t|\mathbf{x}, \mathcal{D}) \, dt \approx \frac{1}{N} \sum_{k=1}^{N} t^{(k)}. \tag{8}$$

Similarly, the samples drawn can be used to compute the variance of $p(t|\mathbf{x}, \mathcal{D})$.

3   Results

As a proof of principle, we applied the approach presented to a simple simulated dataset first, where the true underlying processes determining survival are known. Then, the method was applied to a publicly available dataset on diffuse large B-cell lymphoma (DLBCL), a cancer of the B-lymphocytes.

3.1   Simulated Data

Data Simulation: Gene expression levels and survival times were simulated for 240 patients and 7400 genes, using the Cox model, with constant baseline hazard $\lambda_0(t) = 0.18$. It was assumed that only the first gene $\mathbf{x}_1^{(j)}$ has an influence on the survival time $t^{(j)}$ of patient j, all other genes were assumed not to be correlated to survival. The expression level $\mathbf{x}_i^{(j)}$ was drawn from a normal distribution with mean zero and standard deviation 0.5 for all genes $i = 2, ..., 7400$ and all patients $j = 1, ..., 240$. The "relevant" gene $\mathbf{x}_1^{(j)}$ was either drawn from a normal distribution with mean -0.6 and standard deviation 0.5, or from a normal distribution with mean +0.6 and standard deviation 0.5, corresponding to two assumed subclasses with different survival characteristics. Patients were randomly assigned to one of these two classes. Survival times were then simulated by sampling from the Cox model. Each patient was subject to random, uniform censoring over the time interval $[0, 10]$.

Gene relevances θ: The data was split in training and validation subsets, with 2/3 of the patients in the training subset, and 1/3 reserved for validation. Parameters for the prior distribution over the regression parameters θ were set to $r = 1.6$ and $a = 40$. HMC sampling was started several times with different starting points, with similar results. 500 initial points of the Markov chain were discarded, 2500 points were sampled from $p(\theta|\mathcal{D})$ and used to compute mean and variance. Gene $\mathbf{x}_1$ is correctly identified as correlated with survival, with mean $E[\theta_1] = 0.7274$ and standard deviation 0.0949. All other weights $\theta_2, ..., \theta_{7400}$ have means smaller than 0.0025 and standard deviation smaller than 0.01. Figure 2 shows the projection of the points θ sampled onto the first two dimensions θ_1 and θ_2.

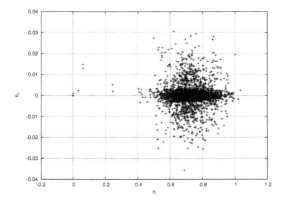

Fig. 2. The 3000 points drawn from the distribution $p(\theta|\mathcal{D})$, projected onto the first two dimensions θ_1 and θ_2, for the simulated dataset

Survival time predictions: Survival time predictions for the validation samples were then made as described. Figure 3 (left) shows the true survivor functions $F(t)$ for the groups predicted to survive up to and including 5 years, and longer than 5 years in the validation subset. Shown along with the survivor functions are two standard deviation error bars on the estimates. A logrank test of the null hypothesis of equality of the survivor functions gives a p-value of 1.7×10^{-6} $(n = 80)$, which is highly significant for rejecting the null hypothesis.

Time dependent receiver operator characteristic [6] curves are useful to evaluate the performance of a predictive algorithm for censored survival times. ROC curves plot all achievable combinations of sensitivity over specificity for a given classifier on two-class classification problems. In the context of survival analysis, a threshold t on the survival time is used to distinguish short from long survivors. The area under the ROC curves (the AUC) can then be plotted for different threshold times t. This value $AUC(t)$ will be in the interval $[0, 1]$, with larger values indicative of better performance of the predictor, and a value of 0.5 equivalent to guessing a survival time.

Figure 3 (right) plots the area under the ROC curve for different threshold-times t in the interval $[0, 20]$, confirming that the sampling algorithm can be used to successfully predict survival times on the simulated dataset.

3.2 Diffuse Large B-Cell Lymphoma

Diffuse large B-cell lymphoma (DLBCL) is a cancer of the lymphocytes, a type of white blood cell. It is the most common type of lymphoma in adults, and can be cured by anthracycline-based chemotherapy in only 35 to 40 percent of patients, the remainder ultimately succumbing to the disease. The dataset used was collected by [12], it consists of 7,399 gene expression values measured in 240 patients, using the lymphochip assay.

The data was preprocessed to reduce dimensionality prior to analysis with the Markov Chain. Several runs on the full dataset were conducted first, but no

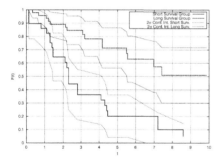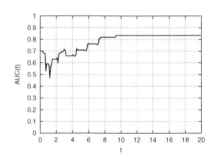

Fig. 3. Left: Survivor functions $F(t)$ estimated from the data for the two subgroups predicted to live short (≤ 5 years) versus long (> 5 years). Validation samples were split in short and long survivors based on the survival time predictions made, plotted are survivor functions for the two subgroups estimated from the true survival times, together with two standard deviation error bars ($p = 1.7 \times 10^{-6}$, $n = 80$). **Right:** Time dependent area under the receiver operator characteristics (ROC) curve for the simulated dataset.

sufficient convergence of the chains was achieved in acceptable running times. Genes with more than 40 percent missing values or with variance less than 0.45 were removed. Genes were then clustered using hierarchical clustering, with Pearson's correlation coefficient as similarity measure. Clusters were represented by their means. Clustering was interrupted when similarity dropped below 0.5 for the first time, at which point the genes were clustered in 1582 distinct clusters. Finally, the data was normalized such that each gene had mean zero and variance one. Patients were then split in training ($n = 160$) and validation ($n = 80$) subsets, using the same split as in the original Rosenwald publication.

Gene relevances θ: 11,000 points $\theta^{(k)}$ were drawn from the distribution $p(\theta|\mathcal{D})$, using the hybrid Monte Carlo algorithm. Parameters used in the prior distribution $p(\theta)$ were $a = 30$ and $r = 1.6$, and constant baseline hazard $\lambda_0(t) = 0.18$. The first 1,000 steps of the Markov Chain were discarded, to allow the chain to converge to its stationary distribution. To check for convergence of the chain, different random starting points were used in different runs, with very similar results.

The genes receiving weights most strongly corresponding to short survival in the runs conducted are LC_25054, E1AF, ETV5, DKFZp586L141, SLCO4A1, MEF2C, CAC-1 and a few others. Genes most strongly associated with long survival are RAFTLIN, RGS16, MMP1, several MHC genes (including DP, DG, DR and DM alpha and beta genes), CD9, TP63, CD47, and a few others. Several of these genes have previously been reported, a biological validation is under way and will be published elsewhere.

Survival time predictions: Survival times for the patients in the validation subset were predicted using the two-step-procedure presented. Figure 4 shows

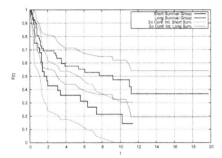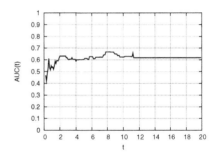

Fig. 4. Left: Survivor functions estimated from the true survival times, for the subgroups of the validation dataset predicted to live long (> 5 years) and short (≤ 5 years), respectively, in the DLBCL dataset ($p = 0.01195$, $n = 80$). **Right:** Area under the ROC curve plotted over time, for the DLBCL dataset.

the survivor functions estimated from the true survival times, for the subsets of patients in the validation subset predicted to live short (≤ 5 years) versus long (> 5 years), along with two standard deviation error bars on the estimate. A logrank test of the null-hypothesis of equality of the survivor functions of the two subgroups yields a p-value of 0.01195 ($n = 80$). Figure 4 shows the area under the time-dependent ROC curves of the survival time predictions, plotted over time.

One of the major advantages of the Bayesian approach presented is the ability of the method to access the full distribution $p(t|\mathbf{x}, \mathcal{D})$. This enables us to compute variances on the predictions made, and makes it possible to visualize the distribution over survival times for the individual patient, providing significantly more information than is available from "classical" regression methods. Figure 5 illustrates this clearly. The plot shows histograms of the predicted survival times for selected patients, generated from the survival time predictions made. Shown are survival time predictions for two representative patients, patient 18 and patient 21.

Patient 18 has a true survival time of only 0.9 years, even though the patient has a good prognosis according to clinical staging using the International Prognostic Index (IPI). The prediction made using the approach presented is very unfavorable, with a mean predicted survival time of merely 1.3 years, and a standard deviation over predictions of 1.4 years. The empirical distribution over survival time predictions is shown in Figure 5 (left), and reflects this unfavorable outcome. The low standard deviation over the predictions indicates a strong confidence in the predictions made, and this distribution would have been an early warning that the patient may be at higher risk than indicated by clinical staging.

Patient 21 is a long survivor, the true survival time is not known. The patient was still alive at 6.4 years, at which time he dropped out of the study. Figure 5 (right) shows relative frequencies of predicted survival times for this patient, the mean of the predictions made is 4.6 years, however, the standard deviation is

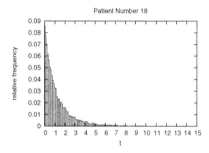 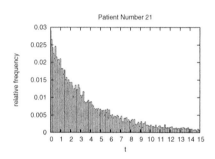

Fig. 5. Left: Distribution over predicted survival times for patient 18, an uncensored patient with true survival time 0.9 years. The mean predicted survival time for the patient is 1.3 years, with a standard deviation of 1.41 years. **Right:**Distribution over predicted survival times for patient 21. The true survival time for this patient is unknown, the patient dropped out of the study after 6.4 years. The mean predicted survival time for the patient is 4.6 years, with a standard deviation of 5.4 years.

5.4 years, indicating less certainty in the predictions made for this patient. The patient falls into the medium risk segment of the International Prognostic Index (IPI), which coincides with the distribution over predicted survival times made by our approach.

4   Discussion

We have analyzed the performance of the method presented on a simulated and a real-world dataset. Results on the simulated data show that the method is able to automatically recognize relevant genes and make predictions of survival directly on high dimensional data, without additional dimension reduction steps. In addition, variances for the survival times predicted can be computed, yielding significantly more information to a clinician than a mere prediction of a risk class. The Kaplan-Meier plots and the receiver operator characteristics analysis in Figure 3 confirm that the method developed has successfully learned the underlying mechanisms influencing survival from the data.

Obviously, an analysis of a real-world dataset is more complex than processing the simulated data. The level of noise in gene expression measurements is unknown, and relevant survival related factors are likely not available. The number of relevant genes affecting survival in DLBCL is unknown, as is the level of redundancy in the data. Last but not least, the Cox regression model used may not appropriately reflect the distribution over survival times, and certainly is a crude approximation to the true distribution at best.

For running time reasons, it turned out to be necessary to carry out some dimension reduction on the DLBCL data prior to analysis by the method presented in this article. Several runs of the Hybrid Monte Carlo algorithm were conducted on the full dataset first, each requiring approximately three days on the machine used (Pentium IV, 3 GHz) to sample 10,000 points from the distribution $p(\theta|\mathcal{D})$. However, during these $10,000$ steps, no sufficient convergence of the Markov Chains was achieved, and much longer sampling appears to be necessary. This is likely due to the high redundancy of microarray gene expression measurements, and correspondingly multiple modes of the distribution $p(\theta|\mathcal{D})$. For this reason, the data was clustered with respect to the genes, reducing dataset dimensionality and removing redundant information. This contrasts to the maximum a-posteriori approach we have previously presented, where the analysis could be carried out on the full data without clustering the genes first [7].

This additional clustering step necessary to reduce dataset dimensionality, and the complications arising from it, may explain the slightly inferior results of the Markov Chain Monte Carlo approach presented when compared to the maximum-a-posteriori (MAP) approach laid out in [7]. From a theoretical point of view, the MCMC approach should be favored, since it makes use of the full distribution $p(\theta|\mathcal{D})$ for predictions, and not just the single parameter vector θ^* maximizing $p(\theta|\mathcal{D})$. In addition, confidence intervals are readily available in the MCMC approach, whereas they cannot directly be computed for the MAP solution since the integrals involved are not analytically tractable, and additional approximations and assumptions would be necessary.

The predictive performance of the MCMC approach is comparable to other, recently published results. For example, [8] have recently published AUC plots for partial Cox regression models, involving repeated least squares fitting of residuals and Cox regression fitting, on the same dataset we used, and report AUC values between 0.62 and 0.67 for the best model in their publication, depending on time t. Similarly, [9] have presented a 6-gene predictor score for survival in DLBCL, applying their method to the Rosenwald data, the area under the time dependent ROC curves is approximately 0.6 for all timepoints t. Results of the MCMC approach we present yields comparable results, but in addition, can provide confidences on the predictions made.

5   Conclusion

From the methodological point of view, this work raises two questions that may deserve further evaluation. Our comparison of results of the MCMC approach and the MAP approach to making predictions indicates that relevant information may get lost in the clustering process. This is an interesting observation in so far that most approaches in use for survival time predictions from gene expression data today make use of extensive dimension reduction steps prior to carrying out the actual data analysis. Such steps are often based on arbitrary cutoff levels determining more or less directly how many genes are used for the prediction. The choice of such cutoff levels is often a difficult problem, but results may depend significantly on them. If too many genes are used, a predictor may be prone to overfitting, and may struggle with multiple modes and redundant information in the data. On the other hand, if too much data reduction is carried out, relevant information may be thrown away. The proper tradeoff is a delicate balance, and difficult to achieve.

The second question raised directly relates to the degree of redundancy in gene expression measurements. Given that some of the genes measured on microarrays are highly correlated, and given the analysis problems stemming from high dimensional data with only few samples in clinical studies, clearly larger studies with more patients and more sophisticated analysis methods are highly desirable. The prior distribution used over the regression parameters in this work aims at dealing with this problem by favoring solutions with only few relevant genes. At the same time, however, a more thorough study of correlation between genes may be useful to guide the design of future clinical studies, and algorithms should exploit the correlation structure in the data. After all, what is the point of measuring expression values of more and more genes, if all we can do with the additional information is to come up with more and more sophisticated methods to reduce dataset dimensionality again?

To summarize, we have presented a novel, Markov-Chain based approach to predict survival times in years and months, together with confidences, from gene expression data. The method provides feedback on the relevance of individual genes by weighting the gene expression measurements, and it considers the full distribution $p(\theta|\mathcal{D})$ over the weights when making predictions, instead of merely using the single vector θ^* maximizing $p(\theta|\mathcal{D})$. By computing first and second order moments of the distribution $p(\theta|\mathcal{D})$, certainties in the gene relevances can be given.

Similarly, the full distribution $p(t|\mathbf{x}, \mathcal{D})$ is evaluated when making survival time predictions for a new patient $\mathbf{x}$, given the training data $\mathcal{D}$. The method allows the computation of moments of the survival distribution, and the full distribution $p(t|\mathbf{x}, \mathcal{D})$ can be visualized for the individual patient, as shown in Figure 5.

From a clinical point of view, the methodology presented permits the analysis of distributions over true survival times in years and months for the individual patient, directly from gene expression data. Such an analysis may aid a clinician in evaluating treatment options, in particular since the method presented provides probabilistic statements along with the prediction made, such as the variance, and full distributions over survival times are available. Patients may benefit from such advanced predictions in the future, and receive more appropriate therapies based on a refined assessment of their individual risk.

Acknowledgement

Parts of this work were conducted while the author was employed at the Center for Applied Computer Sciences, University of Cologne. The author acknowledges support from the BMBF through the Cologne University Bioinformatics Center (CUBIC) during this time.

References

1. Beer,D.G., Kardia,S.L.R., Huang,C.-C., et al, (2002) Gene-expression profiles predict survival of patients with lung adenocarcinoma. *Nature Medicine* 8, 816–24.

2. Chen,L., Qin,Z., Lius,J.S. (2001) Exploring Hybrid Monte Carlo in Bayesian Computation, *Proceedings of ISBA 2000 – The Sixth World Meeting of the Interational Society for Bayesian Analysis.*

3. Cox,D.R. (1972) Regression models and life tables. *J. Roy. Statist. Soc. Ser. B Metho.* 34, 187–220.

4. Duane,S., Kennedy,A.D., Pendleton,B.J., Roweth,D. (1987) Hybrid Monte Carlo, *Physics Letters B*, 195:216–222.

5. Golub,T.R., Slonim,D.K., Tamayo,P. et al. (1999) Molecular classification of cancer: Class discovery and class prediction by gene expression monitoring. *Science* 286, 531–537.

6. Heagerty,P.J., Lumley,T., Pepe,M.S. (2000) Time-dependent ROC curves for censored survival data and a dignostic marker. *Biometrics* 56, 337–344.

7. Kaderali,L., Zander,T., Faigle,U., Wolf,J., Schultze,J.L., Schrader,R. (2006) CASPAR: A Hierarchical Bayesian Approach to predict Survival Times in Cancer from Gene Expression Data, *Bioinformatics* 22, 1495–1502..

8. Li,H., Gui,J. (2004) Partial Cox regression analysis for high-dimensional microarray gene expression data. *Bioinformatics* 20 Suppl. 1, i208–215.

9. Lossos,I.S., Czerwinski,D.K., Alizadeh,A.A. et al. (2004) Prediction of survival in diffuse large B-cell lymphoma based on the expression of six genes. *NEJM* 350, 1828–37.

10. Neal,R.M. (1999) Bayesian Learning for Neural Networks. *Springer-Verlag*, New York.

11. Reid,J.F., Lusa,L., De Cecco,L., et al. (2005) Limits of Predictive Models using Microarray Data for Breast Cancer Clinical Treatment Outcome. *JNCI* 97, 927–930.

12. Rosenwald,A., Wright,G., et al. (2002) The use of molecular profiling to predict survival after chemotherapy for diffuse large-B-cell lymphoma. *NEJM* 346, 1937–47.

13. Shipp,M.A., Ross,K.N., Tamayo,P., Weng,A.P., et al. (2002) Diffuse large B-cell lymphoma outcome prediction by gene-expression profiling and supervised machine learning. *Nature Medicine* 8, 68–74.

14. van De Vijver,M.J., He,Y.D., van't Veer,L.J. et al. (2002) A Gene-Expression Signature as a Predictor of Survival in Breast Cancer. *NEJM* 347, 1999–2009.

Comparing Logic Regression Based Methods for Identifying SNP Interactions

Arno Fritsch and Katja Ickstadt

Universität Dortmund, 44221 Dortmund, Germany
fritsch@statistik.uni-dortmund.de, ickstadt@statistik.uni-dortmund.de

Abstract. In single-nucleotide polymorphism (SNP) association studies interactions are often of main interest. Logic regression is a regression methodology that can identify complex Boolean interactions of binary variables. It has been applied successfully to SNP data but only identifies a single best model, while usually there is a number of models that are almost as good. Extensions of logic regression that consider several plausible models are Monte Carlo logic regression (MCLR) and a full Bayesian version of logic regression (FBLR) proposed in this paper. FBLR allows the incorporation of biological knowledge such as known pathways. We compare the performance in identifying SNP interactions associated with the case-control status of the three logic regression based methods and stepwise logistic regression in a simulation study and in a study of breast cancer.

1 Introduction

In recent years developments in sequencing technology extended the possibilities for researchers to study the effect of genetic variations on the risk of developing a complex disease, such as cancer. The most common sequence variations are single-nucleotide polymorphisms (SNPs), which consist of a single base pair difference in the DNA. Some SNPs have been shown to be associated with diseases, e.g. SNPs in N-acetyltransferase 2 (NAT2), which codes for an enzyme that transforms certain carcinogens, influence the risk of getting colon and bladder cancer [1]. In SNP association studies interest focuses not only on the effect of individual SNPs but also on their interactions. One way of incorporating such interactions is by estimating haplotypes from the genotype data. However, if only a few SNPs from each gene are included in the study it is reasonable to use the individual SNPs in the analysis.

A method to identify complex interactions in situations with many binary predictors is logic regression. As described in [2] the method has been applied successfully to SNP data. A potential problem is that logic regression identifies a single best-fitting model, while in practice usually a number of models fit almost equally well. This problem arises from the large number of possible predictors and can be overcome by considering several plausible models. In [3] it is proposed to incorporate logic regression into a Bayesian framework and to sample from the a posteriori distribution of logic models. The taken approach

S. Hochreiter and R. Wagner (Eds.): BIRD 2007, LNBI 4414, pp. 90–103, 2007.
© Springer-Verlag Berlin Heidelberg 2007

is not fully Bayesian as no prior is set on the regression parameters, they are instead estimated using maximum likelihood. This requires setting a fixed maximum number of predictors and a fixed penalty parameter on model size, which influences the results that are obtained.

In this article we propose a full Bayesian version of logic regression for SNP data that uses only simple AND-combinations of binary variables and their complements as predictors. This allows for an easier interpretation of the resulting models and facilitates the inclusion of available prior information, such as known pathways. We compare how the different methods based on logic regression perform in identifying SNP interactions both in a simulation study and in an application to a case-control study on breast cancer. In both situations we also experiment with the inclusion of a priori pathway information.

2   Methods

2.1   Logic Regression

Logic regression is a regression method designed for data with many binary predictor variables, especially if interactions between these variables are of main interest [4]. The method constructs new predictors as logic (Boolean) combinations of the binary variables X_i, e.g. $X_3 \wedge (X_2^C \vee X_4)$. Logic regression models are of the form

$$g[E(Y)] = \beta_0 + \sum_{i=1}^{k} \beta_i L_i \ , \tag{1}$$

where each L_i represents a logic combination of binaries, g is an appropriate link function and β_i, $i = 0, 1, \ldots, k$, are unknown regression parameters. This is a generalized linear model with binary predictors, as each L_i itself is a binary variable. For case-control data g is taken to be $\mathrm{logit}(x) = \log(x/(1-x))$, where $x \in (0, 1)$. A logic combination can be represented by a logic tree, where each "leaf" contains a variable or its complement and each knot contains one of the operators AND and OR. Figure 1 shows such a logic tree. The number of possible logic trees is huge even for a moderate number of binaries, and studies involving SNPs usually comprise at least 100 of them.

To find a good-fitting model in the large space of possible models simulated annealing, a stochastic search algorithm, is used. Starting from a logic model, at each iteration a candidate model is proposed by randomly changing one of the logic trees, e.g. by replacing a leaf or growing/pruning a branch. A better fitting model is always accepted, a worse fitting one only with a certain probability. The fit is measured by the deviance, i. e. $-2 \cdot \max_\beta \ell(\beta)$, where ℓ is the log-likelihood and β the vector of regression parameters. The algorithm starts with a high probability of accepting worse models to avoid getting stuck in a local minimum. The acceptance probability is then lowered consecutively. More on simulated annealing and its convergence properties can be found in [5]. The best model will likely overfit, as the deviance does not penalize model size which is defined as the total number of leaves on all trees in the model. To avoid

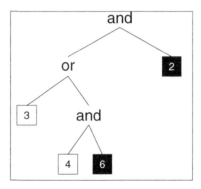

Fig. 1. A logic tree. White numbers on black background denote the conjugate of a variable.

overfitting cross-validation is used, choosing the model size as the one with the lowest mean cross-validation deviance.

To apply logic regression to SNP data, it is necessary to recode the variables to make them binary. A SNP can be regarded as a variable with the possible values 0, 1 and 2, representing the number of variant alleles. In [2] it is proposed to code each SNP variable into two binary variables, $X_{i,d} = 1$ if SNP_i has at least one variant allele and $X_{i,r} = 1$ if it has two variant alleles, both being 0 otherwise. These binary variables have a biologic interpretation: X_d stands for the dominant and X_r for the recessive effect of the SNP.

2.2   Monte Carlo Logic Regression

A problem with logic regression is that usually there are many other models that fit the data almost as good as the best. In Fig. 2 a typical cross-validation result is displayed. The best model would have three leaves on three trees, but for many other model sizes and numbers of trees the mean cross-validation deviance is similar. Therefore it is reasonable to consider more than one plausible model, especially if a SNP association study has exploratory character, i.e. if the focus is on obtaining possible associations rather than one best model.

The uncertainty regarding model choice can be taken into account by using the framework of Bayesian statistics, in which all parameters of a model are treated as random variables. A prior distribution is set on all parameters. It can be uninformative or incorporate prior knowledge about the parameters. The prior distribution $p(\theta)$ is updated by the data vector Y resulting in the posterior distribution $p(\theta|Y)$ using Bayes formula

$$p(\theta|Y) = \frac{p(Y|\theta) \cdot p(\theta)}{p(Y)} \quad , \tag{2}$$

where θ is the parameter of interest and $p(Y|\theta)$ is the likelihood. It is possible to assign a prior distribution to a whole space of models. Equation (2) usually

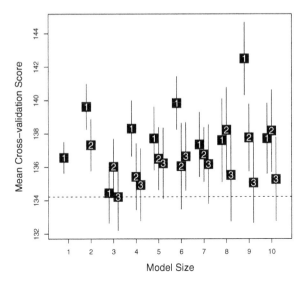

Fig. 2. Cross-validation result. Labels show number of trees in the model, lines indicate ± 1 standard error. Dashed line shows mean cross-validation error of best model.

cannot be solved analytically. Instead a Markov Chain Monte Carlo (MCMC) algorithm can be used to obtain a sample from the posterior distribution by constructing a Markov chain that has the posterior as its stationary distribution. To sample from a model space with changing dimensionality a special case of MCMC, the reversible jump algorithm, is needed, see [6] for details. The method that results from replacing the simulated annealing algorithm in logic regression by an MCMC scheme is called Monte Carlo Logic Regression (MCLR) in [3]. As in simulated annealing, in the reversible jump algorithm a candidate model is proposed and accepted with a certain probability. One difference is that the formula for the acceptance probability does not change with the number of iterations.

In MCLR the prior distribution consists of a geometric prior on model size and a uniform distribution on all models of the same size. The prior for a model M is thus given by

$$p(S) \cdot p(M|S) \propto a^S \cdot \frac{1}{N(S)} \quad a \in (0,1) \; , \tag{3}$$

where $N(S)$ is the number of models of size S. The parameter a from the geometric part of the prior has the effect of penalizing larger models as they have less prior weight. The penalty parameter a and a maximum number of trees K have to be preset. The parameter β is not assigned a prior distribution, it is estimated by maximum likelihood, making MCLR not a full Bayesian method.

The MCMC algorithm is run and after an initial burn-in the found models are stored. Summary statistics of these models can then be computed, such as the proportion of models containing a particular SNP and the proportion of models containing two SNPs together in one logic tree. This gives an idea of which SNPs and SNP interactions might be associated with the outcome of interest.

Despite the uniform distribution on models of the same size not all logic expressions of the same size receive the same prior weight because some expressions have more representations as logic trees than others. An example is given in Appendix A.

2.3 Full Bayesian Logic Regression

In MCLR the penalty parameter a and the maximum number of trees K have to be set. It is difficult to choose these parameters from prior knowledge and their values can have a large influence on the results that are obtained, as will be seen in the simulation study. By assigning a prior distribution to the β parameters and thus extending MCLR to a full Bayesian method, the problem of setting fixed values for these parameters can be overcome.

A large number of sampled models can be summarized better if each of them is easy to interpret. For the proposed Full Bayesian Logic Regression (FBLR) we thus use only AND-combinations of binaries and their complements as predictors. Any AND-combination can be uniquely represented by a sorted tuple of indices of the included binaries, for example $X_4 \wedge X_1^C \wedge X_7$ by $(1C, 4, 7)$. So there is no problem with an unequal prior weight for logic expressions. By de Morgans rule, $(X_i \wedge X_j)^C = X_i^C \vee X_j^C$, the complement of any AND-combination is an OR-combination. It does not matter for the fit whether L_i or L_i^C is included in a model, only the sign of the corresponding β_i changes, so that all possible OR-combinations are also implicitly included as predictors.

A prior distribution on β leads to an automatic penalization of larger models, because for a β of higher dimension the prior mass is spread out more widely, leading to a lower posterior probability than the one of a smaller model with an equal likelihood $p(Y|\theta)$, as computed by (2). Here, the prior on β is chosen to be normal with expectation 0 and covariance matrix $v \cdot I$. As v has an influence on how much larger models are penalized it is not set to a fixed value, but is assigned its own distribution. The prior on all unknown parameters can then be written in factorized form as

$$p(\theta) = p(k) \cdot p(v) \cdot p(\beta|v,k) \cdot \prod_{i=1}^{k} p(s_i) \cdot p(bin|s_i) \ , \tag{4}$$

where k is the number of predictors in the model, s_i is the size of the ith predictor and bin is the vector of binaries making up that predictor. With the automatic penalization of larger model, $p(k)$ can be set to be uniform on $\{0, \ldots, K\}$, where K is chosen high enough not to be influential, usually $K = 10$ suffices. $p(v)$ is an inverse gamma with parameters 0.001 and 0.1. The distribution on a single predictor is chosen similar to the prior in MCLR, setting $p(s_i)$ to a (truncated) geometric distribution with parameter a' and $p(bin|s_i)$ to a uniform distribution on all interactions of size s_i. The search for interactions can be restricted within specific groups of SNPs, making it possible to include prior information, e.g. about known pathways, so that only biologically plausible interactions are considered. For interactions of maximum order 2 we simply set a' to 1, i.e. we do not put an extra penalty on interactions. If higher interactions are included a' should be set to a value below 1, e.g. 0.7, to prevent predictors from becoming to large.

With the distributions of β and v being normal and inverse gamma, samples from the posterior distribution of models can be obtained by adapting a sampling algorithm from the Bayesian MARS classification model described in [7]. The algorithm uses latent variables that make it possible to integrate out β analytically, so that the marginal likelihood $p(Y|M)$ for a model M can be used in the acceptance probability instead of $p(Y|M, \beta^*)$ with a specific value β^*. This leads to higher acceptance rates and thus makes the algorithm more efficient. For details on the algorithm the reader is referred to the original paper. The moves to candidate models for FBLR and how the acceptance probability has to be modified is described in Appendix B.

After a sample from the posterior has been generated it can be computed in what fraction of models a predictor is included and predictions can be made, e.g. by averaging over the predictions by the models in the sample.

Recently another independently developed approach inserting logic regression in a full Bayesian framework has been published [8]. The approach is based on haplotype data and the search is constrained to all logic trees consistent with a perfect phylogeny. We agree that reducing the large number of possible logic regression models by restricting the model space to biologically plausible ones is a good idea. In our algorithm the possibility to search for SNP interactions only within known pathways is in the same spirit.

3   Results

3.1   Simulation Study

The performance of identifying SNP interactions of the methods described in the previous section is compared in a simulation study. The simulated data sets are designed to be simplified versions of SNP data sets used in current research. Each consists of 1000 persons with 50 SNPs each, where each SNP is drawn from a binomial distribution with parameters 2 and 0.3. The assumption of a binomial distribution is equivalent biologically to the SNP being in Hardy-Weinberg equilibrium. The case-control status Y of each person is determined by drawing from a Bernoulli distribution with the logit of the case probability $\eta = \log\left(\frac{P(Y=1)}{P(Y=0)}\right)$ given by

- Model 1: $\eta = -0.70 + 1.0 \cdot (X_{1d} \wedge X_{2d}^C) + 1.0 \cdot (X_{3d} \wedge X_{4d}) + 1.0 \cdot (X_{5d} \wedge X_{6d})$
- Model 2: $\eta = -0.45 + 0.6 \cdot (X_{1d} \wedge X_{2d}^C) + 0.6 \cdot (X_{3d} \wedge X_{4d}) + 0.6 \cdot (X_{5d} \wedge X_{6d})$
- Model 3: $\eta = -0.40 + 1.0 \cdot (X_{1d} \wedge (X_{2d}^C \vee X_{3d}))$
- Model 4: $\eta = -0.70 + 1.0 \cdot (X_{3d} \wedge X_{11d}^C) + 1.0 \cdot (X_{8d} \wedge X_{12d}) + 1.0 \cdot (X_{13d} \wedge X_{14d})$

Models 1 and 2 represent situations where several independent small SNP interactions influence the probability of being a case, each interaction having a moderate or low effect, respectively. The third model contains a more complicated interaction with a moderate effect. A model with a complicated interaction and a low effect is not included as then none of the investigated methods is likely to find any interaction. In model 1 to 3 the SNPs are independent whereas in model 4 some SNPs are correlated. With model 4 it is thus possible to evaluate the effect of linkage disequilibrium, which is often observed in practice. For this model SNPs 1-5 and 6-10 form correlated groups, where each SNP has a correlation of 0.9 with its direct neighbors within the group. Note that the interactions $(X_{3d} \wedge X_{11d}^C)$ and $(X_{8d} \wedge X_{12d})$ in the model contain SNPs from the correlated groups. As mentioned in [9] SNP association studies are usually concerned with low penetrance genes and therefore no high effects are used for the simulation. The intercept β_0 is chosen so that each data set contains about 500 cases and 500 controls. Note that for model 3 the true model can only be approximated by FBLR as the complicated interaction is not in its search space.

For each of the models ten data sets are created and the methods described in the last section as well as a stepwise logistic regression algorithm are applied to them. Stepwise logistic regression is implemented by polyclass [10], using the BIC for model selection. For logic regression we employ 10-fold cross-validation for models having model size 1 to 10 and containing 1 to 3 trees using 25,000 iterations in each cross-validation step. MCLR is run with a burn-in of 10,000 and 100,000 iterations, once with each of the two standard settings of [3], which are $K = 2, a = \frac{1}{2}$ and $K = 3, a = \frac{1}{\sqrt{2}}$. The proposed full Bayesian version is based on at most two-factor interactions with a burn-in of 10,000 and 100,000 iterations. We run it once with and once without pathway information, where it is assumed to be known that SNPs 1 to 20 and 21 to 50 belong to distinct pathways and interactions are only modelled within them.

All computations are performed with the statistical software environment R [11], using the package LogicReg for logic regression and MCLR and the package polspline for stepwise logistic regression. Code for the full Bayesian algorithm is available from the first author upon request.

It is counted how many of the two-SNP interactions of the true model are correctly identified, how many SNP interactions not in the true model are identified and how many of these incorrect interactions include SNPs that are not in the true model. For the third model the three two-factor interactions included in the three-factor interaction are counted. For the fourth model the number of identified interactions in which a SNP from the true model is replaced by one correlated to it is also evaluated. For MCLR and FBLR the following algorithm is used to determine which interactions are counted as identified. First all two-factor interactions included in the sampled model are sorted according to their inclusion proportion. Then the largest jump in the inclusion proportion is determined. An example is given in Fig. 3, where the largest jump in the inclusion probability is after the fourth most common interaction. If there is another jump afterwards that is at least two thirds of the height of the largest, then all interactions above that second jump are deemed identified, otherwise all above the largest jump are used. In the situation of Fig. 3 the jump after the sixth interaction is only slightly less than the one after the fourth, therefore the algorithm identifies the first six interactions. If the largest jump is not at least 2.5% or the most frequent interaction is not present in at least 10% of models no interaction is counted as identified.

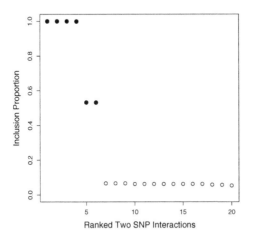

Fig. 3. Identification of interactions. Black dots show interactions identified by the algorithm.

Table 1 shows the results for the first three models averaged over the ten data sets. For model 1 with several interactions having moderate effects logic regression performs worst. It has the fewest correct interactions and the highest number of incorrect interactions. One reason for the large number of incorrect interactions is that logic regression identifies incorrect three-SNP interactions several times which are counted as three incorrect two-SNP interactions. The results of MCLR with $K = 2$ are not very good either, which is due to the fact that the true model would need three trees to be fitted. This shows that the results of MCLR can differ a lot for different values of K. Both settings of MCLR identify a high number of incorrect interactions between the SNPs of the true model. FBLR performs very well, finding on average 2.6 of the 3 correct interactions with no false interactions. The inclusion of pathway information slightly improves the results. The stepwise algorithm also shows good performance. The results for model 2 with several interactions with low effects show a similar pattern, the main difference to model 1 being that stepwise logistic regression does not produce good results anymore. For model 3 MCLR identifies the most correct interactions with similar results for both settings. The search for complicated models seems to be beneficial here, although logic regression again does not perform well. FBLR has fewer correct identifications than MCLR, which reflects that the true model is not in the space of models

Table 1. Results of the simulation study for models 1 to 3

Model 1	Correct interactions	Incorrect interactions	Incorrect with SNPs not in true model
Logic regression	1.3	2.9	2.6
MCLR $K = 2, a = \frac{1}{2}$	1.8	1.4	0.2
MCLR $K = 3, a = \frac{1}{\sqrt{2}}$	2.6	1.3	0.5
FBLR w/o pathways	2.6	0.0	0.0
FBLR with pathways	2.7	0.0	0.0
Stepwise logistic	2.4	0.2	0.0
Model 2			
Logic regression	0.0	1.3	1.3
MCLR $K = 2, a = \frac{1}{2}$	0.5	0.3	0.3
MCLR $K = 3, a = \frac{1}{\sqrt{2}}$	0.9	0.7	0.5
FBLR w/o pathways	1.0	0.3	0.3
FBLR with pathways	1.1	0.1	0.0
Stepwise logistic	0.5	0.7	0.7
Model 3			
Logic regression	0.1	2.1	2.1
MCLR $K = 2, a = \frac{1}{2}$	1.5	0.3	0.3
MCLR $K = 3, a = \frac{1}{\sqrt{2}}$	1.4	0.4	0.4
FBLR w/o pathways	0.7	0.0	0.0
FBLR with pathways	0.9	0.2	0.2
Stepwise logistic	0.4	0.0	0.0

searched through. Nevertheless it also has fewer incorrect interactions and outperforms stepwise logistic regression. With included pathway information it has a few more correct interactions, but also more incorrect ones. Table 2 contains the simulation results for model 4 where the data sets include correlated SNPs. In addition to the correct interactions the table shows how many interactions containing a SNP correlated to the true one are identified. All methods identify less correct interactions, as often a correlated interaction is found instead. Compared to the results of model 1 logic regression and MCLR identify less incorrect interactions. These are probably also replaced by the correlated interactions. The inclusion of pathway information into FBLR has the largest effect for this model. Otherwise the overall pattern of results is similar to the one of model 1. For MCLR and FBLR in some cases a true and a corresponding correlated interaction are both identified. This is an advantage over methods that identify a single best model as it can be concluded that the data does not contain enough information to distinguish between the two interactions, whereas a best model would probably contain only one of them.

3.2 Application: Breast Cancer Study

We apply the methods for identifying SNP interactions to data from the Interdisciplinary Study Group on Gene Environment Interaction and Breast Cancer

Table 2. Results of the simulation study for model 4

Model 4	Correct/ correlated interactions	Incorrect interactions	Incorrect with SNPs not in true model
Logic regression	0.4 / 0.4	2.2	1.4
MCLR $K = 2, a = \frac{1}{2}$	1.5 / 0.5	0.8	0.0
MCLR $K = 3, a = \frac{1}{\sqrt{2}}$	2.0 / 0.5	0.3	0.1
FBLR w/o pathways	1.9 / 0.6	0.0	0.0
FBLR with pathways	2.4 / 0.7	0.0	0.0
Stepwise logistic	1.3 / 0.7	0.0	0.0

in Germany, short GENICA. Data are available for 1191 participants, which include 561 cases and 630 controls. Cases are women of European descent, that were first diagnosed with breast cancer within six months prior to study begin and were not older than 80 years. Controls came from the same study region and were frequency matched according to birth year into 5-year intervals. Sequence information on 68 SNPs are available for the participants. The SNPs come from different pathways including xenobiotic and drug, steroid hormone metabolism and DNA repair. Some SNPs belong to more than one pathway. A number of epidemiological covariates is also available, but for the present analysis we focus on the SNP data alone. After converting the SNP variables to binary ones all variables having less than 10 observations equal to 1 are removed, leading to 113 binary predictor variables used in the analysis. For a more detailed description of the GENICA study see [12] and for a statistical analysis using a variety of methods [13].

All methods are applied with the same settings as in the simulation study. The strongest influential factor found by FBLR is

$$\text{ERCC2\_6540}_d^C \wedge \text{ERCC2\_18880}_d \ , \tag{5}$$

an interaction between two SNPs in the ERCC2 gene from the DNA repair pathway. It is included in 96.8% of the models, the second most common interaction is the same interaction without the complement on ERRCC2_6540, which is included in 3.4%. No other interaction is included in more than 1% of the models. The odds ratio for fulfilling (5) is estimated to be 2.8. Main effect variables included in more than 1% of the models are CYP1B1_1358$_r$ (10.9%), ERCC2_18880$_d$ (3.7%), MTHFR_1$_r$ (1.8%) and ERCC2_6540$_d$ (1.7%). CYP1B1 plays a role both in the xenobiotic and drug and steroid metabolism pathways and MTHFR is a factor relevant for nutrition. Using pathway information and restricting the algorithm to search for interactions only within pathways did not change the results as the only relevant interaction found is the one in ERCC2, which is within a gene and therefore within a pathway.

The best logic regression model found contains only the interaction (5) with an estimated odds ratio of 3.1. The model found by stepwise logistic regression is

$$\text{logit}(p(Y = 1)) = 0.20 - 0.96 \cdot X_{10d} + 0.15 \cdot X_{11d} - 1.10 \cdot X_{10d} \times X_{11d} - 0.96 \cdot X_{16r} \ ,$$

where $X_{10} = $ ERCC2_6540, $X_{11} = $ ERCC2_18880 and $X_{16} = $ CYP1B1_1358. The main effects of the ERCC2 SNPs need to be included as the stepwise model cannot choose an interaction including the complement of one of the variables. So an interaction as in (5) has to be described in more complicated terms. It is an advantage of all logic regression based methods that biologically meaningful variables like (5) can be included directly as predictors.

The tables 3 and 4 show the SNPs and two SNP combinations included most often in the MCLR models. The most important variables are again the ERCC2 SNPs and their interaction, followed by the CYP1B1 SNP. Note that the specific form of the interaction of the two ERCC2 SNPs cannot be seen directly in table 4, as MCLR only counts how often two SNPs occur together in a logic tree without regarding complements. MCLR suggests an interaction of the ERCC2 SNPs with CYP1B1_1358. This interaction is not very plausible as the genes come from pathways with a different biologic function. It is more likely that MCLR suggests an incorrect interaction between SNPs that are associated with the outcome, which also has happened frequently in the simulation study. The gene PGR plays a role in signal transduction, therefore the next frequently found interaction is not very plausible either.

Table 3. The five SNPs that occur most often in MCLR models

SNP	Included in fraction of models	
	$K = 2, a = \frac{1}{2}$	$K = 3, a = \frac{1}{\sqrt{2}}$
ERCC2_6540$_d$	1.0	1.0
ERCC2_18880$_d$	1.0	1.0
CYP1B1_1358$_r$	0.278	0.397
MTHFR_1$_r$	0.040	0.102
PGR_1$_d$	0.036	0.092

Table 4. The five two SNP combinations that occur most often together in a logic tree in MCLR models

SNP1	SNP2	Included in fraction of models	
		$K = 2, a = \frac{1}{2}$	$K = 3, a = \frac{1}{\sqrt{2}}$
ERCC2_6540$_d$	ERCC2_18880$_d$	0.971	0.981
ERCC2_6540$_d$	CYP1B1_1358$_r$	0.200	0.255
ERCC2_18880$_d$	CYP1B1_1358$_r$	0.195	0.255
ERCC2_18880$_d$	PGR_1$_d$	0.030	0.066
ERCC2_6540$_d$	PGR_1$_d$	0.030	0.063

4   Summary and Conclusions

In this paper we compared several methods based on logic regression for identifying SNP interactions that modify the risk of developing a disease: logic regression,

Monte Carlo logic regression and full Bayesian logic regression. All these methods have the advantage over logistic regression, that biologically meaningful interactions involving complements of one of the variables can be directly identified with the method. This leads to better interpretable models in the GENICA study.

While the results of logic regression itself can probably be improved by carefully examining cross-validation plots and looking at several of the best models, its use as an automatic method can definitely not be recommended, because of the high number of false positive SNPs and SNP interactions found in the simulation study.

In the simulation study it was shown that considering more than one model improves the identification of SNP interactions. Bayesian statistics provides a coherent framework to do so. Monte Carlo logic regression can possibly identify complex interactions, but it often gives the impression that SNPs associated with the outcome interact, even if they are in fact independent. In practice it would also be necessary to run MCLR with different settings of the maximum number of trees K and the penalty parameter a since these influence the results that are obtained. Another drawback is that from the output of MCLR it is not possible to directly infer whether an interaction between two variables involves a complement, as in the ERCC2 interaction found in the breast cancer study.

The proposed full Bayesian logic regression does not require the setting of fixed parameters of crucial importance. Since it uses only AND-combinations of binary predictors the method can only approximate more complex interactions. However, a clear advantage of the method is that identified two-SNP interactions can be relied upon, as only very few false positive interactions were found in the simulation study. For the GENICA study only the interaction in ERCC2 has been identified, which corresponds to the results found in [12]. The use of biological knowledge via the inclusion of pathway information improved the results considerably for the data sets containing correlated SNPs. Including such information is also likely to reduce the number of iterations needed, as it reduces the space of models that has to be considered.

Acknowledgments. The authors would like to thank the GENICA network for letting us use their data. Thanks to Holger Schwender and Anna Gärtner for helpful discussions.

References

1. Golka, K., Prior, V., Blaszkewicz, M. and Bolt, H.M.: The Enhanced Bladder Cancer Susceptibility of NAT2 Slow Acetylators Towards Aromatic Amines: a Review Considering Ethnic Differences. Tox. Lett. **128** (2002) 229-241
2. Kooperberg, C., Ruczinski, I., LeBlanc, M.L. and Hsu, L.: Sequence Analysis Using Logic Regression. Genet. Epidemiol. **21** (2001) Suppl. 1 626-631
3. Kooperberg, C. and Ruczinski, I.: Identifying Interacting SNPs Using Monte Carlo Logic Regression. Genet. Epidemiol. **28** (2005) 157-170

4. Ruczinski, I., Kooperberg, C. and LeBlanc, M.: Logic Regression. J. Comp. Graph. Stat. **12** (2003) 475-511
5. Otten, R. H., and Ginneken, L. P.: The Annealing Algorithm. Kluwer Academic Publishers, Boston (1989)
6. Green, P.J.: Reversible Jump Markov Chain Monte Carlo Computation and Bayesian Model Determination. Biometrika **82** (1995) 711-732
7. Holmes, C.C. and Denison, D.G.T.: Classification with Bayesian MARS. Mach. Learn. **50** (2003) 159-173
8. Clark, T.G., de Iorio, M. and Griffiths, R.C.: Bayesian Logistic Regression Using a Perfect Phylogeny. Biostatistics **8** (2007) 32-52.
9. Erichsen, H.C. and Chanock, S.J.: SNPs in Cancer Research and Treatment. Brit. J. Cancer **90** (2004) 747-751
10. Kooperberg, C., Bose, S. and Stone C.J.: Polychotomous Regression. J. Amer. Stat. Assoc. **92** (1997) 117-127
11. R Development Core Team: R: A Language and Environment for Statistical Computing. R Foundation for Statistical Computing, Vienna, Austria (2006) ISBN 3-900051-07-0, URL http://www.R-project.org.
12. Justenhoven, C., Hamann, U., Pesch, B., Harth, V., Rabstein, S., Baisch, C., Vollmert, C., Illig, T., Ko, Y., Brüning, T. and Brauch, H. for the GENICA network: ERCC2 Genotypes and a Corresponding Haplotype Are Linked with Breast Cancer Risk in a German Population. Cancer Epid. Biomark. Prevent. **13** (2004) 2059-2064.
13. Ickstadt, K., Müller, T. and Schwender, H.: Analyzing SNPs: Are There Needles in the Haystack? Chance **19** Number 3 (2006) 21-26.

A   Unequal Prior Weight for Logic Expressions in MCLR

MCLR assigns a uniform prior distribution for all models of the same size. Still not all logic expressions of the same size receive the same prior weight, as they can have a different number of representations as logic trees. For example, there are twelve trees of size three equivalent to the tree in the left panel of Fig. 4 but only four equivalent to the tree in the right panel. This can be seen by first noting that the two branches can be interchanged in both trees. In addition the leaves in the left tree can all switch position without changing the expression the tree represents. For the right tree only the leaves X_1 and X_2^C can be interchanged. That makes $2 \cdot 3! = 12$ or in the other case $2 \cdot 2! = 4$ trees. Thus the first expression has a prior weight four times that of the second.

B   Proposing a Candidate Model in FBLR

To propose a candidate model in FBLR, one of the following moves is carried out on the current model

- *Birth*: Adding a randomly chosen predictor L_i of size one.
- *Death*: Removing a randomly chosen predictor L_i of size one.
- *Change-birth*: Adding a randomly chosen binary X_j to a predictor.

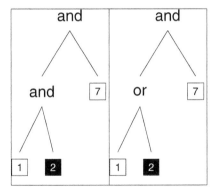

Fig. 4. Two logic trees with a different number of representations

- *Change-death*: Removing a randomly chosen binary X_j from a predictor.
- *Change*: Replacing a randomly chosen binary X_j in a predictor.

To pick a move, first the choice is made, whether to add, remove or change a predictor with the respective probabilities 0.25, 0.25 and 0.5. If a predictor is to be changed, a binary X_j is added to it, removed from it or changed with equal probability.

As described in [7] the acceptance probability depends on the Bayes factor of the two models and on a factor

$$R = \frac{p(\theta^*) \cdot q(\theta^{(t-1)}|\theta^*)}{p(\theta^{(t-1)}) \cdot q(\theta^*|\theta^{(t-1)})} \ , \tag{6}$$

where $p(.)$ is the prior given by (4) and $q(.)$ is the proposal distribution. The value that R takes for the moves of FBLR is shown in Table 5.

Table 5. Value of R for the different move types. N_1 is the number of predictors of size 1 currently in the model, a' is from the geometric prior on predictor size, int is the number of possible interactions for the binary being removed from respectively staying in the predictor to be changed and $\overline{int}$ is the average number of interactions for a SNP.

Move	R
Birth	$1/(N_1 + 1)$
Death	N_1
Change-birth	$1/a' \cdot int/\overline{int}$
Change-death	$a' \cdot \overline{int}/int$
Change	1
Birth from null	$1/4$
Death to null	4

Stochastical Analysis of Finite Point Sampling of 3D Chromatin Fiber in Interphase Cell Nuclei

E. Gladilin[1], S. Goetze[2], J. Mateos-Langerak[2], R. van Driel[2],
K. Rohr[1], and R. Eils[1]

[1] German Cancer Research Center, Theoretical Bioinformatics,
Im Neuenheimer Feld 580, D-69120 Heidelberg, Germany
[2] Swammerdam Institute for Life Sciences, University of Amsterdam,
BioCentrum Amsterdam, Kruislaan 318, 1098 SM Amsterdam, The Netherlands

Abstract. Investigation of 3D chromatin structure in interphase cell nuclei is important for the understanding of genome function. For a reconstruction of the 3D architecture of the human genome, systematic fluorescent in situ hybridization in combination with 3D confocal laser scanning microscopy is applied. The position of two or three genomic loci plus the overall nuclear shape were simultaneously recorded, resulting in statistical series of pair and triple loci combinations probed along the human chromosome 1 q-arm. For interpretation of statistical distributions of geometrical features (e.g. distances, angles, etc.) resulting from finite point sampling experiments, a Monte-Carlo-based approach to numerical computation of geometrical probability density functions (PDFs) for arbitrarily-shaped confined spatial domains is developed. Simulated PDFs are used as bench marks for evaluation of experimental PDFs and quantitative analysis of dimension and shape of probed 3D chromatin regions. Preliminary results of our numerical simulations show that the proposed numerical model is capable to reproduce experimental observations, and support the assumption of confined random folding of 3D chromatin fiber in interphase cell nuclei.

Keywords: 3D genome structure, chromatin folding, fluorescent in situ hybridization (FISH), confocal laser scanning microscopy (CLSM), finite point sampling, confined random point distribution, stochastical analysis, Monte-Carlo simulation.

1 Motivation

The dynamic 3D folding of the chromatin fiber in the interphase nucleus is a key element in the epigenetic regulation of gene expression in eukaryotes [2], [7]. Chromatin structure can be studied by 3D confocal laser scanning microscopy (CLSM) after fluorescent in situ hybridization (FISH) labeling of specific sequence elements in the genome under conditions that preserve biological structure. Using Bacterial Artificial Chromosomes (BACs), FISH enables selective

S. Hochreiter and R. Wagner (Eds.): BIRD 2007, LNBI 4414, pp. 104–118, 2007.

visualization of complete individual chromosomes or subchromosomal domains, or smaller genomic regions of only few hundreds kilobase pairs (Kb $= 10^{-3}$ Mb). In the present study four CSLM imaging channels were used for the simultaneous visualization of three genomic sequence elements (using three spectrally differently labeled BACs) and the overall size and shape of the interphase nucleus (using DAPI labeling of all nuclear DNA). For probing large chromatin regions with a finite number of sampling points, multiple measurements in different cells have to be performed. However, reconstruction of 3D chromatin structure in a 'piece-by-piece' manner from such series of finite point samplings assumes the existence of a mechanically conservative 3D structure that exhibits only small topological cell-to-cell variations. In contrast, distance measurements in different cells show extensive cell-to-cell variations that are not due to measuring errors and therefore indicate a flexible and dynamic structure of interphase chromatin [5]. Repetitive measurements in many otherwise identical cells yield statistical series of simultaneously labeled pairs or triplets of BACs, i.e. coordinates of their mass center points, and the corresponding geometrical features (e.g. pairwise distances) that are used for quantitative analysis of probed chromatin regions. In a number of previous studies [4], [5], [8], statistical models of 3D chromatin folding were proposed. These models provide a qualitative description for some basic experimental observations, for example the $D(L) \approx \sqrt{L}$ relationship between 3D physical distance (D) and genomic length (L) for each two probes along the DNA on a small $L < 2$ Mb genomic scale. However, the saturation plateau of $D(L) \approx const$ observed on larger genomic scales $L > 2$ Mb is not yet satisfactory explained in the literature. The distribution of fluorescent probes for a random folded DNA fiber and probability densities for the corresponding geometrical features (e.g. BAC distances) are derived by authors on the basis of some general thermodynamic principles, e.g. equilibrium state, resulting in a general integral form for sought probability density function (PDF). Closed form solutions of such integrals can only be obtained for some particularly simple geometries of spatial domains, e.g. a spherical ball. However, for the validation of some structural hypotheses and comparative analysis of experimental data, PDFs for a wide range of geometrical features and domain shapes may be of interest. In this article, we present a general approach for numerical computation of arbitrary geometrical PDFs using Monte-Carlo simulations, which is applied for interpretation of statistical series resulting from 4 channel CLSM of human fibroblast cell nuclei. Exemplarily, we restrict our simulations to a natural assumption of uniform confined point distributions, although any arbitrary non-uniform point density function can be used instead. Our simulation results show that the proposed model is capable to reproduce experimental PDFs and provides numerical bench marks for quantitative analysis of probed spatial confinements in terms of dimension and shape. Furthermore, we present a confined random folding model of 3D chromatin fiber, which gives a qualitative explanation for biphasic behavior of the $D(L)$ relationship.

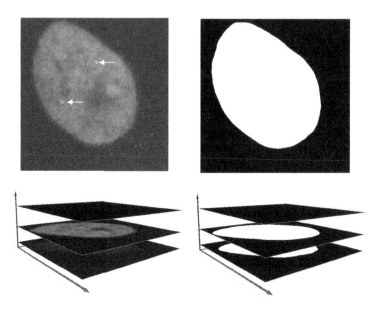

Fig. 1. Left column: 3D CLSM image of a DAPI-stained human fibroblast nucleus (bottom), single slice of 3D image with one BAC in red-color channel (top). Arrows point to two spots of the same BAC corresponding to two sister-chromosomes. Right column: 3D view (bottom) and single 2D slice (top) of the binarized image.

Fig. 2. Segmented 3D CLSM image. Numbered nodes indicate positions of mass centers of the whole nucleus (0) and three BACs in two sister-chromosomes (1-6), respectively.

2   Methods

2.1   Image Acquisition and Preprocessing

3D confocal laser scanning microscopy images of DAPI-stained human fibroblast nuclei are used for geometrical reconstruction of 3D chromatin structure in interphase nuclei of human fibroblasts, see Fig. 1. Besides overall nucleus shape, up to 3 BAC regions are simultaneously labeled via the FISH technique resulting in 4 channel 3D image of the nucleus. 3D images of all channels are consistently preprocessed using following basic steps [3]:

   – Fourier band-pass image smoothing,
   – threshold-based image segmentation.

After thresholding and segmentation, target structures are represented in the binarized images by clearly bounded white regions, see Fig. 1 (right column). Pointwise representation of segmented nuclear and BAC regions is obtained by computing their mass centers:

$$x_i^{mc} = \frac{1}{N} \sum_{j=1}^{N} x_i^j, \tag{1}$$

where x_i^j is the i-th coordinate of the j-th of totally N voxels of one segmented region (e.g. particular BAC or entire nucleus domain). Since each BAC label produces two signals corresponding to two sister-chromosomes, totally 7 mass center points are localized after preprocessing 4 channel 3D images of the nucleus, see Fig. 2.

2.2   Strategies of Finite Point Sampling of 3D Chromatin Fiber

Due to the limited number of independent color-channels, only few gene loci along the chromatin fiber can simultaneously be sampled in one single cell. For continuous dense sampling of larger chromosome regions and assessment of statistical cell-to-cell variations in chromatin structure, each combination of BACs is probed in 50 different cells resulting in statistical series of coordinates of mass center points. On the basis of these statistical series, probability distributions for geometrical features (distances, angles, etc.) are calculated. Two strategies for placement of sampling probes along the DNA fiber and assessment of statistical series are applied:

   1. series of BAC pairs with increasing genomic distances,
   2. series of BAC triplets.

Series of BAC pairs with increasing genomic distance are used for assessment of the $D(L)$ relationship between physical D and genomic L distances for each BACs at small $L = 0.1 - 3$ Mb and large $L = 3 - 28$ Mb genomic scales, see Fig. 3 (left).

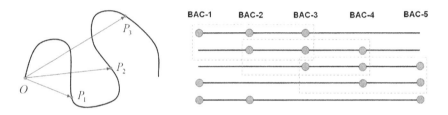

Fig. 3. Left: the relationship $D(L)$ between physical distances $D_i = |OP_i|$ and genomic lengths $L_i = \int_O^{P_i} dl$ for a finite number of sampling points ($i = 1...N$) sequentially placed along the DNA provides insights into 3D chromatin folding. Right: schematic view of the *moving mask* approach in case of 5 BAC triplets, green frames show overlapping pairs of BACs.

Series of BAC triplets are measured for assessment of geometrical features of 3-point combinations (e.g. triangle angles, etc.), which indicate structural variability of probed 3D domains along the DNA fiber. Stochastical analysis of statistical distributions of geometrical features of BAC triplets is applied for the estimation of cross section dimension and shape of sampled chromosome regions. For dense sampling of larger genetic regions in a piece-by-piece manner, a *moving mask* technique is applied. That is three BACs are placed along the DNA in a way that each next triplet has an overlap with the previous one by two common BACs, see Fig. 3 (right). This approach is applied for 3D visualization of finite point sets using multidimensional scaling of cross-distance matrices, see Section 3.4.

2.3   Geometrical Features of BAC Combinations

Spatial distribution of BAC pairs and triplets is analyzed using following geometrical features:

- pairwise distances between BACs,
- radial distances of BACs w.r.t. the nucleus center,
- angles of triangles spanned by BAC triplets,
- spatial orientation of BAC triplets w.r.t. the nucleus center.

Fig. 4 gives an overview over geometrical features of BAC triplets used for numerical computation of PDFs via the Monte-Carlo approach as described in Section 2.5.

Pairwise distances between BACs are computed as Euclidean distances for all 3 pairs of BACs in each triplet $d_{ij} = |B_i B_j| \in [0, S_{max}]$, $\forall i = 1..3, j > i$, where S_{max} is the maximum cross section of the cell nucleus.

Radial distances of BACs w.r.t. the mass center of the nucleus are computed as Euclidean distances between mass centers of the nucleus O and each BAC region B_i, i.e. $R_{OB_i} = |OB_i| \in [0, \frac{S_{max}}{2}]$.

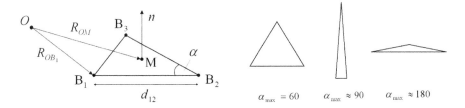

Fig. 4. Left: geometrical features of BAC triplets B_i: pairwise distances $d_{ij} = |B_iB_j|$, radial distances w.r.t. a fixed point $\mathbf{R}_{OB_1}$, triangle angles α, orientation of BAC triangle w.r.t. a fixed point $(\mathbf{n} \cdot \mathbf{R}_{OM})$. Right: maximum triangle angle $\alpha_{\max}$ as a feature of triangle shape.

Maximum angle of BAC triangle $\alpha_{\max}$ serves as a feature of triangle shape, see Fig. 4 (right). The PDF of $\alpha_{\max} \in [60, 180]$ provides insights in the overall shape of the probed domain.

Spatial orientation of BAC triplets w.r.t. the nucleus center is characterized by a scalar product $s = (\mathbf{n} \cdot \mathbf{R}_{OM}) \in [-1, 1]$, where $\mathbf{n}$ is the triangle normal and $\mathbf{R}_{OM}$ is the vector pointing from center of the nucleus to the middle of BAC triangle. $s = 0$ means that $\mathbf{R}_{OM}$ lies in the triangle plain, while $|s| = 1$ indicates that $\mathbf{R}_{OM}$ is perpendicular to the triangle plain.

2.4   Stochastical Analysis of Finite Point Sampling of a 3D Domain

Series of finite point samplings of 3D chromatin fiber in different cells yield statistical distributions of invariant geometrical features (e.g. pairwise distances, angles etc.), which are analyzed using geometrical probability techniques. Stochastical analysis of such statistical series is aimed at

- investigation of the order of randomness and
- quantification of dimension and shape of probed chromatin regions,

and is based on construction of geometrical PDF.

Formally, the PDF of a probability distribution is defined as a non-negative function $p(x) > 0$ of a statistically distributed variable x such that the integral

$$P(A \leq x \leq B) = \int_A^B p(x)dx \leq 1, \tag{2}$$

gives the probability $P(A \leq x \leq B)$ for the variable x being found in the interval $A \leq x \leq B$. From (2) it immediately follows that

$$\int_{-\infty}^{+\infty} p(x)dx = 1. \tag{3}$$

For a discrete distribution of x_i ranging in the interval $x_i \in [A, B]$, $\forall i = 1...N$, the PDF can be constructed using the histogram function $h_j(x_i)$, which is defined as an array of tabulated frequencies of x_i being found within the j-th of totally n intervals $\frac{j}{n}(B - A) \le x_i \le \frac{(j+1)}{n}(B - A)$:

$$p_j(x_i) = C \frac{h_j(x_i)}{N}, \tag{4}$$

where C is the normalization constant resulting from the condition (3)

$$C = \left(\sum_{j=1}^{n} \frac{h_j(x_i)}{N} \frac{(B - A)}{n} \right)^{-1}. \tag{5}$$

For a uniform random distribution of N sampling points $\mathbf{r}_i$ with Cartesian coordinates x_j in a spherical k-dimensional ball $\mathbf{B}_k = \{\mathbf{r}_i(x_j) : \sum_{j=1}^{k} x_j^2 \le R^2\}$, where R denotes the radius of $\mathbf{B}_k$, the PDFs $p_k(d)$ for pairwise distances $d = ||\mathbf{r}_{p=1..N} - \mathbf{r}_{q=(p+1)..N}||$ between the points can be obtained in a closed form [6]:

$$
\begin{aligned}
\mathbf{B}_1 : \quad & p_1(d) = \frac{1}{R}(1 - \frac{d}{2R}) \\
\mathbf{B}_2 : \quad & p_2(d) = \frac{2d}{R^2} - \frac{d^2}{\pi R^4}\sqrt{4R^2 - d^2} - \frac{4d}{\pi R^2}\arcsin\left(\frac{d}{2R}\right) \\
\mathbf{B}_3 : \quad & p_3(d) = \frac{3d^2}{R^3} - \frac{9d^3}{4R^4} + \frac{3d^5}{16R^6}
\end{aligned}
\tag{6}
$$

Plots of $p_k(d)$ in case of unit 1D/2D/3D-balls (i.e. $2R = 1$) are shown in Fig. 5. As one can see, PDFs for pairwise distances essentially depend on the spatial dimension k of $\mathbf{B}_k$. These PDFs can be seen as characteristic signatures of random point distributions for 1D, 2D and 3D isotropic spherical confinements, respectively. Interestingly, the distance with the highest probability ($d_m \in [0, 1]$: $\max(p(d)) = p(d_m)$) for a unit spherical ball of dimension higher than 1 is not the smallest-possible, but some intermediate one:

$$
\begin{aligned}
\mathbf{B}_1 : \quad & d_m = 0 \\
\mathbf{B}_2 : \quad & d_m = 0.42 \\
\mathbf{B}_3 : \quad & d_m = 0.52
\end{aligned}
\tag{7}
$$

These key-values together with further standard PDF features such as first four statistical moments M_i:

$$
\begin{aligned}
M_1 &= \frac{1}{n} \sum_{i=1}^{n} p(d_i) \\
M_2 &= \sqrt{\frac{1}{n-1} \sum_{i=1}^{n} (p(d_i) - M_1)^2} \\
M_3 &= \frac{1}{n} \sum_{i=1}^{n} \left(\frac{p(d_i) - M_1}{M_2} \right)^3 \\
M_4 &= \left(\frac{1}{n} \sum_{i=1}^{n} \left(\frac{p(d_i) - M_1}{M_2} \right)^4 \right) - 3
\end{aligned}
\tag{8}
$$

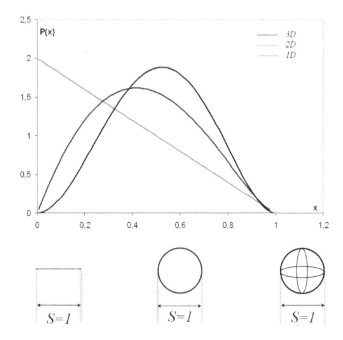

Fig. 5. Plots of theoretical probability density functions $p_k(d)$ for pairwise distances between randomly distributed points in case of spherical k-dimensional balls ($k = 1, 2, 3$) with a unit cross section dimension ($S = 2R = 1$)

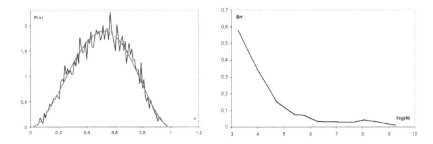

Fig. 6. Left: results of numerical computation of PDFs of pairwise distances for a unit 3D spherical domain with $N = 270$ (blue) and $N = 5219$ (green) random points vs theoretical solution (red). Right: L_2 error norm of numerically computed PDFs w.r.t. theoretical solution as a function of $\log(N)$, where N is the number of random points in the Monte-Carlo simulation.

can be used for quantification of the order of randomness of BAC distributions on the basis of experimental PDFs of pairwise distances. If, for instance, we would observe in our experiment a PDF for pairwise distances, which behaves very much the same as $p_3(d)$ in (6), we would have a strong evidence for a completely random distribution of measured points within a probed spatial domain. And the other way round: if we assumed that the underlying 3D point distribution is confined, random and isotropic, statistical features of the experimental PDF $\bar{p}(d)$ such as (7) gave us an estimate for the upper bound of the cross section dimension $\bar{S}$ of the probed spherical domain, namely

$$\bar{S} = \left(\frac{S}{d_m} \right) \bar{d}_m \approx 1.92\, \bar{d}_m, \tag{9}$$

where $S = 1$ is the cross section dimension of the reference unit ball and $\bar{d}_m$: $\max(\bar{p}(d)) = \bar{p}(\bar{d}_m)$.

2.5   Numerical Computation of PDFs for an Arbitrary Domain

The approach for bench marking experimentally observed PDFs of pairwise distances vs theoretically predicted PDFs can be extended to the case of an arbitrary geometrical feature and arbitrarily-shaped k-dimensional confinement $\Omega \subset \mathbb{R}^k$ with the boundary $\Gamma \subset \Omega$. For more complex geometries, the PDF for pairwise point distances or any other geometrical feature can not be derived in a closed form. However, it can be computed numerically, for example, via a Monte-Carlo simulation. The basic steps of our simulation scheme are as follows

1. specify the domain Ω in a suitable parametric form (e.g. surface or volumetric meshes, point clouds, etc.),
2. generate sufficiently large number of random points $\mathbf{r}_i$ within the bounding box of Ω and select only the points lying inside the domain $\mathbf{r}_i \in \Omega$
3. compute geometrical features (distances, angles, etc.) for all pairs and triplets of $\mathbf{r}_i \in \Omega$,
4. compute corresponding histogram and PDF for simulated statistical series of geometrical features.

At the end, the PDF for a geometrical feature and spatial confinement is given by an $(n-1)$-array of tabulated values corresponding to n intervals of the histogram function. Further details on computation of PDFs for some special cases of 3D domain geometry are in Section 3.

2.6   Confined Random Folding Model

For the interpretation of the $D(L)$ relationship between physical and genomic distances (see Fig. 3 (left)), a confined random folding model of 3D chromatin fiber is proposed. We consider 3D chromatin fiber or one of its fragments to be randomly folded within a confined spatial domain $\Omega \subset \mathbb{R}^3 : \sum_{i=1}^{3}(x_i - x_i^{mc})^2 \leq R^2$, $\forall x_i \in \Omega$, where x_i are coordinates of points, x_i^{mc} is the mass center of Ω, and R is a characteristic dimension of the confinement Ω. The simulation of a randomly folded 3D fiber begins with the generation of a confined random point distribution for Ω as described in Section 2.5. These points are understood as vertex nodes of a 3D fiber randomly folded in Ω, see Fig. 11. An algorithm is developed to perform the reconstruction of a non-closed, loop-free 3D fiber connecting all points of Ω pairwise using the closest neighborhood connectivity. Details on the validation of the confined random folding model are in Section 3.5.

3 Experimental Results

In this section, we present the results of stochastical simulations of confined random point distributions and 3D chromatin fiber folding carried out for the interpretation of experimental observations.

3.1 Numerical Simulation vs Theoretical Solution

First, numerical algorithms for computation of geometrical PDFs are validated by a direct comparison with closed form solutions (6). Fig. 6 (left) shows the results of numerical computations of the PDF of pairwise distances $p_3(d)$ for $N = 113$ and $N = 5219$ sampling points of Monte-Carlo simulations. Plot 6 (right) illustrates the reduction of the numerical error vs theoretical solution with increasing number of sampling points. From numerical point of view, $N = 350$ sampling points is an acceptable lower bound for accurate computation of smooth PDFs. However, one should take into account that $N = 350$ sampling points correspond to $\frac{N^2 - N}{2} = 61750$ pairwise distances, i.e. single distance measurements! One can also reversely calculate the number N of virtual sampling points corresponding to N_d distance measurements: $N \approx \sqrt{2N_d}$. This means that $N_d = 500$ distance measurements correspond to only $N = 31$ virtual sampling points, and, in order to simulate $N = 100$ sampling points, $N_d = 9900$ distance measurements are required.

3.2 Impact of Domain Geometry

In order to investigate impact of domain geometry on the PDF pattern, numerical simulations are carried out Fig. 7 show the simulation results for PDFs of different geometrical features discussed above and several values of Z-scaling factor. As one can see, domain geometry has a strong impact on PDF patterns, which can be exploited for the interpretation of experimental PDF curves.

3.3 Geometrical PDFs of Experimental Series

Our measurements did focus on two regions of the q-arm of human chromosome 1 in G1-phase human primary fibroblasts. The human transcriptome map [1] shows the presence of a gene-dense region of highly expressed genes (named a region of increased gene expression (ridge)) and a gene-sparse region (named antiridge), each comprising several Mb. One ridge (R) and one antiridge (AR) region along the q-arm of human chromosome 1 were probed on a scale 0.7 - 3.3 Mb, see Fig. 9 (left). Distances for each BAC triplet were measured in about 50 clonally grown fibroblast cells and for each of these measurements all geometrical

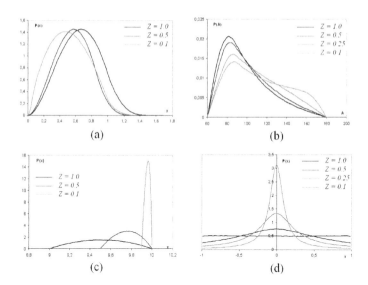

Fig. 7. Simulated PDFs of geometrical features for isotropic ($Z = 1.0$) an Z-scaled spatial domains: (a) pairwise distances, (b) maximum angles of point triplets, (c) radial distances w.r.t. a fixed point, (d) triangle orientation ratio

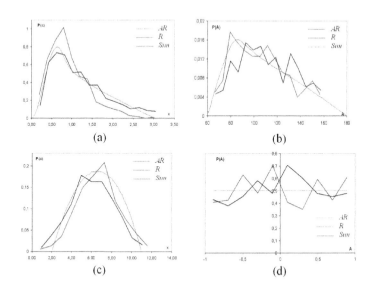

Fig. 8. Geometrical PDFs of statistical series of probing five ridge (blue curves) and antiridge (green curves) regions of human chromosome 1: (a) pairwise BAC distances, (b) maximum angles of BAC triplets, (c) radial distances of BACs w.r.t. the nucleus center, (d) orientation ratio of BAC triangles. Red curves denote simulated PDFs.

features described in Section 3.2 were computed.Altogether, R/AR chromatin regions sampled by five BAC triplets were totally probed $N_t = 200 - 272$ times, which corresponds to $N_d = 3N_t = 600 - 816$ pairwise distances R_{ij}/AR_{ij} and, recalling the discussion in Section 3.1, $\sqrt{2N_d} \approx 35 - 40$ virtual sampling points, respectively. Experimentally assessed geometrical PDFs for entire R/AR series of five BAC probes vs simulated PDFs are plotted in Fig 8. Quantitative comparison between experimental p_i^{ex} and simulated p_i^{sim} PDF patterns carried out on the basis of least square norms $||p_i^{ex} - p_i^{sim}||$ and statistical moments M_3 (skew) and M_4 (kurtosis) indicates a random distribution of BACs within anisotropically shaped 3D confinements, whereas some significant differences in spatial structure of R and AR regions were observed:

- AR domain probed by five BACs is more compact and smaller in size ($1.45 \times 0.42 \times 0.28 \ \mu$m) than R domain ($3.16 \times 1.62 \times 0.26 \ \mu$m),
- AR BACs have in average larger radial distances ($\bar{r}_{AR} = 6.62 \pm 1.98 \ \mu$m) w.r.t. the nucleus center compared to R BACs ($\bar{r}_R = 5.98 \pm 2.23 \ \mu$m)

3.4   3D Visualization of Average Cross-Distance Matrix

Ridge and antiridge BAC triplets have been placed according to the scheme shown in Fig. 3 (right). This BAC placement strategy was introduced for consistent 3D visualization of finite point probes of chromatin regions using a multidimensional scaling (MDS) approach [9], which requires a matrix of cross-distances d_{ij} between all BACs, see Fig. 9 (left). The result of 3D visualization of average R/AR distance matrices after decomposing d_{ij} via the MDS is shown in Fig. 9 (right). In view of large statistical deviations in cell-to-cell distance measurements which order of magnitude lies in the range of $\pm 50\%$ of the average distances, the probed chromatin regions can not be regarded as rigid objects with a same constant shape. Thus, 3D loops depicted in Fig. 9 (right) represent an *average* shape of probed R/AR regions resulting from a statistical series of distance measurements.

3.5   Validation of Confined Random Folding Model

A pure geometrical approach has been applied for qualitative validation of a confined random folding model of 3D chromatin fiber. A synthetic 3D fiber is constructed using the basic computational steps described in Section 2.6. The initial random point cloud consisting of $N = 528$ vertices and the corresponding 3D fiber are shown in Fig. 10. The $D(L)$ relationship between Euclidean distance D and genomic length L of all vertices w.r.t. a fixed starting point of the fiber is shown in Fig. 11 (right, red curve). This is a typical behavior of $D(L)$ function with large periodic oscillations appearing after the $\sqrt{L}$ regime for small L. At this point, we want to draw attention to the fact that the result of numerical computation of $D(L)$ essentially depends on the choice of the starting point for the construction of 3D fiber. For example, if the starting point is selected on the boundary of a spherical confinement with the radius R, the maximum possible

AR	8		19		28		35	
DIS	GEN	EUC	GEN	EUC	GEN	EUC	GEN	EUC
1	0.81	0.65	1.61	0.90	2.49	1.30	3.26	1.37
8			0.80	0.61	1.68	1.21	2.45	1.27
19					0.86	0.74	1.65	1.08
28							0.77	0.76

R	12		21		28		39	
DIS	GEN	EUC	GEN	EUC	GEN	EUC	GEN	EUC
5	0.84	0.72	1.58	1.13	2.39	1.32	3.20	2.12
12			0.74	0.52	1.55	1.23	2.36	2.18
21					0.81	0.92	1.62	2.51
28							0.81	2.57

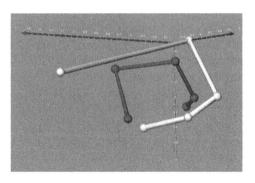

Fig. 9. Left: Genomic (Mb) and Euclidean (μm) distances for five AR/R BACs. Right: 3D visualization of average cross-distance matrices for five ridge (white polygon) and antiridge BACs (red polygon) probed along human chromosome 1 on a genomic scale $0.7 - 3.3$ Mb.

Euclidean distance is $D_{\max} = 2R$, whereas if the starting point is selected in the center of the sphere, the maximum value is $D_{\max} = R$. Also, periodic patterns of $D(L)$ for different starting points are shifted w.r.t. each other with a random phase. From experimental point of view it has following consequences: since experimental $D(L)$ curves are constructed from sequential measurements of the same genetic region in different cells, the physical position of the starting BAC, as well as overall 3D folding of the probed genomic region are varying from cell to cell. This means that an experimentally obtained $D(L)$ relationship is, in fact, the result of averaging N single $D(L)$ curves, where N is the number of measurements in different cells. The result of numerical simulation of such average $D(L)$ is shown in Fig. 11 (right, black curve). As one can see it exhibits a biphasic behavior very much similar to experimentally assessed $D(L)$ curve in Fig. 11 (left). Obviously, the saturation plateau of experimentally assessed $D(L)$ relationships results from statistical smoothing of single $D(L)$ curves due to measurements in different cells. However, one can still recognize remaining quasi-periodic oscillations of single random $D(L)$ curves in the average $D(L)$ on large genomic scale $L > 2$ Mb.

4   Conclusion

In this article, we have presented a novel approach for stochastical analysis and visualization of finite point sampling of 3D chromatin in interphase cell nuclei. The core idea of our approach consists in application of geometrical probability techniques for interpretation of statistical series of finite point sampling of chromatin regions. Numerically computed probability density functions (PDFs) serve as bench marks for the validation of experimentally observed statistical distributions of canonic geometrical features of two- and three-point combinations, e.g. pairwise and radial distances, angles, etc. We have introduced a general Monte-

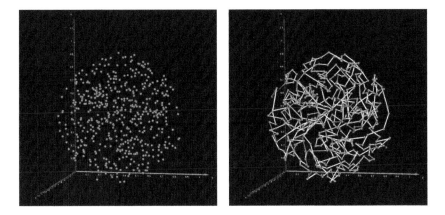

Fig. 10. Left: initial random distribution of $N = 528$ points for a unit spherical confinement. Right: randomly folded 3D fiber computed on the basis of the initial point cloud.

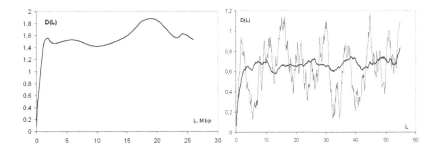

Fig. 11. Left: experimentally assessed relationship $D(L)$ between physical and genomic distances on a scale $0.1 - 28$ Mb for human chromosome 1. Right: numerically computed $D(L)$ for a synthetic fiber randomly folded within a unit 3D spherical confinement (cf. Fig. 10): red curve corresponds to $D(L)$ for a single (starting point dependent) simulation, black curve shows an average $D(L)$ for $N/2 = 264$ simulation runs with varying starting points.

Carlo-based simulation scheme for computation of PDFs of geometrical features of random point distributions for arbitrarily-shaped confined 3D domains, and derived numerical criterions for the estimation of the order of randomness of observed statistical distribution as well as dimension and shape of probed chromosome regions. Preliminary experimental results of sampling human chromosome 1 in primary human fibroblasts in G1 cell cycle phase by five overlapping ridge and antiridge BAC triplets on a genomic scale $0.7 - 3.3$ Mb support the assumption of confined random folding of 3D DNA fiber in interphase cell nuclei. We have proposed a geometrical model of confined random chromatin folding, which is capable to reproduce experimentally observed relationship between physical and genomic distances on a large genomic scale. Further sampling experiments with 4 simultaneous BAC labels and larger statistical series ($\approx 10^4$ distance measurements) are required to provide a more consistent source of geometrical information for distinctive analysis of 3D chromatin structure.

Acknowledgements

This work is part of the 3DGENOME program, which is supported by the European Commission: contract LSHG-CT-2003-503441.

References

1. H. Caron et al. The human transcriptome map: clustering of highly expressed genes in chromosomal domains. *Science*, 291:1289–1292, 2001.
2. T. Cremer et al. Higher order chromatin architecture in the cell nucleus: on the way from structure to function. *Biol. Cell.*, 96:555–567, 2004.
3. E. Gladilin et al. Topological analysis of 3D cell nuclei using finite element template-based spherical mapping. In *Proc. SPIE MI*, volume 6144, pages 1557–1566, 2006.
4. P. Hahnfeldt et al. Polymer models for interphase chromosomes. *Proc. Natl. Acad. Sci. USA, Biophysics*, 90:7854–7858, 1993.
5. R. K. Sachs et al. A random-walk/giant-loop model for interphase chromosomes. *Proc. Natl. Acad. Sci. USA, Biochemistry*, 92:2710–2714, 1995.
6. S. J. Tu et al. Random distance distribution of spherical objects: general theory and applications to n-dimensional physics. *Math. Phys.*, 35(31):6557–6570, 2002.
7. R. van Driel et al. The eucariotic genome: a system regulated at different hierarchical levels. *J Cell. Sci. 116*, pages 4067–4075, 2003.
8. H. Yokota et al. Evidence for the organization of chromatin in megabase pair-sized loops arranged along a random walk path in the human G0/G1 interphase nucleus. *J Cell. Biol.*, 130(6):1239–1249, 1995.
9. F. M. Young et al. *Theory and Application of Multidimensional Scaling*. Eribaum Associates, Hillsdale, NJ, 1994.

Structural Screening of HIV-1 Protease/Inhibitor Docking by Non-parametric Binomial Distribution Test

Piyarat Nimmanpipug[1], Vannajan Sanghiran Lee[1], Jeerayut Chaijaruwanich[2], Sukon Prasitwattanaseree[3], and Patrinee Traisathit[3]

[1] Deparment of Chemistry, Faculty of Science, Chiang Mai University, Chiang Mai 50200, Thailand
`npiyarat@chiangmai.ac.th, vannajan@chiangmai.ac.th`
[2] Department of Computer Science, Faculty of Science, Chiang Mai University, Chiang Mai, 50200, Thailand
`jeerayut@cs.science.cmu.ac.th`
[3] Department of Statistics, Faculty of Science, Chiang Mai University, Chiang Mai, 50200, Thailand
`s.prasit@chiangmai.ac.th, patrinee@chiangmai.ac.th`

Abstract. Attempts have been made to predict the binding structures of the human immunodeficiency virus-1 protease (HIV-1Pr) with various inhibitors within the shortest simulation time consuming. The purpose here is to improve the structural prediction by using statistical approach. We use a combination of molecular docking and non-parametric binomial distribution test considering the combination of binding energy, hydrogen bonding, and hydrophobic-hydrophilic interaction in term of binding residues to select the most probable binding structure. In this study, the binding of HIV-1Pr and two inhibitors: Saquinavir and *Litchi chinensis* extracts (3-oxotrirucalla-7, 24-dien-21-oic acid) were investigated. Each inhibitor was positioned in the active site of HIV-1Pr in many different ways using Lamarckian genetic algorithm and then score each orientation by applying a reasonable evaluation function by AutoDock3.0 program. The results from search methods were screened out using non-parametric binomial distribution test and compared with the binding structure from explicit molecular dynamic simulation. Both complexes from statistical selected docking simulation were found to be comparable with those from X-ray diffraction analysis and explicit molecular dynamic simulation structures.

Keywords: HIV-1 protease, Docking, Binomial Distribution Test, Saquinavir, 3-Oxotirucalla-7, 24-Dien-21-Oic Acid.

1 Introduction

Human Immunodeficiency Virus-1 Protease (HIV-1Pr) has an important role in the replication of HIV-1 by processing two precursor polyproteins, Pr55gag and Pr160gag-pol, into structural proteins and replication enzymes. This enzyme is a type of aspartic protease, and has a $C2$ symmetric homodimer. Each monomer consists of 99 amino acid residues and contributes a loop structure containing the active site triad

S. Hochreiter and R. Wagner (Eds.): BIRD 2007, LNBI 4414, pp. 119–130, 2007.

Asp25(25')-Thr26(26')-Gly27(27'). A cavity for the insertion of the substrate is formed by the loop structures containing the active site triads and flap regions (flaps) which are presumably related to the entry and affinity of the substrate to the enzyme [1-3].

Up to now, much research effort has been focused on computational methods for the prediction of this difficult-to-obtain structural information. In general, docking study has been widely used to generate the enzyme-substrate structures. The candidate from docking typically selected from the lowest energy structure. However, this criteria of energy screening is not always give the correct binding structure, especially for the highly flexible ligand [4]. The purpose here is to improve the structural prediction by using statistical approach; non-parametric binomial distribution test to include the essential amino acids in the binding pocket.

For preliminary study, the prediction was done for the known structure from X-ray diffraction analysis in the case of HIV-1Pr – Saquinavir complex, a peptidomimetic inhibitor of HIV protease in order to proof the efficiency of this method. In addition, since several research groups develop a number of HIV-1Pr inhibitors from natural product but its HIV-1Pr complex is still unknown, efforts have been also done for the unknown HIV-1Pr – inhibitor complex. For our investigation, the binding structure of HIV-1Pr and *Litchi chinensis* seed extracts (3-oxotrirucalla-7, 24-dien-21-oic acid) was predicted and compared with the simulated from molecular dynamic (MD). The inhibitory activity of 3-oxotrirucalla-7, 24-dien-21-oic acid is reported in IC_{50} of 20 mg/L by Chao-me Ma and co-workers in 2000 [5] and this compound was found in chemical constituents in seed of *Litchi chinensis* by Pengfei Tu and co-workers in 2002 [6]. The extracts from such a natural product still have relatively high IC_{50}, in order to get the lower IC_{50} value the molecular modeling is, therefore, a very useful tool.

2   Methods

2.1   Preparation of HIV-1Pr

The initial HIV-1Pr structure was obtained from HIV-1Pr with saquinavir at 2.3 Å resolution (1HXB entry in PDB database) [7]. The structural water and inhibitor from the selected PDB database were then removed in the preparation of the HIV-1Pr. Hydrogen atoms were added in the structure and a short minimization run was performed to remove any potentially bad contacts with the program package AMBER, Version 7 [8,9]. A cutoff distance at 12 Å for non-bonded pair interaction was used in the minimizations. Then the obtained structure model was protonated at catalytic Asp25.

2.2   Preparation of Inhibitors: Saquinavir and 3-Oxotirucalla-7, 24-Dien-21-Oic Acid

Geometry of saquinavir (compound 1) and 3-oxotirucalla-7, 24-dien-21-oic acid (compound 2), as shown in Fig. 1, were optimized using semi-empirical calculation, AM1 in the program package Spartan'04.

saquinavir	3-oxotirucalla-7,24-dien-21-oic acid

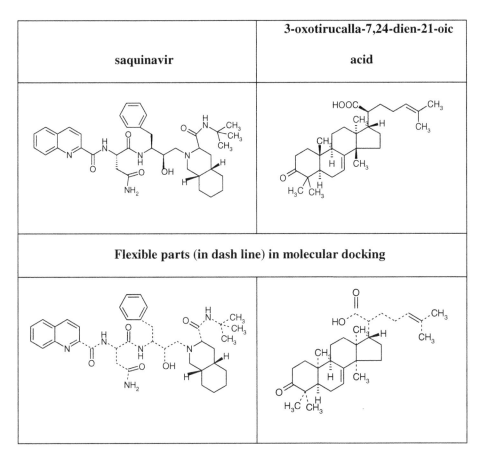

Flexible parts (in dash line) in molecular docking

Fig. 1. Chemical structures of saquinavir (compound 1) and 3-oxotirucalla-7, 24-dien-21-oic acid (compound 2)

2.3 Preparation of HIV-1Pr – Inhibitor Complex by Molecular Docking and Molecular Mechanics Methods

Each structure of HIV-1Pr – inhibitor was obtained by docking compound 1 and compound 2 to HIV-1Pr, respectively. HIV-1Pr was kept rigid while inhibitors are flexible (21 flexible torsions in compound 1 and 13 flexible torsions in compound 2), as shown in dash line in Fig.1, and Gasteiger charges were used. Grid maps have been calculated using the module AutoGrid in AutoDock 3.0 program [10-12] for protease structure. The center of grid was assigned at the center of the cavity, between two catalytic aspartics. A number of grid points in x y z, is 60 60 60 with the spacing 0.375 Å. This parameter set covers the active site extensively and let the ligand move and explored the enzyme active site without any constraints regarding the box size. The inhibitor was positioned in the active site of HIV-1 protease in many different ways using Lamarckian genetic algorithm (LGA). The solvation effect was also included in this docking study via atomic solvation parameters for each atom in the macromolecule of AutoGrid 3.0 module. Each hydrogen bonded to N and O atom in hydrogen grid maps was calculated using self-consistent hydrogen bonding 12-10 parameters in the AutoGrid.

2.4 Molecular Dynamic Simulations

The energy minimized conformation of HIV-1Pr – compound 2 from previous docking calculation was used as starting structure. The molecular mechanics potential energy minimizations and MD simulations were carried out with the program package AMBER, Version 7 [8,9]. Calculations were performed using the parm99 force field reference for HIV-1Pr and compound 2. The atom types for compound 2 were assigned by mapping their chemical properties (element, hybridization, bonding schemes) to AMBER atom type library and the Gasteiger charges were used.

The enzyme-inhibitor complex was solvated with WATBOX216 water model (9298 water molecules) in cell dimension 61.06 x 66.56 x 75.88 Å [3] and treated in the simulation under periodic boundary conditions. All of MD simulations reported here were done under an isobaric-isothermal ensemble (NPT) using constant pressure of 1 atm and constant temperature of 298 K. The volume was chosen to maintain a density of 1 g/cm^3. A cutoff distance (12 Å) was applied for the non-bonded pair interaction. Three sodium and eight chloride ions were added to neutralize and buffer the system. The potential energy minimizations were performed on the systems using the steepest descent method and followed by conjugate gradient method. After a short minimization, molecular dynamic simulations were performed to get an equilibrium structure. The temperature of the whole system was gradually increased by heating to 298 K for the first 60 ps, and then it was kept at 298 K from 61-1800 ps. The temperature was kept constant according to the Berendsen algorithm [13]. All trajectories were kept and analyzed in detail.

2.5 Non-parametric Binomial Distribution Test

The enzyme-inhibitor (compound 1 and compound 2) complexes from 100 runs molecular docking were collected in order to explore all probable binding structures. The vicinity residues in the binding pocket within a trial distance (3 Å) measured from the inhibitor were selected as vital amino acids for enzyme-inhibitor complex formation. As shown in Fig. 2, the average hydrogen bond formation of total 100 runs is only 0-1 bond but 1-2 hydrogen bonds are found for the 25% lower of energy ranking. Therefore, only the lower quartile structures will be taken into consideration here. The present or absent of amino acid was treated as binomial variable x. In this case, the probability of present or absent of amino acid was assumed to be 50% equally. Binomial Distribution Test–value (BDT-value), $P^+(x)$ and $P^-(x)$, for the observed number of *present* (r^+) and *absent* (r^-) can be calculated directly from this binomial distribution:

$$P^+(x) = P(x >= r^+ \text{ when } p = 1/2)$$

$$= \sum_{r^+}^{N} \binom{N}{r^+} (0.5)^{r^+} (0.5)^{N-r^+} \tag{1}$$

$$P^-(x) = P(x >= r^- \text{ when } p = 1/2)$$

If $P^+(x) < \alpha$ amino acid x will be present, on the other hand, if $P^-(x) < \alpha$ amino acid x will be absent. In this case, $\alpha = 0.05$ indicating that if more than 95% of cases were found then the present of absent of the residue will be accepted.

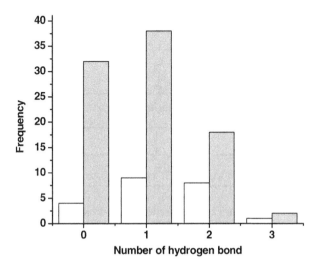

Fig. 2. Comparison of hydrogen bond distribution between total docking structures (gray) and the 25% lower of energy ranking structures (white)

3 Results and Discussion

3.1 Non-parametric Binomial Distribution Test of HIV-1Pr – Inhibitor Complexes

As in most of the existing methods, the protein-ligand complex was composed of a rigid protein structure and a flexible ligand. A flexible ligand has three translational degrees of freedom, three rotational degrees of freedom and one dihedral rotation for each 2 rotable bond. The docking search is computed over a 6+n dimensional space where n is the number of rotable bonds in the ligand. Fig. 3 shows the combined statistical-docking algorithm with the docking simulations. One docking trial allows a protein and a ligand to dock into its binding site, a docking attempt consists of a series of independent structures or the so-called clusters. For each cluster in the lower quartile, all vicinity residues (acidic, basic, polar, and non-polar groups) were tested

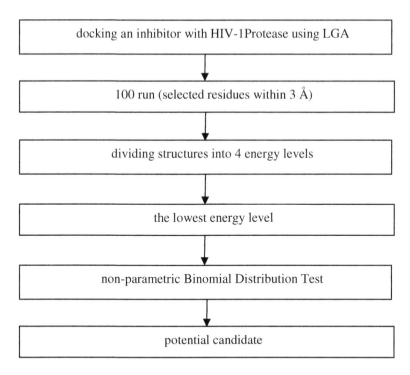

Fig. 3. The statistical-based protein-ligand docking procedure

Table 1. Criteria for screening crucial residues of HIV-1Pr-compound 1 complex

Present residues	Chain A: Asp25, Ala28, Gly48, Gly49, Ile50	Chain B: Asp25', Gly27', Ala28', Asp29', Gly48', Gly49', Ile50'
Absent residues	Chain A: Leu23, Met46, Leu76, Thr80, Val82, Asn83, Ile84	Chain B: Thr26', Leu76', Thr80', Arg87'

for the possibility of appearance. The criteria for essential-inessential residues were concluded and reported in Table 1 (HIV-1Pr – compound 1 complex) and Table 2 (HIV-1Pr – compound 2 complex). Using these criteria (Table 1-2), the inessential residues will be neglected.

Table 2. Criteria for screening crucial residues of HIV-1Pr-compound 2 complex

Present residues	Chain A: Asp25, Gly27, Gly48	Chain B: Asp25', Gly27', Ala28', Asp29', Asp30'
Absent residues	Chain A: Val32, Ile47, Val82, Ile84	Chain B: Val32', Ile47'

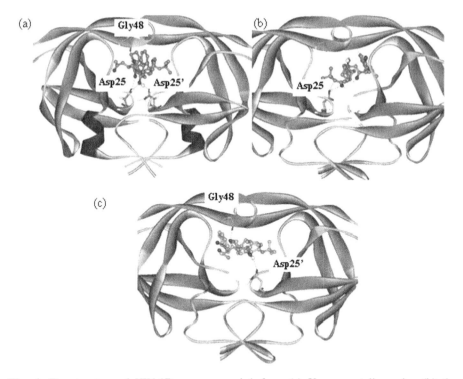

Fig. 4. The structure of HIV-1Pr – compound 1 from (a) X-ray crystallography, (b) the minimized structure after molecular docking, and (c) the statistical selected docking structure. The binding residues were shown in stick.

3.2 Structural Comparison Between Docking Statistical Test, and X-ray Crystallographic Structure of HIV-1Pr – Compound 1 Complex

The X-ray studied structure of HIV-1Pr complexed with compound 1 is shown in Fig. 4a. The inhibitor was bound with protease enzyme using both catalytic and flap regions at Asp25, Asp25', and Gly48 respectively. Compared with experimental investigation, the lowest energy structure from docking (Fig. 4b) give the inappropriate binding structure in both its orientation and the binding sites. After screening using amino acid residues in our non-parametric binomial distribution test criteria, the more appropriate binding residues (Asp25' and Gly48) were observed in Fig. 4c and the more similar inhibitor structure orientation was also found as shown from the superimposition of all atoms in enzyme–compound 1 complexes in Fig. 5.

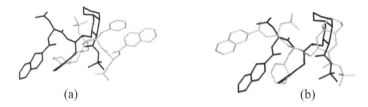

(a)   (b)

Fig. 5. Superimposition of all atoms in enzyme–compound 1 complexes (a) between the X-ray crystallographic structure (black) and the lowest energy docking structure (gray) (b) between the X-ray crystallographic structure (black) and the statistical selected docking structure (gray)

3.3 Structural Comparison Between Docking and Explicit Water MD Simulation of HIV-1Pr – Compound 2 Complex

Firstly, the 100 run docking calculations were used in a prior prediction of binding affinities and to simulate a crystal geometry as a candidate of the ligand/protein complex. However, from the obtained lowest energy minimized structure, the direction of OH group in compound 2 points to the flap site (O:Gly48') instead of catalytic site of enzyme. The reason for this observation may due to the highly flexibility of ligand and rigidity of HIV-1Pr leading to the inappropriate binding structure from docking study. One way to solve problem is to take the flexibility of both enzyme and ligand into account. Therefore MD simulation was performed using this docking structure. The total energy (ET), Potential Energy (EP) and Kinetic energy (EK) over simulations from 0-1800 ps were investigated and the equilibrium was obtained after 600 ps. After the equilibrium stage, compound 2 was found to bind to the enzyme at catalytic site with more than 89% hydrogen bond formation with Asp25 and Asp29. As shown in Table 3, the energy minimized structure obtained from molecular dynamics simulations directs CO and OH group of compound 2 to the catalytic site of enzyme, O:Asp25 and N:Asp29, respectively. The structure of

Table 3. Possible hydrogen bond between compound 2 and HIV-1Pr after reach equilibrium in explicit MD simulation

	% H-bond formation	Average distance / Å
Case 1: HIV-1Pr as donor and compound 1 as acceptor		
Asp25:OD2H - compound1:O78	1.98	3.067
Asp29:NH - compound 1:O21	94.70	2.968
Asp30:NH - compound1:O21	1.73	3.211
Ala28':NH - compound 1:O78	3.27	3.054
Case 2: compound 1 as donor and HIV-1Pr as acceptor		
compound 1:O78H79 – Asp25:OD1	0.04	3.168
compound 1:O78H79 – Asp25:OD2	89.18	2.946
compound 1:O78H79 – Asp25':OD1	12.98	2.929
compound 1:O78H79 –Asp25':OD2	2.00	2.949

HIV-1Pr – compound 2 from docking method and MD simulation were compared in Fig. 6. The binding residues as discussed above were labeled and shown in stick.

Root mean square deviation (RMSD) of all atoms for complex of enzyme-compound 1 in docked - MD structure and statistical docked-MD structure are 3.63 Å and 3.84 Å respectively, indicating somehow conformational difference and molecular displacement in inhibitor structure after both docking results (Fig.7). However binding site of the statistical docked structure is the same as equilibrated MD structure as shown in Fig. 8; both structures bound to Asp25 and Asp29.

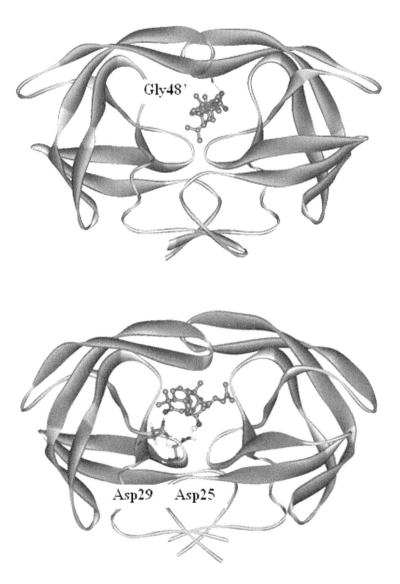

Fig. 6. The structure of HIV-1Pr – compound 2 from docking method (top) and the minimized final structure after explicit MD simulation (below). The binding residues were shown in stick.

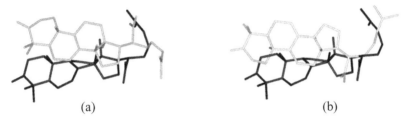

(a)   (b)

Fig. 7. Superimposition of all atoms in enzyme–compound 2 complexes (a) between the final MD simulations structure (black) and the lowest energy docking structure (gray) (b) between the final MD simulations structure (black) and the statistical selected docking structure (gray)

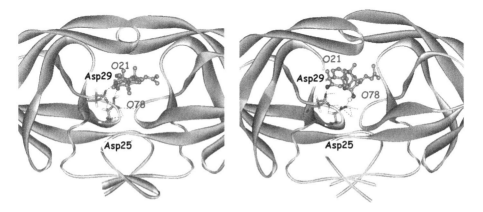

Fig. 8. Binding structure of enzyme-compound 2 complex from statistical selected docking structure (left) compare to that from MD simulation (right). The binding residues were shown in stick.

4 Conclusions

A combination of molecular docking and non-parametric binomial distribution test considering the binding energy, hydrogen bonding, hydrophobic-hydrophilic interaction in term of residue binding to select the most probable binding structure can reduce simulation time consuming. In this study, binding mode and orientation in HIV-1Pr – inhibitors (saquinavir and 3-oxotrirucalla-7, 24-dien-21-oic acid) complexes were found to be comparable with those from X-ray diffraction and explicit molecular dynamic simulation structures. Our statistical selected docking simulation provides the significant improvement of enzyme-inhibitor binding prediction.

Acknowledgements. We gratefully acknowledge a generous grant of computer time from the Computational Simulation and Modeling Laboratory (CSML), Department of Chemistry, Faculty of Science, Chiang Mai University, Chiang Mai, Thailand, and Computational Chemistry Unit Cell (CCUC), Department of Chemistry, Faculty of

Science, Chulalongkorn University, Bangkok, Thailand, where the simulations were performed. This project was supported by the Commission on Higher Education and Thailand Research Fund (TRF), Thailand.

References

1. Lappatto, R., Blundell, T., Hemmings, A., Overington, J., Wilderspin, A., Wood, S., Merson, J. R., Whittle, P. J., Danley, D. E., Geoghegan, K. F., Hawrylik, S. K., Lee, S. E., Scheld, K. G., Hobart, P. M.: X-ray analysis of HIV-1 proteinase at 2.7 A resolution confirms structural homology among retroviral enzymes. Nature, Vol. 342. (1989) 299-302
2. Navia, M. A., Fitzgerald, P. M.: D., McKeever, B. M., Leu, C. T., Heimbach, J. C., Herber, W. K., Sigal, I. S., Darke, P. L., Springer, J. P.: Three-dimensional structure of aspartyl protease from human immunodeficiency virus HIV-1. Nature, Vol. 337. (1989) 20-615
3. Okimoto, N., Tsukui, T., Kitayama, K., Hata, M., Hoshino, T., Tsuda, M., J. Am. Chem. Soc., 122, 5613 (2000)
4. Hassan, S. A., Gracia, L., Vasudevan, G., Steinbach, P. J.: Methods Computer simulation of protein-ligand interactions: challenges and application. Methods Mol. Biol, Vol. 305. (2005) 92-451
5. Ma, C.-m., Nakamura, N., Hattori, M., Kakuda, H., Qiao, J.-c., Yu, H.-l.: Inhibitory effects on HIV-1 protease of constituents from the wood of Xanthoceras sorbifolia. . J. Nat. Prod., Vol. 63. (2000) 42-238
6. Tu, P., Luo, Q., Zheng, J., Zhongcaoyao, 33, 300 (2002)
7. Krohn, A., Redshaw, S., Ritchie, J. C., Graves, B. J., Hatada, M. H.: Novel binding mode of highly potent HIV-proteinase inhibitors incorporating the (R)-hydroxyethylamine isostere. J. Med. Chem, Vol. 34. (1991) 2-3340
8. Case, D. A., Pearlman, D. A., Caldwell, J. W., Cheatham III, T. E., Wang, J., Ross, W. S., Simmerling, C. L., Darden, T. A., Merz, K. M., Stanton, R. V., Cheng, A. L., Vincent, J. J., Crow-ley, M., Tsui, V., Gohlke, H., Radmer, R. H., Duan, Y., Pitera, J., Massova, I., Seibel, G. L., Singh, U. C., Weiner, P. K., Kollman, P. A.: AMBER 7. University of California, San Francisco (2002)
9. Pearlman, D. A., Case, D. A., Caldwell, J. W., Ross, W. S., Cheatham III, T. E., DeBolt, S., Ferguson, D., Seibel, G., Kollman, P. A..: AMBER, a package of computer programs for applying molecular mechanics, normal mode analysis, molecular dynamics and free energy calculations to simulate the structural and energetic properties of molecules. Comp. Phys. Commun, Vol. 91. (1995) 1-41
10. Goodsell, D. S., Olson, A. J.: Automated docking of substrates to proteins by simulated annealing. Proteins: Structure, Function, and Genetics, Vol. 8. (1990) 195-202
11. Morris, G. M., Goodsell, D. S., Huey, R., Olson, A. J.: Distributed automated docking of flexible ligands to proteins: parallel applications of AutoDock 2.4. J Computer-Aided Molecular Design, Vol. 10. (1996) 293-304
12. Morris, G. M., Goodsell, D. S., Halliday, R. S., Huey, R., Hart, W. E., Belew, R. K., Olson, A. J., J. Comp. Chem., 19, 1639 (1998)
13. Berendsen, H. J. C., Postma, J. P. M., Gunsteren, W. F. van, DiNola, A., Haak, J. R., J. Chem. Phys., 81, 3684 (1984)

satDNA Analyzer 1.2 as a Valuable Computing Tool for Evolutionary Analysis of Satellite-DNA Families: Revisiting Y-Linked Satellite-DNA Sequences of *Rumex* (Polygonaceae)

Rafael Navajas-Pérez[1], Manuel Ruiz Rejón[1], Manuel Garrido-Ramos[1], José Luis Aznarte[2], and Cristina Rubio-Escudero[2]

[1] Departamento de Genética, Facultad de Ciencias, Universidad de Granada
Campus de Fuentenueva s/n 18071, Granada, Spain
[2] Departamento de Ciencias de la Computación e Inteligencia Artificial,
Universidad de Granada
C/ Periodista Daniel Saucedo Aranda s/n 18071, Granada, Spain

Abstract. In a previous paper [1] we showed that Y-linked satellite-DNA sequences of Rumex (Polygonaceae) present reduced rates of evolution in relation to other autosomal satellite-DNA sequences. In the present paper, we re-analyze the same set of sequences by using the satDNA Analyzer 1.2 software, specifically developed by us for analysis of satellite DNA evolution. We do not only confirm our previous findings but also prove that the satDNA Analyzer 1.2 package constitutes a powerful tool for users interested in evolutionary analysis on satellite-DNA sequences. In fact, we are able to gather more accurate calculations regarding location of Strachan positions and evolutionary rates calculations, among others useful statistics. All results are displayed in a very comprehensive multicoloured graphic representation easy to use as an html file. Furthermore, satDNA Analyzer 1.2 is a time saving feature since every utility is automatized and collected in a single software package, so the user does not need to use different programs. Additionally, it significantly reduces the rate of data miscalculations due to human errors, very prone to occur specially in large files.

1 Introduction

Despite of sex chromosomes having evolved independently in several different groups of organisms (such as fishes-[2]-, reptiles -[3]- birds – [4]- , mammals –[5]-, insects – [6] - or plants – [7]), they seem to share some common evolutionary features [8]. The commonality is the presence of a pair of heteromorphic sex chromosomes in males (XY) consequence of differentiation and degeneration of Y chromosome. In fact, sex chromosomes undergo a process of gradual suppression of recombination that converts the Y chromosome in a relict chromosome with no counterpart to recombine with. Thus, this process leads to progressive divergence and to the erosion of the Y chromosome [9]. The final outcome of this process is the accumulation of mutations in dispensable regions of Y architecture (high rates of mutation have been

S. Hochreiter and R. Wagner (Eds.): BIRD 2007, LNBI 4414, pp. 131–139, 2007.

described in Y-linked genes- [10]; [11]) and the subsequent loss of function of many genes within the Y chromosome [12]. Y-chromosome degeneration is also accompanied by the accumulation of a set of diverse repetitive sequences such as mobile elements and satellite DNAs [13]; [14];[15]; [16]).

In the present work, we want to emphasize the role of satellite-DNA sequences in the Y degeneration process. Models of evolutionary dynamics for satellite DNA predict its accumulation in chromosomal regions where recombination rates are low [17]. Good examples of this are the non-recombining Y chromosomes ([16]; [18]). However, little is known about how this occurs or about how the absence of recombination affects the subsequent evolutionary fate of the repetitive sequences in the Y chromosome. In the present study, we focus on satellite-DNA sequences accumulation and evolution using as models the dioecious species of Genus *Rumex*, *R. acetosa* and *R. papillaris,* and by means of new computing utilities gathered together in satDNA Analyzer 1.2 package (http://satdna.sourceforge.net).

2   Antecedents and Motivation

Males of *R. acetosa* and *R. papillaris* have a karyotype with 15 chromosomes including a complex XX/XY$_1$Y$_2$ sex-chromosome system, while females have 14 chromosomes, being 2n= 12 + XX. During meiosis, the two Y chromosomes pair only with the ends of each X arm (own observations). All the data indicate that the Ys and the X chromosomes are highly differentiated and that the Y chromosomes are degenerated, as they are heterochromatic and rich in satellite-DNA sequences [19]. In fact, two satellite-DNA families have been found in both species to be massively accumulated in the Y chromosomes, the RAYSI family [20]; [21] and the RAE180 family [22]. Additionally, other satellite-DNA family, RAE730, has been described in heterochromatic segments of some autosome pairs [23].

To elucidate evolutionary rates of Y-linked sequences in relation to autosomal ones, we performed a comparative analysis between *R. acetosa* and *R. papillaris* sequences belonging to three different satellite-DNA families separately. Basically, we performed distance calculations according to the Jukes-Cantor method [24] and from these, we estimated evolutionary rates for both Y and autosomal-linked families. We found that Y-linked satellite sequences evolve two-fold to five-fold slower than autosomal-linked ones. Additionally, we proposed that shared polymorphisms should be removed when analyzing closely related species for more accurate calculations, since they might indicate ancestral variation before splitting of both species but not true divergence. In contrast, non-shared polymorphisms would be automorphies and represent different transitional stages in the process of intraspecific homogenization and interspecific divergence (for full details see [1]).

This study was reinforced by analyzing concerted evolution status (see [25]) of every three satellite-DNA sequences. We followed the method described in [26] which allows to analyze the pattern of variation at each nucleotide site between a pair of species for every of three marker studied (see Materials and Methods for further details).

Our aim is to confirm our previous findings by using the software satDNA Analyzer 1.2 and analyzing the same set of sequences described before. This supposes a non-time demanding method since every step is automated (location of different Strachan positions, removing shared polymorphisms from alignment and all statistics such as average consensus sequences, the average base pair contents, the distribution of variant sites, the transition to transversion rate and different estimates of intra and inter-specific variation) and collected in an unique package, so the user does not need to resort to different softwares. Additionally, the use of the software prevents from data miscalculations due to human errors, very prone to occur specially in large files.

3 Materials and Methods

3.1 Biological Material and Laboratory Procedures

Sequences analyzed in the present work were taken from the EMBL database (http://www.ebi.ac.uk/embl/) with accession numbers AJ580328 to AJ580343, AJ580382 to AJ580398, AJ580457 to AJ580463, AJ580468 to AJ580485, AJ580494 to AJ580496, AJ634478 to AJ634526, AJ634533 to AJ63456 and AJ639709 to AJ639741. These sequences belong to three different satellite-DNA families (RAE180, RAE730 and RAYSI) that we previously isolated in *R. acetosa* and *R. papillaris*. Biological material procedence and laboratory methodologies are fully described in [1].

3.2 Sequence Analysis

For the present work, we have revisited the sequences described above by using a new computing tool , satDNA Analyzer 1.2, a software package for the analysis of satellite-DNA sequences from aligned DNA sequence data implemented in C++. It allows fast and easy analysis of patterns of variation at each nucleotide position considered independently amongst all units of a given satellite-DNA family when comparing sets of sequences belonging to two different species. The program classifies each site as monomorphic or polymorphic, discriminates shared from non-shared polymorphisms and classifies each non-shared polymorphism according to the model proposed by [26] in six different stages of transition during the spread of a variant repeat unit toward its fixation (for a detailed explanation of this method, see also [27]). Briefly described, the classs 1 site represents complete homogeneity between two species, whereas classes 2 to 4 represent intermediate stages in which one of the species shows polymorphism. The frequency of the new nucleotide variant at the site considered is low in stage 2 and intermediate in stage 3, while class 4 comprises sites in which a mutation has replaced the progenitor base in most members of repetitive family in the other species (almost fully homogenized site). Class 5 represents diagnostic sites in which a new variant is fully homogenized and fixed in all members of one of the species while the other species retain the progenitor nucleotide. Class 6 represents an additional step over stage 5 (new variants appear in some of the members of the repetitive family at a site fully divergent between two species). Furthermore, this program implements several other utilities for satellite-DNA analysis evolution such as the design of the average consensus sequences, the average base pair contents, the distribution of variant sites, the transition to transversion rate, and different estimates of intra and inter-specific variation.

Aprioristic hypotheses on factors influencing the molecular drive process and the rates and biases of concerted evolution can be tested with this program. Additionally, satDNA Analyzer generates an output file containing an alignment to be used for further evolutionary analysis by using different phylogenetic softwares. The novelty of this feature is that it allows to discard the shared polymorphisms for the analysis, which as we have demonstrated in [1], can interfere with the results when analyzing closely related species.

satDNA Analyzer 1.2 is freely available at http://satdna.sourceforge.net where supplementary documentation can be also found. satDNA Analyzer 1.2 has been designed to operate under Windows, Linux and MAC operating systems.

4   Results and Discussion

4.1   Subfamilies Detection

One of the main problems researchers have to face up is the recurrent formation of subfamilies in satellite-DNA sets of sequences, due to differential regions within the repeats or by the presence of diagnostic positions specifically fixed in one or another species ([28], [29]). The non-detection of these types of sequences before carrying out further evolutionary analysis can lead to the comparison of non-orthologous sequences and then to subsequent miscalculations. In this work we test the ability of satDNA Analyzer 1.2 to detect such cases. We previously described two paralogous RAYSI subfamilies in *R. acetosa*, called RAYSI-S and RAYSI-J [1]; [21]. We have used as input for our software a set of sequences of RAYSI isolated from *R. acetosa* genome. The study of Strachan stages included as a feature of our software reveals the existence of 72 diagnostic (fixed or almost fixed) positions, what shows the capacity of satDNA Analyzer 1.2 to discriminate both subfamilies (see Figure 1A). This approximately corresponds with our previous estimation of 83 of such as sites. Additionally, 20 sites are in transition stage 6, indicating the beginning of a new cycle of mutation-homogenization. This is supported by the fact that the mean inter-family divergence between both types of sequences is around 18% (see Figure 1B) while the mean intra-family percentage of differences is 4.2% and 5.1% for RAYSI-S and RAYSI-J respectively. Both subfamilies additionally have diagnostic deletions found at different positions in the RAYSI monomers also recognized by satDNA Analyzer as irrelevant positions due to indels.

We then have divided the RAYSI sequences of *R. acetosa* in two different files for RAYSI-J and RAYSI-S respectively. For further analysis, we combined both sets of sequences with RAYSI sequences belonging to *R. papillaris*. The study of diagnostic sites shows that RAYSI sequences of *R. papillaris* belong to the RAYSI-J subfamily since they share more diagnostic positions with this subfamily than with RAYSI-S (see also [21]).

4.2   Evolutionary Analysis and Concerted Evolution

In the present work, we have analyzed the rate of concerted evolution of the three satellite-DNA families studied in *R. acetosa* and *R. papillaris*: the Y-linked RAYSI and RAE180 families and the autosomal RAE730 family. Essentially, we wanted to address the problem of differences in the evolutionary patterns of sequences accumulating in Y chromosomes with respect to those accumulating in autosomes. It is particularly interesting taking into account the non-recombining nature of Y chromosomes. For that task, we used three sets of sequence alignments previously analyzed for us [1] as inputs for satDNA Analyzer 1.2. Specifically, for RAYSI analysis we used a set including RAYSI-J sequences which is the only subfamily present in both *R. acetosa* and *R. papillaris*.

In relation to interspecific divergence, satDNA Analyzer 1.2 reveals that the variability of the Y-associated satellite DNAs, RAYSI-J and RAE180 is much higher than in the autosomic RAE730 sequences. We pointed before that these results might indicate ancestral variation in Y-linked sequences, but not true divergence, due to the significant presence of shared polymorphic positions. We assumed that these sites are ancestral and appeared prior to the split between *R. acetosa* and *R. papillaris*. In this sense, satDNA Analyzer 1.2 includes a utility to discard shared polymorphisms from the analysis for statistics calculations and for further phylogenetic analysis (see supplementary information at http://satdna.sourceforge.net). Then, we performed a second analysis excluding shared polymorphisms. This latter analysis shows that the mean genetic distance for RAE730 sequences between *R. acetosa* and *R. papillaris* are two-fold to three-fold higher than intraspecific variation. Considering that *R. acetosa* and *R. papillaris* diverged 2 million years ago [30], we estimated a rate of sequence change for these three families using other utility of satDNA Analyzer 1.2 (Table 1). This rate of sequence change for RAE730 satellite DNA is around two-fold higher than the rates for the RAYSI and the RAE180 satellite DNAs. These results perfectly correlate with those gathered before manually (see Table 1).

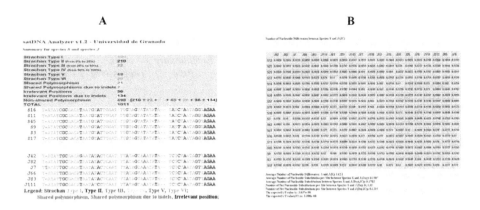

Fig. 1. (A) Partial alignment of RAYSI (RAYSI-S and RAYSI-J) sequences of *R. acetosa*, displaying a summary of different positions and the legend. These are screenshots captured from a much larger output file. Note the significant presence of transition stages 5. **(B)** Example of a table representation in the output.html file generated by satDNA Analyzer 1.2, showing the number of nucleotide differences between RAYSI subfamilies S and J.

We have also studied the transitional stages in the process of concerted evolution according to the Strachan model [26]. SatDNA Analyzer 1.2 distinguishes Strachan stages from 1 to 6. For practical purposes we grouped together stages 2 and 3 in *Initial Stages Class* (**ISC**) and stages 4 and 5 in *Fully or almost Fully Homogenized Class* (**FHC**). RAE730 sequences show higher percentage of **FHC** sites (47 sites) in relation to Y-linked RAYSI-J and RAE180 (4 and 3 respectively). In fact, most positions in Y-linked sequences seem to be in **ISC** yet (312 for RAYSI-J and 33 for RAE180). As shown in Table 1, these calculations differ slightly to our previous results gathered manually. It is probably due to the fact that we considered some indels as positions in the manual calculations. However, this reveals more accuracy in results obtained by satDNA Analyzer 1.2, especially significant in long satellite-DNAs (see Table 1). The mean length of satellite-DNA families has been suggested to be 165 bp in plants [31], but significantly longer sequences have been described in both plants and animals (as the case of RAYSI in *Rumex* with 930-bp repeats- [20]- or some mammals- [32] described a satellite-DNA family with repeats of 2600 bp in bovids) for which satDNA Analyzer 1.2 would be especially suitable.

We have gathered data that correlate significantly with those in [1]. Particularly, we found that within the RAE180 repeat units approximately 59% of the sites represent shared polymorphisms between *R. acetosa* and *R. papillaris*. However, we detected only one nearly fixed difference (0.5% of the sites) between these two species and 17% of polymorphic transitional stages. These data contrast with those found for the RAE730 sequences. In this case, 5.5% of nucleotide sites represent

Table 1. Comparative between data from Navajas-Perez et al., 2005a (**stated as previous data**) and data gathered by satDNA Analyzer in this paper. (**Top**) Mean intraspecific variability and interspecific divergence of three satellite-DNA families considering shared polymorphisms (**SP**), (**Down**) Analysis after excluding shared polymorphic sites (see text for details). Notes: (**ISC**) Initial Stages Class and (**FHC**) Fully or Almost Fully Homogenized Class.

	Mean distance			Differences between species	
With SP	*Intraspecific* (*R.acetosa/R.papillaris*)	*Intersp.*	*Evolutionary Rate*	*FHC* (*stages 4+5*)	*ISC* (*stages 2+3*)
RAE730					
previous data	0.055/0.036	0.099			
satDNA Analyzer	**0.055/0.036**	**0.099**			
RAYSI-J					
previous data	0.048/0.054	0.063			
satDNA Analyzer	**0.051/0.056**	**0.065**			
RAE180					
previous data	0.195/0.203	0.228			
satDNA Analyzer	**0.199/0.211**	**0.235**			
Without SP					
RAE730					
previous data	0.046/0.029	0.088	22×10^{-9}	47	281
satDNA Analyzer	**0.046/0.028**	**0.087**	**21.65×10^{-9}**	**28**	**222**
RAYSI-J					
previous data	0.037/0.042	0.047	11.74×10^{-9}	3	407
satDNA Analyzer	**0.036/0.043**	**0.047**	**11.63×10^{-9}**	**4**	**312**
RAE180					
previous data	0.036/0.037	0.045	11.25×10^{-9}	3	74
satDNA Analyzer	**0.028/0.029**	**0.033**	**8.25×10^{-9}**	**1**	**33**

shared polymorphic sites, while 4% are fixed differences between *R. acetosa* and *R. papillaris* and 30% transitional stages. Clearly, the data support the contention that the rate of concerted evolution is lower for the RAE180 satellite DNA at the Y chromosomes than for the RAE730 autosomic satellite DNA, as it is for RAYSI sequences, since *R. acetosa* and *R. papillaris* differ by only 0.4% of the sites and show 33% transitional stages. However, as opposed to RAE180, RAYSI sequences of the two species share only 6% of polymorphisms. This difference in the number of shared polymorphisms could be explained by the fact that RAE180 sequences are older than RAYSI, and therefore have accumulated a higher number of ancestral polymorphisms. Recent data gathered using Southern-blot hybridization may indicate that RAE180 sequences indeed have an older origin than do RAYSI sequences (own observations).

Additionally, we have tested this software with different sets of sequences gathering same and satisfactory results. However, these results were out of the purposes of this paper and are not shown. In the present work, we do demonstrate the utility of satDNA Analyzer in sets of sequences with main problems when carrying out evolutionary analysis on satellite DNAs, which are: low rates of concerted evolution and subfamilies formation. To summarize, satDNA Analyzer 1.2 constitutes a unique tool for evolutionary analysis of satellite-DNA. In this work we have proved that aprioristic hypotheses on factors influencing the molecular drive process and the rates and biases of concerted evolution can be tested with this program, as comparative analysis of rate between Y-linked and autosomal sequences or subfamily detection. Furthermore, satDNA Analyzer 1.2 supposes a non-time demanding method since every utility is automatized and collected in an unique package, so the user does not need to resort to different softwares. The results are displayed in a very comprehensive multicoloured graphic representation easy to use as an html file (see Figures 1A and 1B). Additionally, the use of the software prevents from data miscalculations due to human errors, very prone to occur specially in large files. Other utilities not shown in this work (as for example design of the average consensus sequences, the average base pair contents, the transition to transversion rate) are included in the software, constituting a complete package for satellite-DNA researchers.

Acknowledgments

This work was supported by Grant CGL2006-00444/BOS awarded by the Ministerio de Educación y Ciencia, Spain. R. N-P was granted by Plan Propio of University of Granada and is currently a Fulbright Postdoctoral Scholar of the Ministerio de Educación y Ciencia, Spain, at Plant Genome Mapping Laboratory in the Genetic Applied Genetic Technologies Center at the University of Georgia, Athens, GA.

References

1. Navajas-Pérez, R., de la Herrán, R., Jamilena, M., Lozano, R., Ruiz Rejón, C., Ruiz Rejón, M. & Garrido-Ramos, MA. (2005a). Reduced rates of sequence evolution of Y-linked satellite DNA in Rumex (Polygonaceae). *J. Mol. Evol.*, 60:391-399.

2. Devlin, RH. & Nagahama, Y. (2002). Sex determination and sex differentiation in fish: an overview of genetic, physiological and environmental influences. *Aquaculture* 208:191-364.
3. Bull, JJ. (1980). Sex determination in Reptiles. *The Quarterly Review of Biology*, 55(1): 3-21.
4. Fridolfsson, AK., Cheng, H., Copeland, NG., Jenkins, NA., Liu, HC., Raudsepp, T., Woodage, T., Chowdhary, B., Halverson, J. & Ellegren, H. (1998). Evolution of the avian sex chromosomes from an ancestral pair of autosomes. *PNAS*, 95(14):8147-8152.
5. Marshall Graves, JA. & Watson, JM. (1991). Mammalian sex chromosomes: Evolution of organization and function. *Chromosoma,* 101(2):63-68.
6. Traut, W. & Marec, F. (1996). Sex chromatin in Lepidoptera. *The Quarterly Review of Biology*, 71(2):239-256.
7. Ruiz Rejón, M. (2004). Sex chromosomes in plants. Encyclopedia of Plant and Crop Sciences Vol. IV of Dekker *Agropedia* (6 Vols.). Marcel Dekker Inc., New York, pp. 1148–1151.
8. Charlesworth, B. (1996). The evolution of chromosomal sex determination and dosage compensation. *Curr. Biol.,* 6(2):149–162.
9. Filatov, DA., Moneger, F., Negrutiu, I. & Charlesworth, D. (2000). Low variability in a Y-linked plant gene and its implications for Y-chromosome evolution. *Nature*, 404, 388–390.
10. Shimmin, LC., Chang, BH. & Li, CH. (1994). Contrasting rates of nucleotide substitution in the X-linked and Y-linked zinc finger genes. *J Mol Evol.,* 39(6):569-578.
11. Filatov, DA. (2005). Substitution rates in a new Silene latifolia sex-linked gene, SlssX/Y. *Mol Biol Evol.* 22(3):402-408. Erratum in: *Mol Biol Evol.,* 22(4):1157.
12. Charlesworth, D. (2002). Plant sex determination and sex chromosomes.*Heredity*, 88: 94–101.
13. Steinemann, M. & Steinemann, S. (1997). The enigma of the Y chromosome degeneration: TRAM, a novel retrotransposon is preferentially located on the neo-Y chromosome of *Drosophila miranda*. *Genetics*, 145:261–266.
14. Bachtrog, D. (2003a). Adaptation shapes patterns of genome evolution on sexual and asexual chromosomes in Drosophila. *Nat. Genet.* 34, 215–219.
15. Bachtrog, D. (2003b). Accumulation of *Spock* and *Worf*, two novel non-LTR retrotransposons, on the neo-Y chromosome of Drosophila miranda. *Mol. Biol. Evol.*, 20:173–181.
16. Skaletsky, H. et al. (2003). The male-specific region of the human Y chromosome is a mosaic of discrete sequence classes. *Nature*, 423:825–837.
17. Charlesworth, B., Sniegowski, P. & Stephan, W. (1994). The evolutionary dynamics of repetitive DNA in eukaryotes. *Nature* 371:215–220.
18. Jobling, MA. & Tyler-Smith, C. (2003) The human Y chromosome: an evolutionary marker comes of age. *Nat. Rev. Genet.*, 4(8):598–612.
19. Ruiz Rejón, C., Jamilena, M., Garrido-Ramos, M.A., Parker, J.S., Ruiz Rejón, M. (1994). Cytogenetic and molecular analysis of the multiple sex chromosome system of *Rumex acetosa*. *Heredity,* 72:209–215.
20. Shibata, F., Hizume, M. & Kuroki, Y. (1999). Chromosome painting of Y chromosomes and isolation of a Y chromosome-specific repetitive sequence in the dioecious plant *Rumex acetosa*. *Chromosoma ,* 108:266–270.
21. Navajas-Pérez, R., Schwarzacher, T., de la Herrán, R., Ruiz Rejón, C., Ruiz Rejón, M. & Garrido-Ramos, MA. (2006). The origin and evolution of the variability in a Y-specific satellite-DNA of Rumex acetosa and its relatives. *Gene*, 368:61-71.
22. Shibata, F., Hizume, M. & Kuroki, Y. (2000a). Molecular cytogenetic analysis of supernumerary heterochromatic segments in *Rumex acetosa*. *Genome*, 43:391–397.

23. Shibata, F., Hizume, M. & Kuroki, Y. (2000b). Differentiation and the polymorphic nature of the Y chromosomes revealed by repetitive sequences in the dioecious plant, *Rumex acetosa*. *Chromosome Res.,* 8: 229–236.
24. Jukes, T.H. & Cantor, C.R. (1969). Evolution of protein molecules. In: Munro, H.D. (ed), Mammalian protein metabolism. Academic press, New york, 21-132.
25. Dover, G. (1982). Molecular drive: a cohesive mode of species evolution. *Nature,* 299, 111–117.
26. Strachan, T., Webb, D. & Dover, G. (1985). Transition stages of molecular drive in multiple-copy DNA families in Drosophila. *Chromosoma,* 108:266–270.
27. Pons, J., Petitpierre, E. & Juan, C. (2002) Evolutionary dynamics of satellite DNA family PIM357 in species of the genus Pimelia (Tenebrionidae, Coleoptera). *Mol. Biol. Evol.,* 19:1329–1340.
28. W illar, HF. (1991). Evolution of alpha satellite. *Current Opinion in Genetics and Development,* 1:509-514.
29. Garrido-Ramos, MA., Jamilena, M., Lozano, R., Ruiz-Rejón, C. & Ruiz-Rejón, M. (1994). Cloning and characterization of a fish centromeric satellite DNA. *Cytogenetic and Cell Genetics,* 65:233-237.
30. Navajas-Pérez, R., de la Herrán, R., López González, G., Jamilena, M., Lozano, R., Ruiz Rejón, C., Ruiz Rejón, M. & Garrido-Ramos, MA. (2005b). The evolution of reproductive systems and sex-determining mechanisms within Rumex (Polygonaceae) inferred from nuclear and chloroplastidial sequence data. *Mol. Biol. Evol.,* 22(9):1929-1939.
31. Macas, J., Meszaros, T. & Nouzova, M. (2002). PlantSat: a specialized database for plant satellite repeats. *Bioinformatics* 18(1):28-35.
32. Singer, MF. (1982). Highly repeated sequences in mammalian genomes. *Int. Rev. Cytol.* 76:67-112.

A Soft Hierarchical Algorithm for the Clustering of Multiple Bioactive Chemical Compounds

Jehan Zeb Shah* and Naomie bt Salim

Faculty of Computer Science & Information Systems, Universiti Teknologi Malaysia,
81310 Skudai, Johor Darul Ta'zim, Malaysia
zebjehan@gmail.com, naomie@fsksm.utm.my

Abstract. Most of the clustering methods used in the clustering of chemical structures such as Ward's, Group Average, K- means and Jarvis-Patrick, are known as hard or crisp as they partition a dataset into strictly disjoint subsets; and thus are not suitable for the clustering of chemical structures exhibiting more than one activity. Although, fuzzy clustering algorithms such as fuzzy c-means provides an inherent mechanism for the clustering of overlapping structures (objects) but this potential of the fuzzy methods which comes from its fuzzy membership functions have not been utilized effectively. In this work a fuzzy hierarchical algorithm is developed which provides a mechanism not only to benefit from the fuzzy clustering process but also to get advantage of the multiple membership function of the fuzzy clustering. The algorithm divides each and every cluster, if its size is larger than a pre-determined threshold, into two sub clusters based on the membership values of each structure. A structure is assigned to one or both the clusters if its membership value is very high or very similar respectively. The performance of the algorithm is evaluated on two bench mark datasets and a large dataset of compound structures derived from MDL's MDDR database. The results of the algorithm show significant improvement in comparison to a similar implementation of the hard c-means algorithm.

Keywords: cluster analysis, chemoinformatics, fuzzy c-means, bioinformatics, chemical information systems.

1 Introduction

The clustering of drug like compound structures is important in many phases of drug discovery and design like the virtual screening, prediction and modeling of structure properties, virtual library generation and enumeration etc. Drug discovery is a complex and costly process, with the main issues being the time and costs of finding, making and testing new chemical entities (NCE) that can prove to be drug candidates. The average cost of creating a NCE in a major pharmaceutical company was estimated at around $7,500/compound [1]. For every 10,000 drug candidate NCE synthesized, probably only one will be a commercial success and there may be 10-12 years after it is first synthesized before it reaches the market [2].

---

* Corresponding author.

S. Hochreiter and R. Wagner (Eds.): BIRD 2007, LNBI 4414, pp. 140 – 153, 2007.
© Springer-Verlag Berlin Heidelberg 2007

Currently, many solution- and solid- phase combinatorial chemistry (CC) strategies are well developed [3]. Millions of new compounds can be created by these CC based technologies but these procedures have failed to yield many drug candidates. Enhancing the chemical diversity of compound libraries would enhance the drug discovery. A diverse set of compounds can increase the chances of discovering various drug leads and optimization of these leads can lead to better drugs. In order to obtain a library of high chemical diversity, a number of structural processing technologies such as diversified compound selections, classification and clustering algorithms have been developed. However, the need for more robust and reliable methods is still seriously felt [4].

The term cluster analysis was first used by Tryon in 1939 that encompasses a number of methods and algorithms for grouping objects of similar kinds into respective categories [5]. The main objective of clustering is to organize a collection of data items into some meaningful clusters, so that items within a cluster are more similar to each other than they are to items in the other clusters. This notion of similarity and dissimilarity may be based on the purpose of the study or domain specific knowledge. There is no pre-notion about the groups present in the data set.

Willett [6] has found that, among the hierarchical methods, the best result was produced by Ward's, Group Average and Complete Linkage hierarchical methods and Jarvis-Patrick was found to be the best method among the non-hierarchical methods tested. They have evaluated almost 30 hierarchical and non hierarchical methods on 10 datasets each containing a group of compounds exhibiting a particular property or biological activity such as anesthetic activity, inhibition activity, molar refractivity, where 2D fingerprints been used as compound descriptors. In another study [7], Barnard and Downs have further investigated Jarvis-Patrick method in more detail using a small dataset of 750 diverse set of compounds from the ECDIN database using 29 physiochemical and toxicological information. Though satisfactory correlations have been obtained yet to obtain the best correlations for different properties and activities different parameter setting was necessary.

In [8], Downs and Willett have analyzed the performance of Ward's, Group Average, Minimum Diameter and Jarvis Patrick methods on two datasets: a small subset of 500 molecules and another one of 6000 molecules from Chemical Abstract Service [9] database. They have incorporated the same 29 physiochemical properties. The performance of Jarvis Patrick's method was very poor. The Minimum diameter method was found to be the most expensive, and the performance of the Ward's method was the best.

Another work on the clustering of chemical dataset was reported by Brown and Martin [9] where Ward's, Jarvis-Patrick's (fixed and variable length nearest neighbor lists), Group Average and Minimum Diameter (fixed and variable diameter) methods have been evaluated on four datasets, each with single activity containing active as well as inactive compounds, summing to a total of 21000 compounds. They have employed a number of descriptors like MACCS 2D structural keys, Unity 2D keys, Unity 2D fingerprints, Daylight 2D fingerprints and Unity 3D pharmacophore screens. The performance of wards was found to be the best across all the descriptors and datasets, whilst the Group average and minimum diameter methods were slightly inferior. The performance of Jarvis Patrick method was very poor for fixed as well as for variable length nearest neighbor lists.

Recently, fuzzy clustering methods have been applied for clustering of chemical datasets. In [10] Rodgers et al have evaluated the performance of fuzzy c-means algorithm in comparison with hard c-mean and Ward's methods using a medium size compound dataset from Starlist database for which LogP values were available. Their results show that fuzzy c-means is better than Ward's and c-means. They have used simulated property prediction method [11] as performance measure, where the property of the cluster is determined by the average property of all the molecules contained in the cluster. This average property of the cluster is called the simulated property of each of the structure in the cluster. The simulated property of each molecule is correlated with the actual property of the compound to find the performance. In [12], we have used fuzzy Gustafson-Kessel, fuzzy c-means, Ward's and Group Average methods to cluster a small size dataset from MDL's MDDR database containing about 15 biologically active groups. Instead of using simulated property prediction method, the active cluster subset method where the proportion of active compounds in active clusters is used as performance measure, was employed. Our results show that the performance of Gustafson-Kessel algorithm is the best for optimal number of clusters. The Ward's, fuzzy c- means and Group Average methods are almost the same for optimal number of clusters.

Bocker et al [13] have revised the hierarchical k-means algorithm and developed an interface to display the resultant hierarchy of compound structures in the form of a very useful colorful dendrogram. The same system has also been used for the display of results for an improved median hierarchical algorithm [14].

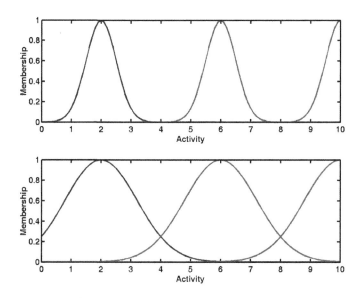

Fig. 1. The upper graph describes the non-overlapping compound structures whereas where as the lower graph describes the overlapping clusters. The vertical axis plots the strength of activity of the compound structure and horizontal axis plots the number of activities.

The main problem of the current clustering methods used in the field of chemoinformatics is their non overlapping nature, where these methods consider the datasets as having very distinct and clearly separable boundaries among the various clusters. It is contradictory to the real problems at hand, where the boundaries are very vague and so it is not always as simple to delineate the clusters as these traditional methods tackle them. In Figure 1, two types of datasets are depicted a) where the boundaries of the clusters are well defined and can easily be separated and b) where the boundaries of the clusters are vague and so difficult to delineate. The second case is a challenge for the current methods where their performance is expected to be not as good as when each data element is not limited to belong to only one cluster.

In case of chemical compounds that are biologically active, it is often the scenario that they exhibit more than one activity simultaneously and grouping such compounds under one cluster is not justified. For example, the MDDR database [15], which currently contains around 16 million compounds, the number of compounds active against multiple targets is considerably large.

Thus the previous researches show the effectiveness of hierarchical methods on one hand and that of the fuzzy methods on the other hand. The fuzzy clustering methods have the promise to care for the overlapping nature of chemical structures. In this work an improved hierarchical fuzzy algorithm is employed for the clustering of chemical structures that exhibit multiple activities. It has also been shown on a dataset sufficiently quantified in terms of the activities that the method result in better clustering when higher overlapping is allowed in the clustering process. The results have also been compared with a similar implementation of the k-means method.

In the next section the dataset and the corresponding structural descriptors used in this work are described. Section 3 discusses the hierarchical implementation of the fuzzy c – means in detail. In section 4 the results are discussed and section 5 concludes the work.

2   Dataset Preparation and Descriptors Generation

In this work three datasets have been utilized to evaluate the performance of the proposed algorithm: two benchmark datasets known as the Fisher's Iris dataset [16] and Golub's Lekuemia datasets [17] and one drug dataset composed of bioactive molecules exhibiting overlapping as well as non-overlapping activities collected from the MDDR database. The MDDR database contain 132000 biologically relevant compounds taken from patent literature, scientific journals, and meeting reports [15]. Each entry of the database contain a 2D molecular structure field, an activity class and an activity class index fields besides many other fields like biological investigation phase, chemical abstract service (CAS) [18] compound identity , and patent information fields. The activity index is a five digit code used to organize the compounds' structures based on biological activity; for example the left most one or two digits describe a major activity group and the next three digits describes sub activities inside the bigger activity. For example, the activity index 31000 shows a large activity of antihypertensive agents and the activity indexes 31250, 31251 show Adrenergic (beta) Blocker, Adrenergic (beta1) Blocker respectively. The dataset used here comprised of 12 major activities where each group can further be divided into a few sub categories. Initially 55000 compounds have been extracted from the database using a number of filtering strategies (as described in the following equations 1 and 2). The number of compounds in dataset1 (DS1) was 29843 and dataset2 (DS2) 6626. The dataset DS1 contain exactly non-overlapping structures where each compound in the dataset can exhibit only and only one activity among the list of activities selected for this work. The dataset DS2 contain bioactive (compounds exhibiting only two activities) such that each compound exhibit two activities only. Let A be the set of activities selected and l be the number of activities in this set, then the DS1 is a superset of sets D_i, a set of compounds exhibiting activity $i \in A$. If $z \in DS1$ is an arbitrary compound, then

$$DS1 = \{z \in D_i \mid i \in A \wedge i \notin A - \{i\}, \quad i = 1,2,...,l\} \tag{1}$$

Similarly, DS2 is a superset of sets D_{ij}, where compound $z \in DS2$ exhibit two activities $i, j \in A$ and so,

$$DS2 = \left\{z \in D_{ij} \mid i, j \in A \wedge i, j \notin A - \{i, j\} \quad \begin{array}{l} i = 1,2,...,l-1 \\ j = i+1,...,l \end{array} \right\} \tag{2}$$

By combining these two datasets, another dataset DS3 has been organized in the same way as depicted in fig 1. It contains single activity compounds from DS1 and in between any two activity groups there are bi-activity molecules from DS2 which will belong to both the groups on its right and left.

The descriptors generation or features extraction is an important step in computational clustering of molecular structures and other problems such as classification and quantitative property/activity relationship modeling. A number of modeling tools are available that can be used to generate structural descriptors. In this work, we use the Dragon software [19] to generate around 99 topological indices for the molecules. Topological indices are a set of features that characterize the arrangement and composition of the vertices, edges and their interconnections in a molecular bonding topology. These indices are calculated from the matrix information of the molecular structure using some mathematical formula. These are real numbers and possess high discriminative power and so are able to distinguish slight variations in molecular structure. This software can generate more than 1600 descriptors which include connectivity indices, topological indices, RDF (radial distribution function) descriptors, 3D-MORSE descriptors and many more.

Scaling of the variables generated is very important in almost all computational analysis problems. If magnitude of one variable is of larger scale and the other one is of smaller scale then the larger scale variable will dominate all the calculations and effect of the smaller magnitude variables will be marginalized. In this work all the variables used were normalized such that the maximum value for any variable is 1 and the minimum is 0.

Table 1. Selected Topological Indices

TI	Description
Gnar	Narumi geometric topological index
Hnar	Narumi harmonic topological index
Xt	Total structure connectivity index
MSD	Mean square distance index (Balaban)
STN	Spanning tree number (log)
PW2	path/walk 2 – Randic shape index
PW3	path/walk 3 – Randic shape index
PW4	path/walk 4 – Randic shape index
PW5	path/walk 5 – Randic shape index
PJI2	2D Petitjean shape index
CSI	Eccentric connectivity index
Lop	Lopping centric index
ICR	Radial centric information index

In order to reduce the descriptor space and to find the more informative and mutually exclusive descriptors a feature selection method principal component analysis (PCA) [20] was used. PCA was carried out using the MVSP 3.13 [21]. It has been found that 13 components can represent more than 98% of the variance in the dataset. The input to our clustering system is thus a 13 X 28003 data matrix. The 13 selected topological indices are shown in Table 1.

3 Methods

Fuzzy clustering is the intrinsic solution to the problem of overlapping data, where the data elements can be member of more than one cluster. The traditional clustering methods do not allow this shared membership by restricting the data elements to belong to only one of the many clusters exclusively. There can be almost three types of partitioning concepts, the traditional hard or crisp one where a compound can belong to only one cluster and so the membership degree of the compound is said to be 1 in any one cluster and zero in the rest of the clusters. Another approach is provided by the fuzzy logic where the membership degree of a compound can be [0, 1] and so the compound can belong with varying degree to more than one cluster [22-23]. In both of these partitioning scenarios, the membership μ_{ik} follow a few conditions such as the sum of the membership values over a range of clusters c is always equal to one:

$$\sum_{i=1}^{c} \mu_{ik} = 1 \quad 1 \le k \le n \tag{3}$$

$$0 < \sum_{k=1}^{n} \mu_{ik} < n \quad 1 \le i \le c \tag{4}$$

Where n is the number of compounds in the dataset, c is the number of clusters and i and k are the indexes for the clusters and data elements respectively.

In the third case, called the possibilistic partitioning [24], this constrain is also relaxed and the sum of the membership degrees is not required to be equal to one, however, clustering algorithms based on this theory are out of the scope of this work.

3.1 Fuzzy Clustering Algorithm

In the literature there are a large number fuzzy based clustering algorithms [22-26] that are variations of the most fundamental and widely used fuzzy c-means, a fuzzy counter part of the ordinary c-means (or k-means) algorithm, first characterized by Dunn [27] and then formalized by Bezdek [28]. The algorithm is based on the iterative minimization of an objective functional and so independent of the initialization conditions and sequence of input presentation. The objective functional is given as:

$$J(C,U,Z) \;=\; \sum_{i=1}^{c}\sum_{k=1}^{n}(\mu_{ik})^{m}\left\|Z_k - V_i\right\|^2 A \tag{5}$$

where Z_k is the kth feature vector representing a molecule of the dataset Z containing a total of n molecules, V_i is the prototype (center or codebook) of ith cluster of the total number of clusters c, $\|.\|^2$ is the square of distance between each molecule and each cluster center, and μ_{ik} is the membership value of molecule Z_k to be part of prototype V_i. A represents a positive definite norm inducing matrix dependent on the type of distance (in case of Euclidean distance it is a unity matrix). The stepwise description of the algorithm is given below:

Step1. Initialize the fuzzification index m, the partition matrix U, the number of clusters c and tolerance ε.

Repeat the following steps

Step2. Compute the cluster prototypes

$$V_i^{(l)} \;=\; \frac{\sum\limits_{k=1}^{n}(\mu_{ik}^{(l-1)})^{m}\,Z_k}{\sum\limits_{k=1}^{n}(\mu_{ik}^{(l-1)})^{m}} \quad 1 \le i \le c \tag{6a}$$

where the subscripts l and $l-1$ represent the current and previous iterations respectively.

Step3. Compute the distances between compound Z_k and cluster center V_i

$$D^2_{ik} = \;(Z_k - V_i^{(l)})^{T}(Z_k - V_i^{(l)}) \quad \begin{array}{l} 1 \le i \le c \\ 1 \le k \le n \end{array} \tag{6b}$$

Step4. Update the partition matrix
If $D_{ikA} > 0$

$$\mu^{(l)}{}_{ik} = \frac{1}{\sum_{j=1}^{c} (D_{ik} / D_{jk})^{2/(m-1)}} \quad \begin{array}{l} 1 \le i \le c \\ 1 \le k \le n \end{array} \quad (6c)$$

else

$$\mu_{ik} = 0 \qquad (6d)$$

Step5. if $\left\| U^{(l)} - U^{(l-1)} \right\| < \varepsilon$ **Stop**
else go to Step 2

The fuzzification parameter m is a measure of the fuzzfication that can have any value from 1.1 to ∞. As the value of m is increased the memberships of molecular structures to the clusters become fuzzier. For a value of $m = 1$, the algorithm will simply becomes the crisp or hard c-means, but it should be avoided as it will result in a divide by zero error in equation 6(c). Many researches suggest a value of 2.0 for m, as the first fuzzy c-means algorithm suggested by Dunn also used the same value [29]. The stopping condition $\varepsilon = 0.001$ is usually enough for convergence, but we have kept it at 0.0001, just to be on the safe side.

3.2 Fuzzy Hierarchical Clustering

The algorithm is a recursive procedure of fuzzy clustering, where each cluster formed is further re-clustered. The number of child clusters in each recursive call can be 2, 3 or any other number greater than 1. However, here in every recursive call the value of c is kept at 2 to obtain binary tree like order on the structures, a fashion more suitable and historical to the chemical structures based on their biological activities. The inputs are a $n \times m$ data matrix Z composed of n (number of structures in each recursive call) rows of feature vectors $Z_k \in \mathfrak{R}^m$ and m columns of features. The output of each recursive call is a c $\times$ n membership matrix U. The two child clusters are formed using the membership matrix U, where a structure Z_k can be a member of either one of the two clusters if the membership of one is greater than the other to some extent, or can be part of both the clusters if their membership degrees do not show much difference. Once a cluster is partitioned into its child clusters, the membership matrix is discarded but the algorithm keeps the necessary global information in the constituent clusters by adding the structures which are closely related to both the clusters. Thus in each recursive call a new membership matrix is generated and optimized based on the local information of the cluster.

This recursive process of clustering continues until every cluster is a singleton (a cluster containing only one structure) or when an optimal partition is obtained. For this purpose the partition validity measure suggested by Bocker et al [13] is adopted. The clustering process is repeated for a number of threshold and at the end of each repetition, the number of singletons, the number of non singleton clusters, and a distance measure D_{max} (Equation 7) are calculated and plotted against the thresholds to find the optimal threshold.

$$D_{max} = \sum_i \max[d(z_k, c_i)], \quad 1 \le k \le n \tag{7}$$

where d is the Euclidean distance, between the structure $z_k \in c_i$ and the prototype of the cluster c_i, and n is the number of structures in each cluster. The value of D_{max} represents the maximum deviation of the clusters from their prototypes.

Once the optimal threshold is obtained from the graph by visual inspection (one shown in Fig 2(a)-(b)), the clustering process is repeated for the last time with the best threshold selected. The main steps of the algorithms are ordered below:

Run1: For finding the optimal threshold
   (i)   A threshold is selected from a range of thresholds
   (ii)   The value c is initialized which is 2 for binary trees, the membership matrix U is initialized
   (iii) The dataset is clustered using the fuzzy –c-means algorithm
   (iv)   Each of the cluster is checked if the size is larger than the Threshold selected, then go to step (ii) for sub-clustering the resultant cluster
   (v)   Plot the number of clusters, singletons and the metric D_{max} against the range of threshold
Run2: (i)   Select the optimal threshold through visual inspection of the graph
       (ii)   Repeat the algorithm for the last time using the optimal threshold.

In clustering a good method is supposed to combine highly similar activity structures together, so large number of singletons is not considered a good gesture. Thus, an appropriate point for a good clustering will be a threshold for which the number of singletons is a minimum.

4   Results and Discussion

Three datasets have been used to evaluate the performance of the clustering process. These include two small size benchmark datasets a leukemia cancer dataset and fisher iris dataset, and a real dataset of chemical structures described earlier in detail. Leukemia dataset is a collection of 72 genes expressions belonging to two types of cancer, acute myeloid leukemia (AML) and acute lymphoblastic leukemia (ALL).Almost 62 of the specimens for this genes expression data were obtained from bone marrow samples of the acute leukemia patients while the rest had been collected from the peripheral blood samples. The fisher's Iris dataset consists of 150 random samples of flowers belonging to the Iris species setosa, versicolor, and virginica. For each of the specie, the dataset contain 50 samples and each sample consists of four variables, sepal length, sepal width, petal length and petal width.

The Iris dataset poses much difficulty to be partitioned into three classes as two of the classes are highly overlapped [30, 31]. However, our method can partition the dataset into three clusters with high accuracy, when a good threshold is selected.

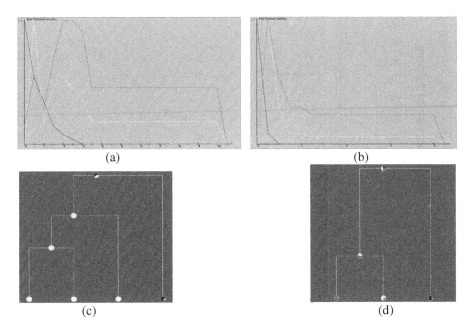

Fig. 2. Clustering results of (a) lekuemia threshold, (b) iris threshold (c) leukemia clustering tree and (d) iris clustering tree

The threshold selection plot is given in figure 2(b) and corresponding dendrogram for the threshold value 0.5, is given in figure 2(d). It can easily be observed that one (shown in yellow color) of the classes could not been separated with much accuracy.

Since these two real datasets are almost non-overlapping but when the number of clusters is decreased, the accuracy (performance) of the clustering degrades. These two results are for the threshold level 0.5 (iris) and 3.0 (leukemia). As the threshold is decreased, the clustering accuracy increases but results in more number of clusters and as the threshold is increased lesser number of clusters and more heterogeneous clusters are obtained.

After confirming the results with the help of benchmark datasets, the methods was applied to the real molecular dataset DS3 developed in section 2. This dataset contain around 12 biologically active and overlapping clusters and the objective of the work is to evaluate the clustering performance of the developed hierarchical fuzzy c-mean (HFCM) algorithm. For evaluation, we use the active cluster subset method [9]. A threshold range of 0.01-0.1 with an increment step of 0.01 was used in this work. For each threshold a number of clusters were obtained. Some of the clusters obtained may be having only actives or inactives structures but many of them will have both. The clusters having at least one active structure are combined to make one super cluster called the active cluster subset.

This subset of the dataset used should not contain any of the singletons, the singletons do not give any clue about the performance of the clustering method, and the clustering method should combine active structures with actives and inactives

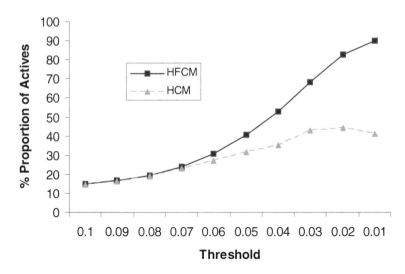

Fig. 3. Performance of the Hierarchical Fuzzy c-means (HFCM) and hierarchical c-means (HCM)

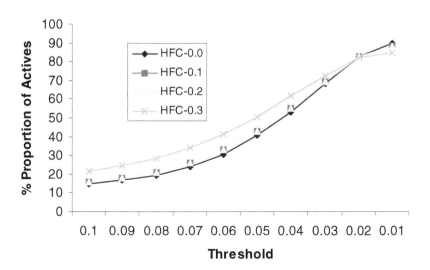

Fig. 4. Performance of the HFCM for various Membership Thresholds. The HFC-0.0 stands for the HFCM with Membership Threshold of 0.0, and so on.

structures with inactives. Thus the singletons are avoided to be included in the active cluster subset. The proportion of actives to inactive structures in the active cluster subset is determined. For each activity group of the dataset, the structures belonging to that activity group were taken as active and the rest of the groups were taken as inactive. The process is repeated for all the 12 bioactivity groups of the dataset for all the clusters obtained for each threshold level and an average proportion was determined.

The fuzzification index of the fuzzy c-means determines the spread in the dataset [24] whose value can range from 1.1 to any finite number, a smaller value means that the data and natural clusters are spread over wide area (volume). As the value of fuzzification index is increased the data and the clusters within becomes more and more compact. The best value of the fuzzification index for which the best clustering can be obtained depends on the dataset used. So, first a fuzzy c-mean method was used to determine the best value for fuzzification index, and was found to be 1.4. The performance of the hierarchical fuzzy c-means is shown in figure 3, for various values of the threshold level, in comparison with the hierarchical hard c-means clustering. Further, to investigate the effect of overlap we repeated a number of experiments. The parent cluster was partitioned into two child clusters based on the membership degree of each compound structure as follows:

If $U_{ik} - Threshold > 0$

   The compound k is assigned to cluster i, 0 <= i <= 1

else

   the compound is assigned to both of the two clusters

Since, $\mu_{ik} \in [0,1]$, so, the value of *Threshold* can be between 0 and 0.5. We have tested for a number of values of *Threshold* and the results are shown in figure 4. As the *Threshold* is increased the compounds are allowed to show more overlap and so we get compounds that go to both of the two child clusters. This permission of higher overlap results in small size and homogeneous clusters, which increases the percentage of active structures in active cluster subset.

5   Conclusions

In this work, an improved hierarchical fuzzy algorithm has been employed for the clustering of chemical structures. The results of the algorithm are very convincing in clustering the multiple activity compounds. A special real dataset have been developed for this purpose where the overlap of activities have been restricted to only two which complements the analysis process for binary tree like clustering. It has been shown that the algorithm have an edge over a similar implementation of the k – means algorithm. Moreover, when higher overlap of activities is allowed, which is incorporated by fuzzy membership as threshold, the results are improved.

References

1. P. Hecht, "High-throughput screening: beating the odds with informatics-driven chemistry," Current Drug Discovery, 2002, pp. 21-24.
2. W.A. Warr, High-Throughput Chemistry: Handbook of Chemoinformatics, Vol. 4, Wiley-VCH, Germany, 2003.

3. D.G. Hall, S. Manku, and F. Wang, Solution- and Solid-Phase Strategies for the Design, Synthesis, and Screening of Libraries Based on Natural Product Templates: A Comprehensive Survey, Journal of combinatorial Chemistry 3 (2001), 125-150.
4. C.N. Parker, C.E. Shamu, B. Kraybill, C.P. Austin, and J. Bajorath, Measure, mine, model, and manipulate: the future for HTS and chemoinformatics? , Drug Discovery Today 11:(19-20) (2006), 863-865
5. R.C. Tryon, Cluster Analysis, MI: Edwards Brothers, 1939.
6. P. Willett, Similarity And Clustering In Chemical Information Systems, Research Studies Press, Letchworth, 1987.
7. G.M. Downs and J.M. Barnard, Clustering of Chemical Structures on the Basis of Two-Dimensional Similarity Measures, Journal of chemical information and computer science 32:(6) (1992).
8. G.M. Downs, P. Willett, and W. Fisanick, Similarity searching and clustering of chemical structure databases using molecular property data, Journal of Chemical Information and Computer Science 34: (1994), 1094-1102.
9. R.D. Brown and Y.C. Martin, Use of structure- Activity data to compare structure based clustering methods and descriptors for use in compound selection, Journal of chemical Information and computer science 36 (1996), 572-584.
10. J.D. Holliday, S.L. Rodgers, and P. Willet, Clustering Files of chemical Structures Using the Fuzzy k-means Clustering Method, Journal of chemical Information and computer science 44 (2004), 894-902.
11. G.W. Adamson and J.A. Bush, A comparison of some similarity and dissimilarity measures in the classification of chemical structures, Journal of chemical Information and computer science 15 (1975), 55-58.
12. J.Z. Shah and N. Salim, FCM and G-K clustering of chemical dataset using topological indices, Proc of the First International Symposium on Bio-Inspired Computing, Johor Bahru, Malaysia, 2005.
13. A. Bocker, S. Derksen, E. Schmidt, A. Teckentrup, and G. Schneider, A Hierarchical Clustering Approach for Large Compound Libraries, Journal of chemical Information and modeling 45:(4) (2005), 807-815.
14. A. Bocker, G. Schneider, and A. Teckentrup, NIPALSTREE: A New Hierarchical Clustering Approach for Large Compound Libraries and Its Application to Virtual Screening, Journal of chemical Information and computer science (2006).
15. "MDL's   Drug   Data   Report,"   Elsevier   MDL. http://www.mdli.com/products/knowledge/drug_data_report/index.jsp
16. R.A. Fisher, The use of multiple measurements in axonomic problems, Annual Eugenics 7 (1936), 179-188.
17. T.R. Golub, D.K. Slonim, P. Tamayo, C. Huard, M. Gaasenbeek, J.P. Mesirov, H. Coller, M.L. Loh, J.R. Downing, M.A. Caligiuri, C.D. Bloomfield, and E.S. Lander, Molecular classification of cancer: Class discovery and class prediction by gene expression monitoring, Science Magazine 285 (1999), 531 - 537.
18. "Chemical Abstract Service," website: http://www.cas.org/.
19. "Dragon," melano chemoinformatics. http://www.talete.mi.it
20. I. Jolife, Principal component analysis, Springer-Verlag, New York, 1986.
21. "MVSP 3.13, Kovach computing services: ." http://www.kovcomp.com/
22. J.C. Bezdek and R.J. Hathaway, Numerical convergence and interpretation of the fuzzy c-shells clustering algorithm, IEEE Transaction on Neural Networks 3, (1992), 787 – 793.
23. R.N. Dave, Fuzzy shell-clustering and applications to circle detection in digital images, International Journal of General Systems 16 (1990), 343-355.

24. F. Hopner, F. Klawonn, R. Kruse, and T. Runkler, Fuzzy Cluster Analysis, John Wiley & Sons, 1999.
25. R. Krishnapurum, O. Nasraoui, and H. Frigui, The Fuzzy C-shells algorithm: A new approach, IEEE Transaction on Neural Networks 3:(5) (1992), 663-671.
26. Y.H. Man and I. Gath, Detection and separation of ring-shaped clusters using fuzzy clustering, IEEE Transaction on pattern analysis and machine intelligence 16:(8) (1994), 855 – 861.
27. J.C. Dunn, A Fuzzy Relative of the ISODATA Process and Its Use in Detecting Compact Well-Separated Clusters, Journal of Cybernetics 3 (1973), 32-57.
28. J.C. Bezdek, R. Ehrlich, and W. Full, FCM: Fuzzy c-means algorithm, Computers and Geoscience (1984).
29. H. Choe and J.B. Jordan, On the optimal choice of parameters in a fuzzy c-means algorithm, Proc of the IEEE Conference on Fuzzy Systems1992, pp. 349 - 354
30. I. Gath and A.B. Geva, Unsupervised optimal fuzzy clustering, IEEE Transaction on pattern analysis and machine intelligence 11:(7) (1989), 773-781.
31. A.B. Geva, Hierarchical unsupervised fuzzy clustering, IEEE Transaction on Fuzzy Systems 7:(6) (1999), 723-733.

A Novel Method for Flux Distribution Computation in Metabolic Networks[*]

Da Jiang[1], Shuigeng Zhou[1], and Jihong Guan[2]

[1] Dept. of Computer Sci. and Eng., Fudan University, Shanghai 200433, China
{jiangda,sgzhou}@fudan.edu.cn
[2] Dept. of Computer Sci. and Tech., Tongji University, Shanghai 200092, China
jhguan@tongji.edu.cn

Abstract. In recent years, the study on metabolic networks has attracted considerable attention from the research community. Though the topological structures of genome-scale metabolic networks of some organisms have been investigated, their metabolic flux distributions still remain unclear. The understanding of flux distributions in metabolic networks, especially when it comes to the gene-knockout mutants, is helpful for suggesting potential ways to improve strain design. The traditional method of flux distribution computation, i.e., flux balance analysis (FBA) method, is based on the idea of maximizing biomass yield. However, this method overestimates the production of biomass. In this paper, we develop a novel approach to overcome the drawback of the FBA method. First, we adopt a series of extended equations to model reaction flux; Second, we build the stoichiometric matrix of a metabolic network by using a more complex but accurate model – carbon mole balance – rather than mass balance used in FBA. Computation results with real-world data of *Escherichia coli* show that our approach outperforms FBA in the accuracy of flux distribution computation.

1 Introduction

Biological function lies in the interactions of metabolites. To understand the biological function of a cell, we require a system viewpoint to analyze its metabolic network. Up to date, many researches have been done on building mathematical models of cell metabolism. Though kinetic models of metabolic networks based on differential equations have been used to predict the networks' capabilities quantitatively, the detailed kinetic mechanism is not quite clear and the parameters in the equations are difficult to determine.

Flux balance analysis (FBA) [1,2], a basic constraint-based approach, is an effective means to analyze biological networks in quantitative manner. It provides

---

[*] This work was supported by National Natural Science Foundation under grants 60373019, 60573183 and 90612007, and the Shuguang Scholar Program of Shanghai Municipal Education Committee. Shuigeng Zhou is the correspondence author, he is also with Shanghai Key Lab of Intelligent Information Processing, Fudan University, Shanghai 200433, China.

S. Hochreiter and R. Wagner (Eds.): BIRD 2007, LNBI 4414, pp. 154–167, 2007.

an appropriate simulation platform for studying the overall phenotypic behavior [3]. This approach is not limited by the availability of kinetic data, and thus can capture the complex, internal interactions of the whole cell. The conventional FBA method uses Linear programming (LP) to solve the flux distribution computation problem, which consists of a linear objective function and some linear constraints. The flux capacities, mass balances, and thermodynamic constraints define the feasible space [4], and the maximization of biomass growth constitutes the typical objective function [2]. Some physiologically feasible cellular states are included in the constrained solution space, while the others excluded by the physicochemical constraints are "infeasible", and thus cannot be held by the cell. The LP method can be used to obtain an optimal solution within the feasible range of cellular capabilities [3].

As shown in the MILP model [5], there are a number of alternate flux distributions that can lead to the same biomass yield. The existence of alternate optimal solutions implies potential drawback of the mathematical models based on FBA method. For simplicity, each of these models usually uses the flux distribution of an alternate optimal solution as input. And different input flux distributions un-avoidably derive different computation results [6]. Consequently, the mathematic methods [7,8,9,10,11,12,13] based on the traditional FBA rely on the alternate optimal solutions for each input flux distribution is just one kind of the numerous distributions in the convex scope of metabolic network [17].

Fong and Palsson [18] calculated the metabolic flux distributions of the wild-type and the gene-knockout strains of *E. coli*, most of their calculated biomass yields are greater than the measured data. There are many instances in which model predictions of metabolic fluxes are not entirely consistent with experimental data, indicating that the reactions in the model do not match the active reactions in the in vivo system [19]. All work mentioned above shows that the traditional FBA method without enough appropriate constraints overestimates the biomass yield.

To obtain more accurate results, we should assign more constraints to the mathematic model. Whereas, the difficulty lies in the fact: we can not obtain so many constraints from the experimental results. The traditional FBA does not aim at finding the accurate solution of the network. Hence, the optimal solution by linear programming should not be the real flux distribution of a metabolic network.

The traditional FBA can obtain only the maximal/minimal value of some metabolite or biomass that we are interested in. However, this method uses only one flux distribution for simplicity, and ignores the alternate solutions of flux distribution. Therefore, the analysis results vary with the used points in reaction ability scope of metabolic network. Furthermore, in the real organism the inner metabolic fluxes are steady unless perturbed by the outer or inner states, and the biomass yield is not as much as calculated result by FBA.

In this paper, we propose an improved method to solve the problem mentioned above. Based on Mahadevan and Schilling's model [6], we add an inequality constraint to calculate the feasible range of each reaction in metabolic network, and obtain the average flux of each reaction in the feasible scope. Then we construct two objective functions to analyze wild-type and gene-knock *E. coli* metabolic networks. Compared with the *in vitro* data, our method can predict metabolic network flux distribution more accurately than the existing models.

The remaining of this paper is organized as follows. Section 2 surveys the related work of this paper. Section 3 introduces the proposed method. Section 4 describes the construction of the *E. coli* metabolic network. Section 5 presents the experimental results. Section 6 concludes this paper and highlights the direction of future work.

2   Related Work

A number of mathematical models have been proposed to extend the FBA method, for depicting the metabolic networks and predicting the outcome of network perturbations more accurately.

In a metabolic network, multiple solutions of flux distribution imply the existence of redundant pathways in the network, and the difference among the multiple solutions is due to the alternate equivalent sets of reactions, which have been investigated in detail by Mahadevan and Schilling [6]. This redundancy renders the network robust against the breakdown of some components. The MILP approach enumerates all the multiple optimal solutions for the given objective function [5]. Recently, a modified MILP approach has been applied to the genome-scale *E. coli* model to generate a limited number of multiple equivalent phenotype states [14].

The MOMA [8] is proposed to analyze the metabolic behavior of mutant strains, while ROOM [7] method calculates the metabolic flux redistribution after certain gene been knocked out with respect to the wild-type strain. The metabolic flux redistribution is an open question. When the wild-type strain is artificially modified with a certain gene knockout, the flux distribution can not be obtained from calculation by using the traditional FBA. The MOMA method, which using the minimum Euclidean distance to the optimal flux point of its wild-type counterpart calculated from FBA, solves a quadratic programme (QP) to find the flux distribution of a mutant. The ROOM method assigns a cost to the expression of a gene and minimizes the number of significant flux changes with respect to the wild-type strain. These two methods predict initial transient behavior after gene manipulation and the post-adaptation flux distribution after a significant period of adapted time, respectively [1].

In recent years, Maranas' group has introduced a series of optimization based frameworks to predict the functions of the metabolic networks [10,11,12,15,16]. An optimization-based framework called ObjFind [10] has been proved to be effective for inferring the most plausible objective function given observed experimental data. It is based on a bi-level optimization problem, where the inner optimization is a traditional FBA problem with undetermined parameters and the outer optimization is for evaluating consistency with the observed fluxes formulated by a quadratic programme. Another bi-level computational framework called OptKnock [11] was developed to suggest reaction deletion strategies that maximize biochemical production. Subsequently, they proposed an integrated framework called OptStrain [15] that extends OptKnock by pinpointing minimal reaction set recombination tasks to confer a desired non-native biochemical production capability on a microbial host. More recently, they developed an approach OptReg [16] to describe the modeling and algorithmic changes required to extend OptKnock to allow for up- and/or down-regulation in addition to gene knockouts to meet a bio-production goal. Their work benefits the geneticists to adopt a systems approach for anticipating the effect of genetic modifications on metabolism.

Markus et. al [19] developed and applied a novel computational method OMNI to identify the genome-scale metabolic network, which imposes measured fluxes as equality constrains. The model identification method uses a bi-level mixed-integer optimization strategy introduced in OptKnock [11] to identify the optimal network structure given one or more sets of experimentally determined metabolic flux data. The OMNI method provides an efficient and flexible way to study and refine genome-scale metabolic network reconstructions using limited amounts of experimental data.

3   Our Method

In this section, we will present our new method to compute flux distribution of metabolic networks. We first introduce the traditional FBA approach, then we extend the traditional FBA model by introducing flux fluctuation and average flux, and derive two new models to compute flux distribution in wild-type strain and gene-knockout strain.

3.1   Linear Programming Based Model

After we obtain the metabolic construction, the mass balance can be defined in terms of the flux through each reaction and the stoichiometry of that reaction as follows.

$$\frac{\mathrm{d}x}{\mathrm{d}t} = S \cdot v \tag{1}$$

where x is the mass vector of metabolites, S is the $m \times n$ stoichiometric matrix of all reactions in the metabolic network, m means the number of metabolites, n is the number of fluxes. Element S_{ij} of the stoichiometric matrix represents the contribution of the j-th reaction to the i-th metabolite. Vector v represents the individual fluxes of the network.

When the metabolic system reaches stable state, the changing of each x's component over time t across all reactions within the system becomes zero. Such handling is proper for most intracellular reactions since they are typically much faster than the rates of changing in the resultant cellular phenotypes such as cell growth [21].

Given a stoichiometric matrix derived from carbon mole balance around all metabolites of a cell, and supposes the metabolic process is at stable state, we get a series of linear equations, which imply the incoming fluxes are balanced by the outgoing fluxes. Formally, we have:

$$S \cdot v = 0 \tag{2}$$

In addition to these linear constraints, there exists thermodynamic constraints on directional flow of the reactions and the capacity constraints. The typical objective function is maximizing the biomass formation. So far, the problem to be solved can be summarized as a LP problem below:

$$\max \ v_{biomass}$$
$$s.t. \ S \cdot v = 0 \tag{3}$$
$$v_{min} \le v \le v_{max}$$

where $v_{biomass}$ is the objective function. $S \cdot v = 0$ means that the majority of metabolites are restricted in an organism and their concentrations are invariable. Besides mass balance equations, reversibility/irreversibility constraints are also imposed on individual internal fluxes ($v_i \ge 0$ for irreversible reactions). v_{max} and v_{min} are the vector containing the maximum and the minimum capacities of the fluxes, respectively. These constraints narrow the spectrum of the possible phenotypes and provide an approach for more specifically characterizing cellular network function. Eq. (3) actually describes the traditional FBA method.

3.2   Flux Fluctuation

For a given optimal state of Eq. (3), the LP problem can give multiple solutions that have the exact same optimal values for the objective function and satisfy all of the constraints. In order to investigate the effects of these alternate optima, Mahadevan and Schilling [6] proposed an alternative strategy to study the issue of alternate optimal solutions and their biological significance. Their LP-based approach focuses on determining the maximum and minimum values of all the fluxes that will satisfy the constraints and achieve the same optimal objective value. This range between the minimum and maximum values is determined for each flux v_i by solving two LP problems. The mathematical formulation of the above approach is described below:

$$\max | \min \ v_i$$
$$s.t. \ S \cdot v = 0$$
$$S^* \cdot v \ge 0 \tag{4}$$
$$v_{biomass} = \alpha \cdot v_{max,biomass}$$
$$v_{min} \le v \le v_{max}$$

where $v_{max,biomass}$ is the value of the objective function calculated previously from Eq. (3), α represents the suboptimal coefficient. So far, the range of the flux of each reaction can be obtained by using Mahadevan and Schilling's method [6]. Considering that some metabolites are not necessary for cellular growth, such as acetic acid, glycerol, lactic acid, ethanol, etc. We add the inequality $S^* \cdot v \geq 0$ to describe some metabolites that can be accumulated and then be excreted out of the cell ultimately.

Note that the above approach provides only the bounds of all solutions, rather than all the possible alternate optimal solutions. It gives the range of the allowable values of each flux component (e.g. v_i illustrated in Fig. 1). This procedure is similar to that of generating the α-spectrum [22,23].

3.3   Average Flux Computation

Using Eq. (4), we can gain the upper and lower flux bounds of each reaction. However, the real metabolic systems are usually not at the extreme states (i.e., having extreme flux values). For some enzymes, their concentrations change with interior diversification (gene-knockout [24]) or environment diversification (different carbon sources [25,26]). This observation inspires us to consider that the metabolic fluxes are robust to inner system and outer environment. Hence, the metabolic fluxes can not be the extremum, otherwise, the robustness of metabolic network can not be guaranteed. In order to achieve maximal robustness, metabolic flux should locate in the middle of the upper and lower bounds. Therefore, we calculate the average flux of the maximal and minimal fluxes of each reaction. Concretely,

$$v_{ave,i} = \frac{v_{max,i} + v_{min,i}}{2} \tag{5}$$

3.4   Balance Flux Computation

Obviously, all of the $v_{ave,i}$ in Eq.(5) are not the balanced flux distribution. In order to calculate the balanced flux, we use the quadratic programming based approach to minimizing the Euclidean distance between the target metabolic flux value and the average metabolic flux value. Thus, the resulting flux will be close to the average flux value as much as possible. In other words, the cell or organism will attempt to maintain a moderate flux distribution rather than the extreme (maximal or minimal) values. So we have the following formulas:

$$
\begin{aligned}
\min \ & \sum (v_i - v_{ave,i})^2 \\
s.t. \ & S \cdot v = 0 \\
& S^* \cdot v \geq 0 \\
& v_{min} \leq v \leq v_{max}
\end{aligned}
\tag{6}
$$

In the computation process of quadratic programming, we use the average flux as the starting point. As a result, v_i will be the closest value to the average value in the feasible solution space, and we can call such flux value *wild-type flux*.

When a certain gene is knockout, in other words, its corresponding reaction is ceased, then the metabolic flux should be redistributed. Flux redistribution is an open problem. For example, the resulting flux distributions computed by MOMA [8] and ROOM [7] are different, but it is difficult to judge which result is more reasonable. In this paper, we partially follow the MOMA method, but consider that the whole cellular reactions participate together to change fluxes. If we still use the Euclidean distance metric, the small values may change considerably while the great values may change trivially. Since all reactions collaboratively take part in the flux redistribution, we impose the average flux $v_{ave,i}$ as the each reaction's weight. As a result, the organism can smoothly adapt to any mutation.

$$\min \ \sum (\frac{v_i - v_{ave,i}}{v_{ave,i}})^2$$
$$s.t. \ \ S \cdot v = 0 \tag{7}$$
$$S^* \cdot v \geq 0$$
$$v_{min} \leq v \leq v_{max}$$
$$v_j = 0$$

where $v_j = 0$ represents that a specific gene (reaction) is knockout.

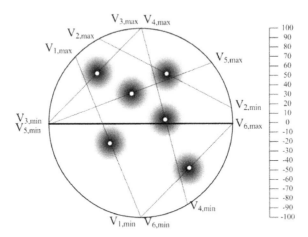

Fig. 1. An illustration of flux distribution

Note: the upside semicircle represents the positive values, and the downside semicircle represents the negative values for reversible reactions. There are six reactions, each reaction's flux value can vary between the minimum and the maximum, which forms a valid flux range. And the real metabolic flux values locate closely to the middle of the corresponding ranges, i.e., the shadowed circle areas.

4 The *E. coli* Metabolic Network

In the literature, the previous work constructs the stoichiometric matrix S with mass balance rather than carbon balance, though mass balance and carbon balance are similar in some aspects. And existing work underestimates the small molecules's mass, such as H_2O, Pi, H, and especially, carbon dioxide. For example, there are about 60 reactions involving carbon dioxide, which are absorbed or discharged in the reconstructed metabolic network [14], and there is around 42% carbon excreted in the form of carbon dioxide [20].

Another point to note is that the mass balance with the traditional FBA is not real balance. For example, the reaction $1F6P \rightarrow 1GAP + 1DHAP$ means to divide F6P into two parts: GAP and DHAP. However, from the stoichiometric matrix S of traditional FBA model, we can see that one unit substrate is converted to two unit products, and thus the mass is doubled. Certainly, this is not acceptable.

Table 1. Reactions of central carbon metabolism in *E. coli*

NO	Reaction	NO	Reaction
R1	1 GLC → 1 G6P	R14	1 MAL → 1 OAA
R2	1 G6P → 1 F6P	R15	1 MAL → 0.75 PYR
R3	1 F6P → 0.5 GAP + 0.5 DHAP	R16	1 PEP ↔ 1.33 OAA
R4	1 GAP → 1 PEP	R17	1 G6P → 0.83 RU5P
R5	1 DHAP → 1 GAP	R18	1 RU5P ↔ 1 R5P
R6	1 PEP → 1 PYR	R19	1 RU5P ↔ 1 X5P
R7	1 PYR → 0.67 AcCoA	R20	0.5 R5P + 0.5 X5P ↔ 0.3 GAP + 0.7 S7P
R8	1 AcCoA → 1 AC	R21	0.3 GAP + 0.7 S7P ↔ 0.4 E4P + 0.6 F6P
R9	0.33 AcCoA + 0.67 OAA → 1 CIT	R22	0.3 GAP + 0.6 F6P ↔ 0.5 X5P + 0.4 E4P
R10	1 CIT → 1 ICI	biom*	6.70 PEP + 2.19 AcCoA + 5.85 αKG
R11	1 ICI → 0.83 αKG		+ 1.68 E4P + 0.24 GAP + 8.04 OAA
R12	1 αKG → 0.8 SUC		+ 9.00 PYR + 0.84 R5P
R13	1 SUC → 1 MAL		→ 1.20 AC + 1 biomass

* The compounds and coefficients of biomass anabolism in this table are based on the
20 amino acids anabolisms in *E. coli* metabolic network [27].

FBA's "balance" means to yield or consume some intermediate metabolites to maintain the mass invariant. But the mass is not balanced in the reaction equation of traditional FBA. Therefore, a reasonable way is to set the reaction equation as $1F6P \rightarrow \beta GAP + \gamma DHAP$, here β and γ are two reaction coefficients to be determined. However, it is hard to evaluate the reaction coefficients such as β and γ by using the mass balance.

Here, we construct the S matric with carbon mole balance. The advantage is that we can assign the coefficient of each metabolite according to its carbon number, and restrict the carbon flux of each reaction in [-100,100] and [0,100] for reversible and irreversible reaction, respectively. Table 1 shows the carbon reaction process, which is different from the traditional reaction. Here, we just

Table 2. Abbreviation lists of metabolites and their carbon numbers

Abb.	Name	C	Abb.	Name	C
AC:	Acetate;	2	ICI:	Isocitrate;	6
AcCoA:	Acetyl-CoA;	2*	MAL:	Malate;	4
αKG:	α-Oxoglutarate;	5	OAA:	Oxaloacetate;	4
CIT:	Citrate;	6	PEP:	Phosphoenolpyruvate;	3
E4P:	Erythrose-4-phosphate;	4	PYR:	Pyruvate;	3
F6P:	Fructose-6-phosphate;	6	R5P:	Ribose 5-phosphate;	5
GAP:	Glyceraldehyde 3-phosphate;	3	RU5P:	Ribulose 5-phosphate;	5
G6P:	Glucose 6-phosphate;	6	S7P:	Sedopheptulose 7-phosphate;	7
DHAP:	Dihydroxyacetone phosphate;	3	SUC:	Succinate;	6
GLC:	Glucose;	6	X5P:	Xylulose 5-phosphate;	5

* There are 2 units of carbon in one AcCoA that participate in carbon flow. Here, CoA is an assistant group and the carbon in CoA will not move to other metabolites.

consider the carbon mole balance. Though we do not consider the carbon dioxide balance, the balance reaction calculates the carbon dioxide indirectly. For example, reaction 1 αKG → 0.8 SUC releases carbon dioxide indirectly, where there is one mole carbon in αKG, and only 0.8 mole carbon is transferred to SUC, while another 0.2 mole carbon is transferred to carbon dioxide.

According to reference [18], the real biomass yield is around 85% of the optimal yield calculated by FBA. Therefore, we set 85% of the maximal biomass yield as the suboptimal biomass yield in Eq. (4). The changing of α value will impact the metabolic flux distribution.

Mahadevan et al. [6] assumed that every metabolite produced in wild-type strain can also be produced in gene-knockout strain. And they did not consider the situation where some metabolites dispensable in metabolic network. To keep the balance state, their model must include all metabolites, even these dispensable metabolites. Obviously, this constraint is so stringent that the system is confined to a narrow range of metabolic ability. They concluded that the effect of the variation of wild-type flux distribution has little effect on the predicted mutant growth rate. However, the real organism is not such a case. For example, the glycerol and acetic acid can be regarded as a substrate or a product according to different environmental states. In this paper, considering that the glucose is the only substrate, the yields of glycerol and acetic acid yields should be equal to or larger than zero. So we do not consider the balances of acetic acid and DHAP (which can yield glycerol).

5   Results and Analyses

In this section, we will give the computational results, which are presented in Table 3 and Fig. 2, and also analyze the results.

There are six reactions whose fluxes are zero for the optimal solution of maximal biomass as shown in Table 3. Under this condition, there is no acetic acid yield (R8=0). And three reactions that can produce carbon dioxide are ceased

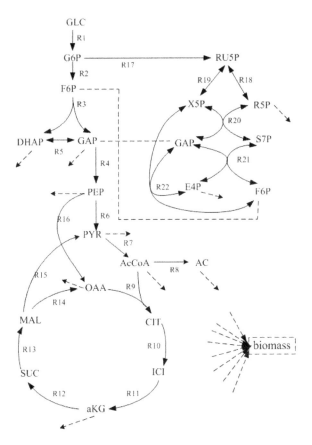

Fig. 2. The central carbon metabolism network of *E. coli* (the corresponding reaction equations are given in Table 1)

(R12, R15 and R17) in order to convert carbon to biomass as much as possible. As soon as the reaction R12 is ceased, the following two reactions (R13 and R14) are also ceased, because there is no substrate input to these two reactions. The resulting flux distribution is not reasonable. Since these reactions do exist in normal organism, they should take on certain functions. That is, their fluxes should not be zero. For the traditional FBA method, in order to maximize biomass yield, some reactions (e.g. R8, R12, R13, R14, R15 and R17) will be stopped. This is unavoidable. However, there has been no direct experimental evidence that supports this process.

Given a specific value of suboptimal coefficient, $\alpha=0.85$, we obtain the flux range of every reaction. The flux range for each reaction is determined by its corresponding minimal and maximal flux values as shown in Table 3. It can be seen that some reactions have large flux ranges (e.g. R2, R17, etc.), while some others have small flux ranges (e.g. R5, R8, etc.). Generally, our method can limit the metabolic network flux values to a relatively small feasible scope.

Table 3. Carbon flux distributions of the *E. coli* metabolic network

NO.	Max(b)	Min	Max	Average	MED	NOM
R1	100.00	100.00	100.00	100.00	100.00	100.00
R2	100.00	10.85	100.00	55.42	55.60	100.00
R3	95.87	66.54	96.49	81.51	81.77	97.07
R4	92.45	78.44	93.60	86.02	86.28	83.10
R5	47.93	33.09	48.25	40.67	40.76	36.99
R6	52.82	44.82	65.05	54.93	53.88	54.66
R7	28.21	23.94	49.88	36.91	39.53	42.80
R8	0.00	0.00	10.15	5.08	2.21	5.02
R9	19.27	16.35	42.30	29.33	29.07	28.98
R10	19.27	16.35	42.30	29.33	29.07	28.98
R11	19.27	16.35	42.30	29.33	29.07	28.98
R12	0.00	0.00	21.54	10.77	11.30	12.71
R13	0.00	0.00	17.23	8.61	9.04	10.17
R14	0.00	0.00	13.44	6.72	3.66	4.59
R15	0.00	0.00	17.23	8.61	5.38	5.58
R16	21.31	13.00	24.52	18.76	17.71	15.46
R17	0.00	0.00	89.15	44.58	44.40	0.00
R18	3.45	2.92	27.59	15.26	15.05	2.44
R19	-3.45	-3.45	46.41	21.48	21.80	-2.44
R20	2.30	1.95	51.28	26.61	26.41	1.63
R21	2.30	1.95	51.28	26.61	26.41	1.63
R22	9.19	-41.54	9.19	-16.17	-17.20	6.51
biom	2.73	2.32	2.73	2.53	2.19	1.94

Note: Max(b) – maximal biomass; Max – the maximal flux value of a reaction; Min – the minimal flux value of a reaction; MED – the minimal Euclidean distance between the target flux value and the average flux value; NOM – the nonlinear optimal method for gene-knockout

The average flux value of each reaction in the metabolic network is not balanced. For example, the sum of the fluxes of R18 and R19 is not equal to that of R17. So, it needs more constraints to achieve balanced flux distribution.

For the MED results, each reaction's flux is near the average value and satisfies the carbon flux balance. The biomass is less than the maximal biomass, concretely, it is about 74% of the maximal biomass. The MED measure provides a way to determine metabolic flux for wild-type strain.

The MED results demonstrate the distribution of metabolic flux,where 55.60% carbon enters the Embden-Meyerhof pathway (EMP), the other 44.40% carbon enters the pentose phosphate pathway (PPP). And 29.07% carbon enters the tricarboxylic acid (TCA) cycle pathway.

When the reaction R17 is ceased (the gene zwf is knocked out), the whole flux will be redistributed. We use the NOM method to calculate flux distributions. We can see that the biomass is less than the MED's biomass, which is about 89% of the latter. This result conforms to *in vitro* experimental value (0.56/0.62) [20]. At that time, all the amount of carbon enters the EMP. And the direction of the flux (R19, R22) through the PPP is reversed.

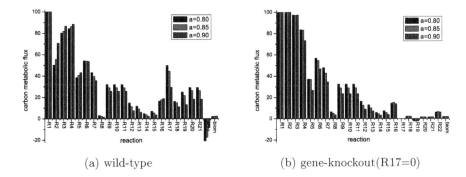

(a) wild-type   (b) gene-knockout(R17=0)

Fig. 3. The flux distributions of wild-type and gene-knockout *E. coli* for different suboptimal coefficient α values

There are two groups of reactions that can produce acetic acid. One is the direct product of the AcCoA, another is the byproduct going with the biomass anabolism, such as citrate anabolism and cysteine anabolism. The acetic acid yields are 4.84 and 7.35 for wild-type and gene-knockout respectively. And the real acetic acid yields [20] are 5.63 and 7.59 for wild-type and gene-knockout respectively (Here, we change the metric and scale the input parameters: change $3.20 \ mmol \ g^{-1} \ h^{-1}$ of glucose to 100 units of carbon for wild-type *E.coli*, and $3.82 \ mmol \ g^{-1} \ h^{-1}$ of glucose to 100 units of carbon for zwf-knockout mutant, and change other experimental data to the corresponding scale for the convenience of comparison with our results). Obviously, the results calculated with our method are quite similar to the real results, which shows the advantage of our method in flux computation accuracy.

Reaction R2 (phosphoglucose isomerase, PGI) is the main offshoot reaction. The activities of PGI are 1277 and 1905 $nmol \ min^{-1}mg^{-1}$ protein in wild-type and zwf-knockout strains, respectively [20]. And the flux values are 55.60 and 100.00. We can see that the ratio of protein activities ($1277/1905 \approx 0.67$) is roughly equal to the ratio of flux values ($55.60/100.00 \approx 0.56$). In summary, all these results conform to the experimental results [20].

Note that different suboptimal coefficient α values in Eq. (4) can result in different flux distributions. We present the results of flux distribution for different α values in Fig. 3. We set the suboptimal coefficient value between 0.8 and 0.9. When the α is 0.8, the acetic acid yields are 5.32 and 8.53 for wild-type and gene-knockout respectively, and the corresponding biomass yields are 2.02 and 1.81. When the α is 0.9, the acetic acid yields are 4.32 and 5.74 for wild-type and gene-knockout respectively, and the corresponding biomass yields are 2.38 and 1.80. As the value of α changes from 0.8 to 0.9 (we compute with three values 0.8, 0.85, and 0.9), for wild-type strain, the acetic acid yield decreases, and biomass yield increases; while for gene-knockout strain, the acetic acid yield increases, but the biomass yield first increases and reaches the maximal value at about 0.85, and then goes down.

6 Conclusion

This paper proposes a new method to compute the flux distribution in the *E. coli* metabolic network. In our method, the biomass production of zwf-knockout strain is less than that of the wild-type strain, and the acetic acid production of zwf-knockout strain is more than that of the wild-type strain. The computation results are all consistent with the *in vitro* data. Our method does not neglect the variety of metabolic flux distributions, but we consider the flux distribution near the average flux value in the feasible scope. Note that in our method we need to set a appropriate α value in Eq. (4) to calculate the flux distribution. When α is set to 0.85, the calculated results conform to the experimental data very well. With our method, it is convenient for the geneticists to select appropriate mutant points for increasing certain useful products or decrease certain useless ones, which suggests potential ways to improve strain design. For future work, we will apply the proposed method to large-scale metabolic networks.

References

1. Lee, J. M., Gianchandani, E. P., Papin, J. A.: Flux balance analysis in the era of metabolomics. Brief Bioinform. **7** (2006) 140–150.
2. Kauffman, K. J., Prakash, P., Edwards, J. S.: Advances in flux balance analysis. Curr. Opin. Biotech. **14** (2003) 491–496.
3. Famili, I., Forster, J., Nielsen, J., et al.: Saccharomyces cerevisiae phenotypes can be predicted by using constraint-based analysis of a genome-scale reconstructed metabolic network. Proc. Natl. Acad. Sci. USA **100** (2003) 13134–13139.
4. Palsson, B. O.: The challenges of in silico biology. Nat. Biotech. **18** (2000) 1147–1150.
5. Lee, S., Phalakornkule, C., Domach, M. M., et al.: Recursive MILP model for finding all alternate optima in LP models for metabolic networks. Comp. Chem. Eng. **24** (2000) 711–716.
6. Mahadevan, R., Schilling, C. H.: The effects of alternate optimal solutions in constraint-based genome-scale metabolic models. Metab. Eng. **5** (2003) 264–276.
7. Shlomi, T., Berkman, O., Ruppin, E.: Regulatory on/off minimization of metabolic flux changes after genetic perturbations. Proc. Natl. Acad. Sci. USA **102** (2005) 7695–7700.
8. Segre, D., Vitkup, D., Church, G. M.: Analysis of optimality in natural and perturbed metabolic networks. Proc. Natl. Acad. Sci. USA **99** (2002) 15112–15117.
9. Fischer, E., Sauer, U.: Large-scale in vivo flux analysis shows rigidity and suboptimal performance of Bacillus subtilis metabolism. Nat. Genet. **37** (2005) 636–640.
10. Burgard, A. P., Maranas, C. D.: Optimization-based framework for inferring and testing hypothesized metabolic objective functions. Biotechnol. Bioeng. **82** (2003) 670–677.
11. Burgard, A. P., Pharkya, P., Maranas, C. D.: Optknock: a bilevel programming framework for identifying gene knockout strategies for microbial strain optimization. Biotechnol. Bioeng. **84** (2003) 647–657.
12. Burgard, A. P., Nikolaev, E. V., Schilling, C. H., Maranas, C. D.: Flux coupling analysis of genome-scale metabolic network reconstructions. Genome. Res. **14** (2004) 301–312.

13. Pal, C., Papp, B., Lercher, M. J.: Chance and necessity in the evolution of minimal metabolic networks. Nature. **440** (2006) 667–670.
14. Reed, J. L., Vo, T. D., Schilling, C. H., Palsson,B.O.: An expanded genome-scale model of Escherichia coli K-12 (iJR904 GSM/GPR). Genome. Biol. **4** (2003) R54.
15. Pharkya, P., Burgard, A.P., Maranas, C.D.: OptSrain: a computational framework for redesign of microbial production networks. Genome Res. **14** (2004) 2367–2376.
16. Pharkya P, Maranas C.D.: An optimization framework for identifying reaction activation/inhibition or elimination candidates for overproduction in microbial systems. Metab. Eng. **8** (2006) 1–13.
17. Price, N.D., Reed, J.L., Palsson, B.O.: Genome-scale models of microbial cells: Evaluating the consequences of constraints. Nat. Rev. Microbiol. **2** (2004) 886–897.
18. Fong, S. S., Palsson, B. O.: Metabolic gene-deletion strains of Escherichia coli evolve to computationally predicted growth phenotypes. Nat. genet. **36** (2004) 1056–1058.
19. Herrgard, M.J., Fong, S.S., Palsson, B.O.: Identification of genome-scale metabolic network models using experimentally measured flux profiles. PLoS Comput Biol. **2** (2006) e72.
20. Zhao, J., Baba, T., Mori, H., et al.: Global metabolic response of Escherichia coli to gnd or zwf gene-knockout, based on 13C-labeling experiments and the measurement of enzyme activities. Appl. Microbiol. Biotechnol. **64** (2004) 91–98.
21. Varma, A., Palsson, B. O.: Stoichiometric flux balance models quantitatively predict growth and metabolic by-product secretion in wild-type Escherichia coli W3110. Appl. Environ. Microbiol. **60** (1994) 3724–3731.
22. Wiback, S. J., Mahadevan, R., Palsson, B. O.: Reconstructing metabolic flux vectors from extreme pathways: Defining the α-spectrum. J. Theor. Bio. **24** (2003) 313–324.
23. Wiback, S. J., Mahadevan, R., Palsson, B. O.: Using metabolic flux data to further constrain the metabolic solution space and predict internal flux patterns: The Escherichia coli α-spectrum. Biotechnol. Bioeng. **86** (2004) 317–331.
24. Siddiquee, K. A. Z., Arauzo-Bravo, M. J., Shimizu, K.: Effect of a pyruvate kinase (pykF-gene) knockout mutation on the control of gene expression and metabolic fluxes in Escherichia coli. FEMS Microb. **235** (2004) 25–33.
25. Daran-Lapujade, P., Jansen, M. L. A., Daran, J. M., et al.: Role of Transcriptional Regulation in Controlling Fluxes in Central Carbon Metabolism of Saccharomyces cerevisiae. J. Biol. Chem. **279** (2004) 9125–9138.
26. Shimizu, K.:Metabolic Flux Analysis Based on 13C-Labeling Experiments and Integration of the Information with Gene and Protein Expression Patterns. Adv. Biochem. Engin.Biotechnol. **91** (2004) 1–49.
27. Stelling, J., Klamt, S., Bettenbrock, K., Schuster, S., Gilles, E. D.: Metabolic network structure determines key aspects of functionality and regulation. Nature. **420** (2002) 190–193.

Inverse Bifurcation Analysis of a Model for the Mammalian G_1/S Regulatory Module

James Lu[1], Heinz W. Engl[1], Rainer Machné[2], and Peter Schuster[2]

[1] Johann Radon Institute for Computational and Applied Mathematics,
Austrian Academy of Sciences
Altenbergerstrasse 69, A-4040 Linz, Austria
james.lu@oeaw.ac.at, heinz.engl@oeaw.ac.at
[2] Theoretical Biochemistry Group,
Institute for Theoretical Chemistry, University of Vienna
Währingerstrasse 17, A-1090 Vienna, Austria
raim@tbi.univie.ac.at, pks@tbi.univie.ac.at

Abstract. Given a large, complex ordinary differential equation model of a gene regulatory network, relating its dynamical properties to its network structure is a challenging task. Biologically important questions include: what network components are responsible for the various dynamical behaviors that arise? can the underlying dynamical behavior be essentially attributed to a small number of modules? In this paper, we demonstrate that inverse bifurcation analysis can be used to address such *inverse problems*. We show that sparsity-promoting regularization strategies, in combination with numerical bifurcation analysis, can be used to identify small sets of "influential" submodules and parameters within a given network. In addition, hierarchical strategies can be used to generate parameter solutions of increasing cardinality of non-zero entries. We apply the proposed methods to analyze a model of the mammalian G_1/S regulatory module.

1   Biological Background

Properties emerging from complex dynamical systems such as *bistability* (the existence of two stable, steady states) or *oscillations* are getting increasing attention as potential design principles of cellular networks of metabolic, gene regulatory or signal transducing systems [1]. While such biological systems form large networks that are difficult to grasp, the theory of dynamical systems provides the means to categorize and reduce them to tangible and understandable core modules governing the overall system properties.

One of the most studied cell biological systems in this regard is the cell cycle. Via iterations of theoretical and experimental analysis it has been successfully analyzed as a modular system consisting of a core oscillator controlled by a series of bistable switches mediating the transition through checkpoints between the different cell cycle phases [2,3,4,5]. While the individual players differ between

S. Hochreiter and R. Wagner (Eds.): BIRD 2007, LNBI 4414, pp. 168–184, 2007.

species, the overall design seems conserved enough to allow researchers to propose a generic cell cycle model that can be mapped onto diverse species [6].

The individual modules are responsible for coordinating each phase of the cell cycle with external resources, such as nutrients or - in a multicellular context - the diverse signaling factors supplied by the organism. The deciding factors include: has the current phase been successfully completed? do external resources still allow the onset of proliferation? The gene regulatory switches have to integrate diverse halt or go signals to form the appropriate decisions. In this methodological paper, we focus on a small part of this vast complexity: a simplified model of the G_1/S transition of the mammalian cell cycle that has recently been proposed by Swat *et al.* [7].

2   A Model of the G_1/S Transition

The model of the mammalian G_1/S transition consists of 9 chemical species, 25 reactions and 40 parameters, representing the transcription factor families AP-1, E2F, pRB, and cyclin/cyclin-dependent kinase complexes cyclin D/Cdk4,6 and cyclin E/Cdk2 [7]. Please see the original publication [7] and the thesis by M. Swat [8] for a description of the model; an SBML version of this model is available for download from [9]. The model may be mathematically expressed as a system of ordinary differential equations (ODEs). Denoting by x and α the biochemical concentrations and parameters respectively, the instantaneous change in $x \in \mathbb{R}^n$ is described by the parametrized vector field $f : \mathbb{R}^n \times \mathbb{R}^m \to \mathbb{R}^n$:

$$\dot{x} = f(x, \alpha). \tag{1}$$

Depending on the parameter α, the parameterized ODE system (1) can exhibit different dynamical behavior. In performing a (forward) *bifurcation analysis*, one examines what different qualitative behavior can arise when various parameters are varied. The computed bifurcation diagrams contain information on the location of parameter regions where the solution behavior stays qualitatively the same, as well as parameter locations where the solution changes its character (see [10] for a mathematical overview of bifurcation analysis).

To understand the significance of bifurcation diagrams for the model system necessitates both a short explanation of the parameters and a rough outline of the biological context, especially the mitogenic stimulation of the G_1/S transition represented here by the bifurcation parameter F_m which is the main focus of our analysis. See Figure 1 for the correspondence between bifurcation points and cell phases. Further details of the model are discussed together with the results of inverse analysis in Section 5.

2.1   Bistable Core

While the model of Swat *et al.* is of a qualitative nature - trying to capture both the basic architecture of interactions and experimentally known dynamics - its underlying equations are frequently used in modeling of gene regulatory networks allowing for straight-forward biological interpretation of results. The model is based on extensive usage of the Hill-Langmuir equation (see e.g. [11,12]) to describe binding equilibria of both the transcription factor E2F1 and its specific DNA binding sites, as well as the transcriptional repressor pRB and this E2F1:DNA complex. The parameters K_{mx}, and J_x (where x is a variable subscript) thus represent dissociation constants in units of concentration, except for the parameter K_{m2}, which falls outside this definition because of an additional factor a used in the core module encoding for the bistability. The parameters k_x represent simple kinetic rates in units of 1/time or 1/(concentration*time) and parameters ϕ_x represent degradation rates in units of 1/time. Maximal transcription rates k_x are then scaled by the temporal averages of the activated $(E2F1/(K_{mx} + E2F1))$ and uninhibited $(J_x/(J_x + pRB))$ gene regulatory sites, according to the Hill-Langmuir equation. The same equations are also used (although with less mechanistic justification) for cyclin autocatalytic activation processes.

The core feature of the model by Swat *et al.* is the autocatalytic activation of the transcription factor E2F1 by binding to its own regulatory site. Two independent binding sites are assumed, both of which need to be occupied for autocatalytic activation. This assumption leads to the necessary nonlinearity in the equation which is the main source of bistability in the model.

2.2   Mitogenic Stimulation

Swat *et al.* realize this classic module of transition through the restriction point and into the S phase of the cell cycle by assuming the following: E2F1 mediated activation of the transcription of all involved genes; a strong inhibition of this process by unphosphorylated pRB; a weakened inhibition by half-phosphorylated pRB (pRB_p); finally no inhibition by fully phosphorylated pRB (pRB_pp). These phosphorylations are sequentially catalyzed by cyclin D/Cdk4,6 and cyclin E/Cdk2 complexes.

The model starts in G_0/G_1 phase and transcription of cyclin D is initiated by a simple linear dependence on the transcription factor AP-1, which itself is modeled as a continuous responder to an input parameter F_m. This latter parameter represents the mitogenic stimulation and the whole model is designed to show bistable dependence on a continuous variation of this mitogenic stimulation parameter. The parameter is used here as the bifurcation parameter. How can biologists relate this to a real life context?

Diverse mitogenic stimuli, usually the growth factors, act on receptors in the cell membrane to activate a complex network of signal transduction. Receptors activate membrane and cytoskeletal modifications in feedback-based switches, with the Ras G protein family as central elements [13]. The pathways at the membrane branch into several intracellular processes, of which two are considered central for cell cycle regulation: intracellular calcium release [14] and the mitogen-activated protein kinase (MAPK) cascades [15] both of which show complex and diverse spatio-temporal variations of intracellular activity but ultimately should converge again in the cellular nucleus to activate further gene transcription via the transcription factors NFAT and AP-1 [16, 17].

Both pathways, calcium release and MAPK activation cascades, have also been intensively studied by theoretical modeling. While calcium increase can show complex oscillatory behavior [18], of which both the frequency and the amplitude might be important information carriers [19, 20, 21, 22, 23, 24], not so much is known about consequences of these spatio-temporal variations for the activation of the cell cycle [14, 25]. In contrast, temporal modulation of MAPK activity is increasingly recognized as the determining signal for cell fate decision [26, 15]. Transient, sustained weak or sustained strong activation have different consequences, either resulting in cell cycle arrest to induce cell differentiation or the onset of the cell cycle. Nuclear feedforward sensors for temporal activity of MAPK that mediate these differences have been identified [27]. From a theoretical perspective, the MAPK cascade has been found again to show bistable dependence on the strength of mitogenic stimulation [28, 29, 30]. Such a bistable, all-or-none switch can, for example, create a discrete border of responding and non-responding cells lined up within continuous concentration gradient of a signal and thus be important for the processes of developmental morphogenesis [31, 32].

Thus, the pathway of a mitogenic stimulus to the activation of nuclear transcription of cell cycle genes, as represented in this model by Swat *et al.* by the simple parameter F_m, involves again a series of bistable switches. However, diverse interacting pathways, be it metabolic or structural states of the cell or opposing signals, can modulate this behavior. In the case of the MAPK cascade's bistability it is known that increasing levels of counteracting phosphatases can render this bistable switch into a continuous responder to a mitogenic signal [33, 34], a process which is often regulated via negative feedback by the MAPK themselves [35]. In such a situation the downstream bistability of the G_1/S would receive increased control.

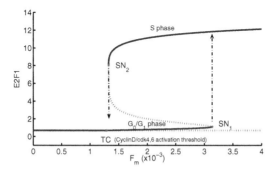

Fig. 1. Bifurcation diagram with respect to mitogenic stimulation F_m, showing various cell phases

3   Inverse Bifurcation Analysis

In the context of molecular biology, the aim of *inverse* bifurcation analysis is to map the geometry of bifurcation diagrams (back) to biochemical parameters (see [36, 37] for applications in biology; refer to [38, 39] for various applications of bifurcation control in engineering). An example of a geometric property important in the study of biological systems is the distance (in parameter space) from reference regions to bifurcation manifolds; this geometric property may be used to quantify the robustness of biological systems. Inverse bifurcation problems may be broadly divided into two classes: those of *design* type, such as finding parameter configurations so that the distance to bifurcation is as large as possible; those of *identification* type, such as finding parameter and network configurations such that the bifurcation diagram exhibits some observed behavior. In this paper, we focus on the latter type: given an ODE model, we would like to infer its properties by computationally mapping bifurcation diagrams of various shapes back to the parameter space.

For gene regulatory systems where the underlying "function" can be related to its bifurcation diagram, the proposed inverse bifurcation analysis can be useful in yielding information on how the regulation process arises from network properties. In Section 3.1, we propose a method to map bifurcation diagrams of different geometric shapes to *sparse* parameters (i.e., parameters with few components of non-zeros). That is, out of the possibly many solutions that result in the desired bifurcation diagram, we select the sparsest ones by adding a sparsity-promoting penalty term to the objective function being minimized. For our model system as described in Section 2, this allows one to identify (minimal) ways to manipulate the qualitative response of the G_1/S transition to varying mitogenic stimulation. In general, even with sparsity constraints the parameter solutions may not be uniquely determined. Hence, in Section 3.2 we propose a *hierarchical* strategy that can be used to identify a sequence of parameters that solve the inverse problem. The combination of sparsity regularization and hierarchical strategy allows for the identification of parameter sets (and the associated submodules of the system) that are influential in generating a variety of dynamical behaviors, thereby providing a way to infer the central non-linear mechanisms. We remark that there exist related methods for model analysis and reduction. For instance, a method based on linear feedback analysis to obtain destabilizing components and interactions has been used to analyze the sources of network behaviors in cell cycle and circadian rhythm models [40]. Model reduction techniques based on Proper Orthogonal Decomposition (POD) have been applied to obtain lower dimensional representations of circadian rhythm models [41]; time-scale decomposition method based on singular perturbation analysis has been used to obtain reduced-order non-linear models of metabolism [42].

3.1   Sparsity-Promoting Regularization

Inverse problems are typically ill-posed, in particular unstable. As a forward problem, bifurcation analysis is typically well-posed: the computed bifurcation diagram exists, is unique and depends smoothly on the problem data (i.e., the vector field in some topology). In contrast, inverse bifurcation analysis is usually ill-posed, in particular: there may be no parameter configurations that can give rise to the desired bifurcation diagram; if solutions exist, they may not be unique or may not depend continuously on the problem data (i.e., the geometric description of the bifurcation diagram). Mathematical techniques, called *regularization methods*, have to be used to cope with this ill-posedness (see [43] for their mathematical theory and applications). While stabilizing ill-posed problems, regularization methods typically bias the solution to some desired behavior. Depending on the mathematical properties of the problem and the application of interest, different regularization techniques may be appropriate. Does one want to obtain a solution of the minimum Euclidean norm? Or is it more desirable to obtain a solution that is sparse, i.e., has as few non-zeros as possible? For biological applications, sparsity is often useful: for instance, one might want to find a network that is as small as possible yet consistent with the experimental data; or, one might wish to identify a small number of parameters whose variation can give rise to a wide range of system behavior. Below, we describe a sparsity-promoting regularization method and apply it in the context of inverse bifurcation analysis.

Consider the following functional, mapping vectors $x = (x_1, \cdots, x_m)$ to $\mathbb{R}$ via:

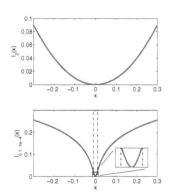

$$l_p(x) = \sum_i x_i^p.$$

For $p = 2$, the above is the square of the Euclidean norm. For the (limiting) case $p = 0$, it measures the number of non-zeros in the vector x; therefore, such a penalty term could be used to obtain sparse solutions [44]. However, $l_0(x)$ is not even a continuous function and hence significant computational effort is needed for its minimization (e.g., using combinatorial methods). In the setting of linearly constrained problems, Donoho and Elad showed that provided the solution allows for a sufficiently sparse representation, the solution of minimum l_0 value can be obtained efficiently by solving the problem with the l_1 penalty term [44]. In our context the problems are nonlinear; to obtain sparse solutions via gradient-based methods, we consider differentiable approximations to $l_0(x)$ of the form

Fig. 2. l_2 and $l_{p,\epsilon}$ functionals

$$l_{p,\epsilon}(x) = \sum_i (x_i^2 + \epsilon)^{p/2}, \quad 1 \geq p > 0, \epsilon > 0,$$

where ϵ is used to remove non-differentiability at $x_i = 0$. We note that for $p < 1$, the above functional is not convex. In particular, the second derivative $d^2/dx_i^2(l_{p,\epsilon}(x))$ changes sign from being positive in the range $|x_i| < \sqrt{\epsilon/(1-p)}$ to being negative for $|x_i| > \sqrt{\epsilon/(1-p)}$. See Figure 2 for a comparison between

l_2 and $l_{p,\epsilon}$ functions; the dotted lines in the latter denote $x = \pm\sqrt{\epsilon}$. We remark that mathematically, l_p with $p \leq 1$ as a regularization method is much less understood than the case of l_2. In the setting of (infinite-dimensional) *linear* inverse problems, Daubechies *et al.* have shown that l_1 is indeed a regularization method [45]. To enforce sparsity constraints using convex penalty terms, Ramlau and Teschke [46] have developed a method for solving *nonlinear* inverse problems using replacement functionals. However, it appears that for $p < 1$, many mathematical questions are still open.

To study the mammalian G_1/S regulatory module, we consider inverse bifurcation problems formulated as constrained optimization problems of the following form: minimize over the set of system parameters α_s,

$$\text{ConMin}(\alpha_s, p, \epsilon) : \min_{\alpha_s} l_{p,\epsilon} \left(\frac{\alpha_s - \alpha_s^*}{\alpha_s^*} \right)$$
$$\text{subject to } \text{SN}_1(\alpha_s) = \text{SN}_1^*$$
$$\text{SN}_2(\alpha_s) = \text{SN}_2^*, \tag{2}$$

where we consider placing the abscissa of saddle nodes at location SN_1^*, SN_2^* different from those for the nominal parameter value α_s^*. In particular, we consider mapping the following 3 modes of bifurcation diagram variation into the parameter space:

- Elongating saddle-node nose: $\text{SN}_1^* \leftarrow \text{SN}_1 + d$
- Moving both saddle-nodes to the right: $\text{SN}_1^* \leftarrow \text{SN}_1 + d$, $\text{SN}_2^* \leftarrow \text{SN}_2 + d$
- Decreasing range of bistability: $\text{SN}_2^* \leftarrow \text{SN}_2 + d$

Using an adjoint-based method for computing the gradient of the saddle-nodes $\text{SN}_{1,2}$ (see [38, 39, 36]), the constrained minimization problem $\text{ConMin}(\alpha_s, p, \epsilon)$ can then be solved using methods such as Sequential Quadratic Programming (SQP) (see [47, 48] for a general overview).

We solve the optimization problem (2) with all parameters being input variables (except, of course, the bifurcation parameter F_m) and taking as α_s^* the nominal parameter values given in the paper [7]. The routine fmincon from the MATLAB [49] Optimization Toolbox is used to solve the problem (2); the underlying algorithm is SQP with approximate Hessian obtained by BFGS-update and line-search is performed (refer to [47] for an overview of optimization algorithms). As the termination criteria, absolute tolerance of 10^{-5} is taken for function value, the constraint violation as well as the optimization variable [1]. Using regularization terms $l_{p,\epsilon}$ with $p = 0.1, \epsilon = 10^{-4}$ and l_2, the top parts of Figure 3 to 5 show the initial and computed bifurcation diagrams in light and dark curves respectively; Table 1 summarizes the number of (forward) bifurcation analyses needed to obtain the results shown in the figures; the large values reflect the tight optimization tolerance used in this study and can be decreased by relaxing the optimality condition. Note that while the bifurcation diagrams shown in Figure 3 and 5 turn out to be qualitatively the same irrespective of

---

[1] Corresponding to the MATLAB settings TolFun, TolCon, TolX.

the choice of regularization term, the respective bifurcation diagrams shown in Figure 4 are qualitatively different; in particular, the transcritical bifurcation point is unconstrained, and it happens that for the parameters obtained using different regularization terms it occurs at different values of the bifurcation parameter F_m.

The bottom part of Figure 3 to 5 compare the computed changes in the parameters using different penalty terms. It can be observed that while l_2 penalty gives rise to maximum parameter changes of smaller magnitude, the computed parameters have many non-zero components. In contrast, the parameters identified using $l_{0.1,10^{-4}}$ regularization are much sparser: for all cases, only 1 or 2 parameters are classified as being identified (i.e., lie outside the range $[-\sqrt{\epsilon}, \sqrt{\epsilon}] = [-0.01, 0.01]$ as denoted by the dotted vertical lines in the figures). The remaining parameters lie within the narrow range about zero where the penalty function is locally convex owing to the smoothing term ϵ. Refer to column 1 of Table 2 for the parameters identified by the algorithm. See Sections 3.2 and 5 for the equations corresponding to the parameters and the biological interpretation of the results respectively.

3.2   Hierarchical Identification Strategy

In general, there are multiple distinct solutions satisfying the constraints in (2). It is useful to obtain a sequence of parameter solutions, allowing for non-zeros of increasing cardinality. An approach towards this goal is to identify parameters in a *hierarchical* manner.

Algorithm 1. <u>HIER-PARAM-IDENT</u>$(\alpha_s^0 \in \mathbb{R}^m$, MaxLev, $p,\epsilon)$

- Initialize: $s \leftarrow \{1, \cdots, m\}$, $I_{\text{identified}} \leftarrow \varnothing$
- FOR $j = 1, \cdots,$ MaxLev
  - $I_{\text{rem}} \leftarrow s \setminus I_{\text{identified}}$
  - Solve $\alpha_{I_{\text{rem}}}^j \leftarrow \text{ConMin}(\alpha_{I_{\text{rem}}}^0, p, \epsilon)$
  - $I_j \leftarrow \{i : |(\alpha_{I_{\text{rem}}}^j)_i| > \sqrt{\epsilon}\}$
  - $I_{\text{identified}} \leftarrow I_{\text{identified}} \cup I_j$

  END
- **Return** $\{\alpha_{I_1}^1, \alpha_{I_2}^2, \alpha_{I_3}^3 \cdots\}$

Table 1. Number of bifurcation runs carried out for inverse analysis

Modification Case \ Regularization Term	$l_{0.1,10^{-4}}$-penalty	l_2-penalty
Elongating SN_1 nose	211	68
Moving $SN_{1,2}$ to right	216	469
Decreasing bistability	197	463

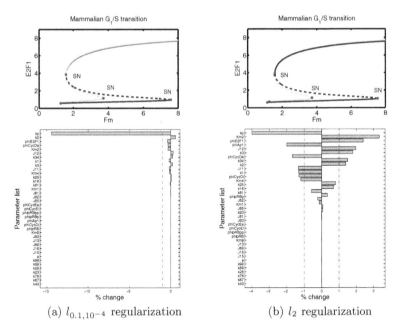

(a) $l_{0.1, 10^{-4}}$ regularization   (b) l_2 regularization

Fig. 3. Elongating saddle-node nose: bifurcation diagrams and solutions obtained using different regularization terms

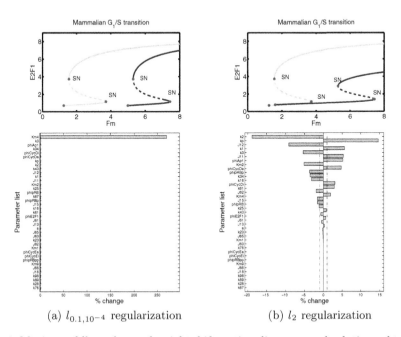

(a) $l_{0.1, 10^{-4}}$ regularization   (b) l_2 regularization

Fig. 4. Moving saddle-nodes to the right: bifurcation diagrams and solutions obtained using different regularization terms

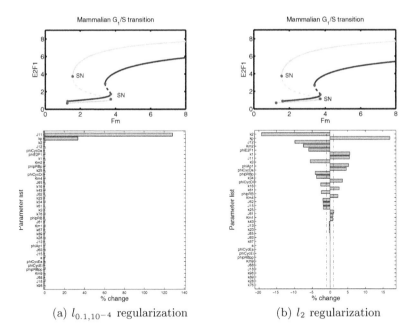

(a) $l_{0.1, 10^{-4}}$ regularization   (b) l_2 regularization

Fig. 5. Decreasing the range of bistability: bifurcation diagrams and solutions obtained using different regularization terms

In the hierarchical approach (see Algorithm 1), parameters are identified in multiple levels. Once a parameter has been identified (using the greater-than-$\sqrt{\epsilon}$ rule), in the subsequent levels it is not permitted to vary from its initial (nominal) value. Thus, it allows for the identification of distinct parameter combinations that all satisfy the same geometric constraints on the bifurcation diagram. Table 2 shows the parameters identified by the hierarchical approach for the same test cases as carried out in Section 3.1. As an illustration, let us look at the case of elongating the nose of saddle-node: the parameter k_p has been identified at level 1; subsequently, it is fixed at its nominal value of 0.05 and only the remaining parameter values $\alpha_s \setminus \{k_p\}$ are allowed to be varied in level 2. It can be observed from Table 2 that the cardinality of the parameter solution exhibits an increasing trend with the level number. Another observation is that of the parameters identified, all except the parameter $\phi_{\text{AP-1}}$ are associated with the following species: the core module,

$$\frac{d}{dt}[\text{pRB}] = k_1 \frac{[\text{E2F1}]}{K_{m1} + [\text{E2F1}]} \frac{J_{11}}{J_{11} + [\text{pRB}]} \frac{J_{61}}{J_{61} + [\text{pRB}_p]}$$
$$- k_{16}[\text{pRB}][\text{CycD}_a] + k_{61}[\text{pRB}_p] - \phi_{\text{pRB}}[\text{pRB}],$$
$$\frac{d}{dt}[\text{E2F1}] = k_p + k_2 \frac{a^2 + [\text{E2F1}]^2}{K_{m2}^2 + [\text{E2F1}]^2} \frac{J_{12}}{J_{12} + [\text{pRB}]} \frac{J_{62}}{J_{62} + [\text{pRB}_p]} - \phi_{\text{E2F1}}[\text{E2F1}]$$

as well as $[\mathrm{CycD}_i], [\mathrm{CycD}_a]$:

$$\frac{d}{dt}[\mathrm{CycD}_i] = -k_{34}[\mathrm{CycD}_i]\frac{[\mathrm{CycD}_a]}{K_{m4} + [\mathrm{CycD}_a]} + \cdots$$
$$\frac{d}{dt}[\mathrm{CycD}_a] = k_{34}[\mathrm{CycD}_i]\frac{[\mathrm{CycD}_a]}{K_{m4} + [\mathrm{CycD}_a]} + \cdots .$$

4   Algorithm Implementation

The numerical results shown here are obtained using the Inverse Bifurcation Toolbox [36]. It is a MATLAB based implementation that utilizes `CL_MATCONT` to perform the underlying bifurcation analysis [50, 51]. Biological models in the Systems Biology Markup Language (SBML) format [52] is read in using the Mathematica [53] package `MathSBML` [54]. The vector fields are symbolically differentiated within Mathematica to obtain the derivatives $f_x, f_\alpha, f_{xx}, f_{x\alpha}$. We remark that while the current implementation is expected to be applicable to systems with dozens of parameters and state variables, it is not geared towards large-scale problems with dimensionality on the order of hundreds or thousands. To up-scale the method, further work on the mathematical as well as informatics aspects of the algorithm are needed. For the mathematical aspects, the regularization aspects needs to studied in more detail since ill-posedness increases with parameter dimension; given the regularized problem, solution procedures can be accelerated by developing appropriate preconditioning strategies for both the inverse problem as well as the bifurcation analysis. For the informatics side, the use of compiled programming languages and efficient data management are necessary to achieve high performance: although high-level, platform-independent languages such as MATLAB and Mathematica allow for rapid algorithm prototyping, the required interpretation can result in significant slow-down of computational speed.

Table 2. Result of hierarchical algorithm with $p = 0.1, \epsilon = 10^{-4}$

Modification Case \ Param. Ident. $(\alpha_{I_j}^j)$	Level $j = 1$	Level $j = 2$	Level $j = 3$
Elongating SN_1 nose	$k_p \downarrow 14.3\%$	$k_{34} \uparrow 31.7\%$ $K_{m2} \uparrow 6.4\%$	$\phi_{\text{AP-1}} \downarrow 20.9\%$ $\phi_{\text{E2F1}} \uparrow 7.3\%$
Moving $SN_{1,2}$ to right	$K_{m4} \uparrow 269.3\%$	$J_{11} \uparrow 191.7\%$ $k_p \uparrow 17.3\%$	$k_2 \downarrow 39.9\%$ $\phi_{\text{E2F1}} \downarrow 11.7\%$ $K_{m2} \downarrow 10.3\%$
Decreasing bistabiliy	$J_{11} \uparrow 128.5\%$ $k_p \uparrow 33.8\%$	$k_1 \uparrow 169.1\%$ $K_{m2} \downarrow 21.7\%$ $J_{12} \downarrow 20.1\%$	$k_2 \downarrow 43.7\%$ $\phi_{\text{E2F1}} \downarrow 28.3\%$

5   Discussion

In this study we attempt to manipulate specific bifurcation diagrams of a simple model of the mammalian G_1/S cell cycle transition in three different ways by inverse bifurcation analysis. Of the parameters identified by sparsity-promoting regularization techniques, most are (perhaps not surprisingly) associated with the reactions of the core module, E2F1 autocatalysis and inhibition by pRB. After all, the model has been designed around a bistable core module [7]. But how can these manipulations and the identified parameters be interpreted in a biological context? While this is only a methodological paper and a simplified model has been chosen on purpose, we can still attempt a biological interpretation of the results. These (admittedly speculative) considerations merely serve to exemplify the general applicability of this approach to situations of specific interest and mechanistically more profound models.

Both the elongation of the nose of the SN_1 saddle-node bifurcation and the shift of the whole bifurcation to the right side would correspond to a generally decreased sensitivity of the cell to growth factors. Such an unresponsiveness is a common situation in embryonic development where cells undergo repeated transitions towards their final differentiation state [55,56]. Intermediately differentiated cells should thus not become fully insensitive. Fully differentiated cells should not undergo anymore cell divisions and choose a different response to growth factor signals. Cell biologists would call this situation a cell cycle arrest, which is often mediated by the E2F1/pRB pair interlinked in another interesting positive feedback cycle with p53/Mdm2 pair [57]. Below, we consider the three modification cases used in the inverse analysis.

Modification 1. Elongating only the right nose of the bifurcation also implies that once activated, the system would stay active also for a subsequent decrease of the input stimulus. Once engaged in the S phase, the cell would continue to do so unless a large decrease in mitogen concentration precedes the activation of the subsequent bistable system in the cell cycle (which is not included in this model). The hierarchical identification strategy reveals three alternative ways to induce such a cell cycle arrest (see Table 2). The easiest (i.e., minimal) way to accomplish this is to increase the basal rate of E2F1 expression k_p and indeed repression of E2F1 expression was one of the two artificial manipulations inducing the expression of differentiation markers of a squamous cell carcinoma cell line [58]. Alternatively, the affinity of E2F1 to its own regulatory site can be decreased (corresponding to an increase of K_{m2}) when at the same time the maximal rate of cyclin D transcription (k_{34}) is increased (Level $j = 2$). In a similar fashion, the AP-1 transcription factor could be stabilized and E2F1 destabilized via modifications of their degradation rates (Level $j = 3$).

Modification 2. In this modification, the whole bifurcation is moved to the right, i.e., to higher mitogen concentrations. This would move the whole window of mitogen responsiveness of the G_1/S transition towards a differentiated state. Level 2 of the hierarchical approach reveals that this can be achieved in a similar manner to the first modification and Level 1, but in opposite direction: increasing k_p, but additionally increasing J_{11}, the dissociation rate of pRB from its own regulatory complex (thus decreasing self-inhibition of pRB). A minimal way to move the bifurcation (computed for Level 1) is to decrease the autocatalysis of cyclin D activation by increasing K_{m4}. As this autocatalysis has been introduced without mechanistic justification (Swat M., personal communication) this could be interpreted as an additional input to the cyclin D activation such as the p21cip1 cyclin inhibitor, which is known to act as an activator of cyclin D/Cdk4,6 at low concentrations and is modulated by MAPK pathways as well [15]. Level 3 of this modification indicates that strengthening the dependence of E2F1 on autocatalytic activation would also yield the same results.

Modification 3. The removal of the bistability of this module, which in some sense corresponds to removing nonlinearity from this feedback system, would render the G_1/S module into a continuous responder to mitogenic signal. This would, for instance, result in the loss of threshold concentration of mitogen. For example, in the MAPK pathways it is known that an initial signal is not enough to enter the cell cycle but can yield opposing responses, depending on whether it is followed by sustained signals or not [26, 15]. A cell population with such a continuous response could loosen the ability to first check for compatibility of the cell state before starting the cell cycle. Thus, one could imagine that removing this bistability at the G_1/S transition could favor the development of cancer, as cells might start cell cycle even in presence of only little mitogenic signal. Let us look at the results of inverse bifurcation and see which parameters could be affected in such pathological conditions. Level 1 of the hierarchical strategy, i.e. the minimal number in the variation of parameters, already reveals a parameter combination that we have met previously but with different magnitudes: increasing both J_{11} (i.e., weakening pRB self-inhibition) and k_p (i.e., increasing the basal rate of E2F1 transcription) would be such deregulation that leads to loss of bistability. In Level 2 we found that an increase in E2F1 autocatalysis (lower K_{m2}), counteracted by higher pRB maximal expression (k_1) and E2F1 inhibition will also yield this result. Level 3 suggests that decreasing the E2F1 degradation rate as well as the maximal rate of auto-activation (k_2) also helps to decrease the non-linear nature of the model. This is actually the same situation as in Level 3 of the second modification case, but without additional enhancement in the E2F1 autocatalysis.

A recent, elegant knock-out study of genes involved in the yeast G_1/S transitions sheds some light on these results. Different yeast knockout strains of G_1/S genes (similar to E2F1 and cyclin D genes in mammals) show either an increased stochasticity in the onset of cell cycle or a significant lengthening of the G_1 phase [59, 60]. Clearly, these results cannot be directly mapped onto the above outlined analysis, but they show that above interpretations could well correlate with such situations. Knock-out of the *swi4* gene, which plays a similar role as our E2F1 transcription factor leads to an increased stochasticity of the G_1/S transition. Could this be a result of a decreased bistability of this control module? A similar result can at least be expected for the third modification, the conversion into a continuous responder to mitogen. Another multiple knock-out involves both the yeast's cyclin D homolog (*cln2*) and another transcription factor with a similar role to E2F1, *mpb1*. In this case, while the G_1 phase of the cells is significantly elongated, the stochasticity of the transition actually decreased. Such a situation might be expected for the first two modifications of this analysis, the shift of the transition point to higher mitogenic stimulation.

To conclude, we remark that the relationship of a specific phenotypic response to perturbation of a cell biological system can be quite complex and often difficult to interpret. Even more challenging is the task of identifying "good" parameters for manipulation to yield a desired phenotype, such as the optimization of a metabolic process or a pharmacological intervention in pathological states. Here we considered the bifurcation diagram of a differential equation model of the mammalian $G1/S$ transition as a representation of the *phenotype*. Rather than intuitive reasoning, we employed a systematic approach to modify this *bifurcation phenotype*. Comparing, for instance, the results of Level 2 of the second modification and Level 3 of the third modification, or Level 3 of the latter two modifications shows how different the consequences can be for very similar manipulations.

Thus our study exemplifies, on a simplified dynamic model of a rather qualitative nature, how the methodology of inverse bifurcation could be used as a tool in the iterative process of experimentation and modeling. A cell biological experiment can usually neither measure nor manipulate all parameters and variables of a system in parallel. Given a mathematical model of a biological system, the inverse bifurcation algorithm helps to gather information about potentially influential species and interactions. The hierarchical application of sparsity promoting constraints further allows to identify minimal sets of core parameters that a cell biologist could change to yield complex manipulations of a phenotype. Subsequently, experiments could concentrate specifically on these parameters to verify or disprove the model's assumptions. This procedure should then be carried out several times until the mathematical analysis yields a result which stands the test of experiment.

Acknowledgements

We gratefully acknowledge Lukas Endler, Stefan Müller, Maciej Swat, Philipp Kügler and Andreas Hofinger for helpful discussions. This work is supported by the WWTF, project number MA05.

References

1. Tyson JJ, Chen KC, Novak B: **Sniffers, buzzers, toggles and blinkers: dynamics of regulatory and signaling pathways in the cell.** *Curr Opin Cell Biol* 2003, **15**(2):221–31.
2. Novak B, Tyson JJ: **Numerical analysis of a comprehensive model of M-phase control in Xenopus oocyte extracts and intact embryos.** *J Cell Sci* 1993, **106** (**Pt 4**):1153–68.
3. Solomon MJ: **Hysteresis meets the cell cycle.** *Proc Natl Acad Sci U S A* 2003, **100**(3):771–2.
4. Pomerening JR, Sontag ED: **Building a cell cycle oscillator: hysteresis and bistability in the activation of Cdc2.** *Nat Cell Biol* 2003, **5**(4):346–51.

5. Pomerening JR, and SYK: **Systems-level dissection of the cell-cycle oscillator: bypassing positive feedback produces damped oscillations.** *Cell* 2005, **122**(4):565–78.

6. Csikasz-Nagy A, Battogtokh D, Chen KC, Novak B, Tyson JJ: **Analysis of a generic model of eukaryotic cell-cycle regulation.** *Biophys J* 2006, **90**(12):4361–79.

7. Swat M, Kel A, Herzel H: **Bifurcation analysis of the regulatory modules of the mammalian G1/S transition.** *Bioinformatics* 2004, **20**(10):1506–1511.

8. Swat MJ: **Bifurcation analysis of regulatory modules in cell biology.** *PhD dissertation*, Humboldt-Universität Berlin 2005, [http://edoc.hu-berlin.de/dissertationen/swat-maciej-2005-11-03/PDF/swat.pdf].

9. **SBML file of the mammalian G_1/S regulatory module proposed by Swat et al. (2004)** [http://www.tbi.univie.ac.at/~raim/models/swat04/].

10. Kuznetsov YA: *Elements of Applied Bifurcation Theory.* New York, USA: Springer-Verlag 2004.

11. Smolen P, Baxter DA, Byrne JH: **Frequency selectivity, multistability, and oscillations emerge from models of genetic regulatory systems.** *Am J Physiol* 1998, **274**(2 Pt 1):C531–42.

12. Hofer T, Nathansen H, Lohning M, Radbruch A, Heinrich R: **GATA-3 transcriptional imprinting in Th2 lymphocytes: a mathematical model.** *Proc Natl Acad Sci U S A* 2002, **99**(14):9364–8.

13. Coleman ML, Marshall CJ, Olson MF: **RAS and RHO GTPases in G1-phase cell-cycle regulation.** *Nat Rev Mol Cell Biol* 2004, **5**(5):355–66.

14. Santella L, Ercolano E, Nusco GA: **The cell cycle: a new entry in the field of Ca2+ signaling.** *Cell Mol Life Sci* 2005, **62**(21):2405–13.

15. Roovers K, Assoian RK: **Integrating the MAP kinase signal into the G1 phase cell cycle machinery.** *Bioessays* 2000, **22**(9):818–26.

16. Macian F, Lopez-Rodriguez C, Rao A: **Partners in transcription: NFAT and AP-1.** *Oncogene* 2001, **20**(19):2476–89.

17. Macian F, Garcia-Cozar F, Im SH, Horton HF, Byrne MC, Rao A: **Transcriptional mechanisms underlying lymphocyte tolerance.** *Cell* 2002, **109**(6):719–31.

18. Schuster S, Marhl M, Hofer T: **Modelling of simple and complex calcium oscillations. From single-cell responses to intercellular signalling.** *Eur J Biochem* 2002, **269**(5):1333–55.

19. Walker SA, Kupzig S, Bouyoucef D, Davies LC, Tsuboi T, Bivona TG, Cozier GE, Lockyer PJ, Buckler A, Rutter GA, Allen MJ, Philips MR, Cullen PJ: **Identification of a Ras GTPase-activating protein regulated by receptor-mediated Ca2+ oscillations.** *EMBO J* 2004, **23**(8):1749–60.

20. Dolmetsch RE, Xu K, Lewis RS: **Calcium oscillations increase the efficiency and specificity of gene expression.** *Nature* 1998, **392**(6679):933–6.

21. Li W, Llopis J, Whitney M, Zlokarnik G, Tsien RY: **Cell-permeant caged InsP3 ester shows that Ca2+ spike frequency can optimize gene expression.** *Nature* 1998, **392**(6679):936–41.

22. Hajnoczky G, Robb-Gaspers LD, Seitz MB, Thomas AP: **Decoding of cytosolic calcium oscillations in the mitochondria.** *Cell* 1995, **82**(3):415–24.

23. Dupont G, Houart G, De Koninck P: **Sensitivity of CaM kinase II to the frequency of Ca2+ oscillations: a simple model.** *Cell Calcium* 2003, **34**(6):485–97.

24. Tomida T, Hirose K, Takizawa A, Shibasaki F, Iino M: **NFAT functions as a working memory of Ca2+ signals in decoding Ca2+ oscillation.** *EMBO J* 2003, **22**(15):3825–32.
25. Nixon VL, McDougall A, Jones KT: **Ca2+ oscillations and the cell cycle at fertilisation of mammalian and ascidian eggs.** *Biol Cell* 2000, **92**(3-4):187–96.
26. Marshall CJ: **Specificity of receptor tyrosine kinase signaling: transient versus sustained extracellular signal-regulated kinase activation.** *Cell* 1995, **80**(2):179–85.
27. Murphy LO, Smith S, Chen RH, Fingar DC, Blenis J: **Molecular interpretation of ERK signal duration by immediate early gene products.** *Nat Cell Biol* 2002, **4**(8):556–64.
28. Huang CY: **Ultrasensitivity in the mitogen-activated protein kinase cascade.** *Proc Natl Acad Sci U S A* 1996, **93**(19):10078–83.
29. Ferrell JEJ, Machleder EM: **The biochemical basis of an all-or-none cell fate switch in Xenopus oocytes.** *Science* 1998, **280**(5365):895–8.
30. Stefanova I, Hemmer B, Vergelli M, Martin R, Biddison WE, Germain RN: **TCR ligand discrimination is enforced by competing ERK positive and SHP-1 negative feedback pathways.** *Nat Immunol* 2003, **4**(3):248–54.
31. Gurdon JB, Bourillot PY: **Morphogen gradient interpretation.** *Nature* 2001, **413**(6858):797–803.
32. Hazzalin CA, Mahadevan LC: **MAPK-regulated transcription: a continuously variable gene switch?** *Nat Rev Mol Cell Biol* 2002, **3**:30–40.
33. Markevich NI, Hoek JB, Kholodenko BN: **Signaling switches and bistability arising from multisite phosphorylation in protein kinase cascades.** *J Cell Biol* 2004, **164**(3):353–9.
34. Hornberg JJ, Bruggeman FJ, Binder B, Geest CR, de Vaate AJ, Lankelma J, Heinrich R, Westerhoff HV: **Principles behind the multifarious control of signal transduction. ERK phosphorylation and kinase/phosphatase control.** *FEBS J* 2005, **272**:244–58.
35. Bhalla US, Ram PT, Iyengar R: **MAP kinase phosphatase as a locus of flexibility in a mitogen-activated protein kinase signaling network.** *Science* 2002, **297**(5583):1018–23.
36. Lu J, Engl HW, Schuster P: **Inverse bifurcation analysis: application to simple gene systems.** *Algorithms for Molecular Biology* 2006, **1**(11).
37. Conrad E: **Bifurcation analysis and qualitative optimization of models in molecular cell biology with applications to the circadian clock.** *PhD dissertation*, Virginia Polytechnic Institute and State University 2006, [http://scholar.lib.vt.edu/theses/available/etd-04272006-1104%09/unrestricted/phd_20060510.pdf].
38. Dobson I: **Computing a closest bifurcation instability in multidimensional parameter space.** *J. Nonlinear Sci.* 1993, **3**(3):307–327.
39. Mönnigmann M, Marquardt W: **Normal vectors on manifolds of critical points for parametric robustness of equilibrium solutions of ODE systems.** *J. Nonlinear Sci.* 2002, **12**(2):85–112.
40. Schmidt H, Jacobsen EW: **Linear systems approach to analysis of complex dynamic behaviours in biochemical networks.** *Systems Biology* 2004, **1**:149–158.
41. Indic P, Gurdziel K, Kronauer RE, Klerman EB: **Development of a two-dimensional manifold to represent high dimension mathematical models of the intracellular mammalian circadian clock.** *Journal of Biological Rhythms* 2006, **21**(3):222–232.

42. Gerdtzen ZP, Daoutidis P, Hu WS: **Non-linear reduction for kinetic models of metabolic reaction networks**. *Metabolic Engineering* 2004, **6**:140–154.

43. Engl HW, Hanke M, Neubauer A: *Regularization of inverse problems, Volume 375 of* Mathematics and its Applications. Dordrecht: Kluwer Academic Publishers Group 1996.

44. Donoho DL, Elad M: **Optimally sparse representation in general (nonorthogonal) dictionaries via l^1 minimization**. *Proc. Natl. Acad. Sci. USA* 2003, **100**(5):2197–2202 (electronic).

45. Daubechies I, Defrise M, De Mol C: **An iterative thresholding algorithm for linear inverse problems with a sparsity constraint**. *Comm. Pure Appl. Math.* 2004, **57**(11):1413–1457.

46. Ramlau R, Teschke G: **Tikhonov replacement functionals for iteratively solving nonlinear operator equations**. *Inverse Problems* 2005, **21**(5):1571–1592.

47. Gill PE, Murray W, Wright MH: *Practical Optimization*. London: Academic Press 1981.

48. Conn AT, Gould NIM, Toint PL: *Trust-Region Methods*. MPS-SIAM Series on Optimization, Philadelphia, USA: SIAM 2000.

49. **MATLAB** [http://www.mathworks.com/products/matlab/].

50. Dhooge A, Govaerts W, Kuznetsov YA: **MATCONT: a MATLAB package for numerical bifurcation analysis of ODEs**. *ACM Trans. Math. Software* 2003, **29**(2):141–164.

51. **MATCONT: continuation software in MATLAB** [http://www.matcont.ugent.be/].

52. **Systems Biology Markup Language** [http://sbml.org/].

53. **Mathematica** [http://www.wolfram.com/products/mathematica/].

54. Shapiro BE, Hucka M, Finney A, Doyle J: **MathSBML: a package for manipulating SBML-based biological models**. *Bioinformatics* 2004, **20**:2829–2831.

55. Perez-Pomares JM, Munoz-Chapuli R: **Epithelial-mesenchymal transitions: a mesodermal cell strategy for evolutive innovation in Metazoans**. *Anat Rec* 2002, **268**(3):343–51.

56. Korenjak M, Brehm A: **E2F-Rb complexes regulating transcription of genes important for differentiation and development**. *Curr Opin Genet Dev* 2005, **15**(5):520–7.

57. Godefroy N, Lemaire C, Mignotte B, Vayssiere JL: **p53 and Retinoblastoma protein (pRb): a complex network of interactions**. *Apoptosis* 2006, **11**(5):659–61.

58. Wong CF, Barnes LM, Dahler AL, Smith L, Popa C, Serewko-Auret MM, Saunders NA: **E2F suppression and Sp1 overexpression are sufficient to induce the differentiation-specific marker, transglutaminase type 1, in a squamous cell carcinoma cell line**. *Oncogene* 2005, **24**(21):3525–34.

59. Bean JM, Siggia ED, Cross FR: **Coherence and timing of cell cycle start examined at single-cell resolution**. *Mol Cell* 2006, **21**:3–14.

60. Ubersax JA: **A noisy 'Start' to the cell cycle**. *Mol Syst Biol* 2006, **2**:2006.0014.

Weighted Cohesiveness for Identification of Functional Modules and Their Interconnectivity

Zelmina Lubovac[1,2], David Corne[2], Jonas Gamalielsson[1,2], and Björn Olsson[1]

[1] University of Skövde, School of Humanities and Informatics, Box 408,
54128 Skövde, Sweden
[2] Heriot-Watt University, School of Mathematical and Computer Sciences,
EH14 4AS Edinburgh, UK
{zelmina.lubovac,bjorn.olsson,jonas.gamalielsson}@his.se,
dwcorne@macs.hw.ac.uk

Abstract. Systems biology offers a holistic perspective where individual proteins are viewed as elements in a network of protein-protein interactions (PPI), in which the proteins have contextual functions within functional modules. In order to facilitate the identification and analysis of such modules, we have previously proposed a Gene Ontology-weighted clustering coefficient for identification of modules in PPI networks and a method, named SWEMODE (Semantic WEights for MODule Elucidation), where this measure is used to identify network modules. Here, we introduce novel aspects of the method that are tested and evaluated. One of the aspects that we consider is to use the k-core graph instead of the original protein-protein interaction graph. Also, by taking the spatial aspect into account, by using the GO cellular component annotation when calculating weighted cohesiveness, we are able to improve the results compared to previous work where only two of the GO aspects (molecular function and biological process) were combined. We here evaluate the predicted modules by calculating their overlap with MIPS functional complexes. In addition, we identify the "most frequent" proteins, i.e. the proteins that most frequently participate in overlapping modules. We also investigate the role of these proteins in the interconnectivity between modules. We find that the majority of identified proteins are involved in the assembly and arrangement of cell structures, such as the cell wall and cell envelope.

Keywords: systems biology, functional modules, Gene Ontology, SWEMODE, interconnectivity.

1 Introduction

At a high level in the complexity pyramid of life [1], protein complexes and proteins interact weakly and transiently with preferred partners to form modules that serve distinct functions. Although modules are often seen as an abstraction of complexes, there is one important distinction between complexes and functional modules. Complexes correspond to groups of proteins that interact with each other at the same

S. Hochreiter and R. Wagner (Eds.): BIRD 2007, LNBI 4414, pp. 185–198, 2007.

time and place, forming a single multimolecular mechanism. Examples of protein complexes include the anaphase-promoting complex, origin recognition complex, protein export and transport complexes, etc. Functional modules, in contrast, do not require physical interaction between all the components at the same point in time, but rather consist of proteins that participate in a particular cellular process while binding to each other at different times and places, such as in different conditions or phases of the cell cycle, in different cellular compartments, etc [2]. Examples of functional modules include the yeast pheromone response pathway, MAP signalling cascades, etc. Furthermore, not all functional modules require a physical interaction between components [2]. A functional module may be conceptualized as a process [3], which does not necessarily correspond to a structure defined in time and in space, like a protein complex.

Consequently, an integrated approach that combines network topology information with knowledge about molecular processes, functions and cellular compartments should be useful for providing new insights about functional modules. The GO Consortium [4] provides three separate ontologies – molecular function, biological process and cellular component – to describe the important attributes of gene products that we seek to integrate with topological information, in order to identify functional modules.

2   Background and Related Work

A series of studies attempting to reveal the modules in cellular networks, ranging from metabolic [5] to protein networks [6, 7], strongly support the proposal that modular architecture is one of the principles underlying biological organisation. The modular nature of the cellular networks, including PPI networks, is reflected in a high tendency towards clustering, which is measured by the clustering coefficient. The clustering coefficient measures the local cohesiveness around a node, and it is defined, for any node i, as the fraction of neighbours of i that are connected to each other [8]. As pointed out in [9], each module may be reduced to a set of triangles, and a high density of such triangles is highly characteristic for PPI networks, thus reflecting the modular nature of such networks.

Although the clustering coefficient is a good measure of the density of interactions in a protein interaction sub-graph, it is strongly dependent on the size of the sub-graph. This makes it very difficult to use clustering coefficient values to discern sub-graphs for which the density is statistically significant. Spirin and Mirny [6] elaborated on this problem by starting from each sub-graph with n proteins and m interactions and computing the probability of obtaining more than m interactions among the same set of proteins in a random graph. They observed that the majority of cliques (fully connected graphs) of size four or greater are statistically significant in PPI networks compared with random graphs. Hence, smaller cliques are likely to appear by chance. Also, we know that data obtained from high-throughput Yeast Two-Hybrid (Y2H) screens is prone to errors, and may contain large numbers of false positives. For example, it is possible that two proteins, although able to interact, and therefore reported as positives in a Y2H screen, are never in close proximity to each other within the cell [10]. Besides this location constraint, there is also a time constraint, meaning that a pair of proteins that interact in the Y2H experiment may be expressed at different points in the cell cycle, and therefore never interact *in vivo*. Therefore, taking into account annotation regarding molecular function of the proteins and their involvement in biological processes or cellular components is likely to increase the reliability of the inferred protein-protein interactions, thereby reducing the number of false positives.

Various methods of network clustering have been applied to reveal modular organisation in protein-protein interaction networks [11-15]. However, those methods have mostly been based on structural properties of the network, such as shortest path distance, mutual clustering coefficient and node degree information. We therefore proposed the method named SWEMODE (Semantic WEights for MODule Elucidation) [16], [17] where topological information was combined with functional annotation, which was found to be a useful approach for identifying functional modules in the network.

3   Materials and Methods

As stated earlier, we here apply and analyse further extensions of SWEMODE, according to two important aspects of biological networks: the overlap between modules and the k-core aspect. In previous work [16], no overlap was allowed between the modules, i.e. proteins were clustered into disjunct modules where one protein could only belong to one module. In this way, modules were treated as isolated functional units, with no possibility to reveal their interconnectivity. However, previous work on the analysis of the yeast filamentation and signalling network indicates that overlapping proteins [18] and highly interconnected proteins [12] in several cases constitute parts of an intermodule path and may play important roles for intermodule communication.

Next, we introduce the cellular component aspect into the calculation of the weighted core-clustering coefficient, and we also add this aspect to a combined weighted core-clustering coefficient that takes into consideration two of the GO aspects – molecular function and biological process.

Besides introducing an overlap aspect and cellular component information, another extension of SWEMODE considers k-cores of the graph. "A subset of interconnected proteins in which each protein has at least k interactions (where k is an integer) forms a k-core. These cores represent proteins that are associated with one another by multiple interactions, as may occur in a molecular complex" [19]. The notion of a core-clustering coefficient has been introduced in previous work [11]. Here, we develop a weighted counterpart, i.e. a weighted core-clustering coefficient, which takes into consideration functional weights and topological properties, i.e. information about the highest k-core for a graph. K-cores have been proposed earlier for detection of protein complexes from protein interaction networks [11], [19]. It has also been found recently that proteins that participate in central cores have more vital functions and a higher probability of being evolutionarily conserved than the proteins that participate in more peripheral cores [20]. This also motivates our attempt to improve SWEMODE by including this aspect.

3.1 Protein Interaction Network

Information on protein interactions was downloaded from the Database of Interacting Proteins (DIP[1]) [21], which contains experimentally determined interactions between proteins in *Saccharomyces cerevisiae*. The majority of the interactions were identified with high-throughput Y2H screens [22]. In Y2H technology, a bait protein, fused to a DNA-binding domain, is used to attract a potential binding protein (prey), fused to a transcriptional activation domain. If the bait and the prey protein interact, their DNA-binding domain and activation domain will combine to form a transcriptional activator, resulting in the expression of a reporter gene. We used a subset of DIP-YEAST denoted CORE, which has been validated in [23]. After removal of 195 self-interactions, the CORE subset contained 6 375 interactions between 2 231 proteins.

3.2 Semantic Similarity Weights

The Gene Ontology (GO) [4] is becoming a *de facto* standard for annotation of gene products. GO consists of three sub-ontologies: molecular function, biological process and cellular component. Based on each of the three sub-ontologies, we use a semantic similarity measure to calculate a weight for each PPI. The weight corresponds to the similarities between the ontology terms assigned to the interacting proteins.

Semantic similarity is calculated as in [24] using the Lin similarity measure [25], which is here calculated using the GO terms assigned to the proteins in the *Saccharomyces* Genome Database (SGD) [26]. To calculate the similarity between two proteins i and j, the similarity between the terms belonging to the GO term sets T_i and T_j that are used to annotate these proteins must first be calculated. Given the ontology terms $t_k \in T_i$ and $t_l \in T_j$, the semantic similarity is defined as [25]:

$$sim(t_k,t_l) = \frac{2\ln p_{ms}(t_k,t_l)}{\ln p(t_k) + \ln p(t_l)} \tag{1}$$

where $p(t_k)$ is the probability of term t_k and $p_{ms}(t_k,t_l)$ is the probability of the minimum subsumer of t_k and t_l, which is defined as the lowest probability found among the parent terms shared by t_k and t_l [27]. We use the average term-term similarity [27] because we are interested in the overall similarity between the pair of proteins rather than between pairs of individual ontology terms. Given two proteins, i and j, with T_i and T_j containing m and n terms, respectively, the protein-protein similarity is defined as the average inter-set similarity between terms from T_i and T_j:

$$ss_{ij} = \frac{1}{m \times n} \sum_{t_k \in T_i, t_l \in T_j} sim(t_k,t_l) \tag{2}$$

where $sim(t_k,t_l)$ is calculated using (1).

---

[1] http://dip.doe-mbi.ucla.edu

3.3 Weighted Clustering Coefficient

As pointed out in previous work [28], the individual edge weights do not provide a general picture of the network's complexity. Therefore, we here consider the sum of all weights between a particular node and its neighbours, also referred to as the protein strength. The strength s_i of node i is defined as:

$$s_i = \sum_{\forall\, j,\, j \in N(i)} ss_{ij} \tag{3}$$

where ss_{ij} is semantic similarity (see Equation 2) between nodes i and j, based on their GO terms, and $N(i)$ is the neighbourhood of node i. Recently, some extensions of the topological clustering coefficient have emerged for weighted networks. Barrat *et al.* [29] introduced a weighted clustering coefficient c^w that combines topological and weighted characteristics. This measure has previously been applied to a world-wide airport network and a scientist collaboration network [29]. We introduced a weighted measure that uses semantic similarity weights [16]. The weighted clustering coefficient c^w is defined as:

$$c_i^w = \frac{1}{s_i(k_i - 1)} \sum_{\forall\, j,h \,|\, \{j,h\} \in K(i)} (ss_{ij} + ss_{ih}) \tag{4}$$

where s_i is the functional strength of node i (see Equation 3), ss_{ij} is the semantic similarity reflecting the functional weight of the interaction, and $K(i)$ is the set of edges connecting neighbours to node i, $K(i) = \{\forall\{j, h\} \mid \{i, j\} \in E \land \{i, h\} \in E \land \{j, h\} \in E\}$. For each triangle formed in the neighbourhood of node i, involving nodes j and h, the semantic similarities ss_{ij} and ss_{ih} are calculated. Hence, not only the number of triangles in the neighbourhood of node i is considered, but also the relative functional similarity between the nodes that form those triangles, with regard to the total functional strength of the node. The normalisation factor $s_i(k_i - 1)$ represents the summed weight of all edges connected to node i, multiplied by the maximum possible number of triangles in which each edge may participate. It also ensures that $0 \le c^w \le 1$.

It should be noted that we calculate three semantic similarity values for each pair of nodes: one based on GO molecular function, the second based on GO biological process, and the third based on GO cellular component. We then use the highest of the three as the final weight of the interaction. This gives the added advantage of taking all three aspects into consideration.

3.4 SWEMODE

In previous work, Bader and Hogue [11] developed an algorithm for finding complexes in large-scale networks, called MCODE, which is based on the weighting of nodes with a core-clustering coefficient. The core-clustering coefficient of a node i is defined as the density of the highest k-core of the closed neighbourhood $N[i]$. The highest k-core of a graph is the central most densely connected sub-graph. Here, we propose a weighted core-clustering coefficient for identifying topologically and functionally cohesive clusters. The weighting scheme, called $core(c^W)$ uses the weighted core-clustering coefficient of node i, which is defined as the weighted clustering coefficient of the highest k-core of the closed neighbourhood $N[i]$ multiplied by the highest core number. The use of weighted core-clustering (instead of the weighted clustering coefficient) is advantageous since it amplifies the importance of tightly interconnected regions, while removing many less connected nodes that are usually present in scale-free networks [11]. The relative weight assigned to node i, based on this measure, is the product of the weighted core-clustering coefficient and the highest k-core number of the immediate neighbourhood of i. By assigning this relative weight to i, the importance of highly interconnected regions is further amplified. There are other functions, such as the density function [11], but these are not evaluated here.

SWEMODE has three options concerning traversal of nodes that are considered for inclusion in a module, as described in [16]. In previous work, we applied immediate neighbour search [16], while we here use depth-first search, i.e. the protein graph is searched starting from the seed node, which is the highest weighted node, followed by recursively traversing the graph outwards from the seed node, identifying new module members according to the given NWP (Node Weight Percentage) criterion. As in [11], the requirement for inclusion of the neighbours in a module is that their weights are higher than a threshold, which is a given NWP of the seed node. At this stage, once a node has been visited and added to the module, it can not be added to another module [16]. However, in the post-processing step, overlap is allowed to some extent. Because we here choose to go further by inspecting the interconnectedness, it is valuable to not only traverse the immediate neighbours but also other indirect neighbours.

In a post-processing step, modules that contain fewer than three members may be removed, both before and after applying a so called "fluffing" step. The degree of "fluffing" is referred to as the "fluff" parameter, and can vary between 0.0 and 1.0 [11]. For every member in the module, its immediate neighbours are added to the module if they have not been visited and if their neighbourhood weighted cohesiveness is higher that the given fluff threshold f.

4 Results

4.1 Overlap Score Evaluation Against MIPS Complexes

We evaluated sets of modules generated across approximately 400 different parameter settings. The NWP parameter was varied from 0 to 0.95 in increments of 0.05 and the fluff threshold parameter was varied between 0 and 1 in increments of 0.1. Modules were post-processed both before and after removing the smallest modules (containing only 1 or 2 members). Resulting modules were compared with the MIPS data set of known protein complexes.

The MIPS[2] protein complex catalogue is a curated set of manually annotated yeast protein complexes derived from literature scanning. After removal of 44 complexes that contain only one member, 212 complexes were left in the data set. In spite of a growing interest in detecting modules, this research area is still in its infancy and lacks a benchmark that could be used for a more through evaluation of prediction accuracy. The MIPS complex data set is incomplete, but it is currently the best available resource for protein complexes that we are aware of. Needless to say, the limitations of MIPS may have affected the presented outcome in terms of the number of matched complexes.

To evaluate the performance of SWEMODE and choose the best parameter settings, we used the overlap score, Ol, defined as [13]:

$$Ol_{ij} = \left| M_i \cap M_j \right| \Big/ \sqrt{\left| M_i \right| \left| M_j \right|} \tag{5}$$

where M_i is the predicted module, and M_j is a module from the MIPS complex data set. The Ol measure assigns a score of 0 to modules that have no intersection with any known complex, whereas modules that exactly match a known complex get the score 1.

We compared the results from the original graph with those from the graph with the highest k-cores graph (Fig. 1). It is obvious that the exclusion of the proteins that belong to the original but not to the highest k-core graphs increased the overlap with MIPS complexes. Interestingly, at the highest threshold value ($Ol > 0.9$), both networks produce the same results, which indicates that the predicted modules that exactly match the MIPS complexes are very robust. This confirms the benefits of using highest k-cores of the graph, because, as pointed out in previous work [20], yeast proteins that participate in the most central cores seem to be evolutionarily conserved and essential to the survival of the organism.

Next, we evaluated the effect of using cellular component information to calculate the weighted core-clustering coefficient, and also adding this aspect to a combined weighted core-clustering coefficient that takes into consideration two of the GO aspects – molecular function and biological process. The result of this comparison is shown in Fig. 2. Both using cellular component as a separate aspect when calculating weights and in combination with the other two aspects generated slightly better results in terms of matched MIPS complexes. This result is interesting, as we found in previous work [16] (when only direct neighbours were considered for inclusion in the modules) that the GO biological process annotation was the most suitable for deriving modules. This may be explained by the fact that we here, in the module prediction step, use another procedure for inclusion of proteins in the modules. In contrast to previous work, when only direct neighbours of each module seed node were considered for inclusion in the modules, the algorithm recursively moves outwards

---

[2] http://www.mips.gsf.de/proj/yeast/

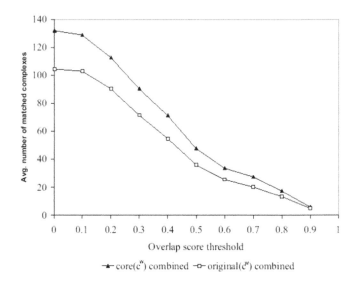

Fig. 1. The result is evaluated in terms of number of matched MIPS complexes at different threshold values. Here, the result using the original protein graph is compared to the result when using the graph with the highest k-cores.

from the seed node (depth-first search), identifying indirect neighbours of the seed node having weights that are higher than a certain threshold, which is given as *NWP* of the seed node. This indicates that the farther we move away from the seed protein, the similarity between the GO terms assigned to the seed protein and the corresponding terms assigned to its indirect neighbours drops faster, when using GO molecular function or GO biological process, compared to the GO cellular component. Simply stated, the seed protein may, for example, be directly or indirectly connected to the neighbours that perform different functional activities or are involved in different processes, but they may still be a part of the same macromolecular complex (which is described in the GO cellular component sub-ontology).

We started by evaluating the importance of introducing the overlap. Generally, we can state that introducing the overlap, i.e. a degree of "fluffing" of the modules, improved our previous results. For example, at the overlap threshold level $Ol > 0.4$, we identified nine more modules on average by using the fluff parameter. The corresponding difference for $Ol > 0.3$ is 12. The best parameter setting, which resulted in the highest number of modules that matched predicted complexes, was obtained with $f > 0$ (i.e. all direct neighbours of the modules with weighted clustering coefficients above zero are added to the module) and $NWP > 0.95$. This parameter setting resulted in 659 modules (471 modules of size three or larger). The module with highest rank, i.e. the module that is generated with seed proteins with the highest weighted core-clustering coefficient, corresponds to the Lsm complex [30]. All eight Lsm-proteins (Lsm1-8) are correctly predicted by the algorithm. Among

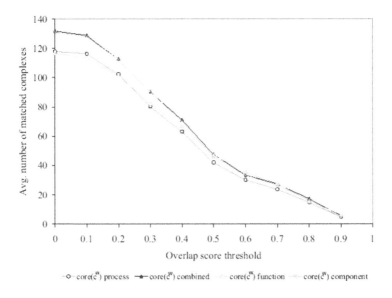

Fig. 2. The results from SWEMODE when using core weighted clustering coefficient based on each separate GO aspect compared to the corresponding measure when all three aspects are combined

other highly ranked modules, we have found the origin recognition complex, the oligosaccharyl transferase (OST) complex [31], the pore complex, etc. The whole list of modules is too large for inclusion in the paper, but may be obtained from the authors upon request.

4.2 Investigating the Interconnectivity Between Modules

To identify topologically and functionally important proteins, we calculated the frequency of each protein across 200 sets of overlapping modules. Those sets have been obtained by varying the *NWP* parameter value (0 to 0.95 in increments of 0.05) and the fluff parameter for each *NWP* value (0 and 0.9 in increments of 0.1). For each seed protein, we calculated the number of times each protein appears in different modules in each module set, divided by the number of module sets it appears in. For example, if protein Nup100 appears in 10 modules in one module set, and 20 modules in another module set, the average frequency of this protein is $(10+20)/2 = 15$.

The majority of the most frequent proteins are annotated with the GO biological process term "cell organization and biogenesis", which has the following GO definition: "the processes involved in the assembly and arrangement of cell structures, including the plasma membrane and any external encapsulation structures such as the cell wall and cell envelope". Table 1 shows the top ten proteins, where 80% (highlighted proteins) belong to the above mentioned class. We have used SGD GO Term Finder[3] to identify the most significantly shared GO term among those proteins.

---

[3] http://www.yeastgenome.org/help/goTermFinder.html

The most significantly shared term is obtained by examining the group of proteins to find the GO term to which the highest fraction of the proteins is associated, compared to the number of times that the term is associated with other yeast proteins. In addition, we have repeated the same evaluation procedure for the 50 most frequent proteins. Still, the majority of the proteins share the GO term "cell organization and biogenesis", which is also the most significant term ($P = 4.01 \cdot 10^{-13}$), but the GO frequency has decreased slightly from 80% to 74%. For comparison, the bottom 50 proteins were evaluated with the same procedure. Here we found that most of the proteins (78%) share the GO biological process term "primary metabolism" ($P = 5.25 \cdot 10^{-7}$). Although those proteins are involved in the anabolic and catabolic processes that take place in all cells, and are very important, they do not seem to have an important role in the interconnectivity of the modules.

Cdc28, which appears most frequently in modules, is one of five different cyclin-dependent protein kinases (CDKs) in yeast and has a fundamental role in the control of the main events of the yeast cell cycle [32]. Topologically, it acts as a hub, i.e., it holds together several functionally related clusters in the interaction network. In previous work, this protein was suggested to be a part of the intramodule path within the yeast filamentation network, because it had the highest intracluster connectivity, i.e. it was the protein with the highest number of interactions with other members of the same cluster [12]. It is therefore highly interesting that we have identified this protein as the most frequent in our modules.

Table 1. Top ten overlapping proteins. Highlighted proteins are annotated with the GO term "cell organization and biogenesis" (8 of 10).

Proteins	Cdc28	Nap1	Prp43	Pre1	Pwp2	Sed5	Tfp1	Nop4	Utp7	Rpc40
Module Frequency	4.2	3.9	2.9	2.7	2.7	2.6	2.6	2.6	2.5	2.5

4.3 Comparison with Other Module Definitions

Recently, a topology-based method for detecting modules from a PPI network has been introduced by Luo and Scheuermann [14]. They proposed a divisive algorithm that uses a new module notion based on the degree definition of the sub-graphs. The approach is based solely on topological properties of the protein sub-graph. It is applied on the same YEAST-CORE data set that we have used here. A total of 99 modules were detected in [14]. For convenience, those modules will be referred to as Feng Luo modules. We have evaluated the Feng Luo modules with the overlap score threshold, and compared them with our modules generated across approximately 400 different parameter settings, and found that our modules show higher agreement with MIPS complexes. The maximum number of modules predicted by our method may seem as a very good result, compared to Feng Luo modules. However, this may be attributable to the fact that some parameter settings generate many small modules of size 2 or 3, which are easily matched with MIPS complexes, while Feng Luo modules are typically large (of size 3 or larger). This is why the average number of matched modules is a more realistic indicator for this comparison. This comparison also indicates that integrating domain knowledge and network topology information seems to be advantageous over using only topology information.

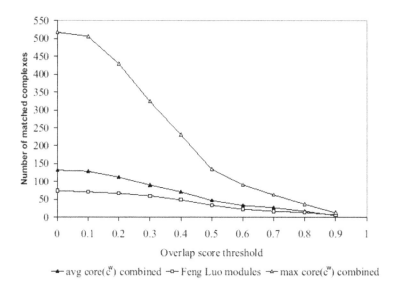

Fig. 3. Comparison between Feng Luo modules and the modules generated with SWEMODE using all three GO aspects

5 Conclusions and Discussion

We have proposed a method for analysis of protein networks using a measure based on a novel combination of topological and functional information of the proteins [16]. The algorithm takes advantage of this combined measure to identify locally dense regions with high functional similarity. In the evaluation of the method, we found many densely connected regions with high functional homogeneity, in many cases corresponding to sets of proteins that constitute known molecular complexes and some additional interacting proteins which share high functional similarity with the complex, but are not part of it. Together, such sets of interacting proteins form functional modules that control or perform particular cellular functions, without necessarily forming a macromolecular complex. Many of the identified modules correspond to the functional subunits of known complexes. Thus, the method may be used for the prediction of unknown proteins which participate in the identified modules. As indicated by the results, the use of a functionally informed measure to generate modules should imply increased confidence in the predicted function.

We have here demonstrated that restricting the analysis to the highest k-core PPI graph instead of the original PPI graph resulted in an improved set of modules, with respect to their overlap with known molecular complexes recorded in MIPS.

We were also able to show that using cellular component as a separate aspect when calculating weights, or in combination with the other two aspects, generated slightly better results in terms of matched MIPS complexes, compared to previous work when only two aspects (molecular function and biological process) were included. One of the main reasons that accounts for this improvement is the inclusion of the indirect neighbours of the seed proteins in the module prediction step. Proteins that are used as seed modules seem to share more similarity with more distant neighbours when cellular component annotation is used, compared to the two other GO aspects. Seed proteins may, for example be connected, directly or indirectly with neighbours that have different functional activities or are involved in different processes, but they may still be part of the same macromolecular complex (which is described in the GO cellular component sub-ontology).

We have also identified topologically and functionally important proteins by calculating the frequency of each protein across 200 sets of overlapping modules. Initial results show that the most frequently appearing proteins that connect several modules are mostly involved in the assembly and arrangement of cell structures, such as the cell wall and cell envelope, which indicates that they are involved in supporting the cell structure rather than in signal transduction.

We will continue our research in this area by investigating functional modules in other organisms, such as *E. coli*. Another line of our research addresses the analysis of the modules in reactome data. Explaining the relationships between structure, function and regulation of molecular networks at different levels of the complexity pyramid of life is one of the main goals in systems biology. We aim to contribute to this goal by integrating the topology, i.e. various structural properties of the networks with the functional knowledge encoded in protein annotations, and also analysing the interconnectivity between modules at different levels of the hierarchy. With the increasing availability of protein interaction data and more fine-grained GO annotations, our approaches will help constructing a more complete view of interconnected functional modules to better understand the organisation of cells.

References

1. Oltvai Z.N., Barabasi A.L.: Systems biology. Life's complexity pyramid. Science 298(5594) (2002) 763-764
2. Pereira-Leal J.B., Levy E.D., Teichmann S.A.: The origins and evolution of functional modules: lessons from protein complexes. Philos. Trans. R. Soc. Lond. B. Biol. Sci. 361(1467) (2006) 507-517
3. Schlosser G.: The role of modules in development and evolution. In: Modularity in development and evolution. Edited by Schlosser G, Wagner A: The Chicago University Press (2004) 519-582
4. Creating the gene ontology resource: design and implementation. Genome Res. 11(8): (2001) 1425-1433
5. Ravasz E., Somera A.L., Mongru D.A., Oltvai Z.N., Barabasi A.L.: Hierarchical organization of modularity in metabolic networks. Science 297(5586) (2002) 1551-1555
6. Spirin V., Mirny L.A.: Protein complexes and functional modules in molecular networks. Proc. Natl. Acad. Sci. U S A 100(21) (2003) 12123-12128

7. Yook S.H., Oltvai Z.N., Barabasi A.L.: Functional and topological characterization of protein interaction networks. Proteomics 4(4) (2004) 928-942
8. Watts D.J., Strogatz S.H.: Collective dynamics of 'small-world' networks. Nature 393(6684) (1998) 440-442
9. Barabasi A.L., Oltvai Z.N.: Network biology: understanding the cell's functional organization. Nat. Rev. Genet. 5(2) (2004) 101-113
10. Van Criekinge W., Beyaert R.: Yeast Two-Hybrid: State of the Art. Biol. Proced. Online, 2 (1999) 1-38
11. Bader G.D., Hogue C.W.: An automated method for finding molecular complexes in large protein interaction networks. BMC Bioinformatics (2003) 4:2
12. Rives A.W., Galitski T.: Modular organization of cellular networks. Proc. Natl. Acad. Sci. U S A 100(3) (2003) 1128-1133
13. Poyatos J.F., Hurst L.D.: How biologically relevant are interaction-based modules in protein networks? Genome Biol. 5(11) (2004) R93
14. Luo F., Scheuermann R.H.: Detecting Functional Modules from Protein Interaction Networks. In: Proceeding of the First International Multi-Symposiums on Computer and Computational Sciences (IMSCCS'06) IEEE Computer Society (2006)
15. Przulj N., Wigle D.A., Jurisica I.: Functional topology in a network of protein interactions. Bioinformatics 20(3) (2004) 340-348
16. Lubovac Z., Gamalielsson J., Olsson B.: Combining functional and topological properties to identify core modules in protein interaction networks. Proteins 64(4) (2006) 948-959
17. Lubovac Z., Olsson B., Gamalielsson J.: Weighted Clustering Coefficient for Identifying Modular Formations in Protein-Protein Interaction Networks. In: Proceedings of the Third International Conference on Bioinformatics and Computational and Systems Biology: August 25-27 Prague, (2006) 122-127
18. Lubovac Z., Gamalielsson J., Olsson B.: Combining topological characteristics and domain knowledge reveals functional modules in protein interaction networks. In: Proceedings of CompBioNets - Algorithms and Computational Methods for Biochemical and Evolutionary Networks, Lyon, France, College Publications (2005) 93-106
19. Tong A.H., Drees B., Nardelli G., Bader G.D., Brannetti B., Castagnoli L., Evangelista M., Ferracuti S., Nelson B., Paoluzi S. et al: A combined experimental and computational strategy to define protein interaction networks for peptide recognition modules. Science 295(5553) (2002) 321-324
20. Wuchty S., Almaas E.: Peeling the yeast protein network. Proteomics 5(2) (2005) 444-449
21. Xenarios I., Rice D.W., Salwinski L., Baron M.K., Marcotte E.M., Eisenberg D.: DIP: the database of interacting proteins. Nucleic Acids Res. 28(1) (2000) 289-291
22. Ito T., Chiba T., Ozawa R., Yoshida M., Hattori M., Sakaki Y.: A comprehensive two-hybrid analysis to explore the yeast protein interactome. Proc. Natl. Acad. Sci. U S A 98(8) (2001) 4569-4574
23. Deane C.M., Salwinski L., Xenarios I., Eisenberg D.: Protein interactions: two methods for assessment of the reliability of high throughput observations. Mol. Cell Proteomics 1(5) (2002) 349-356
24. Lord P.W., Stevens R.D., Brass A., Goble C.A.: Semantic similarity measures as tools for exploring the gene ontology. Pac. Symp. Biocomput. (2003) 601-612
25. Lin D.: An information-theoretic definition of similarity. In: The 15th International Conference on Machine Learning, Madison, WI, (1998) 296-304
26. Dwight S.S., Harris M.A., Dolinski K., Ball C.A., Binkley G., Christie K.R., Fisk D.G., Issel-Tarver L., Schroeder M., Sherlock G. et al: Saccharomyces Genome Database (SGD) provides secondary gene annotation using the Gene Ontology (GO). Nucleic Acids Res. 30(1) (2002) 69-72

27. Lord P.W., Stevens R.D., Brass A., Goble C.A.: Investigating semantic similarity measures across the Gene Ontology: the relationship between sequence and annotation. Bioinformatics 19(10) (2003) 1275-1283
28. Yook S.H., Jeong H., Barabasi A.L., Tu Y.: Weighted evolving networks. Phys. Rev. Lett. 86(25) (2001) 5835-5838
29. Barrat A., Barthelemy M., Pastor-Satorras R., Vespignani A.: The architecture of complex weighted networks. Proc. Natl. Acad. Sci. U S A 101(11) (2004) 3747-3752
30. He W., Parker R.: Functions of Lsm proteins in mRNA degradation and splicing. Curr. Opin. Cell. Biol. 12(3) (2000) 346-350
31. Knauer R., Lehle L.: The oligosaccharyltransferase complex from Saccharomyces cerevisiae. Isolation of the OST6 gene, its synthetic interaction with OST3, and analysis of the native complex. J. Biol. Chem. 274(24) (1999) 17249-17256
32. Mendenhall M.D., Hodge A.E.: Regulation of Cdc28 cyclin-dependent protein kinase activity during the cell cycle of the yeast Saccharomyces cerevisiae. Microbiol. Mol. Biol. Rev. 62(4) (1998) 1191-1243

Modelling and Simulation of the Genetic Phenomena of Additivity and Dominance via Gene Networks of Parallel Aggregation Processes

Elena Tsiporkova[1] and Veselka Boeva[2]

[1] Department of Molecular Genetics, Ghent University
Technologiepark 927, 9052 Ghent, Belgium
elena.tsiporkova@UGent.be
[2] Department of Computer Systems, Technical University of Plovdiv
Tsanko Dyustabanov 25, 4000 Plovdiv, Bulgaria
vboeva@tu-plovdiv.bg

Abstract. The contribution develops a mathematical model allowing interpretation and simulation of the phenomenon of additive-dominance heterosis as a network of interacting parallel aggregation processes. Initially, the overall heterosis potential has been expressed in terms of the heterosis potentials of each of the individual genes controlling the trait of interest. Further, the individual genes controlling the trait of interest are viewed as interacting agents involved in the process of achieving a trade-off between their individual contributions to the overall heterosis potential. Each agent is initially assigned a vector of interaction coefficients, representing the relative degrees of influence the agent is prepared to accept from the other agents. Then the individual heterosis potentials of the different agents are combined in parallel with weighted mean aggregations, one for each agent. Consequently, a new heterosis potential is obtained for each agent. The above parallel aggregations are repeated again and again until a consensus between the agents is attained.

1 Introduction

Heterosis ('hybrid vigour') refers to an improved performance of F1 hybrids with respect to the parents. It has been observed that a cross between quasi-homozygous parents can in some cases lead to an offspring (F1) that is better in terms of yield, stress resistance, speed of development, etc. as compared to the parents. Heterosis is of great commercial importance since it enables the breeder to generate a product (F1 hybrid seed) with preserved values which in turn, allows the farmer to grow uniform plants expressing these heterosis features. Besides a commercial interest there is a more fundamental scientific interest associated with the biological phenomenon of heterosis performance, as an excellent example of what complex genetic interactions can lead to.

S. Hochreiter and R. Wagner (Eds.): BIRD 2007, LNBI 4414, pp. 199–211, 2007.

Although the phenomenon of heterosis has already been studied for many years no complete genetic explanation has been found. Two main models have been considered in attempt to explain heterosis [1]:

- *Additive-Dominance Model* : Heterosis in F1 is due to increased number of loci of the genome where the heterozygote state is superior than any of the parental homozygote states.
- *Epistatic Model* : Heterosis in F1 is due to complex interaction between different genes on different loci of the genome.

Omholt *et al.* have shown in [4] how the phenomena of genetic dominance, over-dominance, additivity, and epistasis are generic features of simple diploid gene regulatory networks. Reflecting the classical genetics perception that the phenomenon of dominance is generated by intra-locus interactions, they have studied two one-locus models, one with a negative autoregulatory feedback loop, and one with a positive autoregulatory feedback loop. In order to include the phenomenon of epistasis and downstream regulatory effects, a model of a tree-locus signal transduction network has been also analysed in [4]. The main finding from this study is that genetic dominance as well as over-dominance may be an intra- as well as inter-locus interaction phenomenon. Moreover according to Omholt *et al.*, it appears that in the intra- as well as the inter-locus case there is considerable room for additive gene action, leading to the conclusion that the latter may explain to some degree the predictive power of quantitative genetic theory, with its emphasis on additive gene variance.

Our approach in the present work is quite different from the one of Omholt *et al.* [4]. We have developed a mathematical formalism allowing interpretation and simulation of additive and dominance phenomena as a gene network of interacting parallel aggregation processes. Thus we have chosen to focus on a more general interpretation of additivity and dominance in terms of gene interaction networks instead of gene regulatory networks of limited size and connectivity.

Initially, we have pursued expressing the overall heterosis potential in terms of the heterosis potentials of each of the individual genes controlling the trait of interest. This has allowed us to gain a better understanding of the biological mechanisms behind the phenomenon of heterosis. According to the additive-dominance model the net heterosis potential can be expressed as a weighted mean of the heterosis potentials of the individual genes weighted with their relative additive effects respectively. Whenever the alleles are dispersed between the parents this weighted mean is further rescaled according to their association-dispersion coefficient.

In a consequent step, the individual genes controlling the trait of interest have been viewed as interacting agents involved in the process of achieving a trade-off between their individual contributions to the overall heterosis potential [5], [6]. Each agent is initially assigned a vector of interacting coefficients, representing the relative degrees of influence this agent accepts from the other agents. Then the individual heterosis potentials of the different agents are combined in parallel with weighted mean aggregations, one for each agent (i.e. taking into account the degrees of influence of each agent). Consequently, a new heterosis potential is obtained for each agent. The above parallel aggregations are repeated again and again until a consensus between the agents is attained.

2   Additive Versus Dominance Effects

2.1   Single Gene Model

Let us consider two parental, quasi-homozygous, inbred lines P_1 and P_2. A cross between them will produce a progeny F_1, which will be heterozygous for all the loci at which the two parental lines contain different alleles. Suppose that the parents and the hybrid have been raised in a replicated experiment and their average phenotypic values for a trait under study have been calculated. These will be denoted as p_1, p_2 and f_1. Then two characteristic quantities can be calculated: the distance between the homozygous parents

$$a = \frac{|p_1 - p_2|}{2}$$

and the deviation of the hybrid value from the mean value of the parents

$$d = f_1 - \frac{p_1 + p_2}{2}.$$

If the two parents differ only in a single gene then a and d represent the estimations of the additive and dominance genetic effects, respectively. The ratio between these effects $hp = d/a$ is referred to as a potential ratio and depending on the values of hp the single gene controlling the trait will exhibit:

- additivity $hp = 0$
- partial dominance $-1 < hp < 0$ or $0 < hp < 1$
- complete dominance $hp = -1$ or $hp = 1$
- over-dominance $hp < -1$ or $hp > 1$.

Practically $hp > 1$ is a quantitative indication of the presence of heterosis for the trait of interest since it implies that the hybrid outperforms both parents.

2.2   Multiple Gene Model

Note that in case of multiple genes involved in the trait of interest, the genetic interpretation of hp is not so straightforward. Suppose that genes $1, \ldots, m$ control the trait of interest. Then the net additive effect can be expressed as a function of the additive effects of the individual genes [3]:

$$a = r_a \cdot \sum_{i=1}^{m} a_i, \tag{1}$$

with $r_a \in [0, 1]$ being the coefficient of gene association-dispersion. Thus $r_a = 1$ means complete association, while $r_a = 0$ indicates complete dispersion of the alleles between the parents. Assuming that J denotes the set of genes with dispersed alleles between the parents then r_a is defined by

$$r_a = 1 - 2 \sum_{j \in J} a_j / \sum_{i=1}^{m} a_i.$$

How the coefficient of gene association-dispersion can be derived is illustrated in Figure 1. For more details also refer to [3].

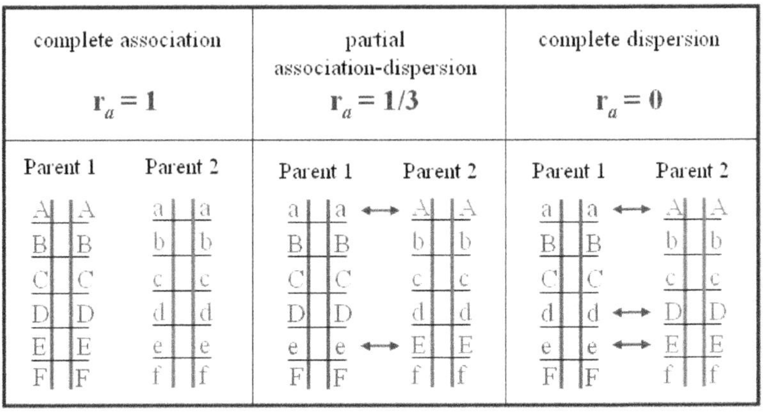

complete association $\mathbf{r}_a = 1$		partial association-dispersion $\mathbf{r}_a = 1/3$		complete dispersion $\mathbf{r}_a = 0$	
Parent 1	Parent 2	Parent 1	Parent 2	Parent 1	Parent 2
A A	a a	a a ⟷ A A	b b	a a ⟷ A A	b b
B B	b b	B B	b b	B B	b b
C C	c c	C C	c c	C C	c c
D D	d d	D D	d d	d d ⟷ D D	D D
E E	e e	e e ⟷ E E	f f	e e ⟷ E E	f f
F F	f f	F F		F F	

Fig. 1. Coefficient of Gene Association-Dispersion (assuming that the additive effects are uniformly distributed between the genes)

Further, taking in view that our considerations are focused on the additive-dominance models (i.e. only inter-allelic interactions between the genes), the net dominance effect can be expressed in terms of the dominance effects of the individual genes [3]:

$$d = \sum_{i=1}^{m} d_i. \tag{2}$$

3   Net Potential Ratio and Heterosis

The net potential ratio is our main concern in this work since it is the true quantitative measure of the heterosis potential for the trait of interest. Hereafter we pursue expressing the net potential ratio or the overall heterosis potential in terms of the heterosis potentials of each of the individual genes. This will allow us to gain a better understanding of the biological mechanisms behind the phenomenon of heterosis.

The net potential ratio or the overall heterosis potential $hp = d/a$ is the ratio of the net dominance effect d in (2) and the net additive effect a in (1)

$$hp = \frac{d}{a} = \sum_{i=1}^{m} d_i / \left(r_a \cdot \sum_{i=1}^{m} a_i \right).$$

After performing some rather straightforward transformations, presented in a detail in Appendix A.1, we obtain that:

$$hp = \frac{1}{r_a} \sum_{i=1}^{m} \lambda_i hp_i, \tag{3}$$

where $hp_i = d_i/a_i$ denotes the heterosis potential of gene i and $\lambda_i = a_i / \sum_{i=1}^{m} a_i > 0$ will be referred as the relative additive effect of gene i. Note that $\sum_{i=1}^{m} \lambda_i = 1$. The coefficient of gene association-dispersion can now be rewritten only in terms of the relative additive effects of the genes dispersed between the parents. Namely

$$r_a = 1 - 2 \sum_{j \in J} \lambda_j$$

and due to $r_a \geq 0$ it follows that $\sum_{j \in J} \lambda_j \leq 1/2$, or in other words, the total additive effect of the dispersed genes can relatively be at most half of the overall additive effect.

Thus we have succeeded in expressing the overall heterosis potential in terms of the heterosis potentials of the individual genes. In case of complete association $r_a = 1$, the net potential ratio can simply be expressed as a weighted mean of the potential ratios of the individual genes weighted with their relative additive effects respectively. Whenever the genes are dispersed between the parents this weighted mean is further rescaled according to their association-dispersion coefficient r_a. Thus according to the additive-dominance model the net heterosis potential is always a linear function of the heterosis potentials of the individual genes assumed to control the trait of interest. Moreover, according to (3), it can easily be seen that

$$\sum_{i=1}^{m} \lambda_i hp_i > r_a \tag{4}$$

is an indication of heterosis, i.e. if the weighted sum of the heterosis potential of the individual genes is greater than their association-dispersion coefficient then heterosis is encountered and vice versa.

Note that the overall heterosis potential for the additive-dominance model is always bounded by

$$\frac{1}{r_a} \min_{i=1}^{m} hp_i \leq hp \leq \frac{1}{r_a} \max_{i=1}^{m} hp_i. \tag{5}$$

Clearly when we have complete association, i.e. $r_a = 1$, the overall heterosis potential will never exceed the extreme heterosis potentials of the individual genes. In case of dispersion, i.e. $r_a < 1$, the above interval can effectively be expanded to infinity.

Further, according to (5), whenever heterosis is encountered we will have that $\max\limits_{i=1}^{m} hp_i > r_a$, which implies that in presence of heterosis there exists at least one gene, among the genes controlling the trait under study, with a heterosis potential greater than the association-dispersion coefficient of these genes. The latter entails a few interesting facts:

- A network of genes, which all exhibit negative heterosis potentials will never lead to an overall heterosis.
- Encountering overall heterosis in case of complete association implies that there must exist at least one positively over-dominant gene.
- Whenever complete association is excluded then at least one gene exhibiting positive partial-dominance must be present when overall heterosis is observed.

4   Heterosis as a Network of Interacting Parallel Aggregation Processes

In this section, we discuss a mathematical model allowing interpretation and simulation of the phenomenon of heterosis as a network of interacting parallel aggregation processes. The sub-processes building up the additive-dominance heterosis are modelled as interacting parallel aggregations combining the heterosis potentials of the individual genes controlling the trait of interest. These genes are viewed as interacting agents involved in the process of achieving a trade-off between their individual contributions to the overall heterosis potential. Thus each agent i is modelled by a vector of interacting coefficients $a_i = [a_{1i}, \ldots, a_{mi}]$, representing the relative degrees of influence this agent is prepared to accept from the rest of the group. It is assumed that the net influence accepted from each agent is 1, or more formally $\sum\limits_{k=1}^{m} a_{ki} = 1$, where a_{ki} is the relative degree of influence agent i accepts from agent k. Clearly, a_{ii} is the corresponding relative degree of self-influence (feedback) of the agent.

Next, the individual heterosis potentials of the different agents are considered as a vector of initial values $[hp_1, \ldots, hp_m]$ that have to be combined in parallel with weighted mean aggregations, one for each agent. Consequently, a new heterosis potential $hp_i^1 = \sum\limits_{k=1}^{m} a_{ki} hp_k$ is calculated for each agent i and a vector of new values $[hp_1^1, \ldots, hp_m^1]$ is obtained. These new values (heterosis potentials) are combined as follows $hp_i^2 = \sum\limits_{k=1}^{m} a_{ki} hp_k^1$. These parallel aggregations are repeated again and again until a consensus between the agents is attained, i.e. at some phase q of the interaction we will have $hp_1^q \approx hp_2^q \ldots \approx hp_m^q$. [1] Assuming

---

[1] The value hp_i^q can be interpreted as a heterosis potential of agent i at interaction phase q.

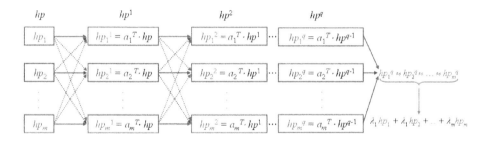

Fig. 2. Interacting Parallel Aggregation Processes

that all interaction coefficients are strictly positive, i.e. $a_{ij} > 0$ for all i and j, is sufficient to guarantee that consensus will be achieved after a final number of interaction steps, or in other words that the interaction process will eventually converge. Moreover, it can be proved that when convergence is attained after q interaction steps then $hp_1^q, hp_2^q, \ldots, hp_m^q$ can be expressed in terms of the original heterosis potentials as follows:

$$\lambda_1 hp_1 + \lambda_2 hp_2 + \ldots + \lambda_m hp_m.$$

Thus according to (3) our network of interacting aggregation processes induces additivity and dominance phenomena. The above aggregation processes are illustrated in Figure 2 and defined in details in Appendix A.2.

The matrix representation of the interacting aggregation processes in (6) (see Appendix A.2) implies that the recursive aggregation is performed not over the vector of initial heterosis potentials but over the interaction coefficients or the so-called interaction matrix. Consequently, a given interaction matrix will uniquely determine a vector of relative additive effects $\lambda_1, \lambda_2, \ldots, \lambda_m$. The latter is demonstrated in Example 4.1 (Figure 4), where two different simulation scenarios are presented. They correspond to parallel aggregation processes performed with one and the same vector of individual heterosis potentials and two slightly different interaction matrices (matrices A and B in Example 4.1). It can be observed in Figure 3 that the induced overall heterosis potentials are quite different for the two interaction matrices. Namely negative partial dominance is generated with matrix A, whereas positive complete dominance is encountered in the second simulation with matrix B. This is not surprising since the two different interaction matrices determine quite different vectors of relative additive effects. A detail description of the above simulations is presented in Example 4.1.

Note that the described above network of parallel aggregation processes can easily be extended into a more complex multilayer topology of interacting networks. Thus one obtains a hierarchical structure with multiple levels of interaction, i.e. there will be nodes at the top level of the network which can further be represented as a network of parallel aggregation processes. Consequently at this lower level of interaction some other nodes will be composed of interacting genes and so on until the bottom level of the hierarchy is reached. The design of our model guarantees that the final aggregation result even from such a complex multilayer gene interaction network can be expressed as a linear combination of the heterosis potentials of the individual genes controlling the trait of interest.

4.1   Simulation Example

Let us consider a set of five genes that supposedly control the trait (biomass, stress resistance, etc.) in which we are interested. The individual heterosis potentials of the genes compose the following vector:

$$hp = \begin{bmatrix} -2 \\ 0.1 \\ 1 \\ 5 \\ -0.5 \end{bmatrix}.$$

In addition, we suppose that these five genes interact with each other according to the network depicted in Figure 3a. In order to perform simulations with parallel aggregation processes we need to construct an interaction matrix. For this purpose, the following assumptions about the network in Figure 3a are made:

- all arrows indicate direct interactions, i.e. an arrow from gene x to gene y means that gene x influences (affects) in some way the activity of gene y;
- all arrows entering the same node have the same weight, i.e. all direct interactions have equal importance;
- each gene accepts 5% (background) influence from the genes that do not have direct interactions with this gene;
- the total amount of interactions accepted by a gene sums to 1.

In view of the above assumptions, the network in Figure 3a defines the following interaction matrix A:

$$A = \begin{bmatrix} 0.8 & 0.425 & 0.3 & 0.05 & 0.3 \\ 0.05 & 0.425 & 0.3 & 0.05 & 0.05 \\ 0.05 & 0.05 & 0.3 & 0.425 & 0.05 \\ 0.05 & 0.05 & 0.05 & 0.425 & 0.3 \\ 0.05 & 0.05 & 0.05 & 0.05 & 0.3 \end{bmatrix}.$$

The evolution of the individual heterosis potentials during the parallel aggregation process performed with the above matrix on the vector hp is illustrated in Figure 3c. After more than 8 iterations the process finally converges to a negative overall heterosis potential of -0.52. The corresponding vector of relative additive effects is $\lambda = (0.58, 0.13, 0.12, 0.1, 0.07)$ (Figure 4). Thus gene 1 delivers the greatest contribution to the overall potential. The latter is certainly due to the fact that, according to Figure 3a, the activities of three of the other four genes are directly influenced by gene 1. The dominating position of gene 1 is also confirmed by the behavior of the individual heterosis potentials in Figure 3c.

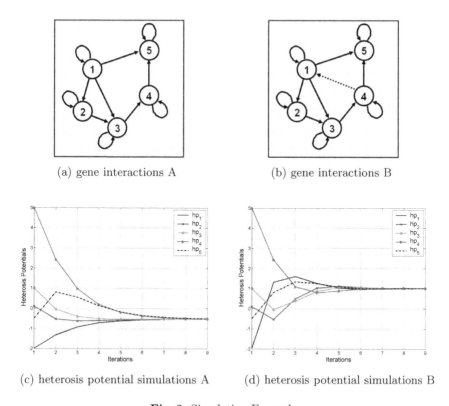

(a) gene interactions A   (b) gene interactions B

(c) heterosis potential simulations A   (d) heterosis potential simulations B

Fig. 3. Simulation Example

Namely the individual potential curves of genes 2 to 5 clearly evolve during the aggregation process toward the curve corresponding to gene 1.

Another interesting observation can be made from the above simulation. According to Figure 3a, the genes that contribute the highest negative and positive initial potentials, genes 1 and 4 respectively, are not involved in a direct interaction. Consequently, the individual potential curves of the other three genes (2, 3 and 5) in Figure 3c are all restricted between the curves of genes 1 and 4.

Assume now that one would like to test a new interaction hypothesis, which implies a direct influence of gene 4 on the activity of gene 1, as depicted in Figure 3b. This assumption entails a new interaction matrix B, which differs from A only in its first column:

$$
B = \begin{bmatrix}
0.425 & 0.425 & 0.3 & 0.05 & 0.3 \\
0.05 & 0.425 & 0.3 & 0.05 & 0.05 \\
0.05 & 0.05 & 0.3 & 0.425 & 0.05 \\
0.425 & 0.05 & 0.05 & 0.425 & 0.3 \\
0.05 & 0.05 & 0.05 & 0.05 & 0.3
\end{bmatrix} .
$$

Analogously, the evolution of the individual heterosis potentials during the parallel aggregation process performed with the above matrix on the vector hp is illustrated in Figure 3d. Around iteration 6-7 the process converges to an overall heterosis potential of 1, i.e. positive complete dominance is attained. The corresponding vector of relative additive effects $\lambda = (0.29, 0.16, 0.2, 0.28, 0.07)$ (Figure 4) indicates that in the present situation gene 1 and gene 4 have almost equal contribution to this complete dominance, i.e. gene 1 has lost its dominant position. The effect of the simulated direct interaction from gene 4 to gene 1 can also be detected in the behavior of the individual heterosis potential curves in Figure 3d, the curves for genes 1 and 4 are clearly attracted to each other in the process of parallel aggregation.

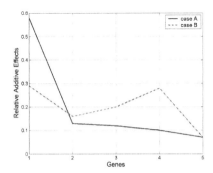

Fig. 4. Relative Additive Effects

When comparing Figure 3c and Figure 3d it becomes evident that adding just one new extra interaction has induced a completely different interaction process leading to a very different overall heterosis potential. The latter is a good indication of the sensitivity potential of our simulation model.

References

1. Birchler, J.A., Auger, D.L., Riddle, N.C.: In Search of the Molecular Basis of Heterosis. The Plant Cell 15 (2003) 2236-2239.
2. J.C. Fodor and M. Roubens, Fuzzy Preference Modelling and Multicriteria Decision Support, Klumer Academic Publishers, Dordrecht, 1994.
3. Kearsey, M.J., Pooni, S.: The Genetical Analysis of Quantative Traits. Chapman & Hall, London (1996).
4. Omholt, S.W., Plahte, E., Oyehauh, L., Xiang K.: Gene Regulatory Networks Generating the Phenomena of Additivity, Dominance and Epistasis. Genetics 155 (2000) 969-979.
5. Tsiporkova, E., Boeva, V.: Nonparametric Recursive Aggregation Process. Kybernetika. J. of the Czech Society for Cybernetics and Inf. Sciences **40** 1 (2004) 51-70.
6. Tsiporkova, E., Boeva, V.: Multi-step ranking of alternatives in a multi-criteria and multi-expert decision making environment. Inf. Sciences **176** 18 (2006) 2673-2697.

A   Appendix

A.1   Net Heterosis Potentials

Consider the ratio of the net dominance effect (2) and the net additive effect (1):

$$hp = \frac{d}{a} = \sum_{i=1}^{m} d_i / \left(r_a \cdot \sum_{i=1}^{m} a_i \right).$$

Next let us multiply and divide d_i in the above expression with a_i and group the expression in the following way:

$$hp = \frac{1}{r_a} \sum_{i=1}^{m} \left(a_i / \sum_{i=1}^{m} a_i \right) \cdot \frac{d_i}{a_i}.$$

Consequently, it is obtained:

$$hp = \left(\sum_{i=1}^{m} \lambda_i hp_i \right) / r_a,$$

where $hp_i = d_i / a_i$ denotes the heterosis potential of gene i and $\lambda_i = \frac{a_i}{\sum\limits_{i=1}^{m} a_i} > 0$

and $\sum\limits_{i=1}^{m} \lambda_i = 1$.

A.2   Interacting Parallel Aggregation Processes

Note that the above parallel aggregation processes can be realized by a recursive aggregation algorithm defined in terms of matrix operations. This algorithm is illustrated in Figure 5 and we proceed with its formulation.

Initialization
Assume that m genes are involved in the trait of interest. Hence, we have m interacting agents and for each agent $i = 1, \ldots, m$ a heterosis potential hp_i is known, i.e. a vector of initial heterosis potential

$$hp^0 = \begin{bmatrix} hp_1 \\ \vdots \\ hp_m \end{bmatrix}$$

is given. Further, each agent $i = 1, \ldots, m$ is modelled by a vector of interacting coefficients $a_i = [a_{1i}, \ldots, a_{mi}]$, where $\sum\limits_{k=1}^{m} a_{ki} = 1$ and a_{ki} is the relative degree of influence agent i accepts from agent k, while a_{ii} is the relative degree of self-influence of agent i in the interaction.

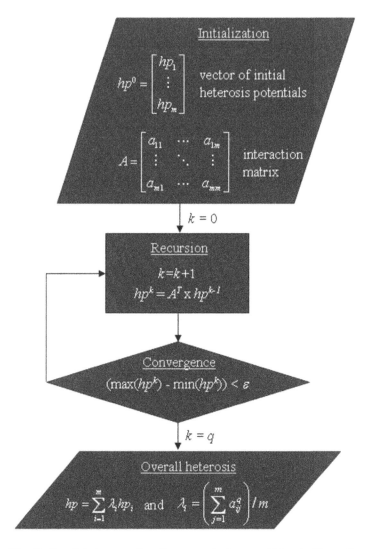

Fig. 5. Parallel Aggregation Processes in terms of Matrix Operations

Next the quadratic interaction matrix $m \times m$ can be composed, as follows:

$$A = [a_1^T, \ldots, a_m^T] = \begin{bmatrix} a_{11} & \cdots & a_{1m} \\ \vdots & \ddots & \vdots \\ a_{m1} & \cdots & a_{mm} \end{bmatrix}.$$

Recursion
Each interaction step is modelled via weighted mean aggregations applied over the results of the previous step, t.e. the vector of heterosis potential at step k $(k = 1, 2, \ldots)$ can be obtained as follows:

$$hp^k = A^T \times hp^{k-1},$$

where the heterosis potential of agent $i = 1, \ldots, m$ at step k is calculated by $hp_i^k = a_i^T hp^{k-1}$.

The above process is repeated again and again until ultimately the difference between the maximum and minimum values in the current vector of heterosis potentials will be small enough to stop further aggregation, i.e. at some q step and for some very small $\epsilon \in R^+$ it will be fulfilled that $(\max(hp^q) - \min(hp^q)) < \epsilon$. Then after q aggregation steps one gets

$$hp^q = A^T \times hp^{q-1} = A^T \times \ldots \times A^T \times hp^0 = (A^q)^T \times hp^0. \qquad (6)$$

Convergence

In [5], we have showed in general that any recursive aggregation process defined via a set of aggregation operators and according to the algorithm described herein, is convergent if the aggregation operators are continuous and strict compensatory. An aggregation operator X is referred to as *strict compensatory* if

$$\min(x_1, \ldots, x_n) < X(x_1, \ldots, x_n) < \max(x_1, \ldots, x_n),$$

for any $(x_1, \ldots, x_n)$, with at least two different values $(n \geq 2)$, [2]. Our aggregation algorithm is dealing with a family of weighted mean operators which are continuous, but not always strict compensatory. However, the latter can be guaranteed when we have non-zero weights, i.e. if $0 < a_{ij} \leq 1$ holds for all $i, j = 1, \ldots, m$.

Notice that the representation in (6) implies that the recursive aggregation is performed not over the vector of initial heterosis potential hp^0 but over the interaction matrix A. Therefore, it can easily be seen that the overall heterosis potential can again be expressed as a weighted mean of the heterosis potentials of the individual agents. Due to the strict compensatory behaviour of the weighted mean operators in case of non-zero weights, we will have that the difference between the maximum and minimum values in each row of A^k decreases at each aggregation step k and at some step q we will have

$$(\max(a_{i1}^q, \ldots, a_{im}^q) - \min(a_{i1}^q, \ldots, a_{im}^q)) < \epsilon.$$

Hence, the overall heterosis can be represented by the expression $hp = \sum_{i=1}^{m} \lambda_i hp_i$, where a consensus weight λ_i can be defined as $\left(\sum_{j=1}^{m} a_{ij}^q \right) / m$, for $i = 1, \ldots, m$.

Protein Remote Homology Detection Based on Binary Profiles

Qiwen Dong, Lei Lin, and Xiaolong Wang

School of Computer Science and Technology, Harbin Institute of Technology, Harbin, China
{qwdong,linl,wangxl}@insun.hit.edu.cn

Abstract. Remote homology detection is a key element of protein structure and function analysis in computational and experimental biology. This paper presents a simple representation of protein sequences, which uses the evolutionary information of profiles for efficient remote homology detection. The frequency profiles are directly calculated from the multiple sequence alignments outputted by PSI-BLAST and converted into binary profiles with a probability threshold. Such binary profiles make up of a new building block for protein sequences. The protein sequences are mapped into high-dimensional vectors by the occurrence times of each binary profile. The resulting vectors are then evaluated by support vector machine to train classifiers that are then used to classify the test protein sequences. The method is further improved by applying an efficient feature extraction algorithm from natural language processing, namely, the latent semantic analysis model. Testing on the SCOP 1.53 database shows that the method based on binary profiles outperforms those based on many other basic building blocks including N-grams, patters and motifs. The ROC50 score is 0.698, which is higher than other methods by nearly 10 percent.

Keywords: remote homology detection; binary profile; latent semantic analysis.

1 Introduction

A central problem in computational biology is the classification of proteins into functional and structural classes given their amino acid sequences. Through evolution, structure is more conserved than sequence, so that detecting even very subtle sequence similarities, or remote homology, is important for predicting structure and function [1].

The major methods for homology detection can be split into three basic groups [3]: pair-wise sequence comparison algorithms, generative models for protein families and discriminative classifiers. Early methods looked for pair-wise similarities between proteins. Among those algorithms, the Smith–Waterman dynamic programming algorithm [4] is one of the most accurate methods, whereas heuristic algorithms such as BLAST [5] and FASTA [6] trade reduced accuracy for improved efficiency. The methods afterwards have obtained higher rates of accuracy by collecting statistical information from a set of similar sequences. PSI-BLAST [7] used BLAST to iteratively build a probabilistic profile of a query sequence and obtained a more sensitive sequence comparison score. Generative models such as profile hidden Markov models (HMM)

S. Hochreiter and R. Wagner (Eds.): BIRD 2007, LNBI 4414, pp. 212–223, 2007.

[8] used positive examples of a protein family, which can be trained iteratively using both positively labeled and unlabeled examples by pulling in close homology and adding them to the positive set [9]. Finally, the discriminative algorithms such as Support Vector Machine (SVM) [10] used both positive and negative examples and provided state-of-the-art performance with appropriate kernel. Many SVM-based methods have been proposed such as SVM-fisher [11], SVM-k-spectrum [12], Mismatch-SVM [13], SVM-pairwise [3], SVM-I-sites [14], SVM-LA and SVM-SW [15]. A comparison of SVM-based methods has been performed by Saigo *et al.* [16].

Sequence homologs are an important source of information about proteins. Multiple sequences alignments of protein sequences contain much information regarding evolutionary processes. This information can be detected by analyzing the output of PSI-BLAST [7, 17]. Since amino acid probability profiles are a richer encoding of protein sequences than the individual sequence themselves, it is necessary to use such evolutionary information for protein remote homology detection.

In this study, the frequency profiles of protein sequences are directly calculated from the multiple sequence alignments and then converted into binary profiles with a cut-off probability for usage. Such binary profiles make up of a new building block for protein sequences. These binary profiles are filtered by an efficient feature selection algorithm, namely, the chi-square algorithm [20]. The protein sequences are mapped into high-dimensional vectors by the occurrence times of each selected binary profile. The resulting vectors can then be inputted to a discriminative learning algorithm, such as SVM. In a previous study [18], we applied Latent Semantic Analysis (LSA), which is an efficient feature extraction technique from natural language processing [19], to protein remote homology detection. Several basic building blocks has been investigated as the "words" of "protein sequence language", including N-grams [12], patterns [20] and motifs [21]. Here, we also demonstrate that the use of latent semantic analysis technology on binary profiles can also improve the performance of protein remote homology detection.

2   Materials and Methods

2.1   Generation of Binary Profiles

The PSI-BLAST [7] is used to generate the profiles of amino acid sequences with the default parameter values except for the number of iterations set to 10. The search is performed against the nrdb90 database (http://www.ebi.ac.uk/~holm/nrdb90/) from EBI [22]. This database has recently been updated (Oct, 2002). The frequency profiles are directly obtained from the multiple sequence alignments outputted by PSI-BLAST. The target frequency reflects the probability of an amino acid in a given position of the sequences. The method of target frequency calculation is similar to that implemented in PSI-BLAST. We use a subset of the multiple sequence alignments with sequence identity less than 98% to calculate the frequency profiles. The sequence weight is assigned by the position-based sequence weight method [23]. Given the observed frequency of amino acid i (f_i) and the background frequency of amino acid i (p_i), the pseudo-count for amino acid i is computed as follows:

$$g_i = \sum_{j=1}^{20} f_i * (q_{ij} / p_j) \tag{1}$$

where q_{ij} is the score of amino acid i being aligned to amino acid j in BLOSUM62 substitution matrix (the default score matrix of PSI-BLAST).

The target frequency is then calculated as:

$$Q_i = (\alpha f_i + \beta g_i) / (\alpha + \beta) \tag{2}$$

where α is the number of different amino acids in a given column minus one and β is a free parameter set to a constant value of 10, the value initially used by PSI-BLAST.

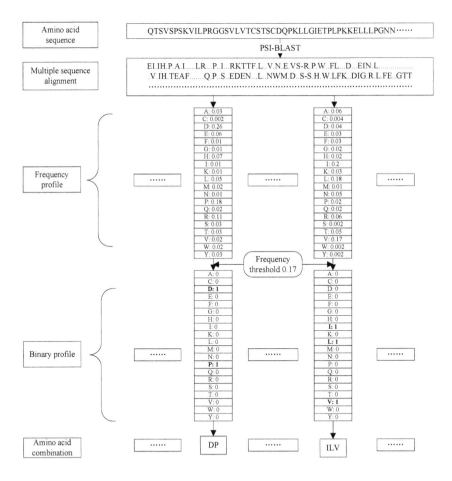

Fig. 1. The flowchart of calculating and converting frequency profiles. The multiple sequence alignment is obtained by PSI-BLAST. The frequency profile is calculated on the multiple sequence alignment and converted to a binary profile with a frequency threshold. The substring of amino acid combination is then collected.

Because the frequency profile is a matrix of frequencies for all amino acids, it cannot be used directly and need to be converted into a binary profile by a probability threshold P_h. When the frequency of an amino acid is larger than p, it is converted into an integral value of 1, which means that the specific amino acid can occur in a given position of the protein sequence during evolution. Otherwise it is converted into 0. A substring of amino acid combination is then obtained by collecting the binary profile with non-zero value for each position of the protein sequences. These substrings have approximately represented the amino acids that possibly occur at a given sequence position during evolution. Each combination of the twenty amino acids corresponds to a binary profile and vice versa. Fig. 1 has shown the process of generating and converting the profiles.

2.2 Chi-square Feature Selection

Most machine-learning algorithms do not scale well to high-dimensional feature spaces [25], and there are too many binary profiles with low probability threshold P_h. Thus, it is desirable to reduce the dimension of the feature space by removing non-informative or redundant features. A large number of feature selection methods have been developed for this task, including document frequency, information gain, mutual information, chi-square and term strength. The chi-square algorithm is selected in this study because it is one of the most effective feature selection methods in document classification task [20].

The chi-square algorithm measures the lack of independence between a feature t and a classification category c and can be compared to the chi-square distribution with one degree of freedom to judge extremeness. The chi-square value of feature t relative to category c is defined to be:

$$\chi^2(t,c) = \frac{N \times (A \times D - C \times B)^2}{(A+C) \times (B+D) \times (A+B) \times (C+D)} \tag{3}$$

where A is the number of times t and c co-occur, B is the number of times the t occurs without c, C is the number of times c occurs without t, D is the number of times neither c nor t occurs and N is the total number of protein sequence.

The chi-square statistic has a natural value of zero if t and c are independent. The category-specific scores of each feature can be combined into two scores

$$\chi^2_{avg}(t) = \sum_{i=1}^{m} P_r(c_i) \chi^2(t,c_i) \tag{4}$$

$$\chi^2_{max}(t) = \max_{i=1,m}\{\chi^2(t,c_i)\} \tag{5}$$

In this paper, the average feature value is used, since its performance is better than the maximum value. A maximum of 8000 binary profiles are selected as the "words" of protein sequence language.

2.3   Remote Homology Detection by SVM

Support Vector Machine (SVM) is a class of supervised learning algorithms first introduced by Vapnik [8]. Given a set of labeled training vectors (positive and negative input examples), SVM can learn a linear decision boundary to discriminate between the two classes. The result is a linear classification rule that can be used to classify new test examples. SVM has exhibited excellent performance in practice and has strong theoretical foundation of statistical learning theory.

Suppose the training set S consists of labeled input vectors $(x_i; y_i)$, $i = 1...m$, where $x_i \in R_n$ and $y_i \in [-1, 1]$. A linear classification rule can be specified:

$$f(x) = < w, x > + b \tag{6}$$

where $w \in R_n$ and $b \in R$. A test example x is then classified as positive (negative) if $f(x) > 0$ ($f(x) < 0$). Such a classification rule corresponds to a hyper-plane decision boundary between positive and negative points with normal vector w and bias term b. When the samples are not linearly separable, the kernel function is used to map the samples from the input space into the feature space in which an appropriate hyper-plane can be found.

In this study, the frequency profiles of protein sequences are converted into binary profiles with a probability threshold P_h. After chi-square feature selection, each protein sequence is represented as a fix-length vector by the occurrence times of each elected binary profile. The protein vectors are then input into SVM to train the classifiers and classify the test protein sequences. The Gist SVM package implemented by Jaakkola et al. [9] is used for protein remote homology detection. The parameters of SVM are used by default of the Gist package except for the kernel function that is the Radius Basis Function (RBF) kernel.

2.4   Latent Semantic Analysis

Latent *semantic* analysis is a theory and method for extracting and representing the contextual-usage meaning of words by statistical computations applied to a large corpus of text [24]. Here, we briefly introduce the basic process of LSA.

The *starting* point of LSA is the construction of a word-document matrix W of co-occurrences between words and documents. The elements of W can be taken as the number of times each word appears in each document, thus the dimension of W is $M \times N$, where M is the total number of words and N is the number of given documents. To compensate for the differences in document lengths and overall counts of different words in the document collection, each word count can be normalized [24].

In the word-document matrix W, each document is expressed as a column vector. However, this representation does not recognize synonymous or related words and the dimensions are too large. In the specific application, singular value decomposition is performed on the word-document matrix. Let K be the total ranks of W, W can be decomposed into three matrices:

$$W = USV^T \tag{7}$$

where U is left singular matrix with dimensions ($M{\times}K$), V is right singular matrix with dimensions ($N{\times}K$), and S is ($K{\times}K$) diagonal matrix of singular values $s_1{\geq}s_2{\geq}...s_K{>}0$. One can reduce the dimensions of the solution simply by deleting the smaller singular values in the diagonal matrix. The corresponding columns of matrix U (rows of matrix V) are also ignored. In practice only the top R ($R{<<}Min\ (M, N)$) dimensions for which the elements in S are greater than a threshold are considered for further processing. Thus, the dimensions of matrices U, S and V are reduced to $M{\times}R$, $R{\times}R$ and $N{\times}R$, leading to data compression and noise removal. Values of R in the range [200, 300] are typically used for information retrieval.

By SVD, the column vectors of the word-document matrix W are projected onto the orthonormal basis formed by the row column vectors of the left singular matrix U. The coordinates of the vectors are given by the columns of SV^T. This in turn means that the column vectors Sv_j^T or, equivalently the row vector v_jS, characterizes the position of document d_j in the R dimensions space. Each of the vector v_jS is referred to a document vector, uniquely associated with the document in the training set.

For a new document that is not in the training set, it is required to add the unseen document to the original training set and the latent semantic analysis model be recomputed. However, SVD is a computationally expensive process, performing SVD every time for a new test document is not suitable. From the mathematical properties of the matrices U, S and V, the new vector t can be approximated as:

$$t = dU \tag{8}$$

where d is the raw vector of the new document, which is similar to the columns of the matrix W.

In this study, the binary profiles are treated as the "words" and the protein sequences can be viewed as the "documents". The word-document matrix is constructed by collecting the weight of each word in the documents. The latent semantic analysis is performed on the matrix to produce the latent semantic representation vectors of protein sequences, leading to noise-removal and smart description of protein sequences. The latent semantic representation vectors are then evaluated by SVM.

2.5 Data Set and Performance Metrics

The standard evaluation data is same as the one used by Li *et al.* [3], which is taken from the Structural Classification of Proteins (SCOP) database [25] version 1.53. Sequences are selected from the ASTRAL database [26]. The data set contains 54 families and 4352 distinct sequences. The sequences with lengths less than 30 are removed because the PSI-BLAST cannot generate profiles on the very short sequences. Remote homology is simulated by holding out all members of a target 1.53 family from a given superfamily. Positive training examples are chosen from the remaining families in the same superfamily and negative test and training examples are chosen from outside of the fold of the target family. The held-out family members serve as positive test examples. The above process is iterated until each family had been tested. Details of the data sets are available at http://www1.cs.columbia.edu/compbio/svm-pairwise/.

Each of the method produces a score for a testing protein sequence, which represents the similarity between the protein and the particular family. Two methods are used to evaluate the experimental results: the Receiver Operating Characteristic (ROC) scores [27] and the Median Rate of False Positives (M-RFP) scores [11]. A ROC score is the normalized area under a curve that is plotted with true positives as a function of false positives for varying classification thresholds. A score of 1 indicates perfect separation of positive samples from negative ones, whereas a score of 0 denotes that none of the sequences selected by the algorithm is positive. The median RFP score is the fraction of negative test sequences that score as high or better than the median score of the positive sequences. Obviously, the smaller the M-RFP is, the better the results are. Hou *et al.* [14] have presented an efficient algorithm to compute these scores.

3   Results and Discussion

3.1   Comparative Results of Various Methods

In a related study [18], we have investigated several basic building blocks as the "words" of "protein sequence language", including N-grams [12], patterns [20] and motifs [21]. Here, the performance of the new building block, binary profile, has been compared with those of other three building blocks as well as other methods, that is PSI-BLAST [7], SVM-pairwise [3] and SVM-LA [15]. For a detail setup procedures of these methods, please refer to [18]. Table 1 summarizes the performance of the various methods in terms of average ROC, ROC50 and M-RFP scores over all families tested. The distributions of ROC and M-RFP scores are plotted in Fig. 2. In each graph, a higher curve corresponds to more accurate homology detection performance.

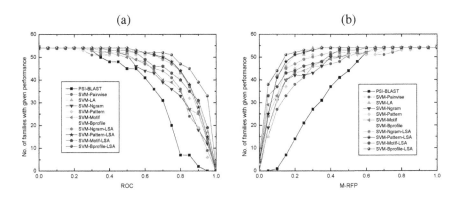

Fig. 2. Comparative performances of homology detection methods. The graph plots the total number of families for which the method exceeds a given performance. Each series corresponds to one of the homology detection methods described in the text. The left part (a) uses the ROC scores and the right part (b) uses the M-RFP scores.

The PSI-BLAST [7] method that is based on sequence alignment and search techniques achieves the lowest performance. The discriminative method, SVM-Pairwise [3], gets improved performance. The SVM-LA method provides state-of-the-art performance with string alignment kernels [15]. The accuracies of the SVM methods based on the basic words are lower than that of SVM-LA. When the LSA model is used, all the SVM methods based on the four basic words achieve higher accuracies. Especially, the ROC50 score of SVM-Bprofile-LSA method is higher than that of SVM-LA method by nearly 10 percent. The SVM-LA method is one of the state-of-the-art methods and outperforms many other methods such as Mismatch-SVM [13] and SVM-Fisher [11] as well as FPS [28] and SAM [29], so the binary profile is an efficient representation of protein sequence for remote homology detection.

Fig. 3 plots the family-by-family comparison of the ROC scores between the methods with LSA and those without LSA. Each point on the graph corresponds to one of the 54 SCOP families. When the families are in the left-upper area, it means that the method labeled by y-axis outperforms the method labeled by x-axis on this family. Obviously, the methods with LSA can outperform the methods without LSA when the binary profiles are taken as the basic building blocks. Such conclusion is also suitable for other building blocks, that is N-grams, patterns and motifs [18].

Table 1. Average ROC and M-RFP scores over all families for different methods

Methods	Mean ROC	Mean ROC50	Mean M-RFP
PSI-BLAST	0.6754	0.330	0.3253
SVM-Pairwise	0.8259	0.446	0.1173
SVM-LA	0.9290	0.600	0.0515
SVM-Ngram	0.7914	0.584	0.1441
SVM-Pattern	0.8354	0.589	0.1349
SVM-Motif	0.8136	0.616	0.1246
SVM-Bprofile	0.9032	0.681	0.0682
SVM-Ngram-LSA	0.8595	0.628	0.1017
SVM-Pattern-LSA	0.8789	0.626	0.0703
SVM-Motif-LSA	0.8592	0.628	0.0995
SVM-Bprofile-LSA	0.9210	0.698	0.0459

SVM-Ngram, SVM-Pattern, SVM-Motif and SVM-Bprofile refer to the SVM-based methods on the four building blocks: N-grams, patterns, motifs and binary profiles respectively. The methods with LSA suffix refer to the corresponding method after latent semantic analysis. Results of SVM-LA method are taken from (Saigo et al. 2004, Bioinformatics, 20: 1682-1689). For the methods based on binary profiles, the probability threshold is taken as the optimal value of 0.13.

3.2 Time Complexity Analysis

One significant characteristic of any homology detection algorithm is its computational efficiency. In this regard, the LSA approaches are better than SVM-pairwise and

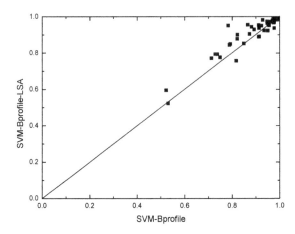

Fig. 3. Family by family comparison of the methods with LSA and those without LSA. The coordinates of each point in the plot are the ROC scores for one SCOP family, obtained by the two methods labeled near the axis.

SVM-LA but a little worse than the methods without LSA and PSI-BLAST [18]. When the four basic building blocks are considered, the binary profiles have the lowest running time because the binary profiles has the least "words" among the building blocks.

Any SVM-based method includes a vectorization step and an optimization step. The vectorization step of SVM-pairwise takes a running time of O (n^2l^2), where n is the number of training examples and l is the length of the longest training sequence. The time complexity of calculation of LA-ekm kernel matrix is same as that of SVM-pairwise [15]. The time complexity of the vectorization step of the method without LSA is O (nml), where m is the total number of words. The main bottleneck of the LSA method is the additional SVD process, which roughly takes O (nmt), where t is the minimum of n and m. The optimization step of SVM-based method takes O (n^2p) time, where p is the length of the latent semantic representation vector. In SVM-pairwise, p is equal to n, yielding a total time of O (n^3). In the method without LSA, p is equal to m. While in the LSA method, p is equal to R. Since $R \ll \text{Min}(n, m)$, the SVM optimization step of LSA method is much faster than those of the other two methods. The time complexity of running PSI-BLAST is O (nN), where N is the size of the database. In the current situation, N is approximately equal to nl.

The time complexity of the method based on binary profiles is better than those based on other building blocks. The running times are dependent on the lengths of the representation vectors, that is, the total number of "words" in each building block. In this regard, there are 8000, more than 10 thousands and 3000 "words" for N-grams, patterns and motifs respectively. For binary profiles, there are only 1087 "words" at the optimal probability threshold of 0.13 (Table 2) and the total number of "words" drops quickly as the probability threshold becomes larger without significantly sacrificing the performance.

3.3 The Probability Threshold Has Not Significant Influence on Remote Homology Detection

The frequency profiles are calculated from the multiple sequence alignments outputted by PSI-BLAST [7] and converted into binary profiles with a probability threshold P_h. The total number of binary profiles is dependent on the size of the database and the value of probability threshold P_h. Since each combination of the twenty amino acids corresponds to a binary profile and vice versa, the total number of binary profiles is 2^20. In fact, only a small fraction of binary profiles appear. These binary profiles constitute novel building blocks of protein sequences. Since the probability threshold P_h is a parameter, it needs to be optimized. The results are shown in table 2. We surprisingly find that the probability threshold P_h has not significant influence on the performance of remote homology detection. A similar conclusion has been derived on the knowledge-based mean force potentials [30]. This results show that a small number of binary profiles contains rich information about evolution.

Table 2. The optimized results of probability threshold

Ph	Number of binary profiles	Mean ROC	Mean ROC50	Mean M-RFP
0.05	18978	0.8771	0.641	0.0764
0.07	10945	0.8890	0.650	0.0683
0.09	4798	0.8892	0.683	0.0794
0.11	2135	0.8931	0.682	0.0736
0.13	1087	0.9032	0.681	0.0682
0.15	573	0.9097	0.654	0.0483
0.17	318	0.8914	0.655	0.0841
0.19	212	0.8805	0.598	0.0932
0.21	175	0.8855	0.583	0.0920
0.23	128	0.8708	0.524	0.1105
0.25	75	0.8470	0.515	0.1315

Given in the table are the number of binary profiles and performances of the SVM-Bprofile method at different probability threshold P_h.

Acknowledgements. The authors would like to thank Xuan Liu for her comments on this work that significantly improve the presentation of the paper. Financial support was provided by the National Natural Science Foundation of China (60673019).

References

[1] Weston, J., Leslie, C., Zhou, D. and Noble, W. S.: Semi-supervised protein classification using cluster kernels. Journal. Cambridge, Mass (Year) 595-602.

[2] Darzentas, N., Rigoutsos, I. and Ouzounis, C. A.: Sensitive detection of sequence similarity using combinatorial pattern discovery: A challenging study of two distantly related protein families. Proteins. 61 (2005) 926-937.

[3] Li, L. and Noble, W. S.: Combining pairwise sequence similarity and support vector machines for detecting remote protein evolutionary and structural relationships. Journal of computational biology. 10 (2003) 857-868.

[4] Smith, T. F. and Waterman, M. S.: Identification of common molecular subsequences. J Mol Biol. 147 (1981) 195-197.

[5] Altschul, S. F., Gish, W., Miller, W., Myers, E. W. and Lipman, D. J.: Basic local alignment search tool. J. Mol. Biol. 215 (1990) 403-410.

[6] Pearson, W. R.: Rapid and sensitive sequence comparison with fastp and fasta. Methods Enzymol. 183 (1990) 63-98.

[7] Altschul, S. F., Madden, T. L., Schaffer, A. A., Zhang, J. H., Zhang, Z., Miller, W. and Lipman, D. J.: Gapped blast and psi-blast: A new generation of protein database search programs. Nucleic Acids Research. 25 (1997) 3389-3402.

[8] Karplus, K., Barrett, C. and Hughey, R.: Hidden markov models for detecting remote protein homologies. Bioinformatics. 14 (1998) 846-856.

[9] Qian, B. and Goldstein, R. A.: Performance of an iterated t-hmm for homology detection. Bioinformatics. 20 (2004) 2175-2180.

[10] Vapnik, V. N.: Statistical learning theory Wiley (1998).

[11] Jaakkola, T., Diekhans, M. and Haussler, D.: A discriminative framework for detecting remote protein homologies. J Comput Biol. 7 (2000) 95-114.

[12] Leslie, C., Eskin, E. and Noble, W. S.: The spectrum kernel: A string kernel for svm protein classification. Journal. (Year) 564-575.

[13] Leslie, C., Eskin, E., Cohen, A., Weston, J. and Noble, S. W.: Mismatch string kernels for discriminative protein classification. Bioinformatics. 20 (2004) 467-476.

[14] Hou, Y., Hsu, W., Lee, M. L. and Bystroff, C.: Efficient remote homology detection using local structure. Bioinformatics. 19 (2003) 2294-2301.

[15] Saigo, H., Vert, J. P., Ueda, N. and Akutsu, T.: Protein homology detection using string alignment kernels. Bioinformatics. 20 (2004) 1682-1689.

[16] Saigo, H., Vert, J. P., Akutsu, T. and Ueda, N.: Comparison of svm-based methods for remote homology detection. Genome Informatics. 13 (2002) 396-397.

[17] Dowd, S. E., Zaragoza, J., Rodriguez, J. R., Oliver, M. J. and Payton, P. R.: Windows.Net network distributed basic local alignment search toolkit (w.Nd-blast). BMC Bioinformatics. 6 (2005) 93.

[18] Dong, Q. W., Wang, X. L. and Lin, L.: Application of latent semantic analysis to protein remote homology detection. Bioinformatics. 22 (2006) 285-290.

[19] Bellegarda, J.: Exploiting latent semantic information in statistical language modeling. Proc. IEEE. 88 (2000) 1279-1296.

[20] Dong, Q. W., Lin, L., Wang, X. L. and Li, M. H.: A pattern-based svm for protein remote homology detection. Journal. 4 Guangzhou, China (Year) 3363-3368.

[21] Ben-Hur, A. and Brutlag, D.: Remote homology detection: A motif based approach. Bioinformatics. 19 Suppl 1 (2003) i26-33.

[22] Holm, L. and Sander, C.: Removing near-neighbour redundancy from large protein sequence collections. Bioinformatics. 14 (1998) 423-429.

[23] Henikoff, S. and Henikoff, J. G.: Position-based sequence weights. J Mol Biol. 243 (1994) 574-578.

[24] Landauer, T. K., Foltz, P. W. and Laham, D.: Introduction to latent semantic analysis. Discourse Processes. 25 (1998) 259-284.

[25] Andreeva, A., Howorth, D., Brenner, S. E., Hubbard, T. J. P., Chothia, C. and Murzin, A. G.: Scop database in 2004: Refinements integrate structure and sequence family data. Nucleic Acids Research. 32 (2004) D226-D229.

[26] Chandonia, J. M., Hon, G., Walker, N. S., Conte, L. L., Koehl, P., Levitt, M. and Brenner, S. E.: The astral compendium in 2004. Nucleic acids research. 32 (2004) 189-192.

[27] Gribskov, M. and Robinson, N. L.: Use of receiver operating characteristic(roc) analysis to evaluate sequence matching. Computers and Chemistry. 20 (1996) 25-33.

[28] Bailey, T. L. and Grundy, W. N.: Classifying proteins by family using the product of correlated p-values. Journal. (Year) 10-14.

[29] Krogh, A., Brown, M., Mian, I. S., Sjolander, K. and Haussler, D.: Hidden markov models in computational biology: Applications to protein modeling. Journal of Molecular Biology. 235 (1994) 1501--1531.

[30] Dong, Q. w., Wang, X. l. and Lin, L.: Novel knowledge-based mean force potential at the profile level. BMC Bioinformatics. 7: (2006) 324.

Biological Sequences Encoding for Supervised Classification

Rabie Saidi[1,3], Mondher Maddouri[2], and Engelbert Mephu Nguifo[1]

[1] CRIL - CNRS FRE 2499 Université d'Artois - IUT de Lens, France
{saidi,mephu}@cril.univ-artois.fr
[2] Computer Science Department, National Institute of Applied Sciences and Technologies,
Tunis-Carthage 2035, Tunisia
mondher.maddouri@fsegt.rnu.tn
[3] FSJEG, University of Jendouba, Tunisia
rabie.saidi@fsjegj.rnu.tn

Abstract. The classification of biological sequences is one of the significant challenges in bioinformatics as well for protein as for nucleic sequences. The presence of these data in huge masses, their ambiguity and especially the high costs of the *in vitro* analysis in terms of time and money, make the use of data mining rather a necessity than a rational choice. However, the data mining techniques, which often process data under the relational format, are confronted with the inappropriate format of the biological sequences. Hence, an inevitable step of pre-processing must be established. This work presents the biological sequences encoding as a preparation step before their classification. We present three existing encoding methods based on the motifs extraction. We also propose to improve one of these methods and we carry out a comparative study which takes into account, of course, the effect of each method on the classification accuracy but also the number of generated attributes and the CPU time.

1 Introduction

The emergence of the bioinformatics that we have witnessed during the last years finds its origin in the technological progress which has helped to conduct large scale research projects. The most remarkable one was the Human Genome Project (HGP) [10] accomplished in 13 years since 1990; a period that seems to be very short compared with the quantity of the collected data on the human genome: 3 billion bases which constitute the human DNA. Thus, several problems are open:

- How does the gene express its protein?
- Where does the gene start and where does it end?
- How do the protein families evolve and how to classify them?
- How to predict the three-dimensional structure of proteins?
- …

S. Hochreiter and R. Wagner (Eds.): BIRD 2007, LNBI 4414, pp. 224–238, 2007.
© Springer-Verlag Berlin Heidelberg 2007

The answer to these questions by the biochemical means and the *in vitro* analysis proves to be very expensive in terms of time and money. Indeed, some tasks, such as the determination of the protein three-dimensional structure, can extend over months and even years whereas the biological sequences quantity generated by the various programs of sequencing knows an exponential growth. Henceforth, the challenge is not the gathering of biological data but rather their exploration in a faster and efficient way making it possible to reveal the cell secrets.

In this context, the need to use the data mining is increasingly pressing. Data mining is particularly essential considering the enormous quantities of the biological data and their ambiguity. The use of mining tools was profitable in several fields also known by their large masses of data such as the commerce, the finance, the information retrieval... However, the data mining techniques, which often process data under the relational format, are confronted with the inappropriate format of the biological sequences. This makes it necessary to apply transformation on these data before their analysis. Our work is within the framework of the biological sequences preprocessing namely their encoding under a standard format appropriate to the analysis which is generally the relational format, frequently used by the data mining tools. We study and compare some existing encoding methods based on the motifs extraction. These methods are implemented in C language and gathered into a DLL enabling their comparison in terms of classification accuracy, number of generated attributes and CPU time.

Introduction to the problem and motivation to the biological sequences encoding can be found in section 2. In section 3, we give an overview on some encoding methods based on the motifs extraction. We propose to improve one of them in section 4. In section 5, we carry out the experimental study. Then we discuss our results in section 6. Section 7 concludes the paper and indicates some possible directions for future work.

2 Protein Sequences Classification by Data Mining Tools

Classification is one of the most significant problems open in bioinformatics. This problem arises, as well for proteins as for DNA. Indeed, the biologists are often interested to identify the family to which belongs a lately sequenced protein. This makes it possible to study the evolution of this protein and to discover its biological functions. For the DNA, the biologists seek, for instance, to classify parts of sequences in coding or non-coding zones [7]. They utilize biochemical means and *in vitro* analysis to perform these tasks which prove to be very expensive in terms of time and money while the biological data quantity is unceasingly growing.

In this context, the use of data mining techniques proves to be a rational response to this problem, since they were efficient in various fields and particularly in supervised classification. However, knowing that biological sequences are represented by strings of characters and that mining tools often process data under the relational format, it will not be possible to apply these tools directly to such data. Hence, the biological sequences have to be encoded into another format. [8] propose a model of a data mining process, illustrated by Fig. 1, to perform this task. The model presents the three main steps of the Knowledge Discovery in Data (KDD) process applied to the problem of the biological sequences classification. It consists of the extraction of a set of motifs from a set of sequences. These motifs will be used as attributes to construct a binary table that contains in row the set of the mentioned sequences. The presence or the absence of an attribute in a sequence is respectively noted by 1 or 0. This binary table is called context. It represents the pre-processing step result and the new sequences encoding format. It will be used as input for the processing step where a classifier will be applied to generate the classification rules. These rules are used to classify other sequences.

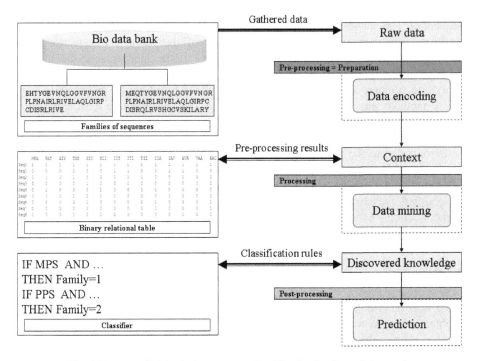

Fig. 1. Process of biological sequences classification by data mining tools

This paper is within the framework of the biological sequences pre-processing. We study how encoding methods affect the discovered knowledge by measuring the classification accuracy. We suppose that a good method will help to increase the accuracy even in hard classification problems.

In the next section we present and describe some encoding methods based on the motifs extraction.

3 Existing Encoding Methods

The nucleic and the protein sequences contain patterns or motifs which have been preserved throughout the evolution, because of their importance in terms of structure and/or function of the molecule. The discovery of these motifs can help to gather the biological sequences in structural or functional families but also to better understand the rules which control the evolution.

The members of a protein family are often characterized by more than one motif: on average each family preserves 3 to 4 regions [11]. The motifs often indicate functional and evolutionary relationships between proteins. Like for proteins, the motifs discovery can be used to determine the function of the nucleic sequences such as the identification of the promoter sites and the junction sites.

We present, hereafter, three methods of motifs extraction which are the methods of the N-Grams (NG), the Active motifs (AM) and the Discriminative Descriptors (DD). Then, we propose, in section 4, a modification of the Discriminative Descriptors method by the use of a substitution matrix (DDSM).

3.1 N-Grams

The simplest approach is that of the N-Grams, known also as *N-Words* or *length N fenestration* [6]. The motifs to be built are length fixed as a preliminary. The N-gram is a sub-sequence composed of N characters, extracted from a larger sequence. For a given sequence, the set of the N-grams which can be generated is obtained by moving a window of N characters on the whole sequence. This movement is carried out character by character. In each movement a sub-sequence of N characters is extracted. This process will be reiterated for all the analyzed sequences (Fig. 2 illustrates this principle). Then, only the distinct N-grams will be kept.

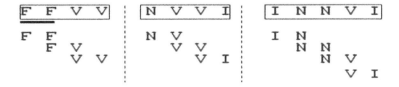

Fig. 2. Extraction of 2-grams from the 3 sequences: FFVV, NVVI and INNVI. For each sequence of length m, the number of extracted N-Grams is: m-N+1.

The N-Grams are widely used in information retrieval and natural languages processing [9]. They are also used in local alignment by several systems such as BLAST [1]. The N-Grams extraction can be done in a $O(m*n*N)$ time, where m is the maximum length of a sequence, n is the number of sequences in question and N is the motif length.

228 R. Saidi, M. Maddouri, and E.M. Nguifo

3.2 Active Motifs

This method is founded on the assumption that the significant regions are better preserved during the evolution and thus they appear more frequently than expected. Indeed, this method enables to extract the commonly occurring motifs whose lengths are longer than a specified length, called *Active Motifs*, in a set of biological sequences. The activity of a motif is the number of sequences which match it within an allowed number of mutations [12].

The motifs extraction is based on the construction of a Generalized Suffix Tree (GST). A GST is an extension of the suffix tree [4] and is dedicated to represent a set of *n* sequences indexed, each one, by $i = 1..n$. Each suffix of a sequence is represented by a leaf (in the shape of a rectangle) labelled by the index i of this sequence. It is composed by the concatenated sub-sequences labelled on the root-to-leaf i path. Each non-terminal node (in the shape of a circle) is labelled by the number of sequences to which belongs its corresponding sub-sequence composed by the concatenation of the sub-sequences labelled on the arcs which bind it to the root (Fig. 3). The candidate motifs are the prefixes of strings labelled on root-to-leaf paths which satisfy the length minimum. Then, only motifs having an acceptable activity will remain.

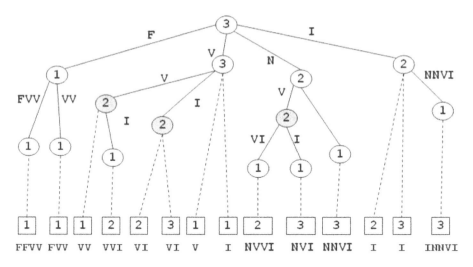

Fig. 3. GST illustration for the 3 sequences: FFVV, NVVI and INNVI. If we suppose that only exactly coinciding segments occurring in at least two sequences and having a minimum length of 2 are considered as active. Then we have 3 active motifs: VV, VI and NV.

There are several algorithms used for the construction of the GST. [12] affirm that the GST can be built in a $O(m*n)$ time, where m the maximum size of a sequence and n the number of sequences in question. To extract the motifs which satisfy the conditions of research, it is necessary to traverse the entire tree. That is to say a complexity of $O((m*n)^2)$.

3.3 Discriminative Descriptors

Given a set of n sequences, assigned to P families/classes F_1 F_2 .., F_P , it is a question of building sub-strings called *Discriminative Descriptors* DD which make it possible to discriminate a family F_i from other families, where $i = 1..P$ [8].

This method is based on an adaptation of the Karp, Miller and Rosenberg (KMR) algorithm [5]. This algorithm can identify the repeats in characters strings, trees or tables. It will be applied here to biological sequences. The extracted repeats are then filtered in order to only keep the discriminative and minimal ones.

The discriminative descriptors are built in a $O(m^2*n^3*N)$ time, where m is the maximum size of a sequence, n is the number of sequences in question and N is the maximum motif length.

Repeats Identification. KMR algorithm is based on the following concept of equivalence: two positions i and j in a character string S of length m are *k-equivalent*, we note $i\ E_k\ j$, if and only if the two sub-strings of length k $S[i; i+k-1]$ and $S[j; j+k-1]$ are identical [5]. We say also that the positions i and j belong to the same class of the equivalence relation E_k. An equivalence relation E_k, where $1<=k<=m$, can be represented by a vector $V_k[1.. m - k +1]$ whose each component $V_k[i]$, with $1<=i<=(m - k + 1)$, represents the class number to which belongs the position i for the equivalence relation E_k. To apply the KMR algorithm to several strings, they are concatenated in one sequence while keeping in memory the information about the positions of the various strings terminals. Fig. 4 shows an example of equivalence applied on 3 concatenated sequences: FFVV, NVVI and INNVI.

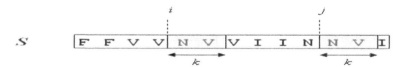

Fig. 4. Illustration of a 2-equivalence between i=5 and j=11. It is about the repeat NV which represents one of the equivalence relation E_2 classes.

Discriminative and Minimal Sub-strings Identification. A sub-string X is considered to be discriminative between the family F_i and of the other families F_j, where $i = 1..P, j = 1..P$ and $i \neq j$ if :

$$\frac{\text{number of sequences of Fi where X appears}}{\text{total number of sequences of Fi}} *100 \geq \alpha \cdot \tag{1}$$

$$\frac{\text{number of sequences of Fj where X appears}}{\text{total number of sequences of Fj}} *100 \leq \beta \cdot \tag{2}$$

A discriminative sub-string is considered minimal if it does not contain any other discriminative sub-string.

4   Proposed Method: Discriminative Descriptors with Substitution Matrix

In the case of protein, the motifs extracted by the Discriminative Descriptors method make it possible to discriminate between various families. But, this method neglects the fact that some amino acids have similar properties and can be thus substituted by each others without changing neither the structure nor the function of the protein [3]. So, we can find in the set of the generated attributes by the Discriminative Descriptors method several motifs which all derive from only one motif. In the same way, during the construction of the context (binary table), we are likely to lose information when we note by 0 the absence of a motif while another one, which can be substituted by it, already exists.

The similarity between the motifs is based, as already mentioned, on the similarity between the amino acids which constitute them. Indeed, there are various degrees of similarity between the amino acids. Since there are 20 amino acids, the mutations between them are scored by a 20x20 matrix called *substitution matrix*.

4.1   Substitution Matrix

In bioinformatics, a substitution matrix estimates the rate that each possible residue in a sequence changes to another residue over time. Substitution matrices are usually seen in the context of amino acid sequence alignment, where the similarity between sequences depends on the mutation rates as represented in the matrix[3].

4.2   Terminology

Let $\mathcal{M}$ be a set of n motifs, noted each one by $\mathcal{M}[p]$, $p = 1.. n$. $\mathcal{M}$ can be divided into m clusters. Each cluster contains a *main motif* M^* and, probably, other motifs which can be substituted by M^*. The *main motif* is the motif which has the highest probability, in its cluster, to mutate to another one. For a motif M of k amino acids, this probability, noted $P_m(M)$ is based on the probability P_i ($i = 1.. k$) that each amino acid $M[i]$ of the motif M does not mutate to any other amino acid. We have:

$$P_m = 1 - \prod_{i=1}^{k} P_i .\tag{3}$$

A P_i is calculated based on the substitution matrix according to the following formula:

$$P_i = S(M[i], M[i]) / \sum_{j=1}^{20} S^+(M[i], AA_j) ; j = 1.. k .\tag{4}$$

$S(x, y)$ is the substitution score of the amino acid y by the amino acid x as it appears in the substitution matrix. We mean by $S^+(x, y)$ a positive substitution score. AA_j is the amino acid of index j among the 20 amino acids.

We consider that a motif M substitutes a motif M' if:

- M and M' have the same length k,
- $S(M[i], M'[i]) >= 0$, $i = 1.. k$,
- $SP(M, M') >= T$, where T is a user-specified threshold: $0 <= T <= 1$.

We note by $SP(M, M')$ the substitution probability of the motif M' by the motif M having the same length k. It measures the possibility that M mutates to M':

$$SP(M, M') = S_m (M, M') / S_m (M, M) . \qquad (5)$$

$S_m (X, Y)$ is the substitution score of the motif Y by the motif X. It is computed according to the following formula:

$$S_m (X, Y) = \sum_{i=1}^{k} S(X[i], Y[i]) \qquad (6)$$

It is clear, according to the substitution matrix, that there is only one best motif which can substitute a motif $M;$ it is obviously itself, since the amino acids which constitute it are better substituted by themselves. This proves that the substitution probability of a motif by another one, if they satisfy the substitution conditions, will be between 0 and 1.

4.3 Methodology

The modification of the Discriminative Descriptors method relates to two aspects. First, the number of the extracted motifs will be, of course, reduced because we will keep only one motif for each cluster of substitutable motifs of the same length. Then, we will modify the context construction rule mentioned in section 2. Indeed, we will note by 1 the presence of the motif or that of one of its substitutes. The first aspect can be also divided into two phases: (1) identify the clusters main motifs and (2) perform the filtering.

Main Motifs Identification and Filtering. The main motif of a cluster is the most likely motif in this cluster to mutate to another one.

To identify all the main motifs, we sort $\mathcal{M}$ in a descending order by motifs lengths then by P_m. For each motif M' of $\mathcal{M}$, we look for the motif M which can substitute M' having the highest P_m. The clustering is based on the computing of the substitution probability between the motifs. We can find a motif which belongs to more than one cluster. In this case, it must be the main motif of one of them.

The filtering consists on keeping only the main motifs and removing all the other ones. The result is a smaller set of motifs which can represent the same information of the initial set.

The main motifs identification and the filtering are performed by the following simplified algorithm:

```
begin
    sort CM in a descending order by (motifs lengths, Pm);
```

```
for each motif CM[i] from i=n to 1
  if Pm(CM[i])=0 then
    CM[i] becomes a main motif;
  else
    x ← position of the first motif having the same
    length as CM[i];
    for each motif CM[j] from j=x to i
      if CM[j] substitutes CM[i] or j=i then
        CM[j] becomes a main motif;

        Break;
      end if
    end for
  end if
end for
for each motif M of CM
  if M is not a main motif then
    delete M;
  end if
end for
end.
```

The time complexity of this algorithm is $O((n^2/2)*k)$, where n is the number of motifs in question and k is the maximum motif length.

Example. Given a BLOSUM62 substitution matrix and the following set of motifs (Table 1) sorted by their lengths and P_m, we assign each motif to a cluster represented by its main motif. We get 5 clusters illustrated by the diagram shown in Fig. 5.

Table 1. Motifs clustering. The third row shows the cluster main motifs

CM	LLK	IMK	VMK	GGP	RI	RV	RF	RA	PP
P_m	0.89	0.87	0.86	0	0.75	0.72	0.72	0.5	0
Main motif	LLK	LLK	LLK	GGP	RI	RI	RI	RV	PP

Context building. The context building is done by noting 1 if a sequence matches a main motif or one of the motifs it can substitute; otherwise we note 0. We use the following algorithm:

```
begin
  for each sequence S
    for each motif M of length k
      repeat
        extract a k-gram M' from S;
        if M substitutes M' then
          note 1 in the context for S and M;
          goto presence;
        end if
```

```
         until the end of S
         note 0 in the context for S and M;
         presence: continue
      end for
   end for
end.
```

The time complexity of this algorithm is $O(m*n*k*l)$, where n is the number of motifs in question, m is the number of sequences, k is the maximum motif length and l is the maximum sequence length.

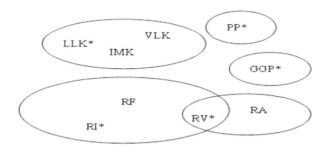

Fig. 5. Motifs clustering. The motif RV belongs to 2 clusters. It is the main motif of one of them.

5 Experiments and Results

The encodings methods are implemented in C language and gathered into a DLL. The accepted formats of the input files are: FASTA format for biological sequences files and the format described by Fig. 6 for the classification files. The DLL generates relational files under various formats such as the ARFF format used by the workbench WEKA [13] and the DAT format used by the system DisClass [7].

```
#classes number
2
#classes names
TLRH
TLRNH
#classes instances
TLRH
TLRNH
TLRNH
TLRNH
```

Fig. 6. Classification file sample. The file describs a FASTA file containing 4 biological sequences belonging to 2 classes: the 1st one belongs to the class «TLRH» and the 3 others ones belong to the class «TLRNH ».

5.1 Experimental Data

To compare the encoding methods we use 3 samples of biological sequences described by Table 2.

Table 2. Experimental data

Data type	Sample	Family/class	Number of sequences	Data source
Protein	Sample S1	High-potential Iron-Sulfur Protein	19	SWISS-PROT
		Hydrogenase Nickel Incorporation Protein HypA	20	
		Hlycine Dehydrogenase	21	
	Sample S2	human TLR	14	Entrepôt de l'université IRVINE
		Non-human TLR	26	
Nucleic	Sample S3	Promoter site	53	
		Non-promoter site	53	

The sample S1 contains three distinct and distant protein families. We suppose that classification in this case will be relatively easy since each family will probably have preserved patterns which are different from those of other families [11]. However, the sample S2 presents a more delicate classification problem. It consists of distinguishing between the human *Toll-like Receptors* (TLR) protein sequences and the non-human ones. The difficulty is due to the structural and functional similarity of the two groups. The sample S3 is the subject of a typical classification problem. It is a question of recognizing the nucleic sequences carrying promoter sites from those which are not. The promoters are short segments of DNA whose identification facilitates the localization of the genes beginnings.

5.2 Experimental Process

In our experiments, we use the 10-fold cross validation technique [2]. Each sample of data is randomly and equitably partitioned to 10 mutually exclusive subsets. The training and the test are carried out 10 times. In each iteration, a subset is reserved for the test and the others are used together for the training. After having built the contexts of training CA and test CT, we start the classification step. Using the classifier C4.5 of the workbench WEKA [13], we generate the classification rules from CA which we test on CT. The classification accuracy is computed as being the average of the 10 iterations accuracies. The encoding and experimental process is illustrated by Fig. 7.

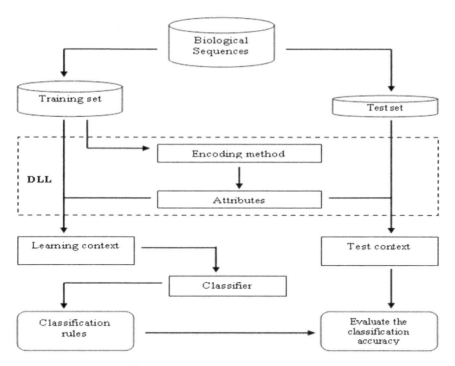

Fig. 7. Encoding and experimental process

5.3 Results

We examine, initially, each method individually while varying its various parameters to seek their optimal values (Table 3). Then, we use the best parameters found to compare them in terms of accuracy (rate of classified sequences), attributes number and CPU time (in seconds). For the DDSM method we use the BLOSUM62 and PAM250 substitution matrix. Results are shown in Table 4.

Table 3. Best parameters

Method	Parameter	Sample		
		S1	S2	S3
NG	N	3	3	4
AM	Min length	3	3	3
	Min activity	50 %	50 %	25 %
DD	Alpha	0	0	0
	Beta	0	0	0
DDSM	Alpha	0	0	-
	Beta	0	0	-
	Substitution matrix	BLOSUM62	BLOSUM62	-
	Threshold	0.7	0.9	-

Table 4. Experimental results

Sample	Method	NG	AM	DD	DDSM
S1	Accuracy	90 %	95 %	95 %	98.33 %
	Attributes	4777	1978	4709	2139
	Time	0.82	38	35	37
S2	Accuracy	60 %	55 %	67.5 %	77.5 %
	Attributes	5340	3458	6839	6562
	Time	1	91	921	954
S3	Accuracy	73.58 %	77.78 %	77.78 %	-
	Attributes	244	314	701	-
	Time	0.05	2	1.57	-

6 Discussion

According to the experimental study, we noticed that there are no optimal and single values for the parameters of the studied methods. In fact, these values depend on the nature of the data in question. So, the adjustment of these parameters requires a preliminary knowledge of the data characteristics such as the lengths of the preserved regions and the mutation rate between the sequences of a family.

The experimental results vary according to the input data. The sample S1 classification was relatively easy since the three protein families are completely distinct. Each one of them probably has its own motifs which characterize it and discriminate it from the others. This explains the high accuracy reached by all the methods especially the Discriminative Descriptors with Substitution Matrix method with which the classification reached a very high accuracy. The methods of the Active Motifs and the Discriminative Descriptors made the best classification accuracy for the sample S3 (this sample does not concern the DDSM method since it contains nucleic sequences). The sample S2 represents a hard classification challenge since the human TLR and the non-human TLR resemble to each other in terms of function and structure. Indeed the two classes share many similar parts what explains the low accuracy with the method of the Active Motifs. Indeed, this method, which extracts motifs based on their occurrences, built attributes which belong at the same time to the two classes, which increases the possibility of confusion. The method of the N-Grams made a better precision, but did not reach the default accepted accuracy which is 65% (if we assign all the sequences to the non-human TLR class). The Discriminative Descriptors method outperforms the two last methods. Since it adopts an approach of discrimination to build the attributes, it allowed a better distinction between the human TLR and the non-human TLR. But, to more improve classification in the sample S2, it is necessary to take into account the phenomenon of mutation and substitution between the amino acids. Indeed, the method DDSM which we proposed made it possible to reach the highest precision while reducing the number of generated attributes.

7   Conclusion

In this paper, we presented the biological sequences encoding as a pre-processing step before their classification. We described three existing encoding methods based on the motifs extraction, which are the methods of the N-Grams (NG), the Active Motifs (AM) and the Discriminative Descriptors (DD). Then, we proposed a modification of the DD method by the use of a substitution matrix.

In order to examine the effect of each encoding method on the classification accuracy, we undertook an experimental study which relates to various biological data comprising protein and nucleic sequences. We also compared the numbers of generated attributes by these methods and their CPU times. Among the existing methods, we noticed that the DD method presents the best accuracy. The modification of this method by the use of a substitution matrix made it possible to improve the classification even in a relatively delicate case. However, we noticed that the DD method like its alternative with the matrix of substitution, are very expensive in term of CPU time compared with the other methods, especially the NG.

Considering this work, several ways are open. It will be interesting to conceive a hybrid encoding method based on the N-Grams, which uses filters like α and β and takes into account the substitution and the order of the extracted motifs in the sequences. We will, also, try to use other amino acids substitution matrices and to extend the application of the modification, which we proposed to the DD method, with the nucleic sequences by using the DNA scores tables.

Acknowledgements

This work was partially supported by the Tunisian Government under a 4-month grant to the first author during his master research training at the Computer Science Research Centre (CRIL-CNRS) in Lens (France).

References

1. Altschul S. F., W. Gish, W. Miller, E. W. Myers, D. J. Lipman. Basic local alignment search tool. Journal of Molecular Biology, Vol. 215(3), pp. 403-413, 1990.
2. Han J., M. Kamber. Data Mining: Concepts and Techniques. ISBN 1-55860-489-8. Morgan Kaufmann Publishers: www.mkp.com, 2001
3. Henikoff S., J. G. Henikoff. *Amino acid substitution matrices from protein blocks*. National Academy of Sciences, USA, 89, pp. 10915-10919, 1992.
4. Hui L. C. K., M. Crochemore, Z. Galil, & U. Manber, (ed.). Combinatorial Pattern Matching. Lecture Notes in Computer Science in Apostolico, Springer-Verlag, 644, 230-243, 1992.
5. Karp R., R. E. Miller, A. L. Rosenberg. Rapid Identification of Repeated Patterns in Strings, Trees and Arrays. 4th Symposium of Theory of Computing, pp.125-136, 1972.
6. Leslie, C., E. Eskin, & W. S. Noble. The spectrum kernel : a string kernel for svm protein classification. Pac Symp Biocomput, 564–575, 2002.

7. Maddouri M, Elloumi M. A data mining approach based on machine learning techniques to classify biological sequences. Knowledge Based Systems, Vol: 15, Issue: 4, pp. 217-223, May 2002.
8. Maddouri M. & M. Elloumi. Encoding of primary structures of biological macromolecules within a data mining perspective. Journal of Computer Science and Technology (JCST); VOL 19, num 1. Allerton Press: 78-88. USA, 2004.
9. Miller E., Shen D., Liu J. & Nicholas C. Performance and scalability of a large-scale N-gram Based Information Retrieval System. Journal of digital information, 1999.
10. National Human Genome Research Institute. National Institute of Health. Available: http://www.nhgri.nih.gov/, June 2006.
11. Nevill-Manning, C. G., Wu, T. D., and Brutlag, D. L. (1998). Highly specfic protein sequence motifs for genome analysis. Proceedings of the National Academy of Sciences of the United States of America, 95(11):5865-5871, 1998.
12. Wang J. T. L., T. G. Marr, D. Shasha, B. A. Shapiro, & G.-W. Chirn. Discovering active motifs in sets of related protein sequences and using them for classification. Nucleic Acids Research, 22(14): 2769-2775, 1994.
13. Witten I. H. & Eibe F. Data Mining: Practical machine learning tools and techniques, 2nd Edition. Morgan Kaufmann, San Francisco, 2005.

Fast Search Algorithms for Position Specific Scoring Matrices[*]

Cinzia Pizzi, Pasi Rastas, and Esko Ukkonen

Department of Computer Science and
Helsinki Institute for Information Technology HIIT
P.O Box 68, FIN-00014 University of Helsinki, Finland
`Firstname.Lastname@cs.helsinki.fi`

Abstract. Fast search algorithms for finding good instances of patterns given as position specific scoring matrices are developed, and some empirical results on their performance on DNA sequences are reported. The algorithms basically generalize the Aho–Corasick, filtration, and superalphabet techniques of string matching to the scoring matrix search. As compared to the naive search, our algorithms can be faster by a factor which is proportional to the length of the pattern. In our experimental comparison of different algorithms the new algorithms were clearly faster than the naive method and also faster than the well-known lookahead scoring algorithm. The Aho–Corasick technique is the fastest for short patterns and high significance thresholds of the search. For longer patterns the filtration method is better while the superalphabet technique is the best for very long patterns and low significance levels. We also observed that the actual speed of all these algorithms is very sensitive to implementation details.

1 Introduction

Position specific scoring matrices [7,20,9] are a popular technique for specifying patterns or motifs such as transcription factor binding sites in DNA and in other biological sequences. So far, most of the effort has been devoted specifically to two aspects: how to define the scores of a matrix to better represent the instances of the modeled motif, and how to define a threshold to detect in the search only the true instances of the motif. These two aspects play a crucial role in the definition and effectiveness of the model itself.

Once the scoring matrices are made available (for example, by synthesizing from alignment blocks databases [12,2,10]), and appropriate thresholds have been fixed, one can search a sequence database for the matches of the matrices that score better than the threshold. This task, which often is in the core of many computational approaches to biological sequence analysis, may become a bottleneck of the computing performance (this is the case for example in the

---

[*] Supported by the Academy of Finland under grant 211496 (From Data to Knowledge) and by EU project Regulatory Genomics.

S. Hochreiter and R. Wagner (Eds.): BIRD 2007, LNBI 4414, pp. 239–250, 2007.

enhancer element prediction system EEL [8]). There has been recent interest in developing improved search algorithms for this search. Most of the widely used algorithms described in literature (e.g. [14,18,22]), use a straightforward trial and error search whose running time is proportional to mn where m is the length of the matrix and n is the length of the sequence database. Recently, a bunch of more advanced algorithms based on score properties [23,11], indexing data structures [3,5], fast Fourier transform [15] and data compression[6] have been proposed to reduce the expected time of computation.

Another possibility to develop improved search algorithms for scoring matrices has been ignored so far, namely generalizing the classic on–line algorithms (i.e., algorithms without database indexing) for exact and approximate key–word matching. This will be done in this paper.

The Aho–Corasick algorithm [1,4,13] is a very fast method for multiple key–word search. To generalize it for scoring matrices, we notice that given a matrix and threshold, we can explicitly generate the sequences that score above the threshold. We use these sequences as the key–word set in the Aho–Corasick algorithm. We develop an economical way of generating the set, based on the lookahead score of [23]. The resulting search algorithm scans the database in time proportional to the database length, independently of the length of the matrix. Unfortunately, the size of the key–word set may become so large that it cannot be accommodated in the fast storage of the computer, hence slowing down the practical performance of this algorithm whose search time is optimal.

Therefore we also propose another algorithm whose storage requirement is easy to control. The algorithm uses 'filtration' technique that is widely used in approximate pattern matching in strings (e.g. [21]). The idea is to evaluate at each database location an upper bound for the actual matching score. The actual score is evaluated only if the bound exceeds the threshold. This gives a fast algorithm if the upper bound is fast to evaluate and the bound is tight enough such that the checking of the actual score is not needed too often. The lookahead scoring search algorithm of [23] is based on filtering idea. We make it faster by using precomputed table of the scores of fixed–length prefixes of the pattern.

We included in the experiments also an algorithm that makes the search in steps that are larger than only one sequence symbol (for example, one IUPAC symbol used in encoding a nucleotide of a DNA sequence). Instead of the standard alphabet we use a 'superalphabet' whose symbols are q–tuples of original symbols. This would give a speed–up by factor q as compared to the naive algorithm. As the speed–up is independent of the threshold, this algorithm may become attractive for low threshold values as then Aho–Corasick and filtration algorithms get slower.

The paper is organized as follows. The position specific scoring matrix search problem is formalized in Section 2. In Section 3 we describe a linear time search algorithm that uses Aho–Corasick technique. Section 4 gives the filtration–based method whose storage requirement is easy to control. Section 5 sketches the superalphabet technique. In Section 6 we compare empirically the performance of the proposed methods on DNA sequences using scoring matrices for transcription factor binding sites from the JASPAR collection [16]. The new algorithms are clearly faster than the still commonly used naive method and also faster than the lookahead scoring algorithm. The Aho–Corasick technique is the fastest for short patterns and high significance thresholds of the search. As compared to the naive search or to the lookahead scoring algorithm, the Aho–Corasick search can be five to ten times faster. For longer patterns the filtration method is better while the superalphabet technique is the best for very long patterns and low significance levels.

2   The Weighted Pattern Search Problem

The problem we consider is to find good occurrences of a positionally weighted pattern in a sequence of symbols in some finite alphabet Σ. A *positionally weighted pattern* (a *pattern*, for short) is a real–valued $m \times |\Sigma|$ matrix $M = (M(i,j))$. In literature, these patterns are also called, e.g., position specific scoring matrices, profiles, and position weight matrices. The matrix gives the weights (scores) for each alphabet symbol occurrence at each position. We call m the *length* and Σ the *alphabet* of M.

Pattern M matches any sequence of length m in its alphabet. The *match score* of any such sequence $u = u_1 \ldots u_m$ with respect to M is defined as

$$W_M(u) = \sum_{i=1}^{m} M(i, u_i).$$ (1)

Let $S = s_1 s_2 \ldots s_n$ be an n symbol long sequence in alphabet Σ. Pattern M gives a match score for any m symbol long segment $s_i \ldots s_{i+m-1}$ (called m-*segment*) of S. We denote the match score of M at location i as

$$w_i = W_M(s_i \ldots s_{i+m-1}).$$ (2)

Given a real–valued *threshold* k, the *weighted pattern search problem* is to find all locations i of sequence S such that $w_i \geq k$.

A special case of this search problem is the exact pattern search problem in which one wants to find within sequence S the exact occurrences of a pattern $P = p_1 p_2 \ldots p_m$ where each p_i is in Σ. The weighted pattern search formalism gives the exact pattern search problem by choosing the weight matrix M as $M(i, p_i) = 1$, and $M(i, v) = 0$ if $v \neq p_i$. Then threshold k selects the locations of S where the symbols of P match the corresponding symbol of S in at least k positions. Choosing $k = m$ gives the locations of the exact occurrences of P in S.

In applications one should fix the threshold value k such that the search would give only relevant findings as accurately as possible. In the experiments (Section6) we will give k indirectly as a p-value or as a MSS value of [14].

The weighted pattern search problem can be solved for given S, M, and k by evaluating w_i from (1) and (2) for each $i = 1, 2, \ldots, n - m + 1$, and reporting locations i such that $w_i \geq k$. We call this the *naive algorithm* (NA). As evaluating each w_i from (1) takes time $O(m)$, the total search time of the naive algorithm becomes $O(mn)$, where m is the length of the pattern and n the length of the sequence S.

In the rest of the paper we develop faster algorithms.

3   Aho–Corasick Expansion Algorithm

Pattern M and threshold k determine the m symbols long sequences whose matching score is $\geq k$. We may explicitly generate all such sequences in a preprocessing phase before the actual search over S, and then use some multipattern search technique of exact string matching to find their occurrences fast from S. The Aho–Corasick multipattern search algorithm can be used here.

The preprocessing phase takes M and k, and generates the Aho–Corasick pattern matching automaton as follows. For each sequence $x \in \Sigma^m$, add x to the set D of the *qualifying sequences* if $W_M(x) \geq k$. In practice we avoid generating and checking the entire Σ^m by using the lookahead technique of [23] to limit the search. We build D by generating the prefixes of the sequences in Σ^m, and expanding a prefix that is still alive only if the expanded prefix is a prefix of some sequence that has score $\geq k$.

More technically, let $u_1 \ldots u_{j-1}$ be a prefix to be expanded by a symbol $u_j \in \Sigma$. We call

$$W_M^{pre}(u_1 \ldots u_{j-1} u_j) = \sum_{i=1}^{j} M(i, u_i) \tag{3}$$

the *prefix score* of $u_1 \ldots u_j$. The maximum possible score of a sequence with prefix $u_1 \ldots u_j$ is given by the *lookahead bound* L_j at j, defined as

$$L_j = \sum_{i=j+1}^{m} \max_{a \in \Sigma} M(i, a). \tag{4}$$

Now, if $W_M^{pre}(u_1 \ldots u_{j-1} u_j) + L_j < k$, we know for sure that no string with prefix $u_1 \ldots u_{j-1} u_j$ can have score $\geq k$ with respect to M, and there is no need to expand $u_1 \ldots u_{j-1}$ to $u_1 \ldots u_{j-1} u_j$. Otherwise there is a string with prefix $u_1 \ldots u_{j-1} u_j$ that has score $\geq k$, and the expansion by u_j is accepted. We organize this process in breadth–first order such that all still alive prefixes of length $j - 1$ are examined and possibly expanded by all $u_j \in \Sigma$ to length j. This is repeated with increasing j, until $j = m$.

For large k, only a small fraction of Σ^m will be included in D. In fact, the above procedure generates D in time $O(|D| + |M|)$ where the $O(|M|) = O(m|\Sigma|)$ time is needed for computing the lookahead bounds L_j from M.

The Aho–Corasick automaton $AC(D)$ for the sequences in D is then constructed [1,4], followed by a postprocessing phase to get the 'advanced' version of the automaton [13] in which the failure transitions have been eliminated. This version is the most efficient for small alphabets like that of DNA. The construction of this automaton needs time and space $O(|D||\Sigma|)$.

The search over S is then accomplished by scanning S with $AC(D)$. The scan will report the occurrences of the members of D in S.

We call this search algorithm the *Aho–Corasick expansion algorithm* (ACE algorithm). The total running time of ACE is $O(|D||\Sigma|+|M|) = O(|\Sigma|^{m+1}+|M|)$ for preprocessing M to obtain $AC(D)$, and $O(|S|)$ for scanning S with $AC(D)$.

4   Lookahead Filtration Algorithm

While the Aho–Corasick expansion algorithm of Section 3 scans S very fast, it may not always be robust enough for practical use. If the pattern length m is large and k is relatively small, set D of the qualifying sequences can become so large that $AC(D)$ cannot be accommodated in the available fast memory. This leads to a considerable slow down of the scanning phase because of secondary memory effects. Moreover, repeating the construction of the Aho–Corasick automaton for different k but the same M seems inefficient.

Therefore we next develop a filtration–based method that preprocesses M only once, and the resulting filtration automaton can be used for any threshold k. Also the size of the automaton can be selected to fit the available memory space, at the cost of weakened filtration capability. A down–side is that the scanning may become slower than with the Aho–Corasick automaton as the filtration method needs to check the findings of the filter to remove false positives. Depending on k, the checking time may even dominate in the total running time.

The general idea of filtration algorithms is to evaluate an *upper bound* for the score of each m–segment of S instead of the accurate score. Whenever the upper bound value is $\geq k$, we know that also the score itself may be $\geq k$. However, to eliminate false positives this must be checked separately as the score may also be $< k$. But if the upper bound is $< k$, we know for sure that the score itself must be less than k, checking is not needed, and we can continue immediately to the next segment. This scheme gives a fast search, provided that the upper bound can be evaluated fast and the bound is strict enough to keep the number of false positives small.

Our *Lookahead filtration algorithm* (LFA) fixes the details of the above scheme as follows. Matrix M is preprocessed to get a finite–state automaton F that reports for each h symbols long segment v of S the prefix score (3) of v. The automaton F has a state for each sequence $u \in \Sigma^h$. It is convenient to denote this state also by u. The prefix score $W_M^{pre}(u)$ of sequence u is associated with state u. The state transition function τ of F is defined for all u in Σ^h and a in Σ as $\tau(u,a) = v$ where $u = bu'$ for some b in Σ and u' in Σ^{h-1}, and $v = u'a$.

As the states and the transition function τ of F have a regular structure, they can be implemented as an implicit data structure without an explicit presentation of τ. The data structure is simply an array of size $|\Sigma|^h$, also denoted by F, whose entry $F(u)$ stores the prefix score of sequence u. Automaton F takes a state transition $\tau(u,a) = v$ by computing the new index v from u and a. This can be done by applying on u a shift operation followed by concatenation with a.

The filtration phase is done by scanning with F the sequence S to obtain, in time $O(n)$, the prefix score of every h symbols long segment of S. Let t_i be the score for such a segment that starts at s_i. (Note that $t_i = F(u)$ where u is the state of F after scanning $s_1 \ldots s_{i+h-1}$.) Now, the upper bound used in filtration is $t_i + L_h$ where L_h is the (also precomputed) lookahead bound (4) for M at h. If $t_i + L_h < k$, then the full occurrence of M at i must score less than k, and we can continue the search immediately, without evaluating the actual score, to the next location. If $t_i + L_h \geq k$, the full score must be evaluated to see if it really is $\geq k$. This is done by adding to t_i the scores of matching the remaining positions $h+1, \ldots, m$ of M against $s_{i+h}, \ldots, s_{i+m-1}$. At each such position, if the prefix score accumulated so far plus the lookahead bound (4) for the rest exceeds k, then this process can be interrupted.

The filtration automaton F and the lookahead bounds can be constructed in time $O(|\Sigma|^h + |M|)$. Scanning S takes time $O(n)$ plus the time for checking the score at locations picked up by the filter. Checking takes $O(m - h)$ time per such a location. Denoting by r the number of locations to be checked, the total checking time becomes $O(r(m-h)) = O(n(m-h))$. Filtration pays off only if r is (much) smaller than $n = |S|$. Increasing k or h would decrease r.

The same filtration automaton F can be used with any threshold k. Hence F needs to be constructed and stored only once. Moreover, the automaton can be extended to multipattern case, to handle several M simultaneously, as follows. Let $M_1, \ldots, M_K$ be the patterns we want to search. The filtration automaton stores in its entry $F(u)$ all the K prefix scores of u that are not smaller than the lookahead bound with respect to the corresponding $M_1, \ldots, M_K$. The above time bounds for initialization should be multiplied by K to get time bounds for the generalized algorithm. The algorithm scans S only once but the scanning becomes slower by a factor which is proportional to the average number of patterns whose prefix score is grater or equal to the lookahead bound. A similar multipattern extension of the ACE algorithm is possible, simply by merging the Aho–Corasick automata. This reduces the scanning time from $O(Kn)$ to $O(n+o)$, where $o \leq Kn$ is the number of occurrences of patterns in S.

To get an idea of what values of h and K may be feasible in practice, assume that $|\Sigma| = 4$. Then array F contains about 2.5×10^8 numbers if $h = 9$ and $K = 1000$ or if $h = 10$ and $K = 250$. Storing them takes about 1 gigabytes. This is a reasonable main memory size of a current PC. To get fastest possible performance, the array should fit into the CPU's cache memory that is typically smaller.

5   Speed–Up by Superalphabet

The naive algorithm NA can be made faster by working in a 'superalphabet' instead of the original one (which is typically small in bioinformatics applications). To define the superalphabet we fix the integer *width* q of the alphabet. Then each q–tuple of the original alphabet is a superalphabet symbol.

Matrix M is preprocessed to obtain equivalent scoring matrix $\hat{M}$ for superalphabet symbols: $\hat{M}$ is an $\lceil m/q \rceil \times |\Sigma|^q$ matrix whose entries are defined as

$$\hat{M}(j, a_1 \ldots a_q) = \sum_{h=1}^{q} M((j-1)q + h, a_h)$$

for $j = 1, \ldots, \lceil m/q \rceil$ and for all q–tuples $a_1 \ldots a_q \in |\Sigma|^q$.

The score of each m symbols long segment of S can now be evaluated on $O(m/q)$ steps using the shift–and–concatenate technique of the previous section to find fast the appropriate entries of $\hat{M}$. We call this method the Naive superalphabet algorithm (NS).

The search time using algorithm NS becomes $O(nm/q)$, giving a (theoretical) speed–up of algorithm NA by factor q, independently of the threshold k. In practice the larger overhead of algorithm NS makes the speed–up smaller. With some care in implementation, matrix $\hat{M}$ can be constructed in time $O(m|\Sigma|^q/q)$.

6   Experimental Comparison of Running Times

We compared the practical performance of the above algorithms NA (naive algorithm), ACE (Aho–Corasick expansion algorithm), LFA (lookahead filtration algorithm), and NS (naive superalphabet algorithm). We also included the lookahead scoring algorithm (LSA), an improved version of NA from [23] that precomputes the lookahead bounds (4) and uses them to stop the score evaluation at a location as soon as the total bound (the prefix score so far plus the corresponding lookahead bound) goes below k.

We collected 123 positionally weighted patterns for DNA from the JASPAR database [16]. The length of these patterns varied from 4 to 30, with average length 10.8.

To test the behaviour of the algorithms on very long patterns, we constructed a collection of 13 patterns, each having a length of 100, by concatenating the JASPAR patterns.

We searched the occurrences of these patterns from a DNA sequence of length 50 megabases. The sequence contained segments taken from the human and the mouse genome.

The threshold k for the experiments was given as a p-value as follows. The original count matrices from the JASPAR database were first transformed into log-odds scoring matrices using uniform background distribution and by adding a pseudo-count 0.1 to every original count. The score that corresponds to a given p-value was then computed using well-known pseudo-polynomial time dynamic programming algorithm [19,23]. As this method requires an integer-valued matrix, the log-odds scores were first multiplied by 100 and then rounded to integers.

For completeness, we also experimented using another way of giving the threshold, namely via the MSS value [14]. This is relative threshold defined as k/k_{max} where k_{max} is the largest possible score given to an m symbols long sequence by the entropy–normalized version of the original count matrix. For example, MSS = 0.90 means that for each matrix we used a threshold that was 0.9 times the maximum score of that matrix.

The algorithms were implemented in C and C++, and the reported times are for 3 GHz Intel Pentium processor with 2 gigabytes of main memory, running under Linux.

Table 1 gives the average running times of different algorithms for JASPAR patterns of length ≤ 15. For longer patterns algorithm ACE would often have required more space than was available (as the set of qualifying sequences becomes large). Therefore we omit ACE in Table 2 that gives the average running times for JASPAR patterns of length ≥ 16. A more detailed view of the run-time dependency on the pattern length is given in Figure 1.

According to Tables 1 and 2, algorithm ACE is the fastest for short patterns and relatively small p-values while for longer patterns algorithm LFA or, for larger p-values, algorithm NS is better. This suggests that a hybrid of the proposed algorithms that chooses the algorithm actually used on the basis of the values of m and p would give the best search time. A possible selection rule for such a hybrid is

if $m \leq \mu$ then run ACE else run LFA

where $\mu = \arg\max_h \{ p * |\Sigma|^h * h \leq 1\,000\,000 \}$. We call this hybrid algorithm LFACE. Its running time is shown in Figure 1 as well as in Table 3 that gives average running times of different algorithms for all JASPAR patterns. In algorithm LFA we used parameter value $h = 7$, and in algorithm NS we used $q = 7$ in the experiments reported in Tables 1, 2, and 3.

Table 1. Average running times per pattern (in seconds, preprocessing included) of different algorithms when searching for 108 JASPAR patterns of length $m \leq 15$, with varying p-values. The average search speed in megabases/s is given in parenthesis.

	$p = 10^{-5}$	$p = 10^{-4}$	$p = 10^{-3}$	$p = 10^{-2}$
NA	1.087(46.1)	1.099(45.6)	1.163(43.1)	1.573(31.8)
LSA	0.627(79.9)	0.815(61.5)	1.190(42.1)	1.845(27.2)
ACE	**0.080(623)**	**0.163(307)**	0.573(87.5)	2.433(20.6)
LFA	0.344(146)	0.375(133)	**0.492(102)**	**1.037(48.3)**
NS	0.649(77.1)	0.695(72.1)	0.740(67.7)	1.172(43.7)

Table 2. Average running times per pattern (in seconds, preprocessing included) of different algorithms when searching for 15 JASPAR patterns of length $m \geq 16$, with varying p-values. The average search speed in megabases/s is given in parenthesis.

	$p = 10^{-5}$	$p = 10^{-4}$	$p = 10^{-3}$	$p = 10^{-2}$
NA	1.987(25.2)	2.181(23.0)	2.145(23.4)	2.674(18.7)
LSA	1.538(32.6)	1.845(27.2)	2.309(21.7)	3.147(15.9)
LFA	**0.623(80.4)**	**0.965(51.9)**	1.313(38.2)	2.252(22.2)
NS	0.939(53.3)	1.079(46.4)	**1.123(44.6)**	**1.709(29.3)**

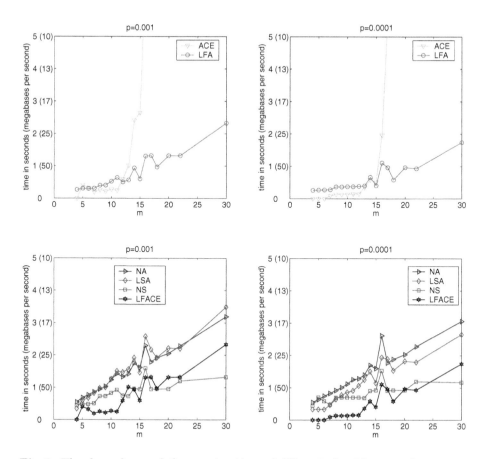

Fig. 1. The dependency of the running time of different algorithms on the pattern length. Average running times for patterns of each length $4, \ldots, 30$ in the JASPAR database are shown. The panels on the left-hand side are for $p = 0.001$ and the panels on the right-hand side for $p = 0.0001$.

Table 3. Average running times per pattern (in seconds, preprocessing included) of different algorithms when searching for all JASPAR patterns, with varying p-values. The average search speed in megabases/s is given in parenthesis.

	$p = 10^{-5}$	$p = 10^{-4}$	$p = 10^{-3}$	$p = 10^{-2}$
NA	1.197(41.8)	1.232(40.7)	1.283(39.0)	1.708(29.3)
LSA	0.739(57.8)	0.941(53.2)	1.327(37.8)	2.004(25.0)
LFA	0.379(132)	0.448(112)	0.593(84.5)	**1.185(42.3)**
NS	0.686(73.0)	0.742(67.5)	0.788(63.6)	1.239(40.4)
LFACE	**0.147(340)**	**0.258(194)**	**0.505(99.3)**	1.230(40.7)

Table 4. Average running times (in seconds, preprocessing included) of different algorithms for long patterns ($m = 100$) and varying p-values

	$p = 10^{-5}$	$p = 10^{-4}$	$p = 10^{-3}$	$p = 10^{-2}$
NA	10.234	10.244	10.434	11.080
LSA	11.835	12.675	13.335	15.118
LFA	9.855	10.347	11.096	12.965
NS	**3.576**	**3.677**	**4.593**	**9.918**

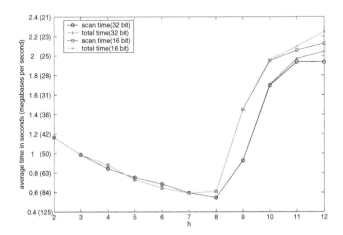

Fig. 2. Average running time of algorithm LFA for the 123 JASPAR patterns when h varies from 2 to 12 and $p = 0.001$. Note the strong increase of running time for $h > 8$. This is because the fast cache memory cannot store the filtration automaton for larger h. The upper curve is for an implementation of F of algorithm LFA in 32-bit integers, and the lower curve is for 16-bit integers.

Table 5. Average running times per pattern (in seconds, preprocessing included) of different algorithms for varying MSS values. The average search speed in megabases/s is given in parenthesis.

	MSS=1.0	MSS=0.95	MSS=0.90	MSS=0.85
NA	1.215(41.2)	1.251(40.0)	1.335(37.5)	1.361(36.8)
LSA	0.568(88.3)	0.863(58.0)	0.965(51.9)	1.103(45.4)
ACE	**0.117(428)**	**0.250(201)**	0.462(108)	0.680(73.7)
LFA	0.366(137)	0.407(123)	**0.461(109)**	**0.504(99.4)**
NS	0.725(69.1)	0.789(63.5)	0.807(62.1)	0.822(60.9)

For the long patterns with $m = 100$, algorithm NS is the best as reported in Table 4. We used $q = 6$ in algorithm NS.

In algorithm LFA, increasing parameter h should improve the filtration performance and make the algorithm faster in principle. Unfortunately it also increases storage requirement, and hence slower memory must be used. This easily

Table 6. Average running times (in seconds, preprocessing included) of different algorithms for the long patterns ($m = 100$) and varying MSS values

	MSS=1.0	MSS=0.95	MSS=0.90	MSS=0.85
NA	9.830	9.894	9.818	10.007
LSA	0.538	2.354	4.072	5.018
LFA	**0.367**	**1.497**	**2.859**	3.788
NS	3.156	3.183	3.164	**3.125**

cancels the speed–up achieved by better filtration. We studied this more carefully by measuring the time of LFA for varying h. Figure 2 shows the result for the JASPAR patterns. The average time first improves as expected, but gets rapidly much worse for $h > 8$. We tried with implementations using 16-bit and 32-bit integers for the filtration automaton F, the 16-bit version being slightly faster as expected.

Finally, Tables 5 and 6 give some average running times for all JASPAR patterns when the threshold is given as the MSS value [14]. Expectedly, the relative speeds of different algorithms show similar pattern as before.

7   Conclusion

Three new algorithms were proposed for searching patterns that are given as position specific scoring matrices. The algorithms give clear improvements in speed as compared to the naive search and its improved variant that uses lookahead scoring. This requires, however, that the implementation is carefully tuned. Moreover, the best alternative among our proposed algorithms depends on the length of the pattern and on the threshold value used in the search.

References

1. Aho, A. V. and Corasick, M.: Efficient string matching: An aid to bibliographic search, *Comm. ACM* 18 (1975), 333–340.
2. Attwood T.K., and Beck, M.E., PRINTS - A Protein Motif Finger-print Database, *Protein Engineering*, 1994; 7(7):841–848.
3. Beckstette M., Homann R., Giegerich R., and Kurtz S., Fast index based algorithms and software for matching position specific scoring matrices, *BMC Bioinformatics*, 2006; 7:389.
4. Crochemore, M. and Rytter, W., *Text Algorithms*, Oxford University Press 1994.
5. Dorohonceanu B., and Neville-Manning C.G., Accelerating Protein Classification Using Suffix Trees, in *Proc. of the 8th International Conference on Intelligent Systems for Molecular Biology (ISMB)*, 2000; 128–133.
6. Freschi V., and Bogliolo A., Using Sequence Compression to Speedup Probabilistic Profile Matching, *Bioinformatics*, 2005; 21(10):2225–2229.
7. Gribskov M., McLachlan A.D., and Eisenberg D., Profile Analysis: Detection of Distantly related Proteins, *Proc. Natl. Acad. Sci.*, 1987; 84(13):4355–8.

8. O. Hallikas, K. Palin, N. Sinjushina, R. Rautiainen, J. Partanen, E. Ukkonen & J. Taipale: Genome-wide prediction of mammalian enhancers based on analysis of transcription-factor binding affinity. *Cell* 124 (January 13, 2006), 47–59.
9. Henikoff S., Wallace J.C., Brown J.P., Finding protein similarities with nucleotide sequence databases, *Methods Enzymol.*, 1990;183:111–132.
10. Henikoff J.G, Greene E.A., Pietrokovski S., and Henikoff S., Increased Coverage of Protein Families with the Blocks Database Servers, *Nucleic Acids Research*, 2000; 28(1): 228–230.
11. Liefhooghe, A., Touzet, H. and Varre, J.: Large Scale Matching for Position Weight Matrices. In: *Proc CPM 2006*, LNCS 4006, pp. 401–412, Springer-Verlag 2006.
12. Matys V., Fricke E., Geffers R., Gossling E., Haubrock M., Hehl R., Hornischer K., Karas D., Kel A.E., Kel-Margoulis O.V., Kloos D.U., Land S., Lewicki-Potapov B., Michael H., Munch R., Reuter I., Rotert S., Saxel H., Scheer M., Thiele S., and Wingender E., TRANSFAC: Transcriptional Regulation, from Patterns to Profiles, *Nucleic Acids Research*, 2003; 31(1):374–378.
13. Navarro, G. and Raffinot, M., *Flexible Pattern Matching in Strings*, Cambridge University Press 2002.
14. Quandt K., Frech K., Karas H., Wingender E., and Werner T., MatInd and MatInspector: New Fast and Versatile Tools for Detection of Consensus Matches in Nucleotide Sequences Data, *Nucleic Acid Research*, 1995; 23(23):4878–4884.
15. Rajasekaran S., Jin X., and Spouge J.L., The Efficient Computation of Position-Specific Match Scores with the Fast Fourier Transform, *Journal of Computational Biology*, 2002; 9(1):23–33.
16. Sandelin, A., Alkema, W., Engstrom, P., Wasserman, W.W. and Lanhard, B., JASPAR: an open-access database for eukaryotic transcription factor binding profiles, *Nucleic Acids Research* 32 (2004), D91–D94.
17. Schneider T.D., Stormo G.D., Gold L. and Ehrenfeucht A., Information Content of Binding Sites on Nucleotide Sequences, *Journal of Molecular Biology*, 1986; 188:415–431.
18. Scordis P., Flower D.R., and Attwood T., FingerPRINTScan: Intelligent Searching of the PRINTS Motif Database, *Bioinformatics*, 1999; 15(10):799–806.
19. Staden R., Methods for calculating the probabilities of finding patterns in sequences, *CABIOS*, 1989; 5(2):89–96.
20. Stormo G.D., Schneider T.D., Gold L.M., and Ehrenfeucht A., Use of the 'Perceptron' Algorithm to Distinguish Translational Initiation Sites in E.coli, *Nucleic Acid Research*, 1982; 10:2997–3012.
21. Ukkonen, E., Approximate string-matching with q-grams and maximal matches. *Theoretical Computer Science* 92 (1992), 191-211.
22. Wallace J.C., and Henikoff S., PATMAT: a Searching and Extraction Program for Sequence, Pattern and Block Queries and Databases, *CABIOS*, 1992; 8(3):249–254.
23. Wu T.D., Neville-Manning C.G., and Brutlag D.L., Fast Probabilistic Analysis of Sequence Function using Scoring Matrices, *Bioinformatics*, 2000; 16(3):233–244.

A Markovian Approach for the Segmentation of Chimpanzee Genome

Christelle Melodelima and Christian Gautier

UMR 5558 CNRS Biométrie et Biologie Evolutive, Université Claude Bernard Lyon 1, 43 boulevard du 11 Novembre 1818, 69622 Villeurbanne Cedex - France and PRABI (Rhône Alpes Bioinformatics Center)
{melo,cgautier}@biomserv.univ-lyon1.fr

Abstract. Hidden Markov models (HMMs) are effective tools to detect series of statistically homogeneous structures, but they are not well suited to analyse complex structures such as DNA sequences. Numerous methodological difficulties are encountered when using HMMs to model non geometric distribution such as exons length, to segregate genes from transposons or retroviruses, or to determine the isochore classes of genes. The aim of this paper is to suggest new tools for the exploration of genome data. We show that HMMs can be used to analyse complex gene structures with bell-shaped length distribution by introducing macrostates. Our HMMs methods take into account many biological properties and were developped to model the isochore organisation of the chimpanzee genome which is considered as a fondamental level of genome organisation. A clear isochore structure in the chimpanzee genome, correlated with the gene density and guanine-cytosine content, has been identified.

Keywords: Hidden Markov model, DNA sequence, isochore modelling.

1 Introduction

The chimpanzee is an excellent model organism in biomedical research due to the similarities between many of its physiological processes and those of human. The availability of the chimpanzee genome sequence has already largely influenced the research in many fields, and more profound impact is certainly to follow. The sequencing of the complete chimpanzee genome led to the knowledge of a sequence of 4.4 billion pairs of nucleotides. Such amounts of data make it impossible to analyse patterns or to provide a biological interpretation analysis unless one relies on automatic data-processing methods. For twenty years, mathematical and computational models have been widely developed in this setting. Numerous methodological efforts have been devoted to multicellular eukaryotes since a large proportion of their genome has no known function. For example, only 1 to 3% of the chimpanzee genome is known to code for proteins. Another difficulty is that the statistical characteristics of the coding region vary dramatically from one specie to the other, and even from one region in a given genome

S. Hochreiter and R. Wagner (Eds.): BIRD 2007, LNBI 4414, pp. 251–262, 2007.

to the other. For example, vertebrate isochores [1], [2] exhibit such a variability in relation to their guanine-cytosine $(G + C)$ frequencies. Thus it is necessary to use different models for different regions if one seeks to detect patterns in genomes.

One way of modelling genomes uses hidden Markov Models (HMMs) [3], [4], [5]. To each type of genomic region (exons, introns, etc.), one associates a state of the hidden process, and the distribution of the stay in a given state, that is, of the length of a region, is geometric. While this is indeed an acceptable constraint as far as intergenic regions and introns are concerned, the empirical distributions of the lengths of exons are clearly bell-shaped [6], [7], [8], hence they cannot be represented by geometrical distributions. Semi-Markov models are one option to overcome this problem [6]. Although these models are widely used, they are very versatile, since they allow to adjust the distribution of the duration of the stay in a given state directly to the empirical distribution. The trade off is a strong increase in the complexity of most algorithms implied by the estimation and the use of these models. For example, the complexities of the main algorithms (forward-backward and Viterbi) are quadratic in the worst case with respect to the length of the sequence for hidden semi-Markov chains and linear for HMMs [6], [9], [10]. This may limit their range of application as far as the analysis of sequences with long homogeneous regions is concerned. Another difficulty is the multiplication of the number of parameters that are needed to describe the empirical distributions of the durations of the states, and which must be estimated, in addition to usual HMM parameters [9]. Thus the estimation problem is more difficult for these variable duration HMMs than for standard HMMs [9]. In other words, semi-Markov models are efficient tools to detect protein coding genes, but they are much more complex than HMMs.

In this paper, HMMs were used to detect isochores which were originally identified as a result of gradient density analysis of fragmented genomes [11]. Mammalian genomes are a mosaic of regions (DNA segments on average more than 300 kb in length) with differing, homogeneous $G + C$ contents. High, Medium and Low-density genomic segments are known as H, M and L isochores in order of decreasing $G + C$ content respectively. The isochore has been classified as a "fundamental level of genome organisation" [12] and this concept has increased our appreciation of the complexity and variability of the composition of eukaryotic genomes [13]. Existing isochore prediction methods only use the overall base composition of the DNA sequence ([14], [15], [16], [17], [18]). The aim of this paper is to suggests a new approach using HMMs and allowing to take into account many biological properties, such as $G + C$ content, gene density, length of the different regions, the reading frame of exons. We suggest to use HMM for modelling the exon length distribution by sum of geometric laws. To do this a state representing a region is replaced by a juxtaposition of states with the same emission probabilities. This juxtaposition of states is called macro-states. Macro-state HMMs models were used for complete genome analysis. Therefore, a method based on a hidden Markov model, which makes it possible to detect the isochore structure has been developped and tested on the chimpanzee genome.

2 Materials

Gene sequences were extracted from Ensembl for the chimpanzee genome. This procedure yielded a set of 22524 genes. The statistical characteristics of the coding and noncoding regions of vertebrates differ dramatically between the different isochore classes [13]. Many important biological properties have been associated with the isochore structure of genomes. In particular, the density of genes has been shown to be higher in H than in L isochores [20]. Genes in H isochores are more compact, with a smaller proportion of intronic sequences, and they code for shorter proteins than the genes in L isochores [16]. The amino-acid content of proteins is also constrained by the isochore class: amino acids encoded by $G+C$ rich codons (alanine, arginine.) being more frequent in H isochores [21] and [22]. Moreover, the insertion process of repeated elements depends on the isochore regions. SINE (short-interspersed nuclear element) sequences, and particularly Alu sequences, tend to be found in H isochores, whereas LINE (long-interspersed nuclear element) sequences are preferentially found in L isochores [23]. Thus, we took into account the isochore organisation of the chimpanzee genome. Three classes were defined and based on the $G + C$ frequencies at the third codon position $(G + C_3)$. The limits were set so that the three classes contained approximately the same number of genes. This yielded classes H=[100%, 70%], M=]58%,70%[and L=[0%,58%], which were used to build a training set. These classes were the same compared with those used by other authors [20], [24] in the human genome. Each class H, L and M, was randomly divided into two equal parts, a training set and a test set. The training sets were used to model the length distributions of the exons and the introns, and to analyse the structure of genes. To test the model, data on all chimpanzee chromosomes were retrieved from ENSEMBL.

3 Method

3.1 Estimation of the HMM Parameters

Estimation of Emission Probabilities
The DNA sequence consists of a succession of different regions, such as gene and intergenic regions. A gene is a succession of coding (exon) and non-coding (intron) region. In this study, HMMs are used to discriminate between these different types of regions. Exons consist of a succession of codons, and each of the three possible positions in a codon (0, 1, 2) has specific statistical properties. Thus, exons were divided into three states [25], [26].

HMMs take into account the dependency between a base and its n preceding neighbours (n defined the order of the model). For our study, n was taken to be equal to 5, as in the studies of Borodovsky [24] and Burge [25]. The emission probabilities of the HMM were therefore estimated from the frequencies of 6-letter words in the different regions (intron, initial exon, internal exons and terminal exon) that made up the training set.

Estimation of the Structure of the Macro-States

We suggest to use sums of a variable number of geometric laws with equal or different parameters in order to model the bell-shaped empirical length distributions of the exons. Thus a "biological state" is represented by a HMM and not by a single Markov state. The emission of probabilities of every state in this HMM are the same. A key property of this macro-state approach is that the conditional independence assumptions within the process are preserved with respect to HMMs. Hence, the HMM algorithms used to estimate the parameters and compute the most likely state sequences still apply [10].

The length distribution of the exons and introns was estimated from the training set (data set sequences are named $x_1...x_n$). Each x_i was considered to be the realization of an independent variable of a given law. We have tested the following laws:

1. the sum of m geometric laws of same parameter Θ (i.e. a binomial negative law):
$$P[X = k] = C_{k-1}^{m-1} \times \Theta^m \times (1 - \Theta)^{k-m} , \tag{1}$$

2. the sum of two geometric laws with different parameters $\Theta_1 > \Theta_2$:
$$P[X = k] = \Theta_1 \times \Theta_2 \frac{(1 - \Theta_2)^{k-1} - (1 - \Theta_1)^{k-1}}{\Theta_1 - \Theta_2,} , \tag{2}$$

3. the sum of three geometric laws with different parameters $\Theta_1 < \Theta_2 < \Theta_3$:
$$P[X = k] = \frac{\Theta_1 \times \Theta_2 \times \Theta_3}{\Theta_2 - \Theta_3} \times$$
$$\left\{ \frac{(1 - \Theta_1)^{k-1} - (1 - \Theta_3)^{k-1}}{\Theta_3 - \Theta_1} - \frac{(1 - \Theta_2)^{k-1} - (1 - \Theta_3)^{k-1}}{\Theta_3 - \Theta_2} \right\} . \tag{3}$$

We define $G_n(D_1, ..., D_n)$ as the distribution of the sum of n random variables of geometric distributions, each with expectation D_i and parameter $\Theta_i = 1/D_i$. Thus the expectation of $G_n(D_1, ..., D_n)$ is $D_1 + ... + D_n$. When $D_i = D$ for every i, this is called a negative binomial distribution with parameters $(n, 1/D)$, which we denote $G_n(1/\Theta)$. Finally $G_n(D)$ is a geometric distribution with expectation D and parameter $\Theta = 1/D$, which we write $G(D)$.

To estimate the parameters of the different laws, we minimised the Kolmogorov Smirnov distance for each law. The law which fits best with the empirical distribution is the law with the smallest Kolmogorov-Smirnov distance.
$$D_{KS} = sup_x|F(x) - G(x)| , \tag{4}$$

where D_{KS} is the Kolmogrorov-Smirnov distance, F is the theorical density distribution, G is the empirical density distribution. However, the classical Newton or gradient algorithm cannot minimise the Kolmogorov-Smirnov distance, since this distance cannot be differentiable. For this reason, we have discretised the parameter space with a step of 10^{-5}. Parameters estimations were not based on the maximum likelihood, which would have matched the end of the exon length

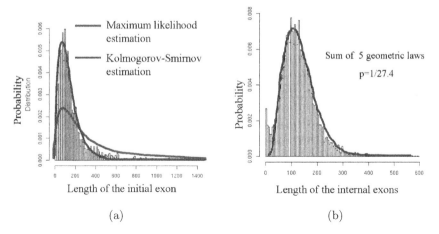

Fig. 1. (a) The histogram shows the empirical distribution of the length of the initial exons in a multi-exons gene. The blue curve shows the theoretical distribution obtained from the Kolmogorov-Smirnov distance. The red curve characterises the binomial distribution, obtained by the maximum likelihood method. (b) The histogram shows the empirical distribution of the length of the internal exons. The blue curve shows the theoretical distribution obtained from the Kolmogorov-Smirnov distance.

distribution while neglecting many small exons (Figure 1a). The definition of the maximum likelihood method is as follow: let x be a discrete variable with probability $P[x|\Theta_1...\Theta_k]$ (where $\Theta_1...\Theta_k$ are k unknown constant parameters which need to be estimated) obtained by an experiment which result in N independant observations, $x_1, ..., x_N$. Then the likelihood function is given by:

$$L(x_1, ..., x_N|\Theta_1...\Theta_k) = \prod_{i=1...N} P[x_i|\Theta_1...\Theta_k] . \qquad (5)$$

The logarithm function is:

$$\wedge = ln(L(x_1, ..., x_N|\Theta_1...\Theta_k)) = \sum_{i=1...N} P[x_i|\Theta_1...\Theta_k] . \qquad (6)$$

The maximum likelihood estimators $\Theta_1...\Theta_k$ are obtained by maximizing L or $\wedge$. Indeed, for a geometrical law or a convolution of geometrical laws, the parameter Θ is estimated by the reverse of the mean ($E[X] = 1/\Theta$) using the maximum likelihood method. The extreme values thus tend to stretch the distribution towards the large ones. We therefore have preferred to use the Kolmogorov-Smirnov distance in order to obtain a better modelling of the chimpanzee gene. Moreover, in order to provide simple but efficient models, equal transitions between states of a macro-state were used when it was possible.

Thus, a region is represented by a hidden state of the HMM. If the length distribution of a region is fitted by a sum of geometric laws, the state representing the region is replaced by a juxtaposition of states with the same emission probabilities, thus leading to macros-states (Figure 2). The state duration is

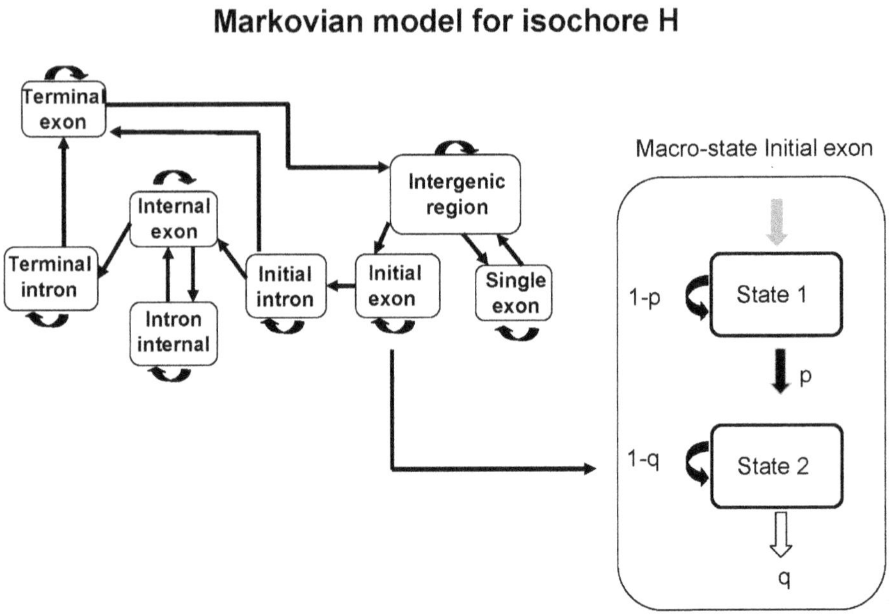

Markovian model for isochore H

Fig. 2. An HMM is composed of states corresponding to different regions within the genome (exons, introns). Each state emits DNA nucleotides (A, C, G and T) with specific emission probabilities. Each exon state is a macro-state. For example, the macro-state initial exon is composed of two smaller macro-states modelling the distribution of the length $G_2(1/p, 1/q)$ of initial exons in H isochore. Black arrows show the transition inside the macro-state. Grey and white arrows show respectively the different input and output of the macro-state.

characterised by the parameters of the sum of these geometric laws. Various studies [6], [27] have shown that the length distribution of the exons depend on their position in the gene. All exon types were taken into account: initial coding exons, internal exons, terminal exons and single-exon genes.

3.2   Modelling of Isochores Organisation

To characterize the three isochore regions (H, L and M) along the chimpanzee genome, three HMM models (H, L and M) were adjusted using the training sets, and then compared on all chimpanzee chromosomes. We divided the DNA of each chimpanzee chromosome into window of 100-kb. Two successive window overlapped by half their length. For each window and for each model (H, L and M), the probability $P[m|S]$ was computed as follows:

$$P(m \mid S) = \frac{P(S \mid m)P(m)}{\sum_{m' \in \{H,M,L\}} P(S \mid m')P(m')}, \qquad (7)$$

m is H, L or M, and S the window that is being tested, $P(S|m)$ is computed by the forward algorithm, and $P(m)$ is estimated by the frequency of genes according to their $G + C_3$ content (due to our definition of $G + C_3$ limits we get $P(H) \approx P(M) \approx P(L) \approx 1/3$, and so our Bayesian approach is numerically very close to a maximum likelihood approach). The computations were realized with the package SARMENT [28]. To be consistent with the preceding definition, we assumed an isochore to be a region consisting of at least 5 consecutive windows associated by the method with the same isochore class. This ensured that all estimated isochore lengths were greater than 300 kb, but meant that some windows can be unassociated to an isochore class.

Several tests were performed in order to check the coherence of isochore prediction: (i) the distribution of isochores was plotted with the distribution of the gene density, and the GC content along the chromosome, (ii) the segmentation allows to define the isochore class of each window along the chimpanzee genome. The isochore repartition of these windows has been compared with a random repartition of these windows. One thousand simulations have been realized. (iii) Furthermore, the ratio of coding regions has been compared between the isochores H and L predicted by our method.

4   Results

4.1   Estimations of the HMMs Parameters

Sums of geometric laws with equal or different parameters were used in order to model the bell-shaped empirical length distributions of exons (Figure 2). The length of an exon depends on its position within the gene. Initial and terminal exons tend to be longer than internal exons (Table 1). The length of introns displays also a noticeable positional variability. The distributions of the lengths of internal and terminal introns are relatively similar. However, internal and terminal introns are both smaller compared with initial introns (Table 1). The lengths of introns depend on their $G + C$ content. Table 1 shows that the $G + C$ frequency at the third codon position is negatively correlated with the length of the introns, i.e., high frequencies correspond to short introns, and vice versa. The length of the exons displays clearly a bell-shaped pattern (Figure 1b), for the three $G + C$ classes. The minimisation of the Kolmogorov-Smirnov distance yields a good fit with the empirical distribution of the length of the exons (Figure 1 and Table 2). Therefore, the Kolmogorov-Smirnov distance was chosen to model their length distribution by sum of geometric laws and to estimate the parameters of these laws (see Method for a comparison with the maximum likelihood approach).

We show here only the results for the modelling of the distributions of the lengths in the H class. However, the distributions of the lengths in the classes M and L were modelled by sums of geometric laws. The estimated distributions are $G_2(52.6, 106.4)$ for initial exons (Figure 1), $G_2(58.8, 108.7)$ for terminal exons, $G_5(27.4)$ for internal exons, $G_3(415.2)$ for intronless genes. The geometric distribution for initial introns was $G(1923.1)$. Other types of introns were also modelled by a geometrical distribution.

Table 1. Length of the exons and of the introns according to their position in the gene and according to the $G + C$ frequency at third codon position in the gene

Position in the gene	Length (bp) in class H		Length (bp) in class M		Length (bp) in class L	
	mean	median	mean	median	mean	meadian
initial coding exon	184	123	167	114	162	112
internal exon	140	101	138	108	139	107
terminal exon	193	138	201	130	202	127
initial intron	2746	2318	3864	3274	4474	4227
internal intron	1192	1045	1446	1437	2841	2322
terminal intron	1247	1012	1534	1479	2691	2136

Table 2. Parameters estimation of different laws obtained for initial exons of class H minimising the Kolmogorov-Smirnov distance

Laws	Parameters p	K-S distance
$G_2(\Theta)$	0.0126	0.05761
$G_3(\Theta)$	0.0197	0.08782
$G_4(\Theta)$	0.0226	0.11243
$G(\Theta_1, \Theta_2)$	0.019 - 0.0094	0.05023

4.2   Modelling of Isochore Organisation

The quality of discrimination between isochore classes for each windows was measured by $max_{H,L,M}(P(m/S))$. For each of the 59075 windows in the chimpanzee genome the maximum value was greater than 0.75, leading to a very clear association between each window and a unique isochore class. A second important criterion was the isochore length, since the method imposes a minimum length of 300 kb, resulting in some unaffected windows. In the chimpanzee genome, unallocated windows represent only 4.75% of the total number of windows. These windows were not considered to constitute an isochore. Along the chimpanzee genome, the distributions of these unallocated windows was random. Figure 3 shows the chimpanzee genome segmentation obtained by the method described in this paper. Figure 3 is available online at http://melodelima.chez-alice.fr/chimpanzee_isochores/chimpanzee_isochore.html.

All the tests performed to verufy our predictions are isochores, were satisfactory. Along the chimpanzee genomes the isochore repartition of windows obtained has been compared with 1000 random repartitions of the same windows. A significant difference between our predictions and random repartitions was observed (respectively the p-value of the χ^2 test were equal to 5.10^{-8}). Furthermore, the percentage of coding region in each isochore class was coherent with the observation obtained along mammals genomes ([20]). The coding regions represent 4.2% of the isochores H and only 1.1% of the isochores L. The p-value of the Wilcoxon test was significant (p= 1.10^{-4}).

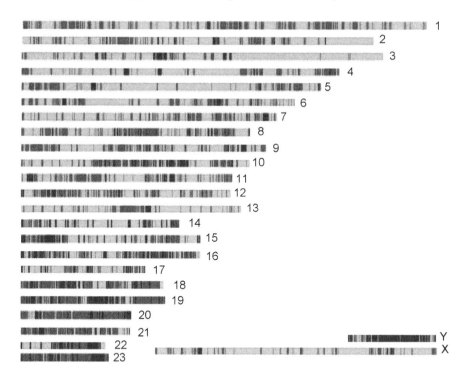

Fig. 3. Distribution of isochores along chimpanzee chromosomes obtained by our method. The detected H, L and M isochores appear respectively in red, green and blue.

The segmentation of the chimpanzee genome correlates well with the $G + C$ content. The mean $G + C$ contents were 0.48 ($\sigma = 0.03$), 0.42 ($\sigma = 0.02$) and 0.38 ($\sigma = 0.02$) respectively for the H, M and L isochore classes as defined by our HMM method. The Kruskal-Wallis non-parametric test was significant (p-value $< 10^{-5}$). The length of the isochores detected by their $G + C$ content is known to depend on the isochore class, with $L > M > H$ [29]. This is what we found here. Although the minimum length of isochores that can be predicted by our method was 300 kb, we were also able to predict much longer isochores. The average length for L isochores was 7.2 Mb, whereas the average length for the H and M isochores was 2 Mb and 4.4 Mb respectively. These lengths were significantly different (Kruskal-Wallis p-value $< 10^{-12}$). Figure 3 shows the relationships between isochore class and gene density. For all chromosomes, the isochore structure is correlated with the gene density distribution along the chromosome. The gene density in the H isochores (8.2 genes per Mb) was higher than the gene density in the L isochores (4.3 genes per Mb), leading to a significant Wilcoxon test (p-value $= 2.10^{-5}$). The same difference was observed when we compared the characteristics of the M isochores (5.6 genes per Mb) with those of the H (p-value $= 4.10^{-3}$) and L isochores (p-value $= 4.10^{-2}$).

5   Discussion

Last year, a large number of genomes were sequenced. This huge amount of data makes it impossible to analyse patterns in order to provide a biological interpretation "by hand". Therefore, mathematical and computational methods have to be used. Our approach, using HMMs, is a very promising method for analysing the organisation of genomes. Our study shows that hidden Markov models could be used to analyse genome organisation. This study was conducted on the chimpanzee genome but our method can be adapted to other eukaryote genomes. To model the bell-shaped length distribution of the exons, we have used sums of a variable number of geometric laws with equal or different parameters. Each region is represented by a macro-state in the HMM. A key property of this macro-state approach is that the conditional independence assumptions within the process are preserved with respect to HMMs. Moreover, we have preferred to use the Kolmogorov-Smirnov distance in order to obtain a better modelling of the chimpanzee genes.

The chimpanzee genomes consists of many nested structures (chromosomes, isochores, genes, exons/introns, reading frame). For the analyses of the isochore organisation of genomes, we have proposed a new method based on HMM, taking into account genes as a local structure. The different approaches already developed for isochore prediction ([14], [15], [16], [17], [18]) use only the overall base composition of the DNA sequence to predict isochores. However, the statistical characteristics of the $G + C$ content differ in the coding and non-coding regions of vertebrate genes. To improve the isochore prediction capacity, we have introduced the idea of using an HMM that takes into account not only the $G + C$ content of the DNA sequence, but also several biological properties associated with the isochore structure of the genome (such as gene density, differences in the $G + C$ content of different regions of the gene, lengths of exons and introns). Therefore, three HMMs were adjusted to each isochore class in order to take into account biological properties associated with H, L and M isochores. In our case, this supplementary information allowed us to determine the precise boundary of the isochores and the structure of a region may be easily analyzed. The segmentation in this paper are linked to an isochore structure of the chimpanzee genome. There was a significant difference between the isochore repartition in our prediction windows and a random repartition of these windows. Furthermore, there was more coding region in isochore H compared with isochores L in the two fishes. Thus, our method has clearly confirmed the existence of an isochore structure in the chimpanzee genome.

In conclusion, The statistical characteristics of the coding and noncoding regions of vertebrates differ dramatically between the different isochore classes [2]. The clarification of the isochore structure is a key to understand the organisation and biological function of the chimpanzee genome and we show here that hidden Markov models were appropriate for each isochore class. One advantage of the model presented in this paper is that the number of basic states for each isochore class (without taking the frame and coding strand into account) in our model is only 7: first, internal and terminal introns and exons and "intergenic regions".

Intergenic region were used to model all non-coding regions of the genome and the introns inserted between two coding exons. This small number of states has made it possible to conduct a complete chimpanzee genome analysis. This method could be easily adapted to other genomes and could be used to study the evolution of isochores among the vertebrate genomes. The comparative genomic analysis have a key role to push our knowledge further in the comprehension of the structure and function of human genes, to study evolutionary changes among organisms and help to identify the genes that are conserved among species.

Acknowledgements. We thanks Marie-France Sagot for assistance in preparing and reviewing the manuscript.

References

1. Thiery, J.P., Macaya, G., Bernardi, G.: An analysis of eukaryotic genomes by density gradient centrifugation. J. Mol. Biol., Vol. 108(1). (1976) 219-35
2. Bernardi, G.: Isochores and the evolutionary genomics of vertebrates. review. Gene, Vol. 241(1). (2000) 3–17
3. Krogh, A.: Two methods for improving performance of an HMM and their application for gene-finding. In Proceedings of the Fifth International Conference on Intelligent Systems for Molecular Biology (1997) 179–186
4. Henderson, J., Salzberg, S., Fasman, K.H.: Finding genes in DNA with a hidden Markov model. Journal of Computational Biology, Vol. 4. (1997) 127–141
5. Lukashin, V.A., Borodovsky, M.: Gene-Mark.hmm: new solutions for gene finding. Nucleic Acids Research, Vol. 26. (1998) 1107–1115
6. Burge, C., Karlin, S.: Prediction of complete gene structure in human genomic DNA. Journal of Molecular Biology, Vol. 268.(1997) 78–94
7. Berget, S.M.: Exon recognition in vertebrate splicing. The Journal of Biological Chemistry, Vol. 270(6). (1995) 2411–414
8. Hawkins, J.D.: A survey on intron and exon lengths. Nucleic Acids Research **16**, (1998) 9893–9908
9. Rabiner, L.: A tutorial on hidden Markov models and selected applications in speech recognition. Poceeding of the IEEE, Vol. 77(2). (1989) 257–286
10. Guédon Y.: Estimating hidden semi-Markov chains from discrete sequences. Journal of Computational and Graphical Statistics, Vol. 12(3). (2003) 604–639
11. Macaya, G., Thiery, J.P., Bernardi, G.: An approach to the organization of eukaryotic genomes at a macromolecular level. J. Mol. Biol., Vol. 108(1). 237–54 (1976)
12. Eyre-Walker, A., Hurst, L.D.: The evolution of isochores. Nat. Rev. Genet., Vol. 2(7). (2001) 549–555 Review
13. Nekrutenko, A., Li, W.H.: Assessment of compositional heterogeneity within and between eukaryotic genomes. Genome Res., Vol. 10(12). (2000) 1986–1995
14. Bernaola-Galvan, P., Carpena, P., Roman-Roldon, R., Oliver, J.L.: Mapping isochores by entropic segmentation of long genome sequences. In: Sankoff D, Lengauer T, RECOMB Proceedings of the Fifth Annual International Conference on Computational Biology. (2001) 217–218
15. Li, W., Bernaola-Galvan, P., Carpena, P., Oliver, J.L.: Isochores merit the prefix 'iso'. Comput. Biol. Chem., Vol. 27(1). (2003) 5–10

16. Oliver, J.L., Carpena, P., Roman-Roldan, R., Mata-Balaguer, T., Mejias-Romero, A., Hackenberg, M., Bernaola-Galvan, P.: Isochore chromosome maps of the human genome. Gene, Vol. 300(1-2). (2002) 117–27

17. Zhang, C.T., Zhang, R.: An isochore map of the human genome based on the Z curve method. Gene, Vol. 317(1-2). (2003) 127–35

18. Costantini, M., Clay, O., Auletta, F., Bernardi, G.: An isochore map of human chromosomes. Genome Research, Vo .16. (2006) 536–541

19. Bernardi, G., Olofsson, B., Filipski, J., Zerial, M., Salinas, J., Cuny, G., Meunier-Rotival, M., Rodier, F.: The mosaic genome of warm-blooded vertabrates. Science, Vol. 228(4702). (1985) 953–958

20. Mouchiroud, D., D'Onofrio, G., Aissani, B., Macaya, G., Gautier, C., Bernardi, G.: The distribution of genes in the human genome. Gene, Vol. 100. (1991)181–187

21. D'Onofrio, G., Mouchiroud, D., Aïssani, B., Gautier, C., Bernardi, B.: Correlations between the compositional properties of human genes, codon usage, and amino acid composition of proteins. J. Mol. Evol., Vol. 32. (1991) 504–510

22. Clay, O., Caccio, S., Zoubak, S., Mouchiroud, D., Bernardi, G.: Human coding and non coding DNA: compositional correlations. Mol. Phyl. Evol. , Vol. 1. (1996) 2–12

23. Jabbari, K., Bernardi, G.: CpG doublets, CpG islands and Alu repeats in long human DNA sequences from different isochore families. Gene, Vol. 224(1-2). (1998) 123–127

24. Zoubak, S., Clay, O., Bernardi, G.: The gene distribution of the human genome. Gene, Vol. 174(1). (1996) 95–102

25. Burge, C., Karlin, S.: Finding the genes in genomic DNA. Curr.Opin.Struc.Biol., Vol. 8. (1998) 346–354

26. Borodovsky, M., McIninch, J.: Recognition of genes in DNA sequences with ambiguities. Biosystems, Vol. 30(1-3). (1993) 161–171

27. Rogic, S., Mackworth, A.K., Ouellette, F.B.: Evaluation of Gene-Finding Programs on Mammalian Sequences. Genome Research, Vol. 11. (2001) 817–832

28. Guéguen, L.: Sarment: Python modules for HMM analysis and partitioning of sequences. Bioinformatics, Vol. 21(16). (2005) 3427–34278

29. De Sario, A., Geigl, E.M., Palmieri, G., D'Urso, M., Bernardi, G.: A compositional map of human chromosome band Xq28. Proc Natl Acad Sci U S A., Vol. 93(3). (1996) 1298–302

Synthetic Protein Sequence Oversampling Method for Classification and Remote Homology Detection in Imbalanced Protein Data

Majid M. Beigi and Andreas Zell

University of Tübingen
Center for Bioinformatics Tübingen (ZBIT)
Sand 1, D-72076 Tübingen, Germany
{majid.beigi, andreas.zell}@uni-tuebingen.de

Abstract. Many classifiers are designed with the assumption of well-balanced datasets. But in real problems, like protein classification and remote homology detection, when using binary classifiers like support vector machine (SVM) and kernel methods, we are facing imbalanced data in which we have a low number of protein sequences as positive data (minor class) compared with negative data (major class). A widely used solution to that issue in protein classification is using a different error cost or decision threshold for positive and negative data to control the sensitivity of the classifiers. Our experiments show that when the datasets are highly imbalanced, and especially with overlapped datasets, the efficiency and stability of that method decreases. This paper shows that a combination of the above method and our suggested oversampling method for protein sequences can increase the sensitivity and also stability of the classifier. **S**ynthetic **P**rotein **S**equence **O**versampling (SPSO) method involves creating synthetic protein sequences of the minor class, considering the distribution of that class and also of the major class, and it operates in data space instead of feature space. We used G-protein-coupled receptors families as real data to classify them at subfamily and sub-subfamily levels (having low number of sequences) and could get better accuracy and Matthew's correlation coefficient than other previously published method. We also made artificial data with different distributions and overlappings of minor and major classes to measure the efficiency of our method. The method was evaluated by the area under the Receiver Operating Curve (ROC).

1 Introduction

A dataset is imbalanced if the classes are not equally represented and the number of examples in one class (major class) greatly outnumbers the other class (minor class). With imbalanced data, the classifiers tend to classify almost all instances as negative. This problem is of great importance, since it appears in a large number of real domains, such as fraud detection, text classification, medical diagnosis and protein classification [1,2]. There have been two types of

S. Hochreiter and R. Wagner (Eds.): BIRD 2007, LNBI 4414, pp. 263–277, 2007.
© Springer-Verlag Berlin Heidelberg 2007

solutions for copping with imbalanced datasets. The first type, as exemplified by different forms of re-sampling techniques, tries to increase the number of minor class examples (oversampling) or decrease the number of major class examples (undersampling) in different ways. The second type adjusts the cost of error or decision thresholds in classification for imbalanced data and tries to control the sensitivity of the classifier [3,4,5,6].

Undersampling techniques involve loss of information but decrease the time of training. With oversampling we do not loose the information but instead it increases the size of the training set and so the training time for classifiers. Furthermore, inserting inappropriate data can lead to overfitting. Some researchers [2] concluded that undersampling can better solve the problem of imbalanced datasets. On the other hand, some other researchers are in favor of oversampling techniques. Wu and Chang [7] showed that with imbalanced datasets, the SVM classifiers learn a boundary that is too close to positive examples. Then if we add positive instances (oversampling), they can push the boundary towards the negative data, and we have increased the accuracy of classifier.

To decide the question of oversampling vs. undersampling, two parameters should be taken into consideration: the *imbalance ratio* and the distribution of data in imbalanced datasets. The *imbalance ratio* ($\frac{NumberOfMinorityData}{NumberOfMajorityData}$) is an important parameter that shows the degree of imbalance. In undersampling we should be sure of the existence of enough information in the minor class and also of not loosing the valuable information in the major class. We found out that the oversampling technique can balance the class distribution and improve that situation. But the distribution of inserted positive instances is of great importance. Chawla et al. [8] developed a method for oversampling named Synthetic Minority Oversampling Technique (SMOTE). In their technique, between each positive instance and its nearest neighbors new synthetic positive instances were created and placed randomly between them. Their approach proved to be successful in different datasets.

On the other hand Veropoulos et al. [6] suggested using different error costs (DEC) for positive and negative classes. So the classifier is more sensitive to the positive instances and gets more feedback about the orientation of the class-separating hyperplane from positive instances than from negative instances.

In protein classification problems the efficiency of that approach (Veropoulos et al. [6]) has been accepted. In kernel based protein classification methods [9,10,1] a class-depending regularization parameter is added to the diagonal of the kernel matrix: $K'(x,x) = K(x,x) + \lambda n/N$, where n and N are the number of positive (or negative) instances and the whole dataset, respectively. But, based on our experiments, if the dataset is highly imbalanced and has overlapping data, choosing a suitable ratio of error costs for positive and negative examples is not always simple and sometimes the values near the optimum value of the error cost ratio give unsatisfying results.

We propose an oversampling technique for protein sequences in which the minority class in the data space is oversampled by creating synthetic examples. Working with protein data in data space instead of feature space allows us to consider the probability distribution of residues of the sequence using a HMM (Hidden Markov Model) profile of the minority class and also one of the majority class and then synthesize protein sequences which can push precisely the boundary towards the negative examples. So we increase the information of the minor class. Our method of oversampling can cause the classifier to build larger decision regions for the minor class without overlapping with the major class. In this work we used real and artificial data with different degrees of overlapping and imbalance ratio to show the efficiency of our methods and we also suggest that our algorithm can be used along with DEC methods to increase the sensitivity and stability of the classifier. As SVM classifiers and kernel methods outperformed other methods in protein classification [9,1,10], we discuss the efficiency of our oversampling technique when used with kernel-based classifiers.

2   SPSO: Synthetic Protein Sequence Oversampling Technique

Given a set of positive training sequences (minor class) S_+ and a set of negative training sequences (major class) S_- we want to create synthetic protein sequences $S_{synthetic}$ as mutated replicas of each sequence of the minor class, provided that those synthetic sequences are created by an HMM profile (Hidden Markov Model profile) of the minor class and are phylogenetically related to that class and far away from the major class. For this, at first we build a multiple alignment of the sequences of the minor class using ClustalW [11] and then we train a hidden Markov model profile with length of the created multiple alignment sequences for each class (positive data and every family belonging to the negative data).

For every sequence in the minor class we create another mutated sequence synthetically. For that, we consider an arbitrary N_m as number of start points for mutation in that sequence. We suppose the $HMMp_+$ (hidden Markov model profile of positive instances) has emitted another sequence identical to the main sequence until the first point of mutation. From that point afterward we assume that $HMMp_+$ emits new residues until the emitted residue is equal to a residue in the same position in the main sequence. From this residue, all residues are the same as residues in the original sequence until the next point of mutation (Fig. 1).

In this way, if the point of mutation belongs to a low entropy area of the HMM profile the emitted residue will be very similar to the main sequence (will have few mutations). We expect the emmitance probability of the synthesized sequence with $HMMp_+$ to be higher than with $HMMp_-$, if not (very rarely), we synthesize another one or we decrease the value of N_m. The N_m parameter can adjust the radius of the neighborhood of the original sequences and the synthesized sequences. With larger values of N_m, the algorithm creates sequences that are phylogenetically farer away from main sequences and vice versa. We used another routine to find a suitable value of N_m. At first, in the minor class, we find

---

Algorithm. SPSO(S_+,S_-)

Input : S_+, set of sequences of minority class; S_-, set of sequences of majority class

Output: $S_{synthetic}$, set of synthetic protein sequences from the minority class

1 Create HMM profile of set S_+, call it $HMMp_+$;
2 Create array of HMM profiles consisting of all families belonging to S_-, call it $HMMp_-$[];
3 Choose an arbitrary number as number of start points for mutation, call it N_m;
4 **for** $i \leftarrow 1$ **to** $|S_+|$ **do**
5     $s = S_+[i]$;
6   **repeat**
7   Create an array of sorted non-repeating random numbers with size of N_m as array of start points for mutation, call it P_m ;
8        $S_{synthetic}$[i]=**newSeq** (s,$HMMp_+$,P_m);
9        $p_+ = P_e(S_{synthetic}[i], HMMp_+)$; (* emmitance probability of synthesized sequence by $HMMp_+$ *)
10       $p_-[] = P_e(S_{synthetic}[i], HMMp_-[])$;
11   **until** $p_+ < \max p_-[]$;
12 **end**
13 **return** $S_{synthetic}$

Function. newSeq(s,$HMMp_+$,P_m)

Input : s, original sequence; $HMMp_+$, HMM profile of set S_+ to which s belongs; P_m, array of start points for mutation

Output: $s_{synthetic}$, synthetic sequence from s

1 $s_{synthetic} = s$;
2 **for** $i \leftarrow 1$ **to** $|P_m|$ **do**
3     $p = P_m[i]$; (* assume that $HMMp_+$ in position p has emitted $s[p]$ *)
4   **repeat**
5        $s_{synthetic}[p+1]$= emitted residue in position $p+1$ by $HMMp_+$;
6        $p = p+1$;
7   **until** (newres $\neq s[p]$) && ($p < |HMMp_+|$) ;
8 **end**
9 **return** $s_{synthetic}$

---

the protein sequence which has the highest emission probability with the HMM profile of the minor class and consider it as root node. Then, we suppose the root node has been mutated to synthesize all other sequences in the minor class through the *newSequence* procedure of our algorithm. It means each sequence is a mutated replica of the root node sequence which is emitted by the HMM profile of the minor class. We gain the value of N_m for each sequence. Then, we get the average of all those values as N_m entry for the SPSO algorithm.

With each call of the SPSO algorithm, we double the minor class. As an example of random synthesizing of sequences, Fig. 1(upper) shows the phylogenetic

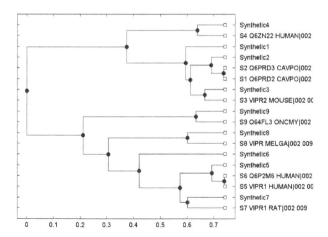

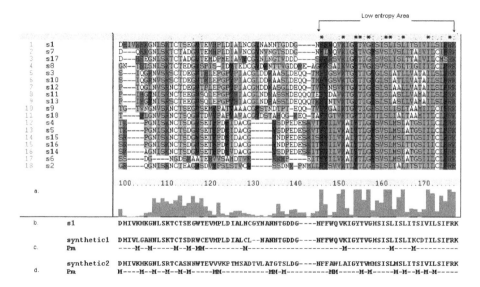

Fig. 1. The phylogenetic tree of the original and the synthesized sequences from the "vasoactive intestinal polypeptide" family of GPCRs (**upper**) and an example of the SPSO algorithm for sequences from the above family (**lower**). **a.** Multiple sequence alignment and low entropy area of that family **b.** A part of sequence $s1$. **c.** Synthetic sequence of $s1$ with N_m=50 . **d.** Synthetic sequence of $s1$ with N_m=100 (P_m: array of start points, shown by M, for mutations).

tree of the original sequences and the synthesized sequences for the vasoactive intestinal polypeptide family of class B (9 out of 18 sequences were randomly selected). It is shown that the synthesized sequences of most original sequences have less distance to them than to other sequences. In that figure (lower) we see two synthetic sequences of $s1$ with different values of N_m. In the low entropy area of the HMM profile of that family we have less mutations.

3   Datasets

To evaluate the performance of our algorithm, we ran our experiments on a series of both real and artificial datasets, whose specification covers different complexity and allows us to fully interpret the results. We want to check its efficiency with different ratio of imbalance and complexity. Fig. 3 shows the pictorial representation of our datasets. In the first one, the distribution of the positive and negative data are completely different and they are separate from each other. With that distribution, we want to see, how the imbalance ratio affects the performance of the classifier by itself. The second one shows datasets in which positive data are closer to negative data and there is an overlap between the minor and major classes. With this distribution, we can consider both the ratio of imbalance and overlap of the datasets in our study. The third one is a case where the minor class completely overlaps with the major class and we have fully overlapping data.

We used the G-protein coupled receptors (GPCRs) family as real data and then created artificial data based on it. G-protein coupled receptors (GPCRs) are a large superfamily of integral membrane proteins that transduce signals across the cell membrane [12]. Through their extracellular and transmembrane domains they respond to a variety of ligands, including neurotransmitters, hormones and odorants. They are characterized by seven hydrophobic regions that pass through the cell membrane (transmembrane regions), as shown in Fig. 2.

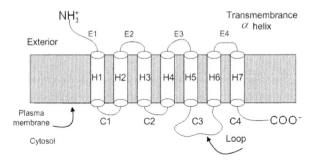

Fig. 2. Schematic representation of GPCR shown as seven transmembrane helices depicted as cylinders along with cytoplasmic and extracellular hydrophilic loops

According to the binding of GPCRs to different ligand types they are classified into different families. Based on GPCRDB (G protein coupled receptor database) [13] all GPCRs have been divided into a hierarchy of 'class', 'subfamily', 'sub-sub-family' and 'type'. The dataset of this study was collected from GPCRDB and we used the latest dataset (June 2005 release). The six main families are: Class A (Rhodopsin like), Class B (Secretin like), Class C (Metabotropic gluta-mate/pheromone), Class D (Fungal pheromone), Class E (cAMP receptors) and Frizzled/Smoothened family. The sequences of proteins in GPCRDB were taken from SWISS-PROT and TrEMBL [14]. All six families of GPCRs (5300 protein sequences) are classified in 43 subfamilies and 99 sub-subfamilies.

If we want to classify GPCRs at the sub-subfamily level, mostly we have only a very low number of protein sequences as positive data (minor class) compared with others (major class). We chose different protein families from that level to cover all states of complexity and imbalance ratio discussed above (Fig. 3). In some experiments we made artificial data using those families and synthesized sequences from them (discussed later). We used numbers to show the level of family, subfamily and sub-subfamily. For example 001-001-002 means the sub-subfamily Adrenoceptors that belongs to subfamily of Amine (001-001) and class A (001).

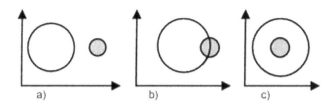

Fig. 3. Pictorial representation of the minor (shaded circle) and major classes of our datasets

4   Experiments

We selected the peptide subfamily (001-002) of Class A (Rhodopsin-like) to classify its 32 families (or sub-subfamily level of class A). We built HMM profiles of all families and measured the probability of emission of sequences belonging to each one by all HMM profiles. We saw that the emission probability of each sequence generated by the HMM profile of its own family is higher than that of almost all other families. So we can conclude that the distribution of the peptide subfamily in a suitable feature map can be considered as in Fig. 3.a. In this study, we used the local alignment kernel (LA kernel) [15] to generate vectors from protein sequences. It has been shown that the local alignment kernel has better performance than other previously suggested kernels for remote homology detection when applied to the standard SCOP test set [10]. It represents a modification of the Smith-Waterman score to incorporate sub alignments by computing the sum (instead of the maximum) over all possible alignments. We build a kernel matrix K for the training data. Each cell of the matrix is a local alignment kernel score between protein i and protein j. Then we normalize the kernel matrix via $K_{ij} \leftarrow K_{ij}/\sqrt{K_{ii}K_{jj}}$. Each family is considered as positive training data and all others as negative training data. After that the SVM algorithm with RBF kernel is used for training. For testing, we created feature vectors by calculating a local alignment kernel between the test sequence and all training data. The number of sequences in the peptide subfamily is in the range of 4 to 251 belonging to (001-002-024) and (001-002-008), respectively. Thus the *imbalance ratio* varies from $\frac{4}{4737}$ to $\frac{251}{4737}$. Fig. 4.**a** shows the result of SPSO oversampling for classification of some of those families. We see that this method can increase the accuracy and sensitivity of the classifier faced with highly imbalanced data without decreasing its specificity. The minority class was oversampled at 100%, 200%, 300%,..., 800% of its original size. We see that the more we increase the synthetic data (oversample) the better result we get, until we get the optimum value. It should be noted that after oversampling, the accuracy of classifiers for the major class didn't decrease.

We compared our method with two other methods. The first one was SMOTE (Synthetic Minority Oversampling Techniques) [8] that operates in the feature space rather than in data space, so it works with all kind of data. The second comparison was done with randomly oversampling, in which we create random sequences by the HMM profile of each family. For this, like our method, we build a multiple alignment of the minor class sequences using ClustalW and then train a hidden Markov model profile with length of the created multiple alignment sequence. Then, we create random sequences by the HMM profile of each family. In this method we don't have enough control on the distribution of created random sequences. We call this method rHMMp in this paper.

In our study, we used the Bioinformatics Toolbox of MATLAB to create the HMM profiles of families and the SVMlight package [16], to perform SVM training and classification.

We used the Receiver Operating Characteristic (ROC) graphs [17] to show the quality of the SPSO oversampling technique. An ROC graph characterizes the performance of a binary classifier across all possible trade-off between the

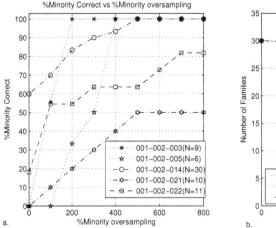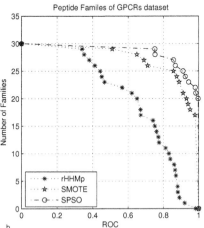

Fig. 4. a. %Minority correct for SPSO oversampling for some families of peptide subfamily (N number of sequences). **b**. Comparison of several methods for oversampling. The graph plots the total number of families for which a given method exceeds an ROC score threshold.

classifier sensitivity (TP_{rate}) and false positive error rates (FP_{rate}) [18]. The closer the ROC score is to 1, the better performance the classifier has. We oversampled each minority class with the three different methods noted above, until we got the optimum performance for one of them. At that point, we calculated the ROC score of all methods.

Fig. 4.b shows the quality of classifiers when using different oversampling methods. This graph plots the total number of families for which a given method exceeds an ROC score threshold. The curve of our method is above the curve of other methods and shows better performance. In our method and in SMOTE, the inserted positive examples have been created more accurately than random oversampling (rHMMp). Our method (SPSO) outperforms the other two methods especially for families in which we have a low number of sequences, although the quality of the SMOTE is comparable to the SPSO method.

To study the second and third representation of the dataset shown in Fig. 3 we had to create some sequences synthetically. At first we built the HMM profile of each family of the peptide families and then computed the probability score of each sequence when emitted not only by the HMM profile of its own family but also from all other families. The average of those scores for sequences of each family when emitted by each HMM profile can be used as a criterion for the closeness of the distribution of that family to other families and how much it can be represented by their HMM profiles. In this way we can find the nearest families to each peptide family. After that we synthesized sequences for each family through the $newSeq$ procedure of the SPSO algorithm, provided that it is emitted by the HMM profile of another near family and not by its own HMM profile. So after each start position for mutation (Fig.1 (lower)) we have residues that are emitted by another HMM profile (we want to have overlap with) instead of its own HMM profile and there is an overlap for the distribution of synthesized sequences between those two families. The degree of overlapping can be tuned by the value of N_m (number of mutations). This dataset (original and new synthesized sequences) can be considered as partially overlapping dataset (Fig. 3.b). If we create more sequences using other HMM profiles the distribution of the dataset is fully overlapping (Fig. 3.c). To study the partially overlapping datasets, we selected 10 families of peptide families and built the synthesized sequences as noted above. To create the fully overlapping dataset, we performed that routine for each family using the HMM profile of three families near to the original family, separately.

We compared our oversampling technique with the SMOTE oversampling technique and the different error cost (DEC) method [6]. Tables 1 and 2 show the results. We see that in general SPSO outperforms the SMOTE and DEC methods, and the performance of the classifier with the SPSO oversampling technique in fully overlapped datasets is more apparent. When there is more overlapping between the minor and major classes, the problem of imbalanced data is more acute. So the position of the inserted data in the minor class is more important and in our algorithm it has been done more accurately than in SMOTE method. With regard to the time needed for each algorithm, DEC has an advantage compared to our method, because in the oversampling technique the minor class, depending on the number of its instances, is oversampled up to 10 times (in our experiments) which increases the dimension of the of kernel matrix. In contrast, in the DEC method choosing the correct cost of error for minority and majority classes is an important issue. One suggested method is to set the error cost ratio equal to the inverse of the imbalance ratio. But, based on our experiments that value is not always the optimum, and especially in partially and fully overlapped datasets we had instability of performance even with values near the optimal value. Based on our experiments in the well-separated imbalanced data the quality of DEC is very near to the SPSO method and for some experiments, even better, and we could find optimum value for error cost ratio simply. So perhaps with this kind of datasets one should prefer the DEC method. But with partially and fully overlapping data, we found that our oversampling method in general has better performance, and if it is used along with the DEC method, it not only increases the performance of the classifier but it also makes finding the value for the error cost ratio simpler. We also have more stability with values close to the optimum value of the error cost ratio. The graphs in Fig. 5.a and Fig. 5.b show the value of the ROC score of classifier for partially overlapped artificial sequences from the family of 001-002-024 ($001 - 002 - 024'$) when the DEC method and DEC along with SPSO (400% oversampling) were applied. We see that when SPSO oversampling is used we have stability in ROC score values and after the optimum value, the ROC score does not change. The drawback is, that we again have to find the best value for the error cost ratio and the rate of oversampling through the experiment by checking different values, but in less time compared to only the DEC method because of the stability that was shown in Fig. 5.b. We used that method for all partially and fully overlapping artificial data (Table 1 and 2). For each experiment we oversampled data in different rates and selected different

Table 1. ROC scores obtained on the partially overlapping classes created from peptide families of GPCR dataset, by various methods. DEC = different error cost

Partially overlapping classes- ROC scores

minority class	# of sequences	SMOTE	DEC	SPSO
$001 - 002 - 015'$	16	0.863	0.943	0.951
$001 - 002 - 016'$	122	0.821	0.912	0.929
$001 - 002 - 017'$	68	0.854	0.892	0.884
$001 - 002 - 018'$	74	0.912	0.871	0.891
$001 - 002 - 020'$	86	0.972	0.975	0.984
$001 - 002 - 021'$	40	0.695	0.739	0.723
$001 - 002 - 022'$	44	0.725	0.762	0.751
$001 - 002 - 023'$	48	0.965	0.982	0.996
$001 - 002 - 024'$	8	0.845	0.834	0.865
$001 - 002 - 025'$	10	0.945	0.972	0.987
overall ROC-score		**0.859**	**0.882**	**0.896**

Table 2. ROC scores obtained on the Fully overlapping classes created from peptide families of GPCR dataset by various methods

Fully overlapping classes- ROC scores

minority class	# of sequences	SMOTE	DEC	SPSO
$001 - 002 - 015"$	32	0.673	0.680	0.724
$001 - 002 - 016"$	244	0.753	0.775	0.821
$001 - 002 - 017"$	136	0.672	0.652	0.643
$001 - 002 - 018"$	148	0.591	0.624	0.672
$001 - 002 - 020"$	172	0.763	0.821	0.858
$001 - 002 - 021"$	80	0.632	0.689	0.681
$001 - 002 - 022"$	88	0.615	0.812	0.854
$001 - 002 - 023"$	96	0.912	0.942	0.968
$001 - 002 - 024"$	16	0.716	0.768	0.819
$001 - 002 - 025"$	20	0.908	0.902	0.921
overall ROC-score		**0.723**	**0.766**	**0.796**

values of error cost ratio until we got the best result. The results in Fig. 5.c show that for those kind of data the ROC scores of SPSO and DEC + SPSO are nearly the same. But in the second method (DEC + SPSO), we needed to oversample data less than in SPSO only method and we could find the best value of the error cost ratio sooner than in DEC only. With less rate of oversampling in SPSO we get less accurate results but we can compensate that with DEC.

For further evaluation of our method, we used our oversampling technique in classification all GPCRs families at sub family and sub-sub family level (mostly we have low number of sequences). In subfamily classification we randomly partitioned the data in two non-overlapping sets and used a two-fold cross validation protocol. The training and testing was carried out twice using one set for training and the other one for testing. To compare with the results of other researchers, the prediction quality was evaluated by Accuracy (ACC), Matthew's correlation coefficient (MCC), overall Accuracy ($\overline{ACC}$)and overall MCC ($\overline{MCC}$) as follows:

$$ACC = \frac{TP + TN}{(TN + FN + TP + FP)} \tag{1}$$

$$MCC = \frac{TP \times TN - FN \times FP}{\sqrt{(TN + FN)(TP + FN)(TN + FP)(TP + FP)}} \tag{2}$$

$$\overline{ACC} = \sum_{i=1}^{N} \frac{ACC(i)}{N} \tag{3}$$

$$\overline{MCC} = \sum_{i=1}^{N} \frac{MCC(i)}{N} \tag{4}$$

(TP = true positive, TN = true negative, FP = false positive , FN = false negative, N=number of subfamily or sub-subfamily)

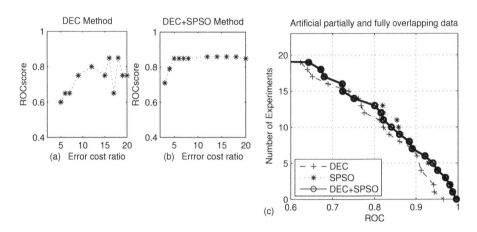

Fig. 5. (a) Comparison of the ROC score at different error cost ratios for artificial sequences of $001 - 002 - 024'$ in classifier with DEC method. (b) classifier with DEC + SPSO mothods (400% oversampled). (c). Comparison of DEC, SPSO and DEC+SPSO methods for imbalanced data. The graph plots the total number of experiments of partially and fully overlapped imbalanced artificial data for which a given method exceeds an ROC score threshold.

The overall accuracy we gained for families A, B and C is 98.94%, 99.94% and 96.95%, respectively, and overall MCC for families A, B and C is 0.98, 0.99 and 0.91, respectively. Almost all of the subfamilies are accurately predicted with our method. Tables 3 shows the results of subfamily classification for classes A of GPCRs. At the subfamily level we compared our method with that of Bhasin et

Table 3. The performance of our method in GPCRs subfamily classification(Class A)

Class A subfamilies	Accuracy (%)	MCC
Amine	99.9	0.99
Peptide	97.8	0.97
Hormone protein	100.0	1.00
(Rhod)opsin	99.6	0.99
Olfactory	99.9	0.99
Prostanoid	99.9	.98
Nucleotide-like	100.0	1.00
Cannabinoid	100.0	1.00
Platelet activating factor	100.0	1.00
Gonadotropin-releasing hormone	100.0	1.00
Thyrotropin-releasing hormone	100.0	1.00
Melatonin	100.0	1.00
Viral	87.0	0.8
Lysosphingolipid	100.0	1.00
Leukotriene	100.0	1.00
Overall	98.4	0.98

Table 4. The performance of our method in GPCRs sub-subfamily classification for Class A,B and C

Class A	subfamilies	Overall Accuracy (%)	Overall MCC
	Amine	97.1	0.91
	Peptide	99.9	0.93
	Hormone protein	100.1	1.00
	(Rhod)opsin	96.6	0.95
	Olfactory	98.9	0.92
	Prostanoid	98.0	0.94
	Gonadotropin-releasing hormone	96.1	0.93
	Thyrotropin-releasing hormone	91.2	0.94
	Lysosphingolipid	98.4	1.00
Class B	Latrophilin	100.0	1.00
Class C	Metabotropic glutamate	98.1	0.96
	Calcium-sensing like	97.2	0.93
	GABA-B	100.0	1.00
Overall		97.93	0.95

al. [19]. They used an SVM-based method with dipeptide composition of protein sequences as input. The accuracy and MCC values of our method outperform theirs. For example in classification of subfamily A, the overall accuracy and MCC of their method were 97.3% and 0.97 but ours are 98.4% and .98, respectively. They did a comparison with other previously published methods like that of Karchin et al. [20] and showed that their method outperformed the others.

For sub-subfamily classification we used 5-fold cross validation. Table 4 shows the results for the sub-subfamily level. We see that in this level also the accuracy is high and we could classify most of GPCRs sub-subfamilies. We could obtain an overall accuracy of 97.93% and a MCC of 0.95 for all sub-subfamilies. At this level we could increase the accuracy, especially when the number of sequences in the positive training data was less than 10, and there was no example in which with our oversampling method the accuracy decreases.

To the best of our knowledge there is only one study which has been done for sub-subfamily classification [21] in GPCRs families. Their approach is based on bagging a classification tree and they achieved 82.4% accuracy for sub-subfamily classification, which is less accurate than ours (97.93% with MCC of 0.95) despite the fact that they had excluded families with less than 10 sequences (we only excluded families with less than 4 sequences). We think our oversampling technique can be widely used for other applications of protein classification with the problem of imbalanced data and it can be used along with the different error cost (DEC) method to overcome the problem of imbalanced data for protein data.

5   Conclusion

In this work, we suggested a new approach of oversampling for the imbalanced protein data in which the minority class in the data space is oversampled by creating synthetic protein sequences, considering the distribution of the minor and major classes. This method can be used for protein classification problems and remote homology detection, where classifiers must detect a remote relation between unknown sequence and training data with an imbalance problem. We think that this kind of oversampling in kernel-based classifiers not only pushes the class separating hyperplane away from the positive data to negative data but also changes the orientation of the hyperplane in a way that increases the accuracy of classifier. We developed a systematic study using a set of real and artificially generated datasets to show the efficiency of our method and how the degree of class overlapping can affect class imbalance. The results show that our SPSO algorithm outperforms other oversampling techniques. In this paper, we also presented evidences suggesting that our oversampling technique can be used along with DEC to increase its sensitivity and stability. For further work, we hope to find an algorithm for finding the suitable rate of oversampling and error cost ratio when DEC and SPSO methods are used together.

References

1. C.Leslie, E. Eskin, A. Cohen, J. Weston, and W.S. Noble. Mismatch string kernel for svm protein classification. *Advances in Neural Information Processing System*, pages 1441–1448, 2003.
2. A. Al-Shahib, R. Breitling, and D. Gilbert D. Feature selection and the class imbalance problem in predicting protein function from sequence. *Appl Bioinformatics*, 4(3):195–203, 2005.
3. M. Pazzini, C. Marz, P. Murphi, K. Ali, T.Hume, and C. Bruk. Reducing misclassification costs. *In proceedings of the Eleventh Int. Conf. on Machine Learning*, pages 217–225, 1994.
4. N. Japkowicz, C.Myers, and M. Gluch. A novelty detection approach to classification. *In Proceeding of the Fourteenth Int. Joint Conf. on Artificial Inteligence*, pages 10–15, 1995.
5. N. Japkowicz. Learning from imbalanved data sets: A comparison of various strategies. *In Proceedings of Learning from Imbalanced Data*, pages 10–15, 2000.
6. K. Veropoulos, C. Campbell, and N. Cristianini. Controlling the sensitivity of support vector machines. *Proceedings of the International Joint Conference on AI*, pages 55–60, 1999.
7. G. Wu and E. Chang. Class-boundary alignment for imbalanced dataset learning. *In ICML 2003 Workshop on Learning from Imbalanced Data Sets II,Washington, DC*, 2003.
8. Nitesh V. Chawla, Kevin W. Bowyer, Lawrence O. Hall, and W. Philip Kegelmeyer. Smote: Synthetic minority over-sampling technique. *Journal of Artificial Intelligence and Research*, 16:321–357, 2002.
9. C. Leslie, E. Eskin, and W. S. Noble. The spectrum kernel: A string kernel for svm protein classification. *Proceedings of the Pacific Symposium on Biocomputing*, page 564575, 2002.
10. H. saigo, J. P. Vert, N. Ueda, and T. akustu. Protein homology detection using string alignment kernels. *Bioinformatics*, 20(11):1682–1689, 2004.

11. J. D. Thompson, D. G. Higgins, and T. J. Gibson. Clustalw: improving the sesitivity of progressive multiple sequence alignment through sequence weighting, positions-specific gap penalties and weight matrix choice. *Nucleic Acids Res.*, 22:4673–4680, 1994.

12. T. K Attwood, M. D. R. Croning, and A. Gaulton. Deriving structural and functional insights from a ligand-based hierarchical classification of g-protein coupled receptors. *Protein Eng.*, 15:7–12, 2002.

13. F. Horn, E. Bettler, L. Oliveira, F. Campagne, F. E. Cohhen, and G. Vriend. Gpcrdb information system for g protein-coupled receptors. *Nucleic Acids Res.*, 31(1):294–297, 2003.

14. A. Bairoch and R. Apweiler. The swiss-prot protein sequence data bank and its supplement trembl. *Nucleic Acids Res.*, 29:346–349, 2001.

15. J.-P.Vert, H. Saigo, and T.Akustu. *Convolution and local alignment kernel In B. Schoelkopf, K. Tsuda, and J.-P.Vert (Eds.), Kernel Methods in Compuatational Biology.* The MIT Press.

16. T. Joachims. Macking large scale svm learning practical. *Technical Report LS8-24*, Universitat Dortmond, 1998.

17. F. Provost and T. Fawcett. Robust classification for imprecise environments. *Machine Learning*, 423:203–231, 2001.

18. J. Swet. Measuring the accuracy of diagnostic systems. *Science.*, 240:1285–1293, 1988.

19. M. Bhasin and G. P. S. Raghava. Gpcrpred: an svm-based method for prediction of families and subfamilies of g-protein coupled receptors. *Nucleaic Acids res.*, 32:383–389, 2004.

20. R. Karchin, K. Karplus, and D. Haussler. Classifying g-protein coupled receptors with support vector machines. *Bioinformatics*, 18(1):147159, 2002.

21. Y. Huang, J. Cai, and Y. D. Li. Classifying g-protein coupled receptors with bagging classification tree. *Computationa Biology and Chemistry*, 28:275–280, 2004.

Stem Kernels for RNA Sequence Analyses

Yasubumi Sakakibara[1], Kiyoshi Asai[2], and Kengo Sato[3]

[1] Department of Biosciences and Informatics, Keio University,
3-14-1 Hiyoshi, Kohoku-ku, Yokohama, 223-8522, Japan
yasu@bio.keio.ac.jp
[2] Department of Computational Biology, University of Tokyo,
CB04 Kiban-tou 5-1-5 Kashiwanoha, Kashiwa, Chiba, 277-8561, Japan
asai@k.u-tokyo.ac.jp
[3] Japan Biological Informatics Consortium,
10F TIME24 Building, 2-45 Aomi, Koto-ku, Tokyo 135-8073, Japan
satoken@bio.keio.ac.jp

Abstract. Several computational methods based on stochastic context-free grammars have been developed for modeling and analyzing functional RNA sequences. These grammatical methods have succeeded in modeling typical secondary structures of RNA and are used for structural alignment of RNA sequences. However, such stochastic models cannot sufficiently discriminate member sequences of an RNA family from non-members and hence detect non-coding RNA regions from genome sequences.

A novel kernel function, *stem kernel*, for the discrimination and detection of functional RNA sequences using support vector machines (SVM) is proposed. The stem kernel is a natural extension of the string kernel, specifically the all-subsequences kernel, and is tailored to measure the similarity of two RNA sequences from the viewpoint of secondary structures. The stem kernel examines all possible common base-pairs and stem structures of arbitrary lengths, including pseudoknots between two RNA sequences and calculates the inner product of common stem structure counts. An efficient algorithm was developed to calculate the stem kernels based on dynamic programming. The stem kernels are then applied to discriminate members of an RNA family from non-members using SVM. The study indicates that the discrimination ability of the stem kernel is strong compared with conventional methods. Further, the potential application of the stem kernel is demonstrated by the detection of remotely homologous RNA families in terms of secondary structures. This is because the string kernel is proven to work for the remote homology detection of protein sequences. These experimental results have convinced us to apply the stem kernel to find novel RNA families from genome sequences.

Keywords: stem kernel, string kernel, SVM, RNA, secondary structure.

1 Introduction

Analysis and detection of functional RNAs are current topics of great significance in molecular biology and bioinformatics research. Since there is a rapidly growing

S. Hochreiter and R. Wagner (Eds.): BIRD 2007, LNBI 4414, pp. 278–291, 2007.
© Springer-Verlag Berlin Heidelberg 2007

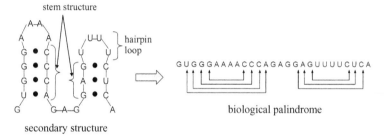

Fig. 1. A typical secondary structure including stem structure of an RNA sequence (left) constitutes the so-called biological palindrome (right)

number of known RNA sequences, structures, and families, computational methods for finding non-protein-coding RNA regions on the genome has garnered much attention and research [2]. Compared with the gene finding difficulties inherent for protein-coding regions, computationally identifying non-coding RNA regions is essentially more problematic as these sequences do not have strong statistical signals. Currently, a general finding algorithm does not exist.

It is commonly known, in RNA sequence analyses, that the specific form of the secondary structures in the cell is an important feature for modeling and detecting RNA sequences. The folding of an RNA sequence into a functional molecule is largely governed by the formation of the standard Watson-Crick base pairs A-U and C-G as well as the wobble pair G-U. Such base pairs constitute "biological palindromes" in the genome (See Figure 1). The secondary structures of RNAs are generally composed of *stems*, *hairpins*, *bulges*, *interior loops*, and *multi-branches*. A stem is a double stranded (paired) region of base-pair stacks (See Figure 1). A hairpin loop occurs when RNA folds back on itself.

To capture such secondary structure features, stochastic context-free grammars (SCFGs) for RNAs have been proposed. SCFGs have been used successfully to model typical secondary structures of RNAs and are also used for structural alignment of RNA sequences [1,9,10,11,13]. However, a serious drawback of the SCFG method is the requirement of prior knowledge. A typical known secondary structure of the target RNA family is needed to design the grammars.

Furthermore, stochastic models such as SCFGs and hidden Markov models (HMMs) have limitations in discriminating member sequences of an RNA family from non-members by only examining the probabilistic scores. Hence, we require stronger discriminative methods to detect and find non-coding RNA sequences.

Recently, the support vector machine (SVM) and kernel function techniques have been actively studied and used to propose solutions to various problems in bioinformatics [5,14]. SVMs are trained from positive and negative samples and have strong and accurate discrimination abilities. Hence, they are better suited for the discrimination tasks. For protein sequence analyses, string kernels [14] have been proposed for the use of SVMs to classify a protein family. In addition, string kernels are proven to work for remote homology detections of protein sequences, i.e. a superfamily.

In this paper, a novel kernel function, called *stem kernel*, is proposed which enhances the ability to measure the similarity of two RNA sequences from the viewpoint of secondary structure. In this study, the similarity features are defined by all possible common base-pairs and stem structures of arbitrary lengths including pseudoknots between two RNA sequences. The proposed stem kernel calculates the inner product of two vectors in the feature space from two RNA sequences. That is, the more stem structures two RNA sequences have in common, the more similar they are. Further, our stem kernel does not assume prior knowledge of the secondary structures in a target RNA family.

Several discrimination task experiments utilizing our stem kernel method were executed. In our experiments on five RNA families, tRNA, miRNA (precursor), 5S rRNA, H/ACA snoRNA, and CD snoRNA, the stem kernel exhibits good discriminative performance even for a smaller number of available sample sequences. In contrast, the string kernel is competitive when an adequate number of sequences are available but the discrimination performance significantly decreases for weakly homologous RNA sequences. Kernel Principal Component Analysis (KPCA) was applied to measure the classification and separation abilities of each kernel function from a mixed RNA sequence data of three different RNA families. In the experiments detecting remote homologous RNA sequences, Tymo_tRNA-like sequences as remote homologies of tRNAs are attempted by using SVMs trained from the positive and negative samples of "tRNA sequences". The stem kernel achieved strong detections of Tymo_tRNA-like sequences while the string kernel failed to adequately detect such sequences. This experimental result convinced us to apply the stem kernel into finding novel RNA families from genome sequences.

2 Methods

This paper proposes a novel kernel function, called *stem kernel*, for the discrimination and detection of functional RNA sequences using SVMs. In the following section, the effectiveness of our stem kernel function in initial experiments for discriminating members of an RNA family from non-members is exhibited.

2.1 String Kernel

First, the string kernel [14] will be briefly reviewed. Special attention is paid to the all-subsequences kernel as our stem kernel is a natural extension of the string kernel in measuring the similarity of two RNA sequences from the viewpoint of secondary structures.

General feature mapping for measuring the similarity between two biological sequences is defined by counting all contiguous or non-contiguous subsequences of the given sequences. For example, two DNA sequences CTG and CAT have 4 common subsequences: ϵ (the empty string), C, T, and C-T. The all-subsequences kernel calculates the inner product of the feature vectors by counting all commonly held non-contiguous subsequences. The inner product of the two sequences CTG and CAT is 4.

2.2   Feature Space for RNA Sequences

First, we need to define a feature mapping the space of RNA sequences to a vector space such that the relative distance in the mapped vector space reflects the similarity between two RNA sequences. We consider a notion of similarity between RNA sequences in terms of common secondary structures. The simplest similarity feature is the count of base pair occurrences that the two RNA sequences have in common. That is, counting four kinds of base pairs: A-U, U-A, C-G, G-C, and the feature space becomes a 4-dimensional vector space. This feature mapping is easily constructed when the secondary structure of a target RNA sequence is available. In this paper, we consider the more general case in which secondary structure information is not available. The strategy is to count all possible base-pair candidates in the RNA sequences.

Example: An RNA sequence AUCGAGUCG contains 3 occurrences of possible A-U base-pairs, 1 occurrence of a U-A base-pair, 4 occurrences of C-G base-pairs, and 2 occurrences of G-C base-pairs. (See Figure 2 as illustration.) The feature space is a 4-dimensional vector space: (# of A-U base-pairs, # of U-A base-pairs, # of C-G base-pairs, # of G-C base-pairs), and the RNA sequence AUCGAGUCG is mapped into a vector $(3, 1, 4, 2)$.

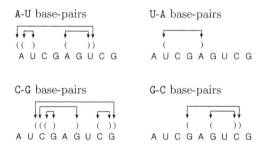

Fig. 2. Occurrences of A-U, U-A, C-G, G-C possible base-pairs contained in AUCGAGUCG

A better suited method for measuring the similarity of secondary structures which two RNA sequences have in common is to count the occurrences of possible stacking base-pairs, called *stem structures*. For example, stems of length 2 constitutes a 16-dimensional vector space: (A(A-U)U), (A(U-A)U), (A(C-G)U), (A(G-C)U), (U(A-U)A), (U(U-A)A), (U(C-G)A), (U(G-C)A), (C(A-U)G), (C(U-A)G), (C(C-G)G), (C(G-C)G), (G(A-U)C), (G(U-A)C), (G(C-G)C), (G(G-C)C).

Example: An RNA sequence AUCGAGUCG is mapped into a 16-dimensional vector space $(0, 1, 2, 0, 0, 0, 1, 0, 1, 0, 0, 2, 1, 0, 0, 0)$ for counting the occurrences of non-contiguous stems of length 2.

In this paper, the similarity feature defined by all possible non-contiguous stems of arbitrary length for an RNA sequence is examined. A novel kernel function is proposed, called *stem kernel*, to calculate the inner product of two vectors in the feature space of two RNA sequences. Hence, the feature space of all possible non-contiguous stems of an arbitrary length is of infinite dimension.

2.3   Stem Kernel: A Novel Kernel Function for RNAs

The dimension of the feature space for counting the occurrences of non-contiguous stems increases exponentially with the length of stem. In addition, the inner product of the feature vectors mapping from two RNA sequences requires a sum over all common stems. Hence, the direct computation of these features are not computationally efficient.

In order to efficiently compute the inner product of the feature vectors for two RNA sequences, we propose the stem kernel function defined in recursive form. For an RNA sequence $v = a_1 a_2 \cdots a_n$ where a_i is a base (nucleotide), we denote a_k by $v[k]$, a contiguous subsequence $a_j \cdots a_k$ by $v[j, k]$, and the length of v by $|v|$. The empty sequence is indicated by ϵ. For base a, the complementary base is denoted as $\bar{a}$. For two RNA sequences v and w, the stem kernel K is defined recursively as follows:

$$K(\epsilon, w) = K(v, \epsilon) = 1, \quad \text{for all } v, w,$$

$$K(va, w) = K(v, w) + \sum_{v[k]=\bar{a}} \sum_{i<j \text{ s.t. } w[i]=\bar{a}, w[j]=a} K(v[k+1, |v|], w[i+1, j-1]).$$

To complete the recursive equations, the following recursive equation is required:

$$K(v, wa) = K(v, w) + \sum_{w[k]=\bar{a}} \sum_{i<j \text{ s.t. } v[i]=\bar{a}, v[j]=a} K(v[i+1, j-1], w[k+1, |v|]).$$

Example: We illustrate a one step calculation of the above recursive equation with two RNA sequences $v = $ AUCCUG and $w = $ CACUAGG.

First, the two RNA sequences vG $= $ AUCCUGG and $w = $ CACUAGG have (A-U), (C-G), and (C-(C-G)-G) in common. Second, we assume that $K(v, w) = 11$, and $K(v[4, 6], w[2, 5]) = 1$, $K(v[4, 6], w[2, 6]) = 2$, $K(v[4, 6], w[4, 5]) = 1$, $K(v[4, 6], w[4, 6]) = 1$, $K(v[5, 6], w[2, 5]) = 1$, $K(v[5, 6], w[2, 6]) = 1$, $K(v[5, 6], w[4, 5]) = 1$, $K(v[5, 6], w[4, 6]) = 1$ are calculated. Then, $K(v$G$, w)$ is inductively calculated as follows:

$$
\begin{aligned}
K(v\text{G}, w) = K(v, w) + \\
K(v[4, 6], w[2, 5]) + K(v[4, 6], w[2, 6]) + K(v[4, 6], w[4, 5]) + \\
K(v[4, 6], w[4, 6]) + K(v[5, 6], w[2, 5]) + K(v[5, 6], w[2, 6]) + \\
K(v[5, 6], w[4, 5]) + K(v[5, 6], w[4, 6]) \\
= 11 + 9 = 20.
\end{aligned}
$$

The computational complexity in calculating the stem kernel function using dynamic programming is estimated. Let the two input RNA sequences v and w be of length $m(= |v|)$ and $n(= |w|)$. The computational complexity of the stem kernel $K(v, w)$ for v and w is equal to calculating a table of $m^2 \times n^2$ elements for the stem kernel function $K(v', w')$ for all subsequences v' of v and w' of w. The calculation of this table is done in time proportional to $m^2 \times n^2$ by using a parsing algorithm and an auxiliary table.

2.4 Variations of the Stem Kernel: Gap, Stack, Loop, Score Matrix

Since the form of the stem kernel introduced in the previous section is the simplest one, the recursive form is adapted in several ways for more practical use.

Gap weight

In order to deal with non-contiguous stems, we introduce a gap weight (decay factor) λ ($0 < \lambda < 1$) for a gap between two non-contiguous stacking base-pairs (See Figure 3 (left)). We need an auxiliary function K' to define the recursive equations:

$$K(va, w) = K(v, w) +$$
$$\sum_{v[k]=\bar{a}} \sum_{i<j \text{ s.t. } w[i]=\bar{a}, w[j]=a} K'(v[k+1, |v|], w[i+1, j-1]),$$
$$K'(\epsilon, w) = \lambda^{|w|}, \quad K'(v, \epsilon) = \lambda^{|v|},$$
$$K'(va, w) = \lambda K'(v, w) +$$
$$\sum_{v[k]=\bar{a}} \sum_{i<j \text{ s.t. } w[i]=\bar{a}, w[j]=a} K'(v[k+1, |v|], w[i+1, j-1])$$
$$\cdot \lambda^{k-1} \lambda^{(i-1)(|w|-j)} \ .$$

Weighting base-pairs with base pairing probability

Any occurrence of A-U, U-A, C-G, G-C pairs in an RNA sequence is a candidate of true base-pairs in the unknown secondary structure. We incorporate the base pairing probability which can be calculated by McCaskill's algorithm [7] to increase the accuracy of counting true base-pairs and stems. The base pairing probability of a pair (a, b) represents the probability of nucleotide a and nucleotide b forming a true base pair in an RNA sequence. We weight an occurrence of a base-pair with the base pairing probability in the calculation of stem kernel function. This weighting significantly improves the accuracy of the discrimination of SVM.

Stacking weight

Since the base-pair stacking contributes to stabilizing its secondary structure, we introduce a stacking weight $(1 + \alpha)$ into the kernel calculation. The stacking weight allocates larger scores for longer stems (See Figure 3 (middle)).

$$K(va, w) = K(v, w) +$$
$$\sum_{v[k]=\bar{a}} \sum_{i<j \text{ s.t. } w[i]=\bar{a}, w[j]=a} K(v[k+1, |v|], w[i+1, j-1]) \cdot (1 + \alpha).$$

This stacking weight parameter α uses the stacking energy scores in the energy-minimization method [15] to predict a secondary structure of RNAs.

Combined with the gap weight, the stacking weight assigns larger scores to contiguous stacking stems of longer length.

Hairpin loop length

The length of a hairpin loop between a closing base pair must satisfy a fixed lower bound L from the biological point of view. We define the least distance L between the closing base-pair such that no base-pairs occur inside of this region.

$$K(va, w) = K(v, w) +$$
$$\sum_{\substack{v[k] = \bar{a} \text{ s.t.} \\ |v| - k \geq L}} \sum_{\substack{i < j \text{ s.t. } w[i] = \bar{a}, w[j] = a, \\ j - i - 1 \geq L}} K(v[k+1, |v|], w[i+1, j-1]).$$

For example, G(U(A(G(G(G(G AAAA C)C)C)C)U)A)G has a hairpin loop between a
$$\underbrace{}_{\geq L}$$
closing base pair G-C which must be of a length longer than the lower bound L (See Figure 3 (right)). The parameter L is needed to satisfy the free energy function of hairpin-loops.

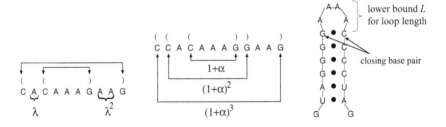

Fig. 3. Gap weight calculation (left), stacking weight calculation (middle), and lower bound for hairpin loop length (right)

Score matrix for base-pairs

It is observed that base-pairs occasionally form secondary structures which are substituted with other base-pairs. For example, a A-U base-pair could be substituted with a U-A or C-G base-pair in a *covariant mutation* substitution. A score (substitution) matrix between pairs of base-pairs is built into the recursive equations. The score matrix must satisfy the positive semi-definite condition to ensure that a valid kernel is defined. In addition, non-canonical G-U wobble base-pairs are taken into account.

$$K(va, w) = K(v, w) +$$
$$\sum_{v[k] = \bar{a}} \sum_{\substack{i < j \text{ s.t.} \\ w[i] = \bar{b}, w[j] = b}} \text{score}((\bar{a}, a), (\bar{b}, b)) \cdot K(v[k+1, |v|], w[i+1, j-1]).$$

3   Results

We tested the abilities of the stem kernel with SVMs to discriminate member sequences of an RNA family from non-members in several varying experiments. The discrimination performances were compared with the string kernel, specifically the all-subsequences kernel with gap weight (decay factor).

3.1   Data

All datasets were taken from the RNA families database "Rfam" at Sanger Institute [3].

   We generated randomly shuffled sequences as the negative sample with the same nucleotide composition as the positive sample sequence for a target RNA family. The shuffling of the sequences was accomplished while preserving the dinucleotide distribution. The discrimination performances of each method was evaluated by the 10-fold cross validation.

3.2   Discriminations of Several RNA Families

In our first experiment, the discrimination abilities of the stem kernel and the string kernel were tested on five RNA families, tRNA, miRNA (precursor), 5S rRNA, H/ACA snoRNA, and CD snoRNA. A 100 member sequence in each RNA family was chosen from the Rfam database as a positive sample and 100 randomly shuffled sequences with the same nucleotide composition were generated as a negative sample. The discrimination performance of both kernels were evaluated by the 10-fold cross validation.

   In addition to calculating the sensitivity and specificity of both kernels, we performed the receiver operating characteristic (ROC) analysis to evaluate the quality of discrimination performances. ROC curves are achieved by changing the threshold of SVM and the quality of a ROC curve can be measured by the area under the curve (AUC). The AUC score measures both sensitivity and specificity by integrating over a ROC curve which plots a true positive rate as a function of a false positive rate.

   The results are summarized in Table 1. Comparisons of the discrimination accuracy of the two kernels are competitive although the stem kernel had greater prediction accuracy overall. The string kernel achieves a better performance for 5S rRNA family because sequence similarity of 5S ribosomal RNA is relatively high.

3.3   Sample Size and Prediction Accuracy

Next, in order to evaluate more accurately the prediction abilities of both kernels and their tolerance to weak homologous sequences, we tested the kernels with only a small number of available sample sequences. The learning and prediction tasks with SVMs on different sample sizes were run from 10 sequences to 100 sequences for the five RNA families. The results of the AUC score plots are shown in Figure 4.

Table 1. Comparison of discriminating abilities of stem kernel and string kernel on five RNA families: tRNA, miRNA, 5S rRNA, H/ACA snoRNA, and CD snoRNA

Stem Kernel				
	AUC	Accuracy	Specificity	Sensitivity
tRNA	0.956	0.905	0.947	0.890
miRNA	0.917	0.750	0.692	0.900
5S rRNA	0.984	0.856	0.986	0.723
H/ACA snoRNA	0.832	0.660	0.848	0.390
CD snoRNA	0.794	0.640	0.606	0.800

String Kernel				
	AUC	Accuracy	Specificity	Sensitivity
tRNA	0.927	0.755	0.918	0.560
miRNA	0.870	0.780	0.900	0.630
5S rRNA	0.998	0.862	1.000	0.723
H/ACA snoRNA	0.809	0.680	0.929	0.390
CD snoRNA	0.851	0.655	0.772	0.440

From this experimental result, it is clear that our stem kernel exhibits a relatively good discrimination performance even if a small number of training sequences are available. Since a small sample size implies that pairwise sequence similarities among sample RNA sequences decreases, the stem kernel shows the ability to tolerate weak sequence similarities. This advantage makes the stem kernel useful in practical situations because only a small number of sequences are currently available for many of the functional RNA families in the Rfam database. On the other hand, the string kernel is sensitive to sequence similarity although the discrimination performance decreased for small samples. It is especially interesting to note in the case of the CD snoRNA family, the discrimination accuracies of the stem kernel and string kernel reverse with the smaller sample size. These results imply that the stem kernel succeeds in capturing the secondary-structure features of RNA sequences for discrimination tasks.

3.4   Kernel Principal Component Analysis

Kernel Principal Component Analysis (KPCA) is used to understand the spread of the input data in the feature space and identify correlations between input vectors and target values. KPCA is performed to visualize similarities and differences of the input RNA sequences in terms of kernel functions. We used KPCA to measure the classification and separation abilities of each kernel function from a mixed RNA sequence data of three different RNA families, 35 sequences of tRNA, 30 sequences of snoRNA, and 35 sequences of 5S rRNA.

The result in Figure 5 shows that the mixed RNA sequences are nearly separated into three clusters using the stem kernel. In contrast, the data is completely mixed using the string kernel.

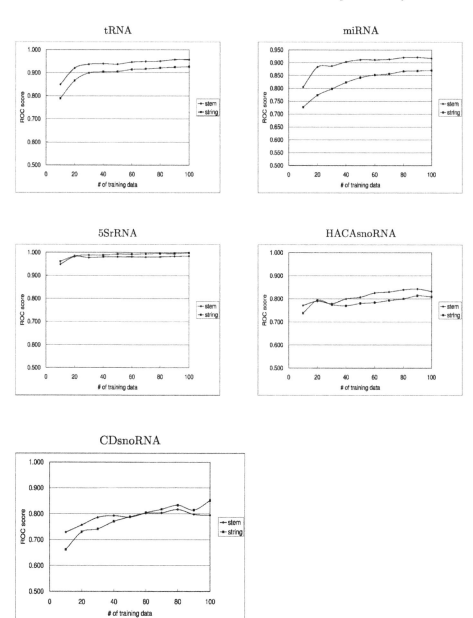

Fig. 4. AUC scores on different sample sizes from 10 sequences to 100 sequences for five RNA families

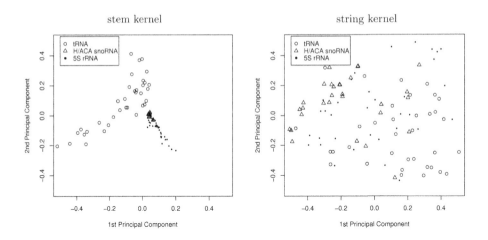

Fig. 5. Results of KPCA for mixed data of tRNA, snoRNA, and 5S rRNA sequences

3.5   Finding a Remote RNA Family

Experiments applying the stem kernel to finding remote homologs of RNA sequences in terms of secondary structures were performed.

The family of Tymovirus/Pomovirus tRNA-like 3'-UTR element (Tymo_tRNA-like) was considered to be remote homologies of tRNAs. The secondary structures of Tymo_tRNA-like elements comprise very similar secondary structures to a clover-leaf structure in tRNAs (See Figure 6 for both secondary structures). This family represents a tRNA-like structure found in the 3' UTR of Tymoviruses and Pomoviruses and are known to enhance translation. The tRNA-like structure is a highly efficient mimic of tRNA, interacting with tRNA-specific proteins as efficiently as tRNA [8].

SVMs trained from the positive and negative samples of "tRNA sequences" were applied to detect 28 Tymo_tRNA-like sequences.

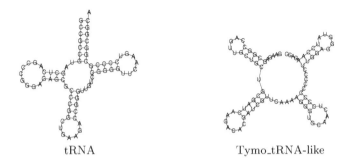

tRNA   Tymo_tRNA-like

Fig. 6. Typical secondary structures of tRNA (left) and Tymo_tRNA-like element (right)

Table 2. Prediction accuracy of detecting Tymo_tRNA-like sequences by using SVMs trained from samples for "tRNA sequences"

	Tymo_tRNA-like		
	AUC	Specificity	Sensitivity
Stem Kernel	0.603	0.614	0.964
String Kernel	0.503	0.222	0.143

The result in Table 2 shows that the stem kernel achieved significantly greater detections of Tymo_tRNA-like sequences. The string kernel, in contrast, failed to adequately detect Tymo_tRNA-like sequences. This experimental result is a strong proponent for application of the stem kernel to discover novel RNA families from genome sequences.

4 Related Works and Discussions

There is very little research on the development of kernel functions and SVMs for the task of discriminating and detecting RNA sequences. The closest study is the marginalized count kernel (MCK) for RNA sequences proposed by Kin et al. [4]. Their kernel calculates *marginalized* count vectors of base-pair features which is counting the occurrences of base pairs with probabilistically estimated hidden states and computing the inner products. The differences between MCK and the proposed stem kernel are: (i) the stem structure length considered with even the 2nd order version MCK is 2 at most. In addition, the marginalized count kernel has a feature space of fixed finite dimension. In contrast, the stem kernel considers stem structures of arbitrary lengths and hence the feature space is of infinite dimension (ii) MCK does not allow substitutions between pairs of base-pairs, (iii) MCK method requires two learning steps for training SVMs, learning the base SCFG model and learning SVMs.

Identical experiments to the previous section were performed to compare the performance of both kernels for five RNA families, tRNA, miRNA, 5S rRNA, H/ACA snoRNA, and CD snoRNA.

The result in Table 3 shows that the stem kernel outperforms the prediction accuracies of the MCK except in the case of miRNA. MCK works well for RNA

Table 3. Comparisons of stem kernel and marginalized count kernel (MCK)

	Stem Kernel (AUC)	MCK (AUC)
tRNA	0.956	0.944
miRNA	0.917	0.960
5S rRNA	0.984	0.953
H/ACA snoRNA	0.832	0.811
CD snoRNA	0.794	0.780

families with solid secondary structures such as miRNA precursor (one hairpin loop structure) and tRNA (cloverleaf structure).

One important issue of the stem kernel is the computational costs. Theoretically, for two input RNA sequences, each of length n, the computational complexity to calculate the recursive equations of the stem kernel is $O(n^4)$ when using a dynamic programming technique. In contrast, the string kernel requires $O(n^2)$ computational time. Efficient implementations of the stem kernel by using various algorithmic techniques, such as hashing, are being developed to relieve this issue.

Acknowledgements

This work is supported in part by Grant-in-Aid for Scientific Research on Priority Area No. 17018029 and grants from the Non-coding RNA Project by New Energy and Industrial Technology Development Organization (NEDO) of Japan.

References

1. Durbin, R., S. Eddy, A. Krogh, and G. Mitchison. *Biological Sequence Analysis.* Cambridge University Press, 1998.
2. Eddy, S. Noncoding RNA genes and the modern RNA world. *Nature Reviews Genetics*, 2, 2001, 919–929.
3. Griffiths-Jones, S., A. Bateman, M. Marshall, A. Khanna, and S. Eddy. Rfam: an RNA family database. *Nucleic Acids Research*, 31, 2003, 439–441. RNA families database of alignments and CMs: http://www.sanger.ac.uk/Software/Rfam/.
4. Kin, T., K. Tsuda, and K. Asai. Marginalized kernels for RNA sequence data analysis. *Genome Informatics*, 13, 2002, 112–122.
5. Lodhi, H., C. Saunders, J. Shawe-Taylor, N. Cristianini, and C. Watkins. Text classification using string kernels. *Journal of Machine Learning Research*, 2, 2002, 419–444.
6. Lowe, T. M. and S. Eddy. A computational screen for methylation guide snoRNAs in yeast. *Science*, 283, 1999, 1168–1171. http://lowelab.ucsc.edu/snoscan/ .
7. McCaskill, J. S. The equilibrium partition function and base pair binding probabilities for RNA secondary structure. *Biopolymers*, 29, 1990, 1105–1119.
8. Matsuda, D. and T. W. Dreher. The tRNA-like structure of Turnip yellow mosaic virus RNA is a 3V-translational enhancer. *Virology*, 321, 2004, 36–46.
9. Matsui, H., K. Sato, and Y. Sakakibara. Pair stochastic tree adjoining grammars for aligning and predicting pseudoknot RNA structures. *Bioinformatics*, 21, 2005, 2611–2617.
10. Sakakibara, Y., M. Brown, R. Hughey, I. S. Mian, K. Sjolander, R. C. Underwood, and D. Haussler. Stochastic context-free grammars for tRNA modeling. *Nucleic Acids Research*, 22, 1994, 5112–5120.
11. Sakakibara, Y. Pair hidden Markov models on tree structures. *Bioinformatics*, 19, 2003, i232–i240.
12. Sankoff, D. Simultaneous solution of the RNA folding, alignment and protosequence problems. *SIAM Journal on Applied Mathematics*, 45, 1985, 810–825.

13. Sato, K. and Y. Sakakibara. RNA secondary structural alignment with conditional random fields. *Bioinformatics*, 21, 2005, ii237–ii242.
14. Shawe-Taylor, J. and N. Cristianini. *Kernel Methods for Pattern Analysis*. Cambridge University Press, 2004.
15. Zuker, M. and P. Stiegler. Optimal computer folding of large RNA sequences using thermodynamics and auxiliary information. *Nucleic Acids Research*, 9, 1981, 133–148.

Prediction of Structurally-Determined Coiled-Coil Domains with Hidden Markov Models

Piero Fariselli[1], Daniele Molinini[1], Rita Casadio[1], and Anders Krogh[2]

[1] Laboratory of Biocomputing, CIRB/Department of Biology, University of Bologna, via Irnerio 42, 40126 Bologna, Italy
[2] The Bioinformatics Centre, Inst. of Molecular Biology and Physiology, University of Copenhagen, Universitetsparken 15, 2100 Copenhagen, Denmark
`piero.fariselli@unibo.it, casadio@alma.unibo.it,`
`krogh@binf.ku.dk.`

Abstract. The coiled-coil protein domain is a widespread structural motif known to be involved in a wealth of key interactions in cells and organisms. Coiled-coil recognition and prediction of their location in a protein sequence are important steps for modeling protein structure and function. Nowadays, thanks to the increasing number of experimentally determined protein structures, a significant number of coiled-coil protein domains is available. This enables the development of methods suited to predict the coiled-coil structural motifs starting from the protein sequence. Several methods have been developed to predict classical heptads using manually annotated coiled-coil domains. In this paper we focus on the prediction structurally-determined coiled-coil segments. We introduce a new method based on hidden Markov models that complement the existing methods and outperforms them in the task of locating structurally-defined coiled-coil segments.

Keywords: Protein structure prediction, Hidden Markov models, coiled-coil domains.

1 Introduction

The coiled-coil is a widespread protein structural motif [1] that has been estimated to be present in 5-10% of the sequences emerging from various genome projects [2]. Coiled-coils have a stabilization function and are frequently involved in protein-protein interaction, cell-activities, signaling and other important cellular processes [1].

Coiled-coils comprise two or more alpha-helices wound around each other in regular, symmetrical fashions to produce rope-like structures [3]. In 1953 Francis Crick and Linus Pauling both proposed models for coiled-coil structures, and although Pauling envisaged a broader set of helix periodicity (4/1, 7/2, 18/5, 15/4, 11/3), the Crick's heptad model gained more popularity, probably because he developed a full mathematical description [3]. The sequence bases of these heptad arrangements are repeating patterns of seven residues, which are labelled from *a* to *g*. A general consensus indicates more hydrophobic residues at *a* and *d* positions, which form a hydrophobic stripe on each helix.

S. Hochreiter and R. Wagner (Eds.): BIRD 2007, LNBI 4414, pp. 292–302, 2007.
© Springer-Verlag Berlin Heidelberg 2007

After 50 years of protein structures determination, we have now structures in the database endowed with the less common periodicities envisaged by Pauling and this enables us to define more general coiled-coil structures. Additionally, we are now in a position where new methods for coiled-coil prediction can be trained on databases containing also structural-derived coiled coil domains.

Coiled-coil segments can be identified in protein structures computationally, particularly with the SOCKET program [2] that was developed to identify general coiled-coil structures. The SOCKET algorithm recognizes the characteristic "knobs-into-holes" side-chain packing of coiled coils, so that it is possible to distinguish coiled-coils from the great majority of helix-helix packing arrangements observed in globular domains. SOCKET is based on the helix-packing structure and therefore coiled-coil domains can be missed in single chains when the coiled-coil is formed with another chain or in half-determined protein structures. Another invaluable source of information is the SCOP classification database [4], in which coiled-coil domains are carefully and manually annotated, and are identified as a specific class (h label). In this paper we use both resources to build a reliable coiled-coil protein database in order to train/test our and other prediction methods.

Several programs for predicting coiled-coil regions in protein sequences have been developed so far, and were parameterized on the basis of the heptad module using manual annotations and sequence similarity inference.

Most of them are based on the notion of position specific score matrices (PSSMs), such as COILS [5], PAIRCOIL [6] and MULTICOIL [7]. Also a machine learning approach (MARCOIL) based on a hidden Markov model was previously described [8]. More recently, PAIRCOIL (PAIRCOIL2 [9]) has been improved so as to include new available data including some structurally derived annotations based on the SOCKET program.

When tested on the long and classical coiled-coil domains, the accuracy of all the programs quoted above is remarkably high, but they are less accurate when they predict short or non classical coiled-coil domains as for example the ones identified by the SOCKET [9]. For this reason, in this paper we specifically focus on the task of predicting the location of structurally-annotated coiled coils domains using new hidden Markov models.

2 Method

2.1 The Protein Database

To build our data set structurally annotated coiled-coil domains, we downloaded the SOCKET pre-computed files from the SOCKET web pages. To weed out homologous pairs, the BLASTCLUST program was adopted (from the NCBI BLAST suite) with default parameters and a similarity threshold of 25%. Only one representative structure was kept from each cluster. This gave 138 sequences (SOCKET138). We also extracted all protein domains from PDB that belong to the coiled coil class according to the SCOP classification. The sequences were filtered to decrease similarity with BLASTCLUST as described above, and this gave a set comprising an additional 111 proteins (SCOP111). These 111 proteins are single representatives of each new cluster generated by BLASTCLUST that did not contain any SOCKET sequence. Our final combined data set (CC249) consists of 249 proteins with a sequence identity of less than 25%.

Furthermore, we ran the BLASTP all-against-all program on CC249 with the low complexity filter turned off. Some 50 protein pairs have sequence identities greater than 25% presumably in low complexity regions, since they were not detected by BLASTCLUST. We included all 249 proteins in our coiled-coil data set; when splitting our set for cross-validation, we made sure that no proteins in the training set had sequence identity greater than 25% with the corresponding test set. As to annotation, we dealt with two different types of files: the files generated by SOCKET and the coiled-coil domains identified in SCOP. Since in this case there is not an explicit indication where the coiled-coil domain starts, we assigned as coiled-coil regions all the helices identified by the DSSP program [10] that fall into a SCOP coiled-coil domain.

A second data set of proteins, not containing coiled-coil domains, was generated using the PAPIA system [11], by removing proteins containing coiled-coil domains. We also checked that no detectable sequence identity with sequences in CC249 were present. The final 'PAPIA' set consists of 2070 protein chains.

Finally for the sake of comparison we used the data set NEWPDB21 (http://paircoil2.csail.mit.edu/supp/new-pdb21.txt) generated for PAIRCOIL2 by McDonnell and coworkers [9]. NEWPDB21 can be regarded as blind set, since contains coiled-coil segments identified only by SOCKET program and not previously recognized using the classical sequence similarity inference and manual annotation.

Data are available at the web page: biocomp.unibo.it/piero/coiled-coils.

2.2 The Hidden Markov Models

The first model we developed and tested was similar to the MARCOIL one (see [8]), and here it is referred to as MChmm. It is endowed with one state modeling the background and 9 groups of 7 states representing the heptad repeats (a,b,c,d,e,f,g). All the states of the same repeat type are tied (they share the same emission probability distributions). This constrains the minimal coiled-coil segment length to nine residues. Contrary to the original MARCOIL model, MChmm has explicit begin and end states, which are silent (non-emitting). Our second model (CChmm1) is quite different from MARCOIL and it is depicted in Figure 1. There is one background state (L) and eight coiled-coil states. The model is fully connected and the heptad order is favored by initializing the transition probabilities, so that the probability to follow the heptad order is close to one (0.94) and that of non-heptad transitions is close to zero (0.01). Moreover, we add one more state called H to the coiled coil model. This state accounts for the deviation from the heptad periodicity, as skips, stutters and stammers [3,12]. Finally, in order to take into account different transition probabilities for sequences that contain one and those that have two or more coiled-coil segments, we introduce a third model (CChmm2) shown in Figure 2.

All training phases were performed using the labeled Baum-Welch algorithm [13] while during testing the maximum accuracy decoding [14] was adopted. In the case of CChmm1, the maximum accuracy decoding converges to the posterior-sum algorithm [13].

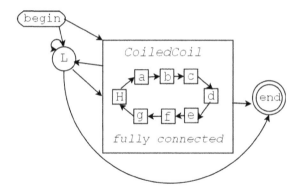

Fig. 1. Automaton representation of the CChmm1 model. The CoiledCoil box represents the coiled-coil states. For sake of clarity only the most probable transitions are indicated.

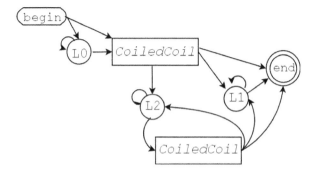

Fig. 2. Automaton representation of the CChmm2 model. The CoiledCoil boxes represent the coiled-coil states as described in Figure 1. The state emission probabilities of two CoiledCoil boxes, as well as those of the L states are tied.

2.3 Scoring the Performance

All the results obtained with our models and other methods are evaluated using the following measures of performance. The fraction of correctly predicted residues is

$$q2 = p/N \tag{1}$$

where p is the total number of correctly predicted residues and N is the total number of residues. This is also used at the sequence level as the fraction of correctly predicted sequences (containing coiled-coil or not), in which case we call it $Q2$. This rule is followed throughout: measures relating to residues are lower case and those relating to complete protein sequences are upper case.

The correlation coefficient for class s is defined as:

$$cor(s) = [p(s)n(s) - u(s)o(s)] / d(s) \tag{2}$$

where $d(s)$ is the normalization factor

$$d(s) = [(p(s)+u(s))(p(s)+o(s))(n(s)+u(s))(n(s)+o(s))]^{1/2} \tag{3}$$

For class s, $p(s)$ and $n(s)$ are the numbers of true positive and negative predictions, respectively, and $o(s)$ and $u(s)$ are the numbers of false positives and negatives, respectively. (Similarly $Cor(s)$ and $D(s)$ are defined for complete sequences of a given class s.)

The coverage or the sensitivity for each class s is

$$sn(s) = p(s)/[p(s)+u(s)] \tag{4}$$

The probability of correct predictions (accuracy or specificity) is computed as:

$$sp(s) = p(s) / [p(s) + o(s)] \tag{5}$$

(Similarly $Sn(s)$ and $Sp(s)$ are defined for complete sequences of a given class s.)

In order to score predictions on a segment basis, i.e. to which extent the predicted coiled-coil segments overlap the experimentally determined ones, we compute a segment-overlap measure introduced before [15]. Specifically, we calculate values of the segment overlap accuracy for the coiled-coil regions (SOVC), for the non-coiled-coil region (SOVN). The SOV index is a measure of the intersection divided by the union of the predicted and observed segments [15].

Finally to compute the Receiver Operating Characteristic curve we measured the True Positive Rate that is equal to $Sn(CC)$ and the False Positive Rate that is equal to 1- $Sn(N)$.

For comparison, we tested the most recently developed programs MARCOIL, and PAIRCOILS2 using their default parameters. Since MARCOIL give predictions with five different thresholds, we show the best performing threshold. For PAIRCOIL2 we used the decision threshold set to 0.025 and we tested two different size of sliding window (21 or 28 residues, respectively). Since the performances of the two window sizes are almost indistinguishable, as also stated previously by the authors (McDonnell et al., 2006), we show only the best performing one.

3 Results

3.1 Locating Coiled-Coil Segments in Protein Sequences

A major problem in protein structure prediction is the location of coiled-coil regions in proteins. A good prediction of this structural motif can also help in protein modelling procedures. The approaches developed so far (COILS, MULTICOIL, PAIRCOIL, PAIRCOIL2 and MARCOIL), have been proved to be very successful in predicting classical manually annotated coiled-coil domains. However they are less suitable to predict structurally-defined coiled-coil segments [9]. Here we tackle the specific problem of predicting structurally-defined coiled-coil segments (CC249 data set) using different hidden Markov models.

We started developing a HMM similar to that previously described in MARCOIL (MChmm), but using our new structurally-annotated data set CC249. Furthermore we developed and implemented other two HMM models: CChmm1, that does not constrain the structural motif length and CChmm2 that distinguish between chains containing one or more coiled-coil motifs (Figure 1 and 2, respectively). All the reported results were obtained using a 5-fold cross validation procedure, in which sequence identity between each training and corresponding testing set has less than 25% identity. From Table 1, we can see that the best performing method is CChmm2. This indicates that the CChmm2 model is more suited to capture the information related to the structurally annotated coiled-coils than other HMMs.

Table 1. Performance of different HMM predictors in locating coiled-coil segments in the protein sequence

Method	q2	sn(CC)	sn(N)	SOVCC	SOVN
MChmm	0.75	0.49	0.80	0.52	0.54
CChmm1	0.80	0.57	0.85	0.55	0.63
CChmm2	0.81	0.59	0.86	0.58	0.66

MChmm, CChmm1 and CChmm2 are scored using a 5-fold cross-validation procedure. CC and N represent the coiled-coil class and the non-coiled-coil class respectively.

Table 2. Performance of different HMM predictors in locating coiled-coil segments in the protein sequences of the SOCKET subset

Method	q2	sn(CC)	sn(N)	SOVCC	SOVN
MChmm	0.78	0.38	0.85	0.38	0.64
CChmm1	0.81	0.45	0.87	0.42	0.69
CChmm2	0.81	0.46	0.87	0.43	0.71

For the legend see Table 1.

Table 3. Performance of different HMM predictors in locating coiled-coil segments in the protein sequences of the DSSP-SCOP subset

Method	q2	sn(CC)	sn(N)	SOVCC	SOVN
MChmm	0.69	0.61	0.73	0.64	0.43
CChmm1	0.80	0.71	0.82	0.71	0.56
CChmm2	0.80	0.76	0.84	0.78	0.61

For the legend see Table 1.

Our CC249 training/testing set contains proteins that have been annotated using either SOCKET or DSSP-SCOP. Since the annotation procedure is different for the two methods (see Introduction) this may affect the performance. We therefore evaluated independently the two protein subsets. In Tables 2 and 3 we list the results. The different HMM predictors score similarly, with the exception of MChmm that shows a drop of performance when tested on the DSSP-SCOP subset. One possible explanation is that the DSSP-SCOP subset contains a larger number of short coiled-coil segments that are not easily detected by MChmm. CChmm2 is apparently the best method on both subsets.

3.2 Scoring the Prediction of Different Numbers of Coiled-Coil Segments

CChmm2 was developed to address the problem that the prediction of coiled-coil segments in proteins is usually more difficult for chains containing more than one coiled-coil region. CChmm2 was implemented with different transition probabilities for paths containing one, two or more coiled-coil segments (see Fig. 2). It turns out that this difference is important for the improvement observed for CChmm2. This is apparent from Table 3, where the small increased accuracy due to the protein sequences that contain more than one coiled-coil segment is shown. These findings support our HMM design.

Table 4. CChmm prediction efficiency for the coiled-coil segment location on different subsets

Subset Containing	Method	q2	sn(CC)	sn(N)	SOVCC	SOVN
1 coiled-coil	CChmm1	0.80	0.68	0.88	0.73	0.65
"	CChmm2	0.80	0.68	0.88	0.73	0.65
2 coiled-coils	CChmm1	0.80	0.42	0.86	0.38	0.64
"	CChmm2	0.81	0.43	0.86	0.40	0.65
3 or more	CChmm1	0.76	0.67	0.75	0.53	0.56
"	CChmm2	0.77	0.67	0.75	0.54	0.58

For the legend see Table 1.

3.3 Comparison with Other Methods

The main goal of this work is to develop a predictor of structurally-defined coiled coil regions to complement the exiting predictor in the task of predicting coiled-coil domains starting from the protein sequence. So that is mandatory to compare our CChmm2 with others previously introduced method specifically developed to predict classical heptad coiled-coil domains. We then compare the performance of our CChmm2 with those obtained with the two most recently introduced methods: MARCOIL [8] and PAIRCOIL2 [9]. In Table 5 we report the results of the different predictor on the NEWPDB21 data set (generated by the PAIRCOIL2 authors). This set is based only on SOCKET annotations and can be considered a perfect structurally-annotated blind test. From Table 6 we can see that our CChmm2 outperforms the existing methods on this particular data set, both on residue bases (6 percentage points of q2) and on the overlap between the predicted and observed coiled-coil segments (more than 20 percentage points on SOVCC). This finding indicates that CChmm2 is to be preferred when the prediction focuses on structurally-defined coiled coil segments.

Table 5. Comparison with other methods on the newPDB21(1) data set

Method	q2	sn(CC)	sn(N)	SOVCC	SOVN
MARCOIL	0.70	0.48	0.74	0.46	0.51
PAIRCOIL2	0.71	0.52	0.80	0.48	0.60
CChmm2*	0.77	0.85	0.66	0.73	0.68

newPDB21 is a new blind set previously generated by [9] using SOCKET algorithm.
(*) The proteins that showed sequence similarity with those of the training set were
predicted using the cross-validation parameters. For the legend see Table 1.

3.4 Discriminating Coiled-Coil Proteins Starting from the Sequence

One of the most important goals in the prediction of protein structure and function is
the classification of a protein sequence into a specific structural (functional) class.

It is interesting therefore to evaluate our new implementation in order to
discriminating coiled-coil proteins from a set of proteins with different structures,
starting from their sequence. This task is very important for structural annotation of
whole genomes. The set of proteins containing coiled-coil domains are the true
positive examples (CC249) and the filtered PAPIA set contains the negative cases
(2070 sequences). To assign a score to each protein sequence with HMMs there are
several possibilities. The most natural one is to adopt the probability of the sequence
given the HMM model ($P(s|HMM)$). However, the $P(s|HMM)$ value is not a good
discriminating function, as discussed before [13]. For this reason as a discriminative
score for our HMMs (only CChmm2 values are shown), we adopted the posterior
probability sum normalized to the protein length. More formally, if $P(\lambda(i)=\Lambda|s)$ is the
posterior probability of emitting the i-th symbol of sequence s in a state whose label is
Λ [13], then our score for that sequence is computed as:

$$D(s)=(\, \Sigma\ P(\lambda(i)=CC|\ s\ )\delta(\ argmax_{\{\Lambda\}}\,(P(\lambda(i)=\Lambda|\ s),CC\,)\,)\,/\,L \qquad (6)$$

where the summation runs over the protein length L, δ is the Kronecker delta, Λ is a
general label and CC is the coiled coil label. This equation gives the sum of the
posterior probability labelling for all the positions predicted to be in a coiled-coil state
and normalized to the protein length. The score is bounded between zero and 1, since
$P(\lambda(i)=CC|\ s\ )$ is always less or equal to 1. In this way, choosing a specific threshold
TH, a given sequence s is assigned to the coiled-coil class when its $D(s)$ score is
greater than TH.

In Figure 3 the ROC curve is obtained with different levels of $D(s)$ using the
CChmm2 model (the curve for CChmm1 is very similar). From the ROC curve it can
be evaluated that CChmm2 scores with a value of Sn(CC) (sensitivity of positive
class) equal to 40% when Sn(N) (sensitivity of negative class) is equal to 99%. In this
case the error rate is 1% (1- Sn(N)). When a larger error is accepted (35%),
sensitivity of the positive class can be as high as 80% (Sn(CC)) (Figure 3).

For comparing with other methods we run the two most recently introduced and
best performing predictors (MARCOIL and PAIRCOIL2) on the same testing set
comprising both CC249 and the PAPIA sequences for a total of 2319 chains (Table 6).
It is worth noticing that the frequency of the coiled-coil proteins in the whole

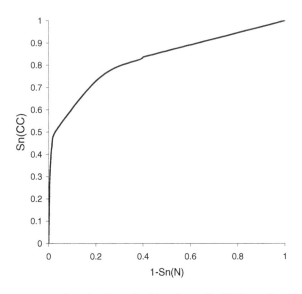

Fig. 3. ROC curve representing the True Positive Rate (Sn(CC)) as function of the False Positive Rate (1-Sn(N)) when CChmm2 is used to discriminate between proteins containing and not containing coiled coil domains. The results are obtained on the non redundant set of globular proteins (PAPIA 2070 proteins) for the False Positive Rate, and in cross validation on the set of 249 coiled coil domains for the corresponding True Positive Rate.

protein set is roughly the same as that estimated in genomes $(249/(2070+249) = 0.10;$ [2]). To compare with the other methods, we report the CChmm2 results using a discriminative threshold set to 0.5 $(D(s) > 0.5)$, which was selected to be a reasonable trade-off between the false positive and false negative rates (Fig. 3). For MARCOIL we report the best discriminating threshold that in this case is TH90 (differently from the previous task in which was TH2, see above). All methods are scoring with similar values of Q2 and values of the correlation coefficient ranging from 0.27 up to 0.56 at the most, indicating that the discriminative power for this specific task is not dependent on the coiled-coil annotation type.

Table 6. Discrimination capability of different predictors for coiled-coil-containing proteins using CC259 and PAPIA sets

Method	Q2	Sn(CC)	Sn(N)	Sp(CC)	Sp(N)	Cor
PAIRCOIL2	0.93	0.41	0.99	0.84	0.93	0.55
MARCOIL	0.92	0.40	0.99	0.79	0.93	0.53
CChmm2*	0.92	0.51	0.97	0.69	0.94	0.56

CC and N represent the coiled-coil class and the non-coiled-coil class respectively.
* Present work with a Dcc(s) threshold set to 0.5 (see Eq. 6).

4   Conclusions

In this paper we derive a database of proteins with structurally annotated coiled-coil segments to train and/or test coiled-coil prediction methods. The coiled-coil annotation does not strictly adhere to the original Crick heptad model, but can contain other shorter knob-into-hole helix-packing as detected by SOCKET or assigned by SCOP (probably closer to the original ideas of Pauling [3]). We introduce new HMMs specifically to predict these general types of coiled-coil structural domains, achieving 81% accuracy per residue and a coiled-coil segment overlap of 58%. We also compare our predictor with the two most recent available methods, which have been proved to be very effective in predicting classical coiled-coil domains [8,9] on a SOCKET-derived data set (NEWPDB21) recently introduced [9] and we showed that our method outperform them of 6 percentage points per residue, and 20 percentage points when measured by coiled-coil segment overlap (SOVCC). This indicates that our HMM (CChmm2) outperforms the existing methods in the prediction of structurally-defined coiled-coil domains, so that CChmm2 can complement the existing to predict a broader types of coiled-coil domains.

Acknowledgments. This work was supported by the following grants: MIUR for a PNR-2003 grant delivered to PF. Biosapiens Network of Excellence project (a grant of the European Union's VI Framework Programme, PNR 2001-2003 (FIRB art.8) and PNR 2003 projects (FIRB art.8) on Bioinformatics for Genomics and Proteomics and LIBI-Laboratorio Internazionale di BioInformatica delivered to RC.

References

1. Lupas A., Coiled coils: new structures and new functions. Trends Biochem Sci. (1996) 21, 375-82.
2. Walshaw J, Woolfson DN. Socket: a program for identifying and analysing coiled-coil motifs within protein structures. J Mol Biol., (2001) 307,1427-1450.
3. Gruber M, Lupas AN. (2003) Historical review: another 50th anniversary--new periodicities in coiled coils. Trends Biochem Sci., (1998) 28, 679-685.
4. Andreeva A, Howorth D, Brenner SE, Hubbard TJ, Chothia C, Murzin SCOP database in 2004: refinements integrate structure and sequence family data. Nucleic Acids Res, (2004) 32, D226-229.
5. Lupas A, Van Dyke M, Stock J. Predicting coiled coils from protein sequences. Science, (1991) 252, 1162-1164.
6. Berger B, Wilson DB, Wolf E, Tonchev T, Milla M, Kim PS. Predicting coiled coils by use of pairwise residue correlations. Proc Natl Acad Sci U S A., (1995) 92, 8259-8263.
7. Wolf E, Kim PS, Berger B. MultiCoil: a program for predicting two- and three-stranded coiled coils. Protein Sci. (1997) 6,1179-1189.
8. Delorenzi,M., and Speed,T. An HMM model for coiled-coil domains and a comparison with PSSM-based predictions. Bioinformatics, (2002) 617-625.
9. McDonnell AV, Jiang T, Keating AE, Berger B Paircoil2: improved prediction of coiled coils from sequence. Bioinformatics. (2006) 22:356-358.
10. Kabsch,W. and Sander,C. Dictionary of protein secondary structure: pattern of hydrogen-bonded and geometrical features. Biopolymers. (1983) 22, 2577-2637.

11. Noguchi,T. and Akiyama,Y. "PDB-REPRDB: a database of representative protein chains from the Protein Data Bank (PDB) in 2003", Nucleic Acids Research, (2003) 31, 492-493.
12. Lupas AN, Gruber M. The structure of alpha-helical coiled coils. Adv Protein Chem. (2005) 70,37-78.
13. Durbin,R., Eddy,S., Krogh,A. and Mitchinson,G. Biological sequence analysis: probabilistic models of proteins and nucleic acids. Cambridge Univ. Press, Cambridge.
14. Kall L, Krogh A, Sonnhammer EL. An HMM posterior decoder for sequence feature prediction that includes homology information. Bioinformatics. (2005)21, i251-i257.
15. Zemla A, Venclovas C, Fidelis K, Rost B. A modified definition of Sov, a segment-based measure for protein secondary structure prediction assessment. Proteins, (1999) 34, 220-223.

Patch Prediction of Protein Interaction Sites: Validation of a Scoring Function for an Online Server

Susan Jones and Yoichi Mukarami

Department of Biochemistry, John Maynard Smith Building, School of Life Sciences, University of Sussex, Falmer, BN1 9RH, UK

Abstract. An online protein interaction server has been designed and implemented to make predictions for 256 nonhomologous protein-protein interaction sites using patch analysis. Predictions of interactions sites are made using a scoring function that ranks four parameters, Solvation Potential, Hydrophobicity, Accessible Surface Area and Residue Interface Propensity, for overlapping patches of surface residues. Using the server, correct predictions were made for 85% of an original hand curated data set of 28 homodimers and for 65% of a new dataset of 256 homodimeric proteins. This is an increased prediction rate over the original algorithm, and proves that the method is valid for a larger set of proteins that includes more diverse interaction sites. In addition, a number of proteins for which predictions are categorized as incorrect, are shown to have alternative protein interaction sites on their surfaces.

Keywords: protein-protein interaction, homodimer, surface patch, acessible surface area, prediction.

1   Introduction

Protein interactions play key roles in a wide range of biological processes within the cell. The mechanisms by which proteins recognise and selectively interact with other proteins are key to processes such as enzyme-substrate binding and immune response. A significant knowledge of the characteristics of protein-protein interaction sites has been built up from analysis of protein complexes solved by X-ray crystallography and deposited in the Brookhaven Protein Databank (PDB)(Berman et al., 2000). An ability to identify potential interaction sites on the surface of a protein has important implications for many areas including drug design, where identification of novel binding sites can lead to new therapeutic targets.

Increasingly large datasets of protein-protein complexes have been analysed (e.g Argos, 1988; Janin et al, 1988; Janin & Chothia, 1990; Jones & Thornton, 1995; Conte et al., 1999; 2004; De et al., 2005). Through such analyses it has become evident that not all protein-protein interactions share the same general chemical and physical characteristics (Ofran and Rost, 2003). Specifically, differences in the interaction site properties of permanent and transient complexes

S. Hochreiter and R. Wagner (Eds.): BIRD 2007, LNBI 4414, pp. 303–313, 2007.

have been observed (e.g. Jones and Thornton, 1996). Permanent complexes do not exist independently, and their interfaces are more hydrophobic and more closely packed than the interfaces of transient complexes. For permanent complexes, the focus of the current work, it has been shown the interfaces are in general more hydrophobic, accessible, and protruding than than other parts of the protein surface (Argos, 1988; Janin et al, 1988; Janin & Chothia, 1990; Jones & Thornton, 1995; Conte et al., 1999).

The observation that interaction sites can be differentiated from the rest of the protein surface has been used to develop prediction methods based on the analysis of patches of surface residues (e.g. Young et a., 1994; Jones & Thornton, 1997a,b). In the original analysis of Jones and Thornton (1997a) patches of surface residues were defined on protomers from small data sets of homodimers, hetero-complexes and antibody-antigen complexes. These patches were then characterised for 6 parameters, solvation potential, residue interface propensity, hydrophobicity, planarity, protrusion and accessible surface area. This study showed that these parameters could differentiate interface patches from other surface patches and the method was developed into a prediction algorithm (Jones & Thornton, 1997b). The predictions involved the calculation of a relative combined score, based on the 6 parameters, which gave the probability that a surface patch will form a protein-protein interaction site. Neural networks have also been applied to the prediction of protein-protein interactions using sequence profiles and accessible surface area as selection parameters (Zhou & Shan, 2001; Fariselli et all, 2002; Ofran & Rost, 2003). In closely related work, support vector machines have also been implemented to predict interaction sites using residue profiles, accessible surface area and hydrophobicity as selection parameters (Yan et al., 2004; Koike & Takagi, 2004).

An alternative method of predicting protein-protein interaction sites has been to dock two protein structures together. In generic terms docking predicts the structure of a complex on the basis of shape and chemical complementarity (e.g. Norel et al, 1999; Smith & Sternberg, 2002; Aloy & Russell, 2003; Comeau et al, 2004). However, this methodology is only applicable when the structures of the two interacting proteins are known. Docking methods generally involve the scoring of combined conformations and the refinement of a combined complex (Smith & Sternberg, 2002). One such method. that predicts all of the possible interactions between homologous proteins has been developed and implemented as a web server (Aloy & Russell, 2002; Aloy & Russell, 2003).

The current work presents an update on the original implementation of patch analysis for protein interaction site prediction (Jones & Thornton, 1997b). This includes the development of an on-line server that has been validated in the prediction of a large dataset of 256 nonhomologous homodimer structures. Improvements in the way in which the combined score is used to rank the patches has led to an increase in prediction rate for the original hand-curated dataset of Jones and Thornton (1997ab), and gives a high prediction rate for a new data set that is ten fold larger. Improvements to the algorithm means that the on-line server provides an efficient and flexible method for the prediction of interaction sites in 3D protein structures.

2   Methods

2.1   Data Set Extraction

A data set of 256 non-homologous homodimeric proteins was extracted from the PDB using search and classification tools available in the Macromolecular Structure Database (MSD)(Brooksbank, 2003; Velankar, 2005), Swiss-Prot (Boeckmann et al, 2003) and the CATH database (Orengo et al, 1997). A data set of homodimers was extracted from MSD by using the MSDpro with key words; assembly class = *homo*, assembly type = *dimeric*, experiment type = *x-ray* and resolution = $< 3.0\text{Å}$.

A data set of homodimers was extracted from Swiss-Prot by using the Sequence Retrieval System (SRS) with key words; comment = *homodimer* and Db-Name = *PDB*. The PDB entry codes extracted from MSD were cross-referenced with those extracted from Swiss-Prot. If a PDB code occurred in both lists it was retained in the dataset, if it only occurred in one list it was removed. This cross-validation was necessary to confirm the classification of assembly class in MSD, as this was known to contain some incorrect classifications.

This MSD:Swissprot cross-validation produced a list of 2158 PDB codes, which included a large number of homologous proteins. To extract a nonhomologous list of proteins the CATH classification database (v2.6.0) was used. This database classifies protein domain structures into a hierarchy which includes four major levels; Class(C), Architecture(A), Topology(T) and Homologous superfamily(H) (Orengo et al, 1997). Each domain is assigned a unique number (CATH domain ID) based on its place in the hierarchy. CATH domain IDs were assigned to each PDB code in the list of homodimers, and where there was more than one domain the IDS were concatenated to form a single ID that represented all the domains in that protein. The allocation of CATH domain IDs to each PDB code allowed the proteins to be grouped into homologous families. The PDB code with the highest resolution was then selected from each homologous family ID. For a homologous family that had multiple PDB codes sharing the same highest resolution, the PDB code of the largest protein complex was selected. In addition, further restrictions were made that excluded any PDB code containing a nucleic acid molecule from the dataset. This process resulted in a data set of 256 nonhomologous homodimeric proteins (Table 1). The dataset from the original work that comprised just 28 homodimeric proteins (Jones and Thornton, 1997(b)) was also used for comparison in the validation of the scoring function.

2.2   Definition of Patch Size

A surface patch was defined as a central surface accessible residue and n nearest neighbours (Jones & Thornton (1997(b))). Estimating the size of the patch

Table 1. Data set of 256 non-homologous homodimeric proteins extracted from the Protein Data Bank (Berman et al., 2000)

1ade	1aih	1aor	1b67	1bg6	1bgf	1ci4	1cmb	1cno	1csh	1ddt	1dek
1e7l	1eej	1eto	1eyv	1f8r	1fx7	1g8m	1gte	1gx5	1hei	1hss	1hw1
1i6m	1itk	1ixm	1jhg	1k3e	1k8u	1kbl	1kbl	1l5o	1lky	1n1b	1nkd
1nkh	1on2	1oq4	1p7n	1q4g	1qi9	1qkm	1qks	1qo8	1qq7	1ses	1tbb
1tw6	1u0e	1uby	1utg	1vsg	2hgs	2pgd	2tct	3sdh	5csm	6pah	1bg5
1brw	1d9c	1e85	1joc	1jqk	1jr8	1k3y	1kny	1l8s	1o17	1qdb	1qjb
1quu	2ilk	3lyn	1dvp	1fp3	1ktz	1mkk	2abx	1cru	1fwx	1flg	1fjr
1pre	1rkd	1flm	1g9o	1jub	1b8a	1bdo	1esr	1fw3	1gvp	1jsg	1l0w
1mvp	1pbo	1uj1	7odc	1cll	1dmh	1ecy	1edh	1ern	1ig0	1k2f	1py9
1q0e	1sox	1x82	2arc	4kbp	1c39	1f3l	1oac	1sii	1lbg	1af5	1by2
1e87	1koq	1mka	1mmi	1nki	1oh0	1oqj	1bow	1e4m	1eye	1fvp	1gve
1h16	1i2k	1jgm	1k87	1n55	1now	1o94	1one	1qpo	12as	1auk	1avv
1b6r	1bol	1c0p	1cbf	1dj0	1eb0	1evl	1fbn	1fi4	1fxd	1g99	1gpe
1hjr	1iho	1j8b	1j98	1jjh	1k0i	1k3s	1kpf	1mu4	1ni9	1o9t	1prx
1puc	1pvg	1q4r	1qmi	1qqq	1t4b	1yer	2hhm	3grs	3ssi	6cro	9wga
1a3a	1a3c	1a4i	1afw	1b73	1bam	1bif	1byi	1byk	1c8k	1cby	1chm
1ctt	1cz3	1d4a	1dbq	1dcf	1djl	1e19	1e59	1e5m	1ev7	1ew2	1ez0
1fcj	1g60	1g8t	1gkd	1gpu	1gu7	1gz0	1hku	1hqs	1hyu	1i24	1i6a
1j9j	1jdn	1jgt	1js3	1jsc	1k0z	1kqp	1l6r	1l8a	1lc5	1lhp	1m2d
1moq	1nox	1oaa	1ooy	1pea	1q92	1qj4	1soa	1ueh	2cmd	3gar	1j7g
1trb	1f89	1jdw	1czj	1d0c	1e9g	1ex2	1fjj	1g57	1gpc	1hyo	1hzt
1i52	1uyr	1cku	1huu								

$(n + 1(\text{central residue}))$ is an important factor in a successful prediction. Jones & Thornton (1995) proposed that there was an approximate correlation between the size of a protein-protein interface and the size of the protomer. Hence, the correlation between the number of residues in a protomer (NR_p) and the number of residues in the observed interface (NR_i) was calculated using the 256 homodimers. A non-linear regression line fitted that had a correlation coefficient of 0.63 (equation 1).

$$NR_i = 1.91 NR_p^{0.55} \tag{1}$$

This was very similar to the regression line used to calculate the size of patches for the dataset of 28 homodimers in the original paper $(NR_i = 1.92 NR_p^{0.56}(r = 0.74).)$ (Jones and Thornton, 1995; Jones and Thornton, 1997b).

2.3 Prediction Algorithm and Combined Score Definition

The patch prediction algorithm includes four major steps

1. The surface of a protomer is divided into a number of overlapping patches of residues, the size of which is determined by a regression line equation
2. A number of parameters are calculated for each patch, and these values are scored on a scale of 1 to 100

3. The multiple parameters calculated for a patch are combined into an individual score that gives a percentage probability using a combined score equation. On this probability scale 1 denotes a very low probability that the patch is a putative interaction site and 100 denotes very a high probability that it is a putative interaction site.
4. Those patches with the highest combined score are selected as the most likely putative interaction sites.

In the original work by Jones and Thornton (1997b) predictions were made using six parameters; *Solvation Potential* (S_{sp}), *Hydrophobicity* (S_{sp}), *Accessible Surface Area* (S_{asa}), *Residue Interface Propensity* (S_{rp}), *Planarity* (S_{pl}) and *Protrusion* (S_{pr}). These were combined into a rank score defined in equation 2.

$$CombinedScorePatch_i(Score\_6) = \frac{S_{sp} + S_{hy} + S_{asa} + S_{rp} + S_{pl} + S_{pi}}{N_p} \quad (2)$$

In the current work the effect of each parameter upon the prediction of the 256 homodimeric dataset were analysed (data not shown) and a new combined score proposed that only included 4 parameters; *solvation potential* (S_{sp}), *hydrophobicity* (S_{hy}), *accessible surface area* (S_{asa}) and *residue interface propensity* (S_{rp}). These parameters were combined into a rank score defined in equation 3, where *Np* is the number of parameters.

$$CombinedScorePatch_i(Score\_4) = \frac{S_{sp} + S_{hy} + S_{asa} + S_{rp}}{N_p} \quad (3)$$

2.4   Online Server for Patch Prediction

The original method developed by Jones & Thornton (1997(b)) has been implemented to incorporate this new combined score definition. In addition an anomaly in the original algorithm that influenced the way in which patches were defined in some proteins, was corrected. This results in an online server which provides more reliable predictions with a shorter execution time. The prediction algorithm is summarised in Figure 1 and can be accessed at URL http://www.bioinformatics.sussex.ac.uk/SHARP2. The prediction server enables users to upload novel protein structures which have not yet been deposited in the PDB, and to obtain the result of predictions via E-mail or web browser. In addition, the user can view the predicted interaction sites on the 3D structure of the protein in a Jmol viewer. Further details of the flexibility of the server have been published elsewhere (Jones and Mukarami, 2006).

3   Results

The predictions for protein interaction sites on the surface of protomers in the data sets of 28 and 256 non-homologous homodimeric proteins were carried out with two different combined score definitions (i) using 4 parameters (*Score_4*)

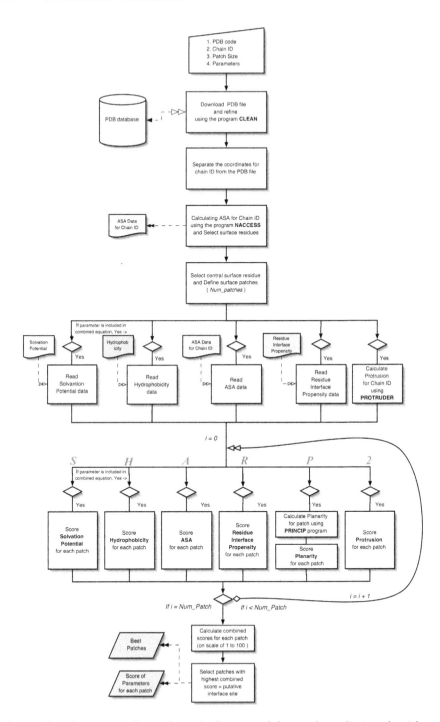

Fig. 1. Flow diagram outlining the main features of the patch prediction algorithm as implemented in the online prediction server

(equation 3) as proposed in the current study, and (ii) using 6 parameters (*Score_6*) (equation 2) as in the original analysis (Jones and Thornton, 1997b). The size of a patch was estimated using two regression lines; $NR_i = 1.92NR_p^{0.56}$ and $NR_i = 1.91NR_p^{0.55}$, the former calculated in the original study (Jones & Thornton, 1997(b)), the latter derived from the 256 proteins extracted here.

Two measures, used by Jones & Thornton (1997(b)), to assess the accuracy of predictions were calculated for each of the top three patches with the highest combined scores in each protomer. The first is the percentage overlap value (*P1*) which evaluates how each patch sampled the observed interface, calculated as

$$PercentageOverlap(P1) = \frac{NrO \bigcap NrC}{NrO} \times 100 \qquad (4)$$

where NrO is the number of residues in the observed interfaces and NrC is the number of residues in the calculated interface patch. The second is the relative overlap (*P2*) value which is calculated as

$$RelativeOverlap(P2) = \frac{PercentageOverlap(P1)}{maximumP1} \qquad (5)$$

where $maximumP1$ is the patch with the highest percentage overlap with the observed interface. As defined by Jones & Thornton (1997(b)), if the relative overlap value (*P2*) was more than 70% for any of the top three patches, the prediction was defined as correct.

In addition, the rank order of the patch with the maximum overlap with the observed interface (*MaximumP1*) was calculated within the total number of patches, (*Rank Max P1*). A rank order of 1 denoted that *MaximumP1* had the highest combined score of all patches, this means that the predictive algorithm provided the best possible result. The results of the 10 best predictions and the 10 worst predictions for the data set of 256 homodimers are shown in Table 2. A summary of the prediction results using different combined scores and different regression lines to calculate the size of the patch is shown in Table 3.

In the original data set of 28 examples, the predictions with *Score_4* and $NR_i = 1.92NR_p^{0.56}$ was successful for 82%(23/28), and the predictions with *Score_6* were successful for 75%(21/28) (Table 3). In the original study (Jones & Thornton (1997(b))), the success rate of the predictions using *Score_6* was 66%(19/28). The increased success rate from 66% to 75% was due to the correction of the method for making distance calculations for nearest neighbour residues, that affected the size of surface patches for 4 proteins in the dataset. The increase to a prediction rate of 82% was due to the exclusion of two of the parameters in the combined score definition (*Score_4*).

In the new data set of 256 examples, the predictions with *Score_4* and $NR_i = 1.91NR_p^{0.55}$ were successful for 65% (166/256), and the predictions with *Score_6* were successful for 61% (157/256) (Table 3). Thus, in the large data set of 256 examples presented here, the regression line $NR_i = 1.91NR_p^{0.55}$ gave the best estimate of patch size. In addition, the prediction with *Score_4* always had a higher success rate than that with *Score_6* (Table 3). Hence, the regression

Table 2. Results of the interaction site predictions for the 10 best and 10 worst proteins in dataset of 256 non-homologous homodimeric proteins, based on the rank of the patch with the max P1 value (last column)

PDB code	Chain size	Patch size	Overlap P1 (%) 1st	2nd	3rd	Max P1 (%)	Relative overlap 1st	2nd	P2 (%) 3rd	No.diff patches	Rank max P1
1lbg	111	27	52.3	34.1	38.6	52.3	100.0	65.2	73.9	2	1
1b73	184	40	76.2	28.6	47.6	76.2	100.0	37.5	62.5	2	1
1dmh	247	45	53.2	53.2	33.8	53.2	100.0	100.0	63.4	2	1
1e19	229	45	54.8	41.9	46.8	54.8	100.0	76.4	85.3	1	1
1e5m	248	52	56.6	44.7	52.6	56.6	100.0	79.0	93.0	1	1
1gu7	269	49	96.0	88.0	88.0	96.0	100.0	91.6	91.7	1	1
1gvp	79	22	79.2	75.0	62.5	79.2	100.0	94.7	78.9	1	1
1hjr	124	31	87.5	0.0	25.0	87.5	100.0	0.00	28.6	3	1
1hku	252	46	53.6	52.1	50.7	53.6	100.0	97.3	94.6	1	1
1ig0	227	45	75.0	75.0	67.5	75.0	100.0	100.0	90.0	2	1
..											
..											
1c8k	543	77	35.4	32.3	21.5	53.8	65.7	60.0	40.0	2	87
1gx5	391	59	25.0	25.0	25.0	77.8	32.1	32.1	32.1	1	91
1ern	170	36	4.8	9.5	0.0	90.5	5.2	10.5	0.0	2	94
1oq4	252	48	26.4	30.6	33.3	55.6	47.5	55.0	60.0	1	96
1qks	369	62	20.5	12.8	15.4	79.5	25.8	16.1	19.3	1	106
1kbl	590	79	0.0	0.0	19.4	67.7	0.0	0.00	28.6	2	122
1q4g	396	62	18.2	21.2	12.1	57.6	31.6	36.8	21.0	1	128
1auk	308	57	0.0	0.0	0.0	92.1	0.0	0.00	0.00	1	132
1fp3	270	52	0.0	0.0	0.0	87.1	0.0	0.00	0.00	2	141
1hei	318	55	2.8	33.3	30.6	77.8	3.57	42.8	39.3	1	143

Table 3. Prediction accuracy for two data sets of non-homologous homodimeric proteins. Prediction rates are shown using two combined score definitions and two regression lines to define the size of the surface patches.

Parameters		Examples	$NR_i = 1.92NR_p{}^{0.56}$	$NR_i = 1.91NR_p{}^{0.55}$
Score_4	S_{sp}, S_{hy}	256	64% (164/256)	**65% (166/256)**
	S_{asa}, S_{rp}	28	**82% (23/28)**	75% (21/28)
Score_6	S_{sp}, S_{hy}, S_{asa}	256	59% (151/256)	**61% (157/256)**
	S_{rp}, S_{pl}, S_{Pi}	28	**75% (21/28)**	67% (19/28)

line $NR_i = 1.91NR_p{}^{0.55}$ and combined score definition that used 4 parameters (*Score_4*) is used as the default settings for the online prediction server.

4 Discussion

This new implementation of the patch prediction algorithm as an online server provides a fast and reliable means of making predictions of protein interaction sites. The implementation increased the prediction rate for the original data set

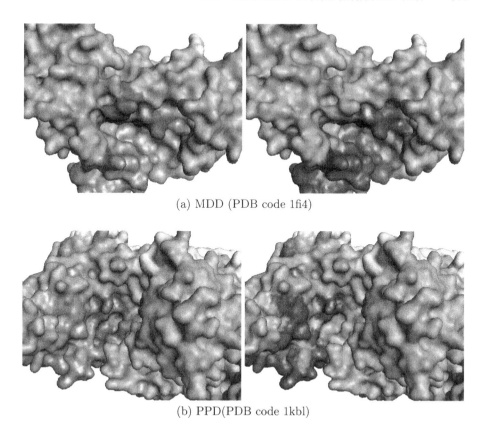

(a) MDD (PDB code 1fi4)

(b) PPD(PDB code 1kbl)

Fig. 2. Alternative binding sites for two "incorrectly" predicted proteins (1kbl, 1fl4) (a) Mevalonate 5-diphosphate decarboxylase (MDD, 1fi4), the binding site of mevalunate is shown in red (right), the predicted best interface is shown in blue (left). (b) Phyruvate phosphate dikinase (PPD, 1kbl), the binding site of pyruvate is shown in red (right), the predicted best interface is shown in blue. (left). In all figures the true interface site is shown in yellow.

of 28 homodimers to 83%, and produced a prediction rate of 65% for the new data set of 256 nonhomologous homodimers. This work proves that the original premise of the algorithm was valid, and holds for much larger datasets that include more diverse protein-protein interaction sites.

Proteins whose binding sites were not predicted correctly were studied in more detail, and a number were observed to have more than one protein interaction site on their surface. Two examples are highlighted here, Mevalonate 5-Diphosphate Decarboxylase (MDD, *PDB entry code*; 1fi4) and Pyruvate Phosphate Dikinase (PPDK, *PDB entry code*; 1kbl). The interaction sites of these proteins were never predicted using combined score *Score_4*, but it was observed that the patches with a highest combined score were placed near or overlapped with substrate binding sites within these proteins (Figure 2).

MDD is a single-domain α/β protein with a deep cleft, which is responsible for the synthesis of isopentenyl diphosphate from mevalonate (Bonanno et al (2001)). Residues Pro113, Thr114, Ala115, Ala116, Gly117, Leu118, Ala119, Ser120, Ser121 and Ala122 that lie within a deep cleft (Figure 2 (a)) are candidates for binding the anionic substrate, mevalonate-5-diphosphate (Bonnanno et al, 2001). In the prediction of 1fi4, the best patch with a highest combined score overlapped part of the substrate binding site within the cleft at the C-terminal end of the chain. A similar case is seen with PPDK (Figure 2(b)), which is responsible for catalysis for interconverting ATP, Pi and pyruvate with AMP, PPi and PEP(phosphoenolpyruvate) (Herzberg et all, 2002). In this enzyme the Arg561, Arg617, Asp620, Glu745, Asn768, Asp769 and Cys831 are known to form the pyruvate/PEP binding site within the centre of the α/β barrel in the C-terminal end of the chain. In the prediction of 1kbl, the best patch with a high combined score overlapped with part of this substate binding site. Hence even those predictions that are classified as "incorrect" actually revealed interesting information.

This new implementation of patch analysis prediction in the form of an on-line server provides a fast means of predicting protein interaction sites with an increased rate of correct predictions. The fact that the server can reveal information about alternative binding sites on the surface of a protein, means that it has the potential to contribute to the identification of novel therapeutic target sites.

References

1. Aloy, P. & Russel, R.B. (2002) Interrogating protein interaction networks through structural biology. *Proc.Natl.Acad.Sci (PNAS)*, vol.99, no.9, pp.5896-5901.
2. Aloy, P. & Russell, R.B. (2003) InterPreTS: protein interaction prediction through tertiary structure. *Bioinformatics.* vol.19, no.1, pp.161-162.
3. Argos, P. (1988) An investigation of protein subunit and domain interface. *Protein Eng.* vol.2, pp.101-113.
4. Berman, H.M., Westbrook, H., Feng, Z., Gilliland, Z., Bhat, T.N., Weissig, H., Shindyalov, I.N. & Bourne, P.E. (2000) The Protein Data Bank. *Nucleic Acids Research*, vol.28 pp. 235-242.
5. Boeckmann, B., Bairoch, A., Apweiler, R., Blatter, M.C., Estreicher, A., Gasteiger, E., Martin, M.J., Michoud, K., O'Donovan, C., Phan, I., Pilbout, S. & Schneider, M. (2003) The Swiss-Prot protein knowledgebase and its supplement TrEMBL in 2003. *Nucl. Acid. Res.*, vol.31, no.1, pp.365-370.
6. Bonanno, J.B., Edo, C., Eswar, N., Pieper, U., Romanowski, M.J., Llyin, V., Gerchman, S.E., Kycia, H., Studier, F.W., Sali,A. & Burley, S.K. (2001) Structural genomics of enzymes involved in sterol / isorenoid biosynthesis. *Proc.Natl.Acad.Sci (PNAS)*, vol.98, pp.12896-12901.
7. Brooksbank, C., Camon, E., Harris, A.M, Magrane, M., Martin, M.J., Mulder, N., O'Donovan, C., Parkinson, H., Tuli, M.A., Apweiler, R., Birney, E., Brazma, A., Henrick, K., Lopez, R., Stoesser, G., Stoehr, P. & Cameron, G. (2003) The European Bioinformatics Institute's data resources, *Nucl. Acid. Res.*, vol.31, no.1, pp.43-50.

8. Comeau, S.R., Gatchell, D.W., Vajda, G. & Camacho, C.J. (2004) ClusPro: an automated docking and discrimination method for the prediction of protein complexes. *Bioinformatics*, vol.20, no.1, pp.45-50.
9. Conte, Lo, Chothia, C. & Janin, J. (1999) The atomic structure of protein-protein recognition sites. *J. Mol. Biol*, vol.285, pp.2177-2198.
10. De S, Krishnadev,O Srinivasan N, & Rekha N. (2005) Interaction preferences across protein-protein interfaces of obligatory and non-obligatory components are different. *BMC Structural Biology* vol. 5, pp56-65.
11. Fariselli, P., Pazos, F., Valencia, A. & Casadio, R. (2002) Prediction of protein-protein interaction sites in heterocomplexes with neural networks. *Eur. J. Biochem*, vol.269, pp.1356-1361.
12. Herzberg, O., Chen, C.C.H., Fiu, S., Tempczyk, A., Howard, A., Wei, M., Ye, D. & Dunaway-Mariano, D. (2001) Pyruvate site of pyruvate phosphate dikinase: crystal structure of the enzyme-phosphonopyruvate complex, and mutant analysis. *Biochem.*, vol.41, pp.780-787.
13. Janin, J., Miller, S. & Chotia, C. (1988) Surface, subunit interfaces and interior of oligomeric proteins. *J.Mol.Biol*, vol.204, pp.155-164.
14. Janin, F & Chothia, C. (1990) The structure of protein-protein recognition sites. *J.Biol.Chem*, vol.265, pp.16027-16030
15. Jones, S. & Thornton, J. M. (1995) Protein-Protein interactions: a review of protein dimer structures. *Prog.Biophys.Mol.Biol*, vol.63, pp.31-65.
16. Jones, S. & Thornton, J.M. (1996) Principles of protein-protein interactions. *Proc.Natl.Acad.Sci (PNAS).* USA, vol.93, pp.13-20.
17. Jones, S. & Thornton, J. M. (1997(a)) Analysis of protein-protein interaction sites using surface patches. *J.Mol.Biol*, vol. 272, pp.121-132.
18. Jones, S. &Thornton, J. M. (1997(b)) Prediction of protein-protein interaction sites using patch analysis. *J.Mol.Biol*, vol.272, pp.133-143.
19. Koike, A. & Takagi, T. (2004) Prediction of protein -protein interaction sites using support vector machines. *Prot. Eng.*, vol.17, no.2, pp.165-173.
20. Murakami, Y. & Jones, S. (2006) SHARP2: protein-protein interaction predictions using patch analysis. *Bioinformatics In Press.*
21. Norel, R., Petrey, D., Wolfson, H.J. & Nussinov, R. (1999) Examination of shape complementarity in docking of unboun d proteins. *Proteins*, vol.36, pp.307-317.
22. Ofran, Y & Rost, B, (2003) Analysing six types of protein-protein interface. ₁J. Mol. Biol. vol. 325 pp.377-387.
23. Orengo, C.A., Michie, A.D., Jones, S., Jones, D.T., Swindells, M.B. & Thornton, J.M. 1997, CATH - a hierarchic classification of protein domain structures, *Structure*, vol.5, no.8, pp.1093-1109.
24. Smith, G.R. & Sternberg, M.J, (2002) Prediction of protein-protein interactions by docking methods, *Current opinion in structural biology*, vol.12, pp.28-35.
25. Velankar, S., McNeil, P., Mittard-Runte, V., Suarez, A., Barrell, D., Apweiler, R. & Henrick, K. (2005) E-MSD: an intergrated data resource for bioinformatics, *Nucleic Acids Research*, vol.33, pp.D262-265.
26. Yan, C., Honavar, V. & Dobbs, D. (2004) Identification of interface residues in protease-inhibitor and antigen-antibody complexes: a support vector machine approach, *Neural Comput & Applic*, vol.13, pp.123-129.
27. Young, L., Jerniga, R.L. & Covell, D. G. (1994) A role for surface hydrophobicity in protein-protein recognition. *Protein Sci.* vol.3, pp.717-729.
28. Zhou, H.X. & Shan, Y. (2001) Prediction of protein interaction sites from sequence profiles and residue neighbour list, *Proteins.*, vol.44, pp.336-343.

Statistical Inference on Distinct RNA Stem-Loops in Genomic Sequences

Shu-Yun Le[1] and Jih-H. Chen[2]

[1] CCR Nanobiology Program, NCI Center for Cancer Research,
National Cancer Institute, NIH,
Bldg. 469, Rm 151, Frederick, Maryland 21702, USA
[2] Advanced Biomedical Computing Center
SAIC-NCI/FCRDC, Frederick, MD 21702, USA

Abstract. Functional RNA elements in post-transcriptional regulation of gene expression are often correlated with distinct RNA stem-loop structures that are both thermodynamically stable and highly well-ordered. Recent Discoveries of microRNA (miRNA) and small regulatory RNAs indicate that there are a large class of small non-coding RNAs having the potential to form a distinct, well-ordered and/or stable stem-loop in numbers of genomes. The distinct RNA structure can be well evaluated by a quantitative measure, the energy difference (E_{diff}) between the optimal structure folded from the segment and its corresponding optimal restrained structure where all base pairings formed in the original optimal structure are forbidden. In this study, we present an efficient algorithm to compute E_{diff} of local segment by scanning a window along a genomic sequence. The complexity of computational time is $O(L \times n^2)$, where L is the length of the genomic sequence and n is the size of the sliding window. Our results indicate that the known stem-loops folded by miRNA precursors have high normalized E_{diff} scores with highly statistical significance. The distinct well-ordered structures related to the known miRNA can be predicted in a genomic sequence by a robust statistical inference. Our computational method StemED can be used as a general approach for the discovery of distinct stem-loops in genomic sequences.

Keywords: well-ordered RNA structure, dynamic programming, statistical inference, microRNA stem-loops.

1 Introduction

Recent advances in studies of non-coding RNAs (ncRNAs) and RNA interference (RNAi) indicate that RNA is more than a messenger between genome and protein. The ncRNAs are involved in various regulatory mechanisms of gene expression at multiple levels. The functional structured RNAs (FSRs) that can perform the regulatory activity comprise transfer RNA, ribosomal RNAs, self-cleavage ribozymes, small microRNAs (miRNAs) and various RNA regulatory elements, such as iron-responsive element (IRE) in the non-coding region (NCR)

S. Hochreiter and R. Wagner (Eds.): BIRD 2007, LNBI 4414, pp. 314–327, 2007.
© Springer-Verlag Berlin Heidelberg 2007

of ferritin mRNAs, internal ribosome entry sequence in the 5' NCR and cis-acting RNA elements involving in nuclear mRNA export, such as Rev response element (RRE) of HIV-1 and constitutive transport element (CTE) of Mason-Pfizer monkey virus [1]. The known biological functions of ncRNA continue to grow and newly discovered miRNA genes are one of the new classes of regulatory genes in animals. The $\sim$ 22 nucleotides (nt) miRNAs can control gene expression by binding to complementary sites in the 3' NCR of target mRNA. It is interesting to note that most of miRNA precursors are of $\sim$ 80 nt in length and form a conserved fold-back stem-loop structure across the divergent species in which the conserved $\sim$ 22 nt miRNA sequences are within one arm containing at least 16 base-pairings. Intriguingly, about 3800 distinct miRNA genes have been determined in various genomes from the plant to animal [2]. It is conceivable that there are a large number of various undiscovered ncRNAs or FSRs in the plant and animal genomes.

A complete understanding of a FSR requires a knowledge of its primary sequence and 3-D structure. Currently, computational methods for analysis and detection of FSRs have made a great progress. A number of tools [3-9] such as tRNAscan-SE, Palingol, RNAMotif, ERPIN, RNAprofile, HomoStRscan, RE-SEARCH and INFERNAL have been developed and have practical applications to the search for FSRs of which primary sequences and higher-ordered structures are well determined, such as tRNA, 5S rRNA genes, signal recognition particle and IRE. In contrast to these successful programs and algorithm, a few computational methods are available for finding structured RNA elements as potential FSRs or regulatory elements in mRNAs of which primary sequences and higher-ordered structures are not well determined [10]. A computational machine learning approach using neural networks has been recently developed to identify genes for FSRs in genomic sequences [11]. Some new computational procedures integrating various methods and experimental data have been developed for identifying miRNAs encoded in various genomes [12-18].

RNA structure comparison and analysis from a number of laboratories show that some specific combinations of base pairings and some conserved loop sequences in stem-loops are more abundant in FSRs [1]. A computational test on tRNAs, RNase P RNAs, TAR and RRE of HIV-1, IRES of HCV and ribozyme showed that the FSRs had well-ordered conformations that were unlikely to occur by chance [19]. The observation and computational test suggested that FSRs possess well-ordered conformations that are both thermodynamically stable and uniquely folded.

Our hypothesis is that the quality of the well-ordered RNA structure can be assessed by the computed energy difference (E_{diff}) between the folded optimal structure (OS) and its corresponding optimal restrained structure (ORS) in which all the previous base pairings in the OS are forbidden. Recent studies indicated that known miRNA precursors and other FSRs were closely related to the well-ordered stem-loops with high statistical significance [10, 18]. The previously developed EDscan [10] can extract the signals that have statistically high normalized scores of E_{diff} computed from these successive segments of length

n by scanning the window of size n along the genomic sequence of size L. The method EDscan needs a computation time of $O(L \times n^3)$. Also, the extracted signals are quite sensitive to the size of sliding window selected in the computation. In order to discover the well-ordered stem-loops in genomic sequence, we present here an efficient algorithm StemED to compute E_{diff} for the local fragments by moving a window along a genomic sequence. The computational time is improved to the order of $O(L \times n^2)$. Furthermore, our tests show that the signals of the identification of miRNA genes in genomic sequence are quite stable even if the variant window sizes are selected. This is especially important when the data mining approach is used to explore those undiscovered FSRs in a genomic sequence in the first step. The predicted candidates of FSRs or miRNA precursors can be verified further by additional information, such as experimental data or the conservation of both the sequence and structure.

2   Methods

In our method StemED, the quality of a locally well-ordered folding sequences (WFS) is evaluated by a quantitative measure E_{diff} that is defined as the difference of free energies between the OS and its corresponding ORS [10]. That is

$$E_{diff} = E_f - E$$

where E and E_f are the lowest free energies of the OS and ORS folded by the local segment (S), respectively. To compute E_f, we first have to determine the secondary structure of the OS folded by the S in addition to compute E. By prohibiting all base-pairings in the folded OS, we then compute the lowest free energy of the local segment S again. Thus, we need to fold the segment S twice under the two specific conditions in order to compute E_{diff} of S.

It is well established that most nucleotide residues in all RNA molecules interact to form short, intramolecular helical regions and unpaired loops, so defining a secondary structure. The helical stem regions consist of a series of uninterrupted Watson-Crick or wobble G:U base-pairs. The loops include hairpin loops, internal loops, bulge loops, and multibranch loops. The energy of a secondary structure is considered to be the sum of the energy contributions from all of the stacked base-pairs and loops it contains. For a given RNA sequence, we use the dynamic programming algorithm to predict a secondary structure with the lowest free energy by the given energy rules [20]. The dynamic programming is a commonly used technique for solving the discrete optimization problems by calculating recursively the minimum energy structure on progressively longer subsequences.

2.1   Computation of the Lowest Free Energy

Let $V(i, j)$ denote the lowest free energy in a segment from i to j given in which i is paired with j denoted by $i : j$. From the definition of the various types of loops in a secondary structure [21], $V(i, j)$ can be computed as follows:

$$V(i,j) = \min\{V_H(i,j), e_s(i,j) + V(i+1,j-1), V_{IB}(i,j), V_M(i,j)\}$$

where $V_H(i,j)$ is the energy of the structure if base-pair $i:j$ closes a hairpin loop, $e_s(i,j)$ is the stacking energy if $i:j$ stacks over $(i+1):(j-1)$, $V_{IB}(i,j)$ is the optimal energy if $i:j$ closes a bulge or interior loop, and $V_M(i,j)$ is the optimal energy if $i:j$ closes a multibranch loop. $V_{IB}(i,j)$ can be computed by

$$V_{IB}(i,j) = \min\{e_{ib}(i,j;i',j') + V(i',j'), i < i' < j' < j\}$$

where $e_{ib}(i,j;i',j')$ is the energy contribution of the bulge or interior loop closed by base-pairs $i:j$ and $i':j'$.

If we assume the energy rule for loops is basically a function of loop size, then it requires a single calculation for each $i:j$ and therefore the total computation time is $O(n^2)$ in the calculation of $V_H(i,j)$ and $e_s(i,j)$ for the fragment of size n. Biologically, a loop of order 2, a loop closed by two base-pairs, is seldom very large. It is reasonable to assume that the size of a bulge or interior loop is less than some fixed number. Under this assumption, the time to search for the optimal bulge or interior loop reduces from $O(n^4)$ to $O(n^2)$. To compute $V_M(i,j)$, we need to consider all possible multibranch loops closed by $i:j$. This implies that the time complexity for computing $V_M(i,j)$ is $O(2^{j-i})$ and therefore the overall execution time for V_M is $O(2^n)$.

Since our aim is to discover those distinct well-ordered stem-loops in a genomic sequence, we exclude the multibranch loop in our method [22], StemED. Furthermore, we only consider a secondary structure with a single stem-loop structure by the local fragment. Under these restrictions, the computation time for V in StemED is $O(n^2)$, where n is the length of the local fragment.

If we let $W(i,j)$ be the optimal energy regardless i and/or j formed a base-pair or not, then, under these restrictions, we have

$$W(i,j) = \min\{V(i,j), V(i+1,j) + d(i), V(i,j-1) + d(j), v(I+1,J-1) + d(i) + d(j), W(i,j-1), W(i+1,j)\}$$

where $d(i)$ is the dangling contribution of base i. The computation for W requires $O(n^2)$. Thus, the energy matrices V and W can be obtained as follow:

> **do** j=1 to n
>   **do** i=j down to 1
> $V(i,j) = \min\{V_H(i,j), e_s(i,j) + V(i+1,j-1), V_{IB}(i,j)\}$ (1)
> $W(i,j) = \min\{V(i,j), V(i+1,j) + d(i), V(i,j-1) + d(j),$
> $V(i+1,j-1) + d(i) + d(j), W(i,j-1), W(i+1,j)\}$ (2)
>   **enddo**
> **enddo**

The lowest free energy, E, of an optimal structure OS, folded by the local fragment (length is n) is the matrix element $W(1,n)$ from the definition of $W(i,j)$. The base-pairs in the optimal stem-loop structure are then determined by the following traceback algorithm:

```
initialize: i = 1, j = n, and e = W(1, n)
do while (e = W(i, j))
    i = i + 1
enddo
do while (e = W(i, j))
    j = j − 1
enddo
if (e = V(i, j)) then
    store i and j in a base-pair list
elseif (e = V(i + 1, j) + d(i)) then
    store i + 1 and j in a base-pair list
    set e = V(i + 1, j) and i = i + 1
elseif (e = V(i, j − 1) + d(j)) then
    store i and j − 1 in a base-pair list
    set e = V(i, j − 1) and j = j − 1
elseif (e = V(i + 1, j − 1) + d(i) + d(j)) then
    store i + 1 and j − 1 in a base-pair list
    set e = V(i + 1, j − 1), i = i + 1 and j = j − 1
endif
do while (e ≠ V_H(i, j))
    if (e = e_s(i, j) + V(i + 1, j − 1)) then
        store i + 1 and j − 1 in a base-pair list
        set e = V(i + 1, j − 1), i = i + 1 and j = j − 1
    elseif (e = V_{IB}(i, j)) then
        store i' and j' in a base-pair list
        set e = V(i', j'), i = i' and j = j'
    else
        write error message
    endif
enddo
```

In order to compute E_f, we fold the fragment again under the condition where all base-pairs in the OS are prohibited by assigning a large positive value to $V(i, j)$ for each base-pair in OS [21] whose base-pair list has been determined above by the traceback algorithm. Using the same algorithm we calculate energy matrices, $V(i, j)$ and $M(i, j)$ by Eq.(1) and Eq.(2).

2.2   Statistical Inference of WFS

One important question in our approach is how to evaluate the statistical significance of the score E_{diff} computed from a local segment. Statistical inference of the computed data needs to quantify uncertainties involved in the analysis. In the statistical analysis we adapt Monte Carlo simulations to estimate the typical behavior of E_{diff} computed in a random sample that is selected to be related to the local segment [23]. For a given segment S_i we generate a large number of randomly shuffled sequences $(RS_{i,1}, RS_{i,2}, \ldots, RS_{i,m})$, where the number m is often determined by the length of the segment S_i. $E_{diff}(S_i)$ and $E_{diff}(RS_{i,j})$ $(1 \leq j \leq m)$ of the m randomly shuffled sequences are then computed. The random variables $E_{diff}(RS_{i,j})$ are expected to follow the Poisson distribution approximately (see Fig. 1). To facilitate statistical inference, a normalized z-score, $SigZscr_e(S_i)$ can be computed by the formula

$$SigZscr_e(S_i) = \frac{E_{diff}(S_i) - E_{diff}(RS_i)}{std(RS_i)}$$

where $E_{diff}(RS_i)$ and $std(RS_i)$ are the sample mean and sample standard deviation (std) computed from the random sample. The distribution of the random variables $SigZscr_e(RS_{i,j})$ $(1 \leq j \leq m)$ are expected to approximately follow a normal distribution. Thus, the statistical significance of $E_{diff}(S_i)$ for the segment S_i can be easily estimated by means of the table of normal distribution.

To extract the signal with statistically high E_{diff} computed in a genomic sequence, we also compute a normalized score $Zscr_e$ of E_{diff}, similarly. The normalized score can be computed by the formula

$$Zscr_e(S_i) = \frac{E_{diff}(S_i) - E_{diff}(w)}{std(w)}$$

where the sample mean $E_{diff}(w)$ and sample standard deviation $std(w)$ are computed from the sample composed of $E_{diff}(S_1), E_{diff}(S_2), \ldots, E_{diff}(S_{L-n+1})$ by sliding the window of size n in steps of a few nt from 5' to 3' along the sequence of length L.

3   Results and Discussion

3.1   Statistical Inference of Known *Arabidopsis Thaliana* miRNAs

92 miRNA stem-loops of *Arabidopsis thaliana* are listed in the database of miRNA Registry (2005 version) [2]. For each of 92 miRNA stem-loops, to assess the statistical significance of E_{diff}, the sample mean $E_{diff}(RS_i)$ and sample standard deviation $std(RS_i)$ were computed from 500 randomly shuffled sequences. On the average, $E_{diff}(RS_i)$ and $std(RS_i)$ were 3.99 and 0.80 (kcal/mol), respectively. In contrast, the mean and std of E_{diff} computed from 92 miRNA precursors were 30.54 and 8.22 (kcal/mol). The average $SigZscr$ computed from these miRNA precursors was 8.48 that was highly significant in a normal distribution. Statistical analysis of computed E_{diff} for these miRNA stem-loops are listed in Table 1. Our results indicate that the E_{diff} computed from these miRNA stem-loops are significantly greater than that computed from their corresponding randomly shuffled sequences. For example, scores E_{diff} computed from the stem-loops of MIR395a, MIR395b and MIR395c are 26.5, 25.9 and 30.8 (kcal/mol), respectively. However, $E_{diff}(RS_{i,j})$ scores are ranged from 0 to 15.1 kcal/mol in the random sample of MIR395a precursor. Similarly, $E_{diff}(RS_{i,j})$ scores are ranged from 0 to 18.7 in the random sample of MIR395b and 0 to 22.4 in the random sample of MIR395c, respectively. The distributions of $E_{diff}(RS_{i,j})$

Table 1. Random test of quantntative measure E_{diff} computed from miRNA precursors of *Arabidopsis thaliana*

miRNA	Size	Natural Seq E_diff	SigZscr	Random E_diff Mean	std	miRNA	Size	Natural Seq E_diff	SigZscr	Random E_diff Mean	std
MIR161	173	38.10	9.35	4.50	3.59	MIR163	331	50.70	9.21	6.76	4.77
MIR170	93	26.90	10.31	3.00	2.32	MIR173	102	32.10	9.98	3.47	2.87
MIR156a	84	26.20	9.54	3.09	2.42	MIR156b	183	37.70	8.05	5.17	4.04
MIR156c	85	25.40	7.60	3.45	2.89	MIR156d	118	34.80	10.27	3.69	3.03
MIR156e	107	38.80	11.52	4.06	3.02	MIR156f	132	24.00	5.95	4.25	3.32
MIR156g	103	24.80	7.98	3.42	2.68	MIR156h	106	20.30	5.72	3.54	2.93
MIR157a	98	24.80	7.75	3.53	2.74	MIR157b	132	32.30	9.27	3.80	3.08
MIR157c	188	20.50	4.17	4.74	3.78	MIR157d	182	28.40	7.37	4.00	3.31
MIR158a	100	24.80	7.89	3.35	2.72	MIR158b	85	17.60	5.73	2.95	2.56
MIR159a	184	41.50	9.97	4.50	3.71	MIR159b	196	50.30	12.16	4.75	3.75
MIR159c	225	29.60	6.39	5.10	3.84	MIR160a	85	27.60	8.34	3.58	2.88
MIR160b	89	33.30	10.56	3.23	2.85	MIR160c	89	30.90	8.26	4.17	3.23
MIR162a	140	33.10	9.67	3.99	3.01	MIR162b	111	27.70	8.16	3.66	2.95
MIR164a	113	33.60	10.39	3.43	2.91	MIR164b	153	24.80	5.87	4.39	3.48
MIR164c	102	17.20	4.99	3.38	2.77	MIR165a	111	19.00	5.14	3.89	2.94
MIR165b	182	29.30	6.88	4.33	3.63	MIR166a	170	25.90	5.88	4.80	3.59
MIR166b	119	27.30	7.19	4.13	3.22	MIR166c	126	14.60	3.83	3.74	2.84
MIR166d	113	15.20	3.82	3.71	3.01	MIR166e	143	22.50	6.30	3.72	2.98
MIR166f	105	25.20	7.31	3.73	2.94	MIR166g	122	26.70	7.15	3.86	3.19
MIR167a	101	29.60	10.06	3.20	2.62	MIR167b	109	29.10	8.39	3.71	3.03
MIR167c	160	38.00	10.85	3.95	3.14	MIR167d	377	29.80	5.52	6.10	4.29
MIR168a	138	30.80	6.93	4.81	3.75	MIR168b	124	25.30	5.87	4.76	3.50
MIR169a	214	35.80	8.93	4.39	3.52	MIR169b	181	43.60	10.03	5.06	3.84
MIR169c	411	50.30	9.04	6.56	4.84	MIR169d	154	37.70	10.23	4.15	3.28
MIR169e	217	31.90	8.32	4.23	3.32	MIR169f	167	33.70	8.21	4.56	3.55
MIR169g	379	37.50	6.08	6.78	5.05	MIR169h	190	44.30	11.71	4.65	3.38
MIR169i	206	46.50	11.93	4.62	3.51	MIR169j	221	30.10	6.95	4.90	3.63
MIR169k	213	36.40	9.18	4.46	3.48	MIR169l	211	26.50	6.33	4.85	3.42
MIR169m	212	48.30	13.13	4.66	3.32	MIR169n	215	31.60	7.79	4.53	3.47
MIR171a	123	27.90	9.82	3.14	2.52	MIR171b	117	29.40	8.69	3.87	2.94
MIR171c	116	24.60	7.27	3.61	2.89	MIR172a	95	20.40	6.46	2.99	2.70
MIR172b	95	25.30	9.02	2.90	2.48	MIR172c	133	29.30	8.53	3.75	2.99
MIR172d	124	25.20	7.61	3.65	2.83	MIR172e	125	31.10	7.94	4.10	3.40
MIR319a	176	43.90	11.92	4.23	3.33	MIR319b	172	55.20	14.35	4.39	3.54
MIR319c	199	40.70	9.80	4.87	3.66	MIR393a	133	27.90	8.32	3.72	2.90
MIR393b	160	37.80	12.07	3.56	2.84	MIR394a	117	28.60	8.05	3.74	3.09
MIR394b	121	21.20	5.89	3.76	2.96	MIR395a	93	26.50	8.45	3.53	2.72
MIR395b	100	25.90	8.10	3.43	2.77	MIR395c	100	30.80	9.11	3.52	3.00
MIR395d	100	28.20	9.25	3.27	2.70	MIR395e	95	27.60	9.38	3.14	2.61
MIR395f	112	30.70	10.01	3.42	2.73	MIR396a	151	34.60	12.46	3.12	2.53
MIR396b	135	25.80	9.28	3.12	2.44	MIR397a	107	27.40	10.13	2.96	2.41
MIR397b	109	19.20	6.44	2.93	2.53	MIR398a	105	28.50	10.01	3.07	2.54
MIR398b	116	28.60	8.72	3.74	2.85	MIR398c	115	26.80	7.78	3.69	2.97
MIR399a	123	33.70	10.80	3.31	2.81	MIR399b	135	41.30	12.67	3.66	2.97
MIR399c	114	29.70	7.65	3.95	3.36	MIR399d	100	16.90	4.44	3.70	2.97
MIR399e	109	33.10	10.53	3.57	2.80	MIR399f	118	25.70	7.53	3.53	2.95

from the three random samples are displayed in Fig. 1. The values of $SigZscr$ computed from the stem-loops of MIR395a, MIR395b and MIR395c are 8.45, 8.10, and 9.11, respectively.

For other known miRNA stem-loops listed in the miRNA database (2005 version), we used the same approach to evaluate the statistical significance of the distinct stem-loop structures folded by miRNA precursors in human, mouse, rat, *Gallus gallus*, fly, *Caenorhabditis elegans*, *Caenorhabditis briggsae*, and *Oryza*

Table 2. Statistical analysis of computed E_{diff} from miRNA precursor and their corresponding randomly shuffled sequences

```
---------------------------------------------------------------
      miRNAs    Sample mean and standard deviation (std) computed from
  Genome Number       Natural Sequence       Randomly Shufflled Sequence
                 E_diff(std)  SigZscr(std)  E_diff(std)  SigZscr(std)
---------------------------------------------------------------
  Human   207   20.52 (5.89)  6.11 (2.25)  3.50 (0.55)   0.0 (1.0)
  Mouse   208   19.14 (6.19)  5.76 (2.34)  3.43 (0.52)   0.0 (1.0)
    Rat   187   20.40 (6.13)  5.93 (2.15)  3.54 (0.46)   0.0 (1.0)
  G.gal   121   19.81 (5.02)  6.13 (1.77)  3.30 (0.36)   0.0 (1.0)
    Fly    78   18.22 (4.68)  5.80 (1.73)  3.14 (0.41)   0.0 (1.0)
 C.eleg   116   20.15 (7.03)  6.09 (2.40)  3.36 (0.41)   0.0 (1.0)
 C.brig    50   21.84 (6.03)  6.77 (2.30)  3.36 (0.34)   0.0 (1.0)
  A.tha    92   30.54 (8.22)  8.48 (2.17)  3.99 (0.80)   0.0 (1.0)
  O.sat   122   30.35 (9.89)  7.72 (2.61)  4.35 (0.66)   0.0 (1.0)
---------------------------------------------------------------
  * All data were based on the 2005 Version of miRBase [2].
```

sativa genomes. These statistical analysis data are summarized in Table 2. It is clear that the folded stem-loops from these miRNA precursors are distinct, well-ordered and can be well characterized by the measure E_{diff}.

3.2   Fold-Back Stem-Loops of the Reported miRNA Precursor Are Coincident with Statistically Significant WFSs

The three miRNAs of MIR395a, MIR395b and MIR395c are clustered and encoded in the chromosome 1 (Accession No. NC_003070.5) of *Arabidopsis thaliana*. Among them, MIR395b and MIR395c are encoded in the positive stranded sequence (PSS) and MIR395a is encoded in the reverse complementary sequence of the listed sequence. Fig. 2 graphically displays the observed distributions of the scores $Zscr_e$ and $SigZscr_e$ that were computed by moving the 100-nt window in steps of 3 nt along the PSS of chromosome 1. It is clear that the distinct WFS associated with the three miRNAs can be explicitly detected by the two scores.

We also searched for miRNA stem-loop structures in *Drosophila melanogaster* by StemED. The chromosome 3R of fly genome has about 27.9 million nts. We computed the $Zscr_e$ distribution by StemED and scanning the 100-nt window with a step of 5 nt along the chromosome sequence. We found some interesting noncoding regions in which the computed WFS elements with very high $Zscr_e$ were clustered. As shown in Fig. 3 we detect those WFSs that are coincident with the well known miRNA genes, mir-277 (5925762 − 5925861), mir-92a (21461593 − 21461687) and mir-279 (25030673 − 25030772). The computed $Zscr_e$ are 6.38, 6.92 and 5.96, respectively. Except for the distinct miRNA genes we also found the other distinct stem-loop structures with high $Zscr_e$ values as seen in Fig. 3.

3.3   Predictions Are Not Sensitive to the Selected Window

We previously reported that *let-7* stem-loop could be precisely predicted by EDscan [5] from a 2460-nt genomic sequence of *C.elegans* (Accession number AF274345). However the accuracy of prediction was closely associated with the

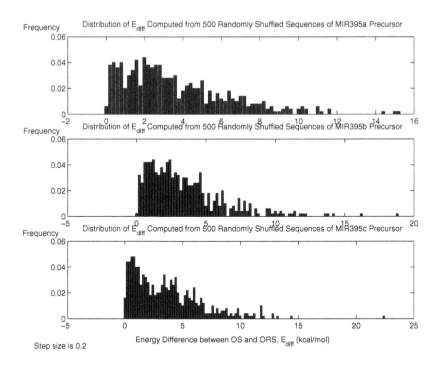

Fig. 1. E_{diff} distributions computed from three random samples corresponding to the three distinct stem-loop structures of MIR395a, MIR395b and MIR395c. Each random sample has 500 randomly shuffled sequences of the precursor sequences of MIR395a (top), MIR395b (middle) and MIR395c (bottom). In the plot, the horizontal axis represents E_{diff} and plotted with step size of 0.2 kcal/mol.

window size selected in the calculation. The shortcoming is inconvenient to search for those miRNA and/or FSRs that are not well established. In contrast to EDscan, the weakness in the prediction of miRNA stem-loops can be improved by using StemED. This is because StemED searches for the distinct stem-loop only without considering those potential multiple-loops and opening bifurcate stem-loops folded in the local segments. To test our hypothesis, we computed $Zscr_e$ by StemED and a set of windows which sizes were 75, 100, 125 and 150-nt, respectively. In all cases, the *let-7* stem-loop were detected precisely (see Fig. 4). Our results indicate that the prediction of miRNA genes by StemED is less sensitive to the window size. StemED is more stable in the prediction of the distinct stem-loop structures than EDscan. As a rule of thumb, we often scan a window of which size is little bigger than the target size of potential FSRs or ncRNAs by StemED.

Using StemED and EDscan [10] we also detect the small regulatory RNAs in *E. coli* genomes. Those small regulatory RNAs do not possess the conserved secondary structure but most of them have statistically high $Zscr_e$ and $SigZscr_e$ (data are not shown). Thus, StemED is generally applicable for finding FSRs

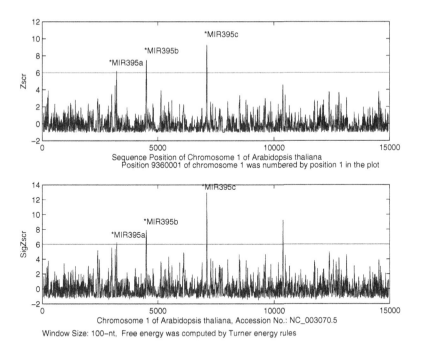

Fig. 2. $Zscr_e$ and $SigZscr_e$ of local segments computed in the region $9360001-9375000$ of *Arabidopsis thaliana* chromosome 1. The quantitative measure $Zscr_e$ was computed by moving a 100-nt window in steps of 3 nt from 5' to 3' along the genomic sequence by StemED. The plot was made by plotting $Zscr_e$ against the position of the middle base of the local overlapping segments. $Zscr_e = 6.0$ is marked by a horizontal line in the top plot and a horizontal line of $SigZscr_e = 6.0$ is marked in the bottom. The peak with the maximal scores in each plot is coincident with the reported miRNA genes, MIR395a, MIR395b and MIR395c that are marked. Among them, MIR395a is encoded in the reverse complementary sequence of the listed sequence in the genome database.

that are associated with their specific stem-loop conformations that are expected to be both significantly more ordered and thermodynamically more stable than anticipated by chance.

Rapid advances in computational biology and bioinformatics are providing new approaches to complex biological systems. For example, several computational approaches for detection of miRNA precursors have been currently developed [10-18, 22-23]. They have been successfully applied to discover a number of miRNAs. Most of these methods use the routine software of RNA secondary structure prediction, such as mfold [20, 21] to detect the most stable RNA structures. From the predicted stable stem-loop pool, the sequence conservation and common structural pattern in the pool are further checked so that the potential miRNA-like stem-loops are predicted. In this study, we present a general

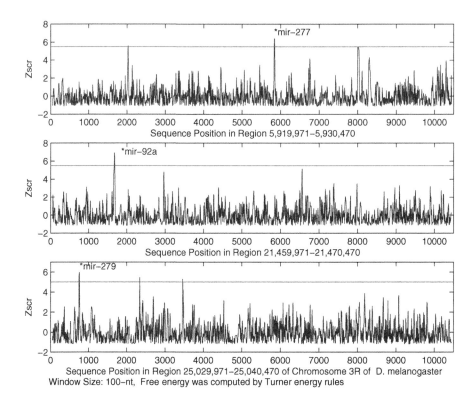

Fig. 3. $Zscr_e$ of local segments computed in the region $5919971 - 5930470$ (top), region $21459971 - 21470470$ (middle) and region $25029971 - 25040470$ (bottom) of *D. melanogaster* chromosome 3R. The quantitative measure $Zscr_e$ was computed by moving a 100-nt window in steps of 5 nt from 5' to 3' along the sequence by StemED. The plot was made by plotting $Zscr_e$ against the position of the middle base of the local overlapping segments. The peak with the maximal $Zscr_e$ in each plot is coincident with the reported miRNA genes, mir-277, mir-92a and mir-279 that are marked. We also detected other distinct stem-loops that were the both thermodynamic stable and highly well-ordered. $Zscr_e = 5.5$ is marked by a horizontal line in the plot.

approach to discover well-ordered stem-loop structures in genomic sequences. The method, StemED is more powerful for finding distinct stem-loops in genome than that of classical mfold. With the improvement of the integrating algorithms of statistical and computational tools of RNA folding, the computational approach can be used to discover ncRNAs and/or FSRs that are associated with important biological properties. The demand is growing in proportion to the size of sequence databases, which are growing exponentially. The existing tools, although already successful in finding interesting structural features of ncRNAs, can be improved further by future development.

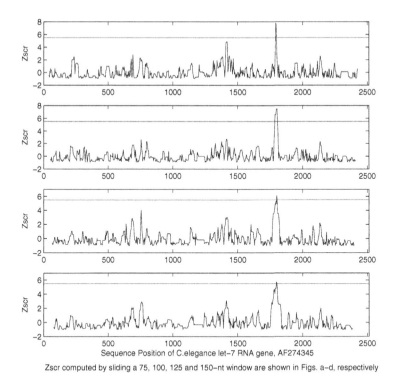

Zscr computed by sliding a 75, 100, 125 and 150-nt window are shown in Figs. a-d, respectively

Fig. 4. $Zscr_e$ of local segments computed in the genomic sequence of *C. elegans* (Accession No. AF274345). $Zscr_e$ values were computed by moving a set of windows whose sizes are 75-nt (shown in row 1), 100-nt (row 2), 125-nt (row 3) and 150-nt (row 4) in steps of 3 nt from 5' to 3' along the sequence by StemED. The plot was made by plotting $Zscr_e$ against the position of the middle base of the local overlapping segments. The reported miRNA stem-loops *let-7* can be easily distinguished in each plot by the maximal $Zscr_e$ as denoted by peak.

Acknowledgments

This research was supported by the Intramural Research Program of the NIH, National Cancer Institute, Center for Cancer Research. The content of this publication does not necessarily reflect the views or policies of the Department of Health and Human Services, nor does mention of trade names, commercial products, or organizations imply endorsement by the U.S. Government.

References

1. Simons, R.W., Grunberg-Manago, M. (eds.): *RNA Structure and Function,* Cold Spring Harbor Lab. Press, New York (1998)
2. Griffiths-Jones, G.R., van Dongen, S., Bateman, A., Enright, A.J.: miRBase: microRNA sequences, targets and gene nomenclature. *Nucleic Acids Res.* **34** (2006) D140-144

3. Macke, T.J., Ecker, D.J., Gutell, R.R., Gautheret, D., Case, D.A., Sampath, R.: RNAMotif, an RNA secondary structure definition and search algorithm. *Nucleic Acids Res.* **29**, (2001) 4724-4735.
4. Le, S.Y., Maizel Jr., J.V., Zhang, K.: An Algorithm for detecting homologues of known structured RNAs in genomes. *Proceedings of the 2004 IEEE Bioinformatics Conference, CSB2004* Stanford, California (2004) 300-310
5. Klein, R.J., Eddy, S.R.: RSEARCH: Finding homologs of single structured RNA sequences. *BMC Bioinformatics* **4** (2003) 44
6. Gautheret, A., Lambert, A.: Direct RNA motif definition and identification from multiple sequence alignments using secondary structure profiles. *J. Mol Biol.* **313** (2001) 1003-1011
7. Grillo, G., Licciulli, F., Liuni, S., Sbisa, E., Pesole, G.: PatSearch: a program for the detection of patterns and structural motifs in nucleotide sequences. *Nucleic Acids Res.* **31** (2003) 3608-3612
8. Lowe T.M., Eddy, S.R.: tRNAscan-SE: a program for improved detection of transfer RNA genes in genomic sequence. *Nucleic Acids Res.* **25** (1997) 955-964
9. Billoud, B., Kontic, M., Viari, A.: Palingol: a declarative programming language to describe nucleic acids' secondary structures and to scan sequence database. *Nucleic Acids Res.* **24** (1996) 1395-1403
10. Le, S.Y., Chen, J.H., Konings, D., Maizel Jr., J.V.: Discovering well-ordered folding patterns in nucleotide sequences. *Bioinformatics* **19** (2003) 354-361.
11. Carter, R.J., Dubchak, I., Holbrook, S.R.: A computational approach to identify genes for functional RNAs in genomic sequences. *Nucl. Acids Res.* **29** (2001) 3928-3938.
12. Lim, L.P., Lau, N.C., Weinstein, E.G., Abdelhakim, A., Yekta, M., Rhoades, M.W., Burge, C.B., Bartel, D.P.: The microRNAs of *Caenorhabditis elegans*. *Genes & Development* **17** (2003) 991-1008.
13. Lai, E.C., Tomancak, P., Williams, R.W., Rubin, G.M.: Computational identification of *Drosophila* microRNA genes. *Genome Biology* **4** (2003) R42.1-R42.20.
14. Nam, J.W., Shin, K.R., Han, J., Lee, Y., Kim, V.N., Zhang, B.T.: Human microRNA prediction through a probabilistic co-learning model of sequence and structure. *Nucl. Acids Res.* **33** (2005) 3570-358
15. Pfeffer, S. Sewer A., Lagos-Quintana, M., Sheridan R. Sander, C., Grasser, F.A., van Dyk, L.F., Ho, C.K., Shuman, S., Chien, M., Russo, J.J., Ju, J., Randall, G., Lindenbach, B.D., Rice, C.M., Simon, V., Ho, D.D., Zavolan, M., Tuschl, T.: Identification of microRNAs of the herpesvirus family. *Nature Methods* **2** (2005) 269-276
16. Berezikov, E., Guryev, V., Jose van de Belt, W. Plasterk, E, Ronald, H.A., Cuppen, E.: Phylogenetic shadowing and computational identification of human microRNA genes. *Cell* **120** (2005) 21-24
17. Wang, X., Zhang, J., Li, F., Gu, J., He, T., Zhang, X., Li, Y.: MicroRNA identification based on sequence and structure alignment. *Bioinformatics* **21** (2005) 3610-3614
18. Le, S.Y., Maizel Jr., J.V., Zhang, K.: Finding conserved well-ordered RNA structures in genomic sequences. *Int. J. Comp. Intelligence and Applications* **4** (2004) 417-430
19. Le, S.Y., Zhang, K., Maizel Jr., J.V.: RNA molecules with structure dependent functions are uniquely folded. *Nucleic Acids Res.* **30** (2002) 3574-3582
20. Mathews, D.H., Sabina, J., Zuker, M., Turner, D.H.: Expanded sequence dependence of thermodynamic parameters improves prediction of RNA secondary structure. *J. Mol. Biol.,* **288** (1999) 911-940

21. Zuker, M., Stiegler, P.: Optimal computer folding of large RNA sequences using thermodynamics and auxiliary information. *Nucleic Acids Res.* **9** (1981) 133-148
22. Bennasser, Y., Le, S.Y., Yeung, M.L., Jeang, K.T.: MicroRNAs in human immunodeficiency virus-1 infection. *Methods Mol. Biol.,* **342** (2006) 241-252
23. Le, S.Y., Chen, J.H., Maizel Jr., J.V.: Statistical inference for well-ordered structures in nucleotide sequences. *Proc. IEEE Comput. Soc. Bioinform. Conf.,* **2** (2003) 190-196

Interpretation of Protein Subcellular Location Patterns in 3D Images Across Cell Types and Resolutions

Xiang Chen[1] and Robert F. Murphy[1,2]

[1] Center for Bioimage Informatics and Departments of Biological Sciences and Machine Learning, Carnegie Mellon University, Pittsburgh, PA 15213
[2] Department of Biomedical Engineering, Carnegie Mellon University, Pittsburgh, PA 15213
xiang.chen@yale.edu, murphy@cmu.edu

Abstract. Detailed knowledge of the subcellular location of all proteins and how they change under various conditions is essential for systems biology efforts to recreate the behavior of cells and organisms. Systematic study of subcellular patterns requires automated methods to determine the location pattern for each protein and how it relates to others. Our group has designed sets of numerical features that characterize the location patterns in high-resolution fluorescence microscope images, has shown that these can be used to distinguish patterns better than visual examination, and has used them to automatically group proteins by their patterns. In the current study, we sought to extend our approaches to images obtained from different cell types, microscopy techniques and resolutions. The results indicate that 1) transformation of subcellular location features can be performed so that similar patterns from different cell types are grouped by automated clustering; and 2) there are several basic location patterns whose recognition is insensitive to image resolution over a wide range. The results suggest strategies to be used for collecting and analyzing images from different cell types and with different resolutions.

Keywords: Location Proteomics, Pattern Recognition.

1 Introduction

The central goal of proteomics is to characterize all proteins in a given cell type, which includes but is not restricted to the characterization of protein sequence, structure, expression, localization, function and regulation. Subcellular location is a critical property of a protein, and knowledge of the location pattern is essential to the complete understanding of its function. However, unlike many other properties for which systematic and automated approaches have been developed [1, 2], the systematic study of protein location patterns has just started [3]. Most of the approaches that exist today are based on assignment by database curators of terms from a restricted vocabulary, such as that generated by the Gene Ontology Consortium (http://geneontology.org). These approaches generally cannot distinguish location patterns beyond the level of major subcellular organelle categories. The development of random tagging techniques [4, 5] enables us to produce on a reasonable time scale a large number of clones from a given cell type that each express a different fluorescently-tagged

S. Hochreiter and R. Wagner (Eds.): BIRD 2007, LNBI 4414, pp. 328–342, 2007.

protein [6, 7]. Advances in microscope technologies have further enabled us to record high resolution, 3D fluorescence images of living cells. These techniques combined together can provide a collection of high resolution images for a large number of proteins in a given cell type. Machine learning techniques have been applied to provide tools for automated and systematic analysis of such images, initially by our group [8-10] and more recently by others [11, 12]. These methods are suitable for automated analysis of subcellular location on a proteome-wide basis.

The core of our approach is the design of Subcellular Location Features (SLF), numerical descriptors that can quantitatively describe the distribution of proteins inside a cell. Cells vary greatly in their size, shape, position and rotation in the field. The total intensity of a cell is also affected by the labeling and imaging techniques. We designed SLFs so that they are able to represent location patterns while at the same time not being too sensitive to changes in cell intensity, position and orientation. Using the SLFs, we have built automated classifiers that are able to distinguish the major subcellular organelles and structures with high accuracy. An important conclusion of this work is that some classifiers built on SLFs are able to distinguish location patterns which are essentially indistinguishable to human visual inspection [13].

Besides automated classifiers, we have also described approaches for objectively partitioning an arbitrary protein collection by subcellular location pattern. Initial trials with a limited set of randomly tagged proteins in the NIH 3T3 cell line demonstrated the feasibility of this approach [14, 15].

Our previous trials have validated the effectiveness of SLFs in describing protein subcellular location patterns in different data sources [9, 10, 15]. Although each of these trials was based on a single dataset, they suggest that SLFs describe general characteristics of location patterns, regardless of the data source. However, we have observed that classifiers trained on one dataset cannot simply be applied to another dataset taken for a different cell type or under different microscopy conditions. We therefore sought features and/or transformations of features that would be robust against variations in cell type or labeling and imaging protocols. This would be an important step towards our ultimate goal of systematically studying protein location patterns across all cell types. It would enable researchers to utilize classification or clustering procedures without first building their own reference set. Furthermore, as more and more published literature becomes available on line, we can potentially acquire and analyze many images without performing a single experiment. Such an effort (the SLIF, or Subcellular Location Image Finder, project) is ongoing in our group [16]. Given a suitable feature set and/or transformation, images retrieved from a range of sources could be analyzed to give a more comprehensive understanding of location.

A separate but related issue is the relationship between the resolution of an image and the resolution of the subcellular location assignment that can be made from it. Our initial experiments on 3D images were performed on a set of high resolution confocal fluorescence microscope images ($0.05 - 0.1$ µm horizontal pixel resolution) [10, 14]. However, even with recent advances in automated imaging technologies, the data acquisition step for high resolution fluorescence images is still the bottleneck for location studies at the proteome level. There is frequently a tradeoff between speed and resolution during image collection yet image resolution is one of the most important factors affecting the discriminating power of any interpretation method. For example, although we have shown that using our feature based classification approaches, two Golgi proteins (giantin and gpp130) could be distinguished with near-perfect accuracy in the high resolution 3D HeLa dataset [17], it is unlikely that these proteins would be distinguishable using images where the Golgi complex is barely separated from other subcellular organelles. Therefore, it is essential to establish guidelines for the maximum discrimination between subcellular patterns that can be expected for images at different resolutions. Strategies for image collection appropriate to a given problem can be properly chosen once such a guideline is established.

We have previously evaluated the effects of several image manipulations on 2D SLFs, including compression, resizing, and intensity rescaling [18]. The results suggested that while most individual features were sensitive to resizing, neural network classifiers trained on resized images were robust to a minor degree of resizing. It is therefore of interest to more fully study the performance of automated interpretation tools using images at different resolutions, and to consider 3D images.

This paper describes initial approaches to the problem of analyzing protein location patterns across different cell types and conditions. We combined 3D image data from two sources: a HeLa cell dataset of 11 different location patterns obtained by immunofluorescence labeling and a 3T3 cell dataset for 90 clones expressing different GFP-tagged proteins obtained by random tagging techniques. We searched for a proper transformation between features calculated for the 3D HeLa dataset and their counterparts for the 3D 3T3 dataset. We also evaluated the effects of pixel resolution on protein location pattern interpretation using the 3D HeLa dataset.

2   Methods

2.1   Datasets

The acquisition of the 3D HeLa dataset has been described previously [10]. To create it, antibodies were chosen for 8 different proteins found in 7 subcellular organelles and structures: ER (endoplasmic reticulum protein p63), Golgi (giantin and gpp130), lysosomes (LAMP2), endosomes (transferrin receptor), mitochondria (a mitochondrial inner membrane antigen), nucleoli (nucleolin), and microtubules (tubulin). A secondary antibody conjugated with Alexa 488 dye was then applied. A ninth protein (F-actin, present in microfilaments) was labeled using phalloidin directly-conjugated to Alexa 488. In parallel, total DNA and total cellular protein were labeled using propidium iodide (PI) and Cy5 reactive dye, respectively. The two Golgi proteins were included in the image dataset to test the ability of any classification system to separate protein location patterns which we have shown are essentially indistinguishable to visual inspection [13]. Images were taken with a three-channel confocal laser scanning microscope with lateral pixel resolution of 0.049 μm and axial pixel resolution of 0.203 μm. The 3D HeLa dataset contains 11 different subcellular location patterns with 50 – 52 cells per pattern (558 cells in total). Each cell is represented as a 3-color stack of from 14 to 24 2D slices each consisting of 1024 × 1024 pixels with 8-bit intensities. Example images are shown in Figure 1.

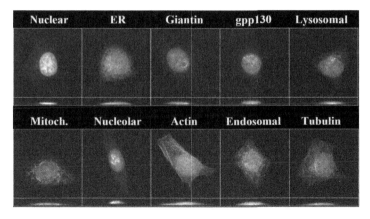

Fig. 1. Typical images from the 3D HeLa image dataset. Red, blue and green colors represent DNA, total protein and target protein staining, respectively. Projections on the X-Y (top) and the X-Z (bottom) planes are shown. Reprinted by permission of Carnegie Mellon University.

The acquisition of the 3D 3T3 dataset has also been described previously [14]. NIH 3T3 cell clones expressing GFP-tagged proteins were generated using CD-tagging techniques [4, 6]. A population of cells was infected with a retrovirus that creates a new GFP exon if inserted into intronic sequence. After the infection, cells expressing GFP-tagged proteins were isolated and subcloned. The identity of the tagged protein was determined by RT-PCR and sequencing. GFP images were taken with a single-channel spinning disk confocal microscope with lateral pixel resolution of 0.11 μm and axial pixel resolution of 0.5 μm. The 3D 3T3 dataset contains 90 different GFP expressing clones with 9 – 33 cells per clone (1554 cells in total). Each cell is represented as a 1-color 1280 × 1024 × 31 stack with 12-bit pixel intensities. Example images are shown in Figure 2.

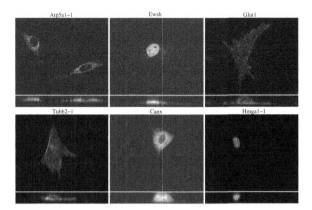

Fig. 2. Selected images from the 3D 3T3 image dataset. Tagged protein names are shown with a hyphen followed by a clone number if the same protein was tagged in more than one clone in the dataset. Projections on the X-Y (top) and the X-Z (bottom) planes are shown. Reprinted from reference 15.

2.2 Image Processing

Before feature extraction, the raw images were first corrected for background fluorescence, segmented into single cell regions, and thresholded. The segmentation for the 3D HeLa dataset was performed using parallel DNA and total protein images and a seeded watershed algorithm as described previously [10]. The segmentation of the 3D 3T3 dataset was achieved by manually defining a cropping mask for each cell since DNA and total protein images were not available.

For experiments on clustering proteins from different sources, the 3D HeLa dataset was down-sampled to the resolution of the 3D 3T3 dataset before feature extraction. When studying effects of pixel resolution, each preprocessed image from the 3D HeLa dataset (after background correction and segmentation) was resized from 2 to 7-fold (corresponding to a reduction in pixel size of from 50% to 14%).

2.3 Feature Extraction

For clustering of images from different data sources, feature set SLF11 [14] was calculated for all images. It consists of 14 morphological features from set SFL9 [10], 2 edge features [14] and 26 Haralick texture features [14, 15]. The 26 texture features in SLF11 consist of mean and range features for the 13 directions of pixel adjacency; the range features were not used in this study. Texture features for both the HeLa and 3T3 datasets were calculated after downsampling to 0.5 µm pixel size and 64 gray levels, which were determined to be the optimal settings for the 3T3 dataset [15].

For assessment of effects of pixel resolutions, feature set SLF19 [19] was calculated for the 3D HeLa dataset for the original images as well as for images downsampled from 2 to 7 fold (downsampling ratios refer to the lateral dimensions; downsampling in the in the axial dimension was only done as needed to match the lateral dimensions). It consists of the 42 SLF11 features plus an additional 14 morphological features calculated with reference to the parallel DNA image. For the original images and the 2-fold down-sampled images, texture features were calculated after additional downsampling to 0.4 µm pixel resolution and 256 gray levels, the optimal settings for the HeLa dataset [17]. For the other down-sampled images, texture features were calculated without further downsampling.

2.4 Feature Transformation Between Cell Lines

To find a set of features and a transformation function that could make HeLa cell features comparable to 3T3 cell features, four sets of subcellular patterns that were expected to be similar between the two cell types were identified. For each set, the mean feature values were calculated for both cell lines (since all values of feature SLF11.28, the average information measure of correlation 1, were negative, we converted them to absolute values). For each feature, we then fit the following transformation:

$$\log(f_{3T3}) = \alpha + \beta \times \log(f_{HeLa}) \tag{1}$$

where α and β are fitted regression coefficients. We chose a simple linear regression model since we have only four data points for each feature. The log form was used since the dynamic range of each feature can be quite large. Once the intercepts and coefficients for each individual feature were determined, we transformed the features calculated for all cells using the formula. Four of the features were eliminated: SLF9.2 (Euler number of objects) had both positive and negative values and could not be log transformed, and SLF9.8, SLF9.17, SLF9.20 were dropped because they were not observed to be discriminatory before transformation.

2.5   Feature Selection

We have previously observed after characterization of a number of methods for feature selection that Stepwise Discriminant Analysis (SDA) performed the best in the domain of subcellular location features [20]. We formed a merged HeLa-3T3 dataset but in which the four classes common to both datasets were joined (as discussed above). This gave a total of 82 classes (90 3T3 classes plus 11 HeLa classes minus 19 3T3 classes that belong to the four common classes). We applied SDA to the 25 features for the 82 classes, resulting in a list of ranked features with descending order of discriminating power. We selected the optimal feature set using classifiers with increasing numbers of the SDA ranked features [15].

A similar procedure was followed for the downsampled HeLa images.

2.6   Implementation of Classification and Clustering Algorithms

A support vector machine was trained and tested with an rbf kernel and max-win strategy and the overall classification accuracy was estimated by 10-fold cross validation.

Two clustering approaches were used as described previously [15]. The first approach started by performing k-means clustering using standardized (z-scored) Euclidean distance on all single cell observations with k increasing from 2 to either 20 for the resampled 3D HeLa dataset or 101 for the combined dataset (90 3T3 clones and 11 HeLa classes). Akaike Information Content (AIC) was used to select the optimal k. For each class, the cluster containing the largest membership was found. If that cluster contained at least 33% of the cells for that class, those cells were retained and the class was assigned to that cluster. Otherwise the class was dropped from further consideration. The second approach started from the cells retained in the first approach. It consisted of building a set of dendrograms using randomly selected halves of the retained dataset, and then constructing a majority consensus tree from the set of dendrograms. AIC was used to select the optimal cutting of the consensus tree into disjoint clusters.

2.7   Creation of Cluster Relationship Maps

Cluster relationship maps were created to visualize the effect of pixel size on the number of distinguishable location classes. The input to create these maps is a list of original classes and a matrix showing which classes are clustered together as a function of degree of downsampling. This matrix was created for the 3D HeLa dataset using k-means/AIC clustering as described above. A graph is then created with a node for each cluster found at each level of downsampling. Edges are created between nodes at different levels that contain the same class. In theory this graph could be quite complex, but in practice it was observed for the 3D HeLa data to be composed of distinct cliques. A cluster relationship map is created from the graph in two steps. The class indices are reindexed so that classes in the same clique have continuous indices. Coordinates for plotting are then assigned to each node: the downsampling level as the ordinate and the average index of its members as the abscissa. To facilitate comparison between levels, the overall accuracy of a classifier across all classes at a given downsampling ratio is noted adjacent to the left axis and the accuracy of a classifier trained on just the number of nodes at each level is noted on the right side. Visual clues are also provided to faciliate interpretation of the map. The width of the edge is drawn proportional to the number of classes shared between two clusters. If classes in a lower level (smaller downsampling ratio, higher resolution) node split into multiple nodes at the next higher level (larger downsampling ratio, lower resolution), the numerical IDs of the classes represented by each edge are written on the edge. If all classes at a lower level node still belong to a single node at the next higher level, the numeric IDs are omitted. Using this labeling scheme, membership of each node as well as individual edge could be inferred from the plot without ambiguity.

3 Results

3.1 Feature Conversion and Selection for the Combined HeLa-3T3 Dataset

The 3D HeLa dataset, consisting of 11 distinct location patterns, and the 3D 3T3 dataset, consisting of 90 randomly tagged protein clones, were derived independently on two different instruments. Table 1 lists the major differences between the two datasets. The two datasets were derived from cell lines that have different cell size, were labeled with different techniques and were imaged with different microscopes and protocols. Consequently, we expect some differences in features calculated for the same location pattern in the two datasets.

Our initial approach was to find an optimal subset of features that are not sensitive to these variations by feature selection rather than transformation (data not shown). Starting from feature set SLF11 (after reduction to 29 features as described in the Methods) and following procedures described previously [15], we obtained a mixed result. Although the nucleolar and nucleus location patterns from 3T3 and HeLa cell lines partitioned into the same clusters, most of the HeLa location patterns formed a distinct cluster that was isolated from the rest of the 3T3 clusters. This result suggested that some (if not most) of the features are dependent on the cell type or methodology used for acquisition.

Table 1. Differences between the datasets used in this study

Name	Labeling Method	Microscopy Method	Objective	Resolution (μm)
3D HeLa	Immunofluor	Laser Scanning Confocal	100×	$0.049 \times 0.049 \times 0.203$
3D 3T3	CD-tagging	Spinning Disk Confocal	60×	$0.11 \times 0.11 \times 0.5$

We therefore sought to find a suitable transformation that could be used to convert features for HeLa cells into their counterparts for 3T3 cells. Given the limited number of classes in common between the two, we sought to fit a linear regression model to the log of the feature values for corresponding pattern classes.

In order to achieve this goal, we started by identifying several sets of proteins that share the same location pattern in the two different cell lines. After visual inspection of the images, we identified four calibration sets, namely cytoskeletal, mitochondrial, nucleolar and uniform (whole cell) patterns. Although both the 3D HeLa dataset and the 3D 3T3 dataset have a nuclear pattern, we did not consider them to be corresponding. This is because the nuclear pattern in the HeLa cell set was generated by DNA staining (Figure 1) while the nuclear pattern in the 3T3 cell dataset was from a specific nuclear protein, Hmga1-1 (Figure 2), and the former pattern shows more uniform staining across the nucleus.

We calculated the mean feature vector for each location pattern of both datasets. For each feature, we used the feature values for the four calibration sets to estimate the intercepts and coefficients for a log linear regression model.

In order to get the optimal subset of features for the combined dataset, we followed the feature selection procedure for clustering described previously [15]. The principle behind this approach is to use increasingly numbers of the features ranked by SDA to train a classifier for all classes and choose the subset that gives the best classification accuracy (recognizing that some classes may not be distinguishable). When we used the optimal feature subset to train a neural network classifier to recognize the 4 combined patterns and the remaining single patterns, the overall classification accuracy was 62%. The classification accuracy for the combined cytoskeletal, mitochondrial, nucleolar and uniform cellular location patterns were 65%, 83% 98% and 97% respectively, suggesting that the classifier could learn decision boundaries for both cell types reasonably well (particularly for nucleolar and uniform cellular patterns). The inspection of classification errors within the cytoskeletal class also revealed that the highest error occurred for the HeLa F-actin class, which can be expected since its pattern is similar but not identical to that of tubulin.

3.2 Clustering of the Combined Dataset

A robust set of features would group the protein images based on their location patterns. We tested this hypothesis by using two different approaches to clustering.

The first approach is the k-means/AIC algorithm. Using the selected optimal subset of transformed features, we found the minimum AIC was achieved at $k = 31$. However, 14 of these 31 clusters contained only minority images (images from a particular class that were not in the plurality cluster for that class) and were eliminated. This left 17 protein clusters that contained at least 33% of the images for at least one class. The number of classes per cluster ranged from 1 to 12 and the 11 HeLa classes were distributed into 8 clusters.

A parallel approach was to perform consensus clustering as we have described previously [15]. Since we had multiple images for each class in our datasets, we constructed a set of dendrograms in which each was built using a random half of the images for each class. A single majority consensus tree, which contains only the frequently-observed structures, was constructed from the set of dendrograms (Figure 3).

A set of disjoint clusters could be selected from the consensus tree by finding that set of cuts that minimized AIC (these are shown as short vertical lines in Figure 3). Eleven clusters were obtained with 3 to 23 clones per cluster.

Comparison of the clusters obtained in the two approaches revealed that the partitioning obtained by consensus tree analysis is contained in the partitioning by k-means/AIC. In other words, while members of any single cluster from k-means/AIC belong to a single cluster from consensus clustering, a cluster from consensus clustering contains one or more clusters from k-means/AIC.

Examination of the consensus tree (Figure 3) revealed that those proteins expected to have similar location patterns were grouped together. For example, we consider the DNA pattern in the HeLa dataset to be similar (but, as discussed above, not identical) to the nuclear proteins in 3T3 (two clones of Hmga1 and Unknown-9), and they formed a cluster in the consensus tree even though we did not use this pair to estimate the feature transformation. Since the 3T3 images did not have parallel DNA images, we could not use any features that use this as a reference point (an important aspect of our HeLa cell work). Without being able to calculate position to the nucleus, the Golgi patterns and nucleolar patterns from the two cell types were all considered similar in the tree (both are small, closely placed objects). Tubulin and actin are both cytoskeletal proteins and in the consensus tree, their patterns from HeLa cells form a cluster with a tubulin protein (Tubb2) from the 3T3 dataset.

3.3   Classification of Images at Different Resolutions

As discussed in the introduction, our second goal was to determine the relationship between pattern resolution and image resolution. For this purpose, we downsampled the 3D HeLa images to varying degrees and compared the performance of classifiers trained to distinguish all 10 classes using an optimal feature set selected for each resolution.

The overall classification accuracies across all 10 classes using images resized to different degrees are summarized in Table 2. As the texture features contained information of optimal resolution, we first considered classification on morphological and edge features but not texture features. A clear descending trend was detected with increasing downsampling. While the classifier using original resolution images achieved an overall accuracy of 94.5%, this number quickly dropped below 90% for a minor downsampling (2-fold) and finally dropped below 80% for downsampling of 6 and 7 fold. These results suggested that the discriminant power of our 3D morphological and edge features is modestly sensitive to the pixel resolution of the images.

Inspection of the confusion matrices from these classifiers revealed that the two Golgi proteins (giantin and gpp130) could only be distinguished with confidence at the original resolution and even a two-fold down-sample operation largely eliminated the separation (data not shown) between them. At a 7-fold down-sample ratio, the classification between these two classes was essentially random (data not shown). This result confirmed that these two location patterns are extremely similar and could only be distinguished at very high resolutions.

Previously we have shown that using a combination of morphological, edge and texture features, we could achieve 98% overall classification accuracy in the same HeLa dataset [17]. Therefore we next included DNA and texture features in the

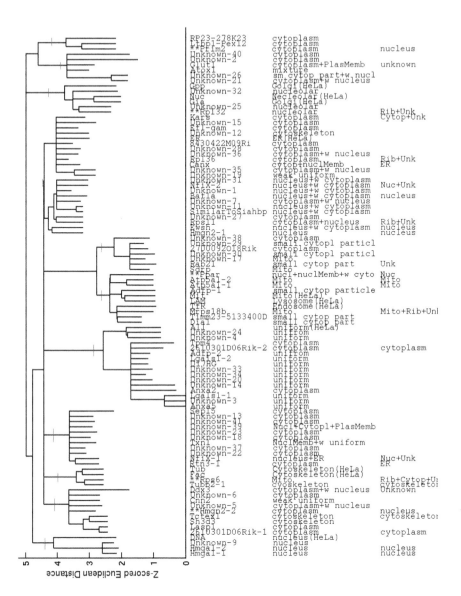

Fig. 3. A consensus Subcellular Location Tree for the combined image dataset using SDA selected features. The columns show the protein names (if known), human observations of subcellular location, and subcellular location inferred from Gene Ontology (GO) annotations. The sum of the lengths of horizontal edges connecting two proteins represents the distance between them in the feature space. Proteins for which the location described by human observation differs significantly from that inferred from GO annotations are marked (**). The vertical lines show the 11 clusters selected by AIC from this tree.

scheme. Firstly, we observed that the inclusion of DNA related features did not further improve the classifier's performance at the original resolution. The same overall classification accuracy (98%) was achieved. Secondly, inclusion of texture features improved the classification performance, confirming the value of texture features in 3D image analysis. Finally, the same descending trend was detected with increasing downsampling ratios (from 98% at original resolution to 82% at 6-fold downsampled images), indicating that the discriminant power of the texture features is also modestly sensitive to the pixel resolution of the images.

3.4   Effects of Pixel Resolution on Cluster Stabilities

As an alternative way to examine what patterns are distinguishable as a function of pixel resolution, we next evaluated the effects of pixel resolution on solutions of clustering approaches.

We partitioned the 10 HeLa classes using both k-means/AIC and consensus tree algorithms at the original resolution as well as 6 downsampled resolutions. Results using only morphological and edge features revealed perfect agreements between the clusters obtained from k-means/AIC algorithm and the clusters obtained from the consensus tree algorithm at each of the 7 resolutions analyzed (data not shown). To graphically show the relationship between patterns and resolution and the stability of clusters, we constructed cluster relationship maps. The goal of these maps is to demonstrate which high-resolution patterns become confused as the resolution is decreased.

Figure 4 shows such a map constructed using only morphological and edge features. As might be expected, the DNA class formed a distinct cluster at all resolutions (as represented by the thin straight line on the left). The two Golgi proteins, giantin and gpp130, formed a cluster separated from all other classes. They were merged at all levels. The LAM class (lysosomal pattern), the Mit class (mitochondrial pattern) and the TfR class (endosomal pattern) formed a single cluster at the original image level, split into two clusters at downsample level 2, and merged again at level 6.

Table 2. Comparison of classification accuracy using images at different resolutions. For images at a specific resolution, the overall classification accuracy using a SVM with 10-fold cross validation and optimal feature subset (with and without texture features) selected by SDA was reported.

		Classification Accuracy (%)	
Downsampling Factor	**Pixel Size (μm)**	**without Texture Features**	**with Texture Features**
1 (None)	$0.05 \times 0.05 \times 0.2$	94.5	97.8
2	$0.10 \times 0.10 \times 0.4$	88.6	95.8
3	$0.15 \times 0.15 \times 0.6$	86.4	92.6
4	$0.20 \times 0.20 \times 0.8$	84.0	85.6
5	$0.25 \times 0.25 \times 1.0$	81.6	84.9
6	$0.30 \times 0.30 \times 1.2$	77.4	82.0
7	$0.35 \times 0.35 \times 1.4$	77.9	84.1

Figure 4 clearly revealed that by morphological and edge features we can distinguish 5 basic clusters within the 3D HeLa dataset down to a lateral pixel resolution of 0.35 µm: a cluster of DNA (uniform nuclear pattern), a cluster of giantin and gpp130 proteins (Golgi complex pattern), a cluster of nucleolin (nucleolar pattern), a cluster of ER, F-Actin and Tubulin proteins (cytoplasmic network pattern) and a cluster of Mit, LAM and TfR proteins (cytoplasmic punctate pattern). Although some basic clusters could be separated into sub-clusters at higher resolutions (for example, the separation of Mit and LAM from TfR at downsampling ratios of 2 to 5), this did not happen in a continuous manner.

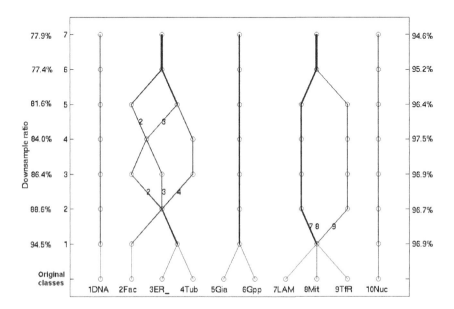

Fig. 4. A cluster relationship map for the 3D HeLa dataset using 3D morphological features (3D SLF 9) and 3D edge features (3D SLF11.15 and 3D SLF11.16). The overall classification accuracies for 10 classes at specific resolutions are shown on the left and the classification accuracies for 5 basic location pattern clusters are shown on the right.

The separation of the 5 basic clusters was further validated by training a classifier to distinguish these 5 clusters. The overall classification accuracy from 10-fold cross validation was labeled on the right side of Figure 4. These 5 clusters could be accurately distinguished with roughly the same accuracy at any resolution.

We repeated this analysis after adding texture features. Results indicated extensive agreement between k-means/AIC algorithm and consensus tree analysis algorithm with some slight differences (mainly relationships among LAM, Mit and TfR patterns). The relationship map using morphological, edge and texture features is shown in Figure 5. It indicated that inclusion of texture features greatly improved the resolution of the clustering solution. Generally at least 8 clusters could be distinguished from the dataset at different pixel resolutions. The stability of these clusters was also

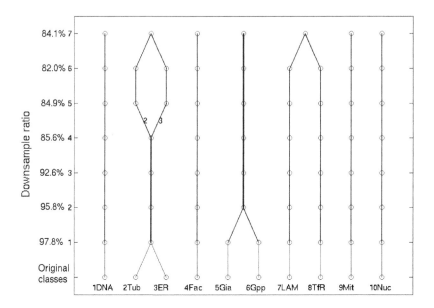

Fig. 5. A cluster relationship map for the 3D HeLa dataset using 3D morphological, edge and texture features. The overall classification accuracies for 10 classes at specific resolutions are shown on the left.

better. For example, the giantin and gpp130 patterns were placed in distinct clusters at the original resolution but were merged into a single cluster at lower resolutions. The LAM and TfR patterns were not merged until the lowest resolution tested.

4   Discussion and Conclusions

We have previously shown that SLFs can be used effectively to numerically describe protein subcellular location patterns using high resolution images in a single cell line. Automated classifiers trained on these features can recognize the location pattern in previously unseen images of HeLa cells with high accuracy. A clustering/partitioning scheme based on SLFs was shown to achieve objective clustering on a dataset of 3D images from 3T3 cells.

The effectiveness of SLFs in different datasets suggested that the SLFs are describing general aspects of the location patterns, which are not dataset dependent. In the current study, we extended our approaches to a multi-source dataset where the input data were from two cell types and two types of microscopes. We proposed a simple method to search for the proper transformation between features from different data sources. Using the transformed features, the partitioning of a combined 3T3 and HeLa dataset was largely based on intrinsic location patterns rather than on the data source. We also studied the effects of image resolution on automated interpretation. Overall classification accuracies for individual classes in the 3D HeLa dataset decreased with decreasing resolution. However, several basic clusters exist and these clusters are insensitive to image resolution, at least down to 0.35 µm pixel size. When texture features were combined with morphological and edge features, the improvement of the resolution as well as the stability of clustering solutions further validated the good discriminant power of texture features.

The results suggest the feasibility of constructing a combined subcellular location tree across many cell types with different resolutions. They also have implications for the collection of future image sets. These are that images of a reasonably large common set of proteins should be acquired for each cell type to be combined, images should be collected at relatively high resolution and that if at all possible parallel images of the DNA distribution should be acquired. The promise of the approach described here is that the similarities and differences of protein distributions in different cell types and different resolutions can be systematically organized and represented.

Acknowledgments. This work was supported in part by NIH grant R01 GM068845 and NSF grant EF-0331657. X. Chen was supported by a Graduate Fellowship from the Merck Computational Biology and Chemistry Program at Carnegie Mellon University established by the Merck Company Foundation.

References

1. Cutler, P.: Protein arrays: The current state-of-the-art. Proteomics **3** (2003) 3-18
2. Sali, A., Glaeser, R., Earnest, T., Baumeister, W.: From words to literature in structural proteomics. Nature **422** (2003) 216-225
3. Huh, W.-K., Falvo, J.V., Gerke, L.C., Carroll, A.S., Howson, R.W., Welssman, J.S., O'Shea, E.K.: Global analysis of protein localization in budding yeast. Nature **425** (2003) 686-691
4. Jarvik, J.W., Adler, S.A., Telmer, C.A., Subramaniam, V., Lopez, A.J.: CD-Tagging: A new approach to gene and protein discovery and analysis. BioTechniques **20** (1996) 896-904
5. Rolls, M.M., Stein, P.A., Taylor, S.S., Ha, E., McKeon, F., Rapoport, T.A.: A visual screen of a GFP-fusion library identifies a new type of nuclear envelope membrane protein. J. Cell Biol. **146** (1999) 29-44
6. Jarvik, J.W., Fisher, G.W., Shi, C., Hennen, L., Hauser, C., Adler, S., Berget, P.B.: In vivo functional proteomics: Mammalian genome annotation using CD-tagging. BioTechniques **33** (2002) 852-867
7. Sigal, A., Milo, R., Cohen, A., Geva-Zatorsky, N., Klein, Y., Alaluf, I., Swerdlin, N., Perzov, N., Danon, T., Liron, Y., Raveh, T., Carpenter, A.E., Lahav, G., Alon, U.: Dynamic proteomics in individual human cells uncovers widespread cell-cycle dependence of nuclear proteins. Nat Methods **3** (2006) 525-531
8. Boland, M.V., Markey, M.K., Murphy, R.F.: Classification of Protein Localization Patterns Obtained via Fluorescence Light Microscopy. 19th Annu. Intl. Conf. IEEE Eng. Med. Biol. Soc. (1997) 594-597
9. Boland, M.V., Murphy, R.F.: A neural network classifier capable of recognizing the patterns of all major subcellular structures in fluorescence microscope images of HeLa cells. Bioinformatics **17** (2001) 1213-1223

10. Velliste, M., Murphy, R.F.: Automated determination of protein subcellular locations from 3D fluorescence microscope images. 2002 IEEE Intl. Symp. Biomed. Imaging (2002) 867-870
11. Danckaert, A., Gonzalez-Couto, E., Bollondi, L., Thompson, N., Hayes, B.: Automated recognition of intracellular organelles in confocal microscope images. Traffic **3** (2002) 66-73
12. Conrad, C., Erfle, H., Warnat, P., Daigle, N., Lorch, T., Ellenberg, J., Pepperkok, R., Eils, R.: Automatic identification of subcellular phenotypes on human cell arrays. Genome Research **14** (2004) 1130-1136
13. Murphy, R.F., Velliste, M., Porreca, G.: Robust numerical features for description and classification of subcellular location patterns in fluorescence microscope images. J VLSI Sig Proc **35** (2003) 311-321
14. Chen, X., Velliste, M., Weinstein, S., Jarvik, J.W., Murphy, R.F.: Location proteomics - Building subcellular location trees from high resolution 3D fluorescence microscope images of randomly-tagged proteins. Proc SPIE **4962** (2003) 298-306
15. Chen, X., Murphy, R.F.: Objective clustering of proteins based on subcellular location patterns. J Biomed Biotechnol **2005** (2005) 87-95
16. Murphy, R.F., Kou, Z., Hua, J., Joffe, M., Cohen, W.W.: Extracting and Structuring Subcellular Location Information from On-Line Journal Articles. IASTED Intl. Conf. Knowl. Sharing Collab. Eng. (2004) 109-114
17. Chen, X., Murphy, R.F.: Robust classification of subcellular location patterns in high resolution 3D fluorescence microscopy Images. 26th Annu. Intl. Conf. IEEE Eng. Med. Biol. Soc. (2004) 1632-1635
18. Murphy, R.F., Velliste, M., Yao, J., Porreca, G.: Searching Online Journals for Fluorescence Microscope Images Depicting Protein Subcellular Locations. 2nd IEEE Intl. Symp. BioInf. Biomed. Eng. (2001) 119-128
19. Nair, P., Schaub, B.E., Huang, K., Chen, X., Murphy, R.F., Griffith, J.M., Geuze, H.J., Rohrer, J.: Characterization of the TGN Exit Signal of the human Mannose 6-Phosphate Uncovering Enzyme. J. Cell Sci. **118** (2005) 2949-2956
20. Huang, K., Velliste, M., Murphy, R.F.: Feature reduction for improved recognition of subcellular location patterns in fluorescence microscope images. Proc SPIE **4962** (2003) 307-318

Bayesian Inference for 2D Gel Electrophoresis Image Analysis

Ji Won Yoon[1], Simon J. Godsill[1], ChulHun Kang[2], and Tae-Seong Kim[3]

[1] Signal Processing Group, Engineering Department, Cambridge University, UK
[2] Graduate School of East-West Medical Science, Kyung Hee University, Korea
[3] Biomedical Engineering Department, Kyung Hee University, Korea

Abstract. Two-dimensional gel electrophoresis (2DGE) is a technique to separate individual proteins in biological samples. The 2DGE technique results in gel images where proteins appear as dark spots on a white background. However, the analysis and inference of these images get complicated due to 1) contamination of gels, 2) superposition of proteins, 3) noisy background, and 4) weak protein spots. Therefore there is a strong need for an automatic analysis technique that is fast, robust, objective, and automatic to find protein spots. In this paper, to find protein spots more accurately and reliably from gel images, we propose Reversible Jump Markov Chain Monte Carlo method (RJMCMC) to search for underlying spots which are assume to have Gaussian-distribution shape. Our statistical method identifies very weak spots, restores noisy spots, and separates mixed spots into several meaningful spots which are likely to be ignored and missed. Our proposed approach estimates the proper number, centre-position, width, and amplitude of the spots and has been successfully applied to the field of projection reconstruction NMR (PR-NMR) processing [15,16]. To obtain a 2DGE image, we peformed 2DGE on the purified mitochondiral protein of liver from an adult Sprague-Dawley rat.

1   Introduction

Recent advances in proteomics play a key role in life science by identifying and characterizing overall proteins, and provide insights of disease and drug interactions. 2DGE is a widely used technique to analyze the protein complexes in proteomics and bioinformatics [11,8]. The two dimensions in 2DGE correspond to isoelectric point and mass: the isoelectric point separates the proteins in terms of a gradient of pHs, and the mass according to the weights of proteins. 2DGE yields an image representing the distribution of protein spots.

The 2DGE image analysis includes spot detection, segmentation, characterization, quantification, and etc. However, such analysis are complicated for the following reasons. First, there may be weak and small spots which are not be detected. Second, spots can be superimposed. These mixed spots are hard to separate by inspection or many deterministic approaches and the mixed spots are often likely to be regarded as one big spot. Even though interesting spots are clearly visible, it is difficult to recognize them if they are mixed with other

S. Hochreiter and R. Wagner (Eds.): BIRD 2007, LNBI 4414, pp. 343–356, 2007.
© Springer-Verlag Berlin Heidelberg 2007

spots. Finally, there are spots not discernible from background noise. Spots in 2DGE image may be corrupted by two kinds of noise: global noise and local noise. Global noise is a background noise which has a specific pattern. Local noise effects the intensity of a pixel or a small area of image. Thus, we may have to restore images and find important spots with careful consideration of noise.

Many researchers have been worked on 2DGE image processing and analysis using several methods such as filtering in the wavelet domain [12], watershed techniques [14] and pixel value collection [10]. However, they are not good enough in noisy images and produce only limited results such as segmentation or quantification. In this paper, we tackle these problems by applying Reversible Jump Markov Chain Monte Carlo (RJMCMC) [2,7]. This method has been successfully applied to Projection-reconstruction NMR (PR-NMR) to reconstruct NMR spots in 2D signals [15,16]. Our application of RJMCMC to 2DGE attempts to subtract background noise from the image to enhance weak spots. That is, RJMCMC searches for weak spots which are likely to be ignored due to their weakness. The method also finds the proper number of spots automatically, restores the noisy images, and unmixes spots into more meaningful ones. In modeling of spots, we assumed Gaussian shape, as applied in other studies [13,9]. The assumption of such a specific shape of a spot has a significant benefit in that, since it is robust against local noise, we can estimate the signals based on interesting areas rather than each pixel. In our proposed RJMCMC method, we have incorporated the following characteristics to meet the requirements of gel image analysis: dimension invariant and moves in birth, death, split, and merge.

This paper consists of three main parts. In the first part, we present the mathematical model for 2D gel electrophoresis. Next, we describe the main algorithms for RJMCMC. Finally, the synthetic and experimental results from RJMCMC are demonstrated.

2   Model for 2D Gel Electrophoresis

As shown in Fig. 3, a typical 2D gel image contains numerous protein spots which might be individual or mixed. Therefore in our study, we model the image as a mixture of spots with Gaussian profile as follows:

$$I(x) = \sum_{k=1}^{K} A_k \phi_k(x; \mu_k, \Sigma_k) + \epsilon_e(x) \tag{1}$$

$$\begin{cases} x = [x_1, x_2]^T \\ \mu_k = [\mu_{k,1}, \mu_{k,2}]^T \\ \phi_k(x; \mu_k, \Sigma_k) \\ \quad = \frac{1}{\sqrt{2\pi\Sigma_k}} \exp\left\{-\frac{1}{2}(x_1 - \mu_k)^T \Sigma_k^{-1}(x - \mu_k)\right\} \\ \Sigma_k = diag(\sigma_{k,1}^2, \sigma_{k,2}^2) \\ \epsilon_e(x) \sim N(\epsilon_e(x); \mu_e, \sigma_e^2) \end{cases}$$

where $I(x)$ is the intensity at position x of the image. A_k is the amplitude of each spot, and $\phi_k(x; \mu_k, \Sigma_k)$ denotes the radial functions with a specific shape such as Gaussian, Lorentzian, or Laplacian shape. In this paper, we use a Gaussian

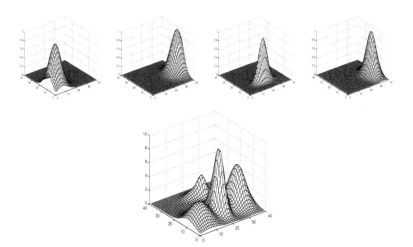

Fig. 1. Combination of radial functions: 4 different radial images with a spot are combined after multiplying their amplitudes - all images are vectorized in the linear model

shape for each spot. The radial function for each spot consists of two components, centre position μ_k and width of the spot σ_k. We assume that all spots have different spot widths and each spot has different spot width in terms of the x_1 and x_2 axes. $\epsilon_e(x)$ is a white noise, at position x, which is generated from normal distribution with a mean μ_e and a standard deviate σ_e. N denotes a normal distribution. RJMCMC estimates the μ_e and σ_e during its simulation automatically. Eq. (1) is well represented by Fig. 1.

In practice, Eq. (1) may be written in the linear model framework as follows:

$$Y = XA_{1:K} + \epsilon_e \text{ where } \epsilon_e \sim N(\epsilon_e; \mu_e, \sigma_e^2 I) \tag{2}$$

$A_{1:K} = [A_1, A_2, \cdots, A_K]$ is a vector for amplitudes of spots. X and Y are built from Eq. (1) by assembling all intensities $I(x)$ into a vector Y. That is, Y is a vector of noisy image intensities. X is defined by $[X_1, X_2, \cdots, X_K]^T$ where X_k is a vector made up from the image profile of spot k.

Denote by $\theta_k \in \Theta_k$, the parameter vector associated with the model indexed by $k \in \kappa$. Then, the priors are defined as

$$\theta_{1:K} = (\mu_{1:K}, A_{1:K}, \sigma_{1:K}) \tag{3}$$

$$\text{where } \begin{cases} \mu_{1:K} = (\mu_{1,1:K}, \mu_{2,1:K}) \\ \sigma_{1:K} = (\sigma_{1,1:K}, \sigma_{2,1:K}) \end{cases}$$

$$A_k \overset{iid}{\sim} N(A_k; \mu_A, \sigma_A^2)$$

$$\mu_{1,k} \overset{iid}{\sim} U(\mu_{1,k}; 0, T_1) \quad \mu_{2,k} \overset{iid}{\sim} U(\mu_{2,k}; 0, T_2)$$

$$\sigma_{1,k} \overset{iid}{\sim} G(\sigma_{1,k}; \alpha, \beta) \quad \sigma_{2,k} \overset{iid}{\sim} G(\sigma_{2,k}; \alpha, \beta)$$

where $k = 1, 2, \cdots, K$ and $K \in \{0, \cdots, K_{max}\}$. T_1 and T_2 are the size of an image. N, U and G stand for the normal, uniform, and gamma distributions respectively. α and β are assumed known. μ_A and σ_A are estimated during simulation.

Now, we may remove nuisance parameters $A_{1:K}$ by linear analytical integration since $A_{1:K}$ are assumed linear Gaussian,

$$P(\mu_{1:K}, \sigma_{1:K}|Y) = \int P(\mu_{1:K}, \sigma_{1:K}, A_{1:K}|Y)dA_{1:K} \qquad (4)$$

The removal of the nuisance parameters makes RJMCMC more efficient (Rao-Blackwellization) [3] . The removed nuisance parameters $A_{1:K}$ are sampled from their full conditional $P(A_{1:K}|\mu_{1:K}, \sigma_{1:K}, Y)$, when required for estimation.

2.1   Likelihood

Marginalising the nuisance parameters, the likelihood is defined as follows:

$$P(Y|X, \mu_A, \Sigma_A, \Sigma_e) = \frac{1}{(2\pi)^{(T_1 T_2)/2}|\Sigma_e|^{1/2}|\Sigma_A|^{1/2}|\Phi|^{1/2}} \qquad (5)$$

$$\times \exp\left\{-\frac{1}{2}(Y'^T \Sigma_e^{-1} Y' + \mu_A^T \Sigma_A^{-1} \mu_A - \Phi^T \hat{A})\right\}$$

$$\text{where} \quad \begin{cases} \hat{A} = \Phi^{-1}\phi \\ \Phi = X^T \Sigma_e^{-1} X + \Sigma_A^{-1} \\ \phi = X^T \Sigma_e^{-1} Y' + \Sigma_A^{-1}\mu_A \\ Y' = Y - \mu_e \end{cases}$$

where $\Sigma_A = \sigma_A^2 I$ and $\Sigma_e = \sigma_e^2 I$ respectively.

3   Algorithms

Since one does not know the exact number of spots in a given image, the number must be estimated during the processing. That is, to calculate the proper number of spots is equivalent to estimate the exact dimension of the parameters as they are proportional to each other. This kind of problem is addressed in trans-dimensional approaches. One of the best known trans-dimensional approaches is a generalization of Markov Chain Monte Carlo method, so called Reversible Jump Markov Chain Monte Carlo method (RJMCMC) [2,7]. RJMCMC proposes a next state given by current state in the time series and it constraints on Markov chains. In this paper, RJMCMC has several moves to find parameters of interests and their dimensions: Birth, Death, Split, Merge, and invariant moves. The next image is proposed from the current image via these moves. The Birth move makes a new spot in the current image randomly in terms of a given proposal distribution. The Death move deletes an existing spot in the current image. Birth and Death moves are designed to satisfy reversibility conditions. The Split move divides a spot in the current iteration into two different spots in the next iteration. Conversely, Merge move makes two selected spots in the current iteration into a single spot in the next iteration. Split and Merge moves have reversibility conditions as for Birth and Death moves [2]. The last move is a dimension invariant move, i.e. it does not change the dimension of the parameters. Instead, each parameter is sampled by a standard Markov Chain Monte Carlo (MCMC) step.

RJMCMC for 2D gel electrophoresis image has the following procedure in this paper:

- Propose a type of move from Birth, Death, Split, Merge, and Dimension invariant.
- If the move type is Dimension invariant, RJMCMC samples parameters using a standard Metropolis-Hastings (MH) algorithm, so that each unknown parameter is updated according to an acceptance probability.

$$\alpha_K = \min\left\{1, \frac{P(Y|\theta'_{1:K})P(\theta'_{1:K})q(\theta_{1:K};\theta'_{1:K})}{P(Y|\theta_{1:K})P(\theta_{1:K})q(\theta'_{1:K};\theta_{1:K})}\right\} \tag{6}$$

- If the move type is one of Birth, Death, Split, and Merge, RJMCMC follows a generalized MH step with an acceptance probability.

$$\alpha_{K'} = \min\left\{1, \frac{P(Y|K',\theta'_{1:K'})P(K')P(\theta'_{1:K'}|K')q_1(K;K')q_2(\theta_{1:K};\theta'_{1:K'})}{P(Y|K,\theta_{1:K})P(K)P(\theta_{1:K}|K)q_1(K';K)q_2(\theta'_{1:K'};\theta_{1:K})}\right\} \tag{7}$$

3.1 Dimension Invariant Move

RJMCMC is the same as standard MCMC in the case of Dimension invariant moves in that the dimension is fixed in Eq.(6). The Dimension invariant move samples two types of parameters, $\mu_{1:K}$ and $\sigma_{1:K}$. The prior structure of $\theta_{1:K}$ is assumed to be $P(\theta_{1:K}) = \prod_{k=1}^{K} P(\mu_k)P(\sigma_k)P(A_k)$, see Eq. (4) The kernel function $q(\theta'_{1:K};\theta_{1:K})$ proposes parameters using a Metropolis Hastings algorithm within the Gibbs method:

Dimension invariant Move

- Propose new parameters $\theta'_{1:K}$.
  - $\mu'_{1:K} \sim N(\mu'_{1:K};\mu_{1:K},\gamma^2 I)$
  - $\sigma'_{1:K} \sim G(\sigma'_{1:K};\alpha,\beta)$ where α and β are known.
- Calculate likelihood from Eq. (5).
- Obtain α_K from Eq. (6).
- $u \sim U(u;0,1)$
- if $\alpha_K > u$, then
  - Accept the proposed parameters and replace them by the current parameters.
- else
  - Reject the proposed parameters and maintain the current parameters.

3.2   Other Moves: Birth, Death, Split, and Merge Moves

In Eq. (7), the prior distribution for the dimensionality, $P(K)$ is assumed uniform. From experience, convergence is rather slow when $\mu_{1:K}$ and $\sigma_{1:K}$ are updated jointly. Thus, only $\mu_{1:K}$ is sampled in the dimension variant moves (but σ_k is also proposed from the prior in the Birth move for a new spot) and $\sigma_{1:K}$ is updated in the dimension invariant move. These four moves assume the prior distribution $P(\mu_{1:K}|K)$ is a uniform distribution.

The kernel functions for dimension, $q_1(K'|K)$, and parameters, $q_2(\theta'_{1:K'}|\theta_{1:K})$ are designed as follows in Birth and Death moves. The Birth move creates a new spot so that $q(K' = K+1|K) = 1$. The Death move however selects a spot randomly to delete among the K existing spots in the current step. Hence, the proposal probability of dimension for the death move is $q(K' = K-1|K) = 1/K$. In the Birth move, the new spot K' is proposed from the prior, i.e. $q(\mu'_k, \sigma'_k) = p(\mu'_k, \sigma'_k)$.

Birth move

  - $\mu_{K'}$ and $\sigma_{K'}$ are proposed for spot position and width.
  - Calculate likelihood from Eq. (5).
  - Obtain $\alpha_{K'}$ in Eq. (7).
  - Let $u \sim U(u; 0, 1)$
  - if $\alpha_{K'} > u$, then
    • Accept the proposed parameters and replace them by the current parameters.
  - else
    • Reject the proposed parameters and maintain the current parameters.

Death move

  - Select one among K spots and remove it.
  - Calculate likelihood from Eq. (5).
  - Obtain $\alpha_{K'}$ in Eq. (7).
  - $u \sim U(u; 0, 1)$
  - if $\alpha_{K'} > u$, then
    • Accept the proposed parameters and replace them by the current parameters.
  - else
    • Reject the proposed parameters and maintain the current parameters.

The Split and Merge moves are related in a similar way. The first transition kernel function $q_1(K'|K)$ is defined to be $1/K$ for both moves. The Split move divides a single spot, randomly chosen from the K existing spots, into two spots. In the Merge move, a single spot is randomly chosen and merged with its closest neighbours. The Split kernel for $q_2(\theta'_{1:K'}|\theta_{1:K})$ divides spot k into spots k and m, as follows:

$$\mu'_m \sim q_2(\mu'_m|\mu_k) = N(\mu'_m; \mu_k, \lambda)$$
$$\mu'_k \sim q_2(\mu'_k|\mu_k) = N(\mu'_k; \mu_k, \lambda)$$

where λ is assumed known. The Merge Kernel for $q_2(\theta'_{1:K'}|\theta_{1:K})$ combines a spot k with its the nearest neighbour m, as follows:

$$\mu'_k \sim q_2(\mu'_k|\mu_m, \mu_k) = N(\mu'_k; \bar{\mu}, \nu)$$
$$\text{where } \begin{cases} \bar{\mu} = \mu_m \times \omega_m + \mu_k \times \omega_k \\ \omega_m = \frac{A_m}{A_m+A_k}, \ \omega_k = \frac{A_k}{A_m+A_k} \end{cases} \tag{8}$$

where ν is assumed known. In both moves, σ_k parameters are proposed from the prior.

Split move

- Select one among K spots and divide it into two spots in Eq. (8)
- Calculate likelihood from Eq. (5).
- Obtain $\alpha_{K'}$ as in Eq. (7).
- $u \sim U(u; 0, 1)$
- if $\alpha_{K'} > u$, then
  - Accept the proposed parameters and replace them by the current parameters.
- else
  - Reject the proposed parameters and maintain the current parameters.

Merge move

- Select one among K spots and search for its closest neighbour.
- Merge the two selected spots from Eq. (8)
- Calculate likelihood from Eq. (5).
- Obtain $\alpha_{K'}$ in Eq. (7).
- $u \sim U(u; 0, 1)$
- if $\alpha_{K'} > u$, then
  - Accept the proposed parameters and replace them by the current parameters.
- else
  - Reject the proposed parameters and maintain the current parameters.

Owing to the non-unique labeling of individual spots in the model Eq. (1), it is likely that spots become re-ordered during sampling, especially in a RJMCMC procedure where spots can be detected or added at each iteration. In order to

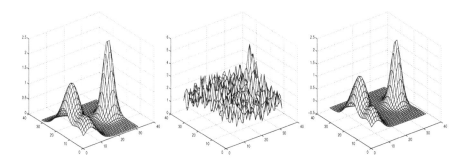

Fig. 2. A synthetic image with 4 spots : a pure image (a), a noisy image (b), an estimated image by RJMCMC (c)

address this labeling problem, we run a fixed dimensional RJMCMC with invariant moves after variant dimensional RJMCMC. That is, the variant dimensional RJMCMC generates the number of spots and the initial parameters for the fixed dimensional RJMCMC (see e.g. [5] for a detailed theoretical treatment of such issues).

4   Results

4.1   Synthetic Data

We tested RJMCMC performance on a synthetic image in Fig. 2. The first synthetic image has 4 spots, one very large spot and three overlapping spots. All spots have a Gaussian shape defined by centre position, width, and amplitude of the spot. The size of this image is 32 by 32. White Gaussian noise is added to the pure image with mean 2 and standard deviation 0.5. 2000 iterations are performed to address this problem including 1000 burn-in and 1000 for estimation. The estimated mean of noise is 2.0364. The left, centre and right figures denote a pure image without noise, corrupted image with noise generated by $\mu = 2$ and $\sigma = 0.5$ and a restored image from RJMCMC. The error is calculated by

$$\epsilon = ||\hat{S} - S||, \tag{9}$$

where $\hat{S}$ and S are the estimated image from RJMCMC and the original image without noise. We obtain $\epsilon = 1.8422$ after simulation for the first synthetic data set.

4.2   Experimental Data

To obtain experimental 2DGE images, we performed 2DGE on the purified mitochondiral protein of liver from an adult Sprague-Dawley rat. Fig. 3 shows that the resultant GE images present numerous protein spots.

We used two different concentrations of the sample and the images (a) and (b) of the figure show the results of $100\mu g$ and $200\mu g$ respectively. As we can

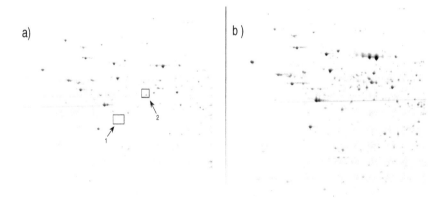

Fig. 3. Two gel images with a protein sample of $100\mu g$ (a) and $200\mu g$ (b): two rectangles (1 and 2) of the left image are used to search for the underlying spots such as weak spots and mixed spots in a mixed shape spot using RJMCMC

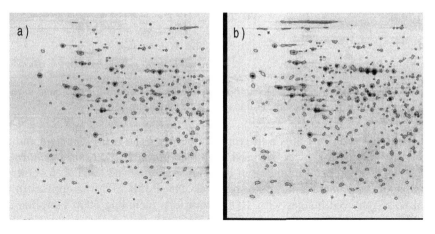

Fig. 4. Detection by commercial software (*Imagemaster 2D elite software v3.1*) for two gel images with a protein sample of $100\mu g$ (a) and $200\mu g$ (b)

see, the GE image (b) reveals more spots and much clearer than the image (a) due to the higher concentration which we intend to use in our evaluation and validation of the proposed methodology for detecting and identifying the spots.

To make a comparison against our proposed technique, the GE images in Fig. 3 were analyzed using a commercial gel image analysis software, Imagemaster 2D elite software v3.1 [6]. The images in Fig. 4 show the detected spots from each image. It is clear that there are more spots detected in (b).

For the RJMCMC image analysis, we selected the identical two sub-images from Fig. 3 (a) with the lower concentration as indicated in the figure with boxes. Before applying the RJMCMC, to remove the local variations or noise in

Fig. 5. Subtracting background noises: raw image (a), subtracted background noise (b), and extracted image (c)

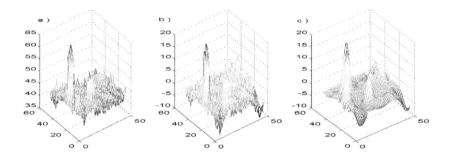

Fig. 6. Three images for the first example: a raw image (a), a filtered image without background noise (b) and a sample image in RJMCMC run (c)

the gel image, which creates a discrepancy between the real GE image and the theoretical model in Eq. (1), we used a simple approach, *local mean removal*, in which the average pixel intensity in local areas is removed [1]. Fig. 5 shows an example of local background noise removal. The RJMCMC was applied to the local noise-removed images with 10000 iterations including 5000 sampling and 5000 burn-in for sub-images. After the RJMCMC simulation, we run the fixed dimensional MCMC with 10000 iterations including 5000 sampling and 5000 burn-in for the labeling of the spots.

Fig. 6 and 8 show the steps of our RJMCMC algorithm. The image (a)s are selected from original 2DGE and the image (b)s are resulant images after removing background noise. The image (c)s are averaged images of samples from RJMCMC analysis. One can clearly notice that the RJMCMC generated images reveal potential spots with much better clarity. Furthermore, the RJMCMC generates statistical inferences to these spots.

Fig. 7 and 9 show the comparisons of the RJMCMC-inferenced spots against the detected spots by the commercial software. We used threshold to plot circles for the spots over 2.5 and 5 of the amplitudes (intensities) for Fig. 7 and 9 respectively.

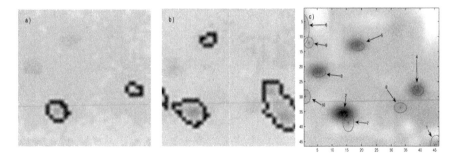

Fig. 7. comparison of detection for the first example: detection by commercial software for $100\mu g$ (a), detection by commercial software for $200\mu g$ (b) and detection by RJMCMC method for $100\mu g$ (c)

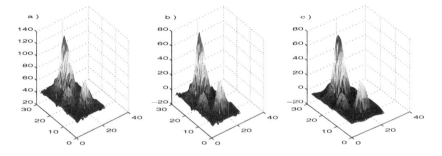

Fig. 8. Three images for the second example: a raw image (a), a filtered image without background noise (b) and a sample image in RJMCMC run (c)

Fig. 7 demonstrates that the RJMCMC can detect even very weak spots which would be unlikely to be detected in the conventional approaches. Note that the RJMCMC was applied to the gel image with a protein sample of $100\mu g$. The RJMCMC inferences numerous spots, as shown in Fig. 7 (c), that are not detected by the conventional software as in Fig. 7 (a). For instance Spot no. 5 and 10 are detected by the RJMCMC, but they are not detected in the (a), the lower concentration GE image. However, the higher concentration GE image confirms the presence of the spot in Fig. 7 (b). For Spot no. 2 and 7, the same spot has been detected in all three images, but the RJMCMC indicates the spots are composed of two spots. Additionally the RJMCMC indicates there could be more spots which are not visible in (a) and (b). It is not clear at the moment whether those RJMCMC inferenced weak spots are real or artificial, but it is clear that RJMCMC provides much higher sensitivity toward the detection of possible protein spots.

The results in Fig. 9 show the analysis of the mixed or overlapped spot in the selected region. The conventional software cannot separate the mixed and cannot give information about the possible individual profiles in both the low

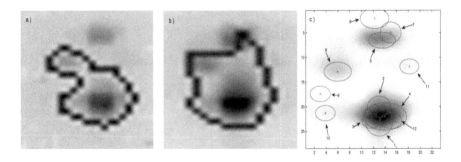

Fig. 9. comparison of detection for the second example: detection by commercial software for $100\mu g$ (a), detection by commercial software for $200\mu g$ (b) and detection by RJMCMC method for $100\mu g$ (c)

Table 1. Spot information for the first sub-image generated from RJMCMC

num	position (row)	position (col)	width (row)	width (col)	amplitude
1	45.5077	46.0000	1.8687	3.1683	3.5181
2	14.1045	35.8273	2.4262	1.5338	17.4910
3	5.6448	21.9955	2.1765	1.5216	10.8062
4	39.3266	28.0185	1.9230	1.6635	10.0152
5	18.8038	13.2272	2.1765	1.5122	9.8300
6	0	7.6177	2.2833	5.0953	15.3704
7	15.5598	38.6679	1.8768	3.0360	4.5647
8	33.2575	33.6518	2.1495	1.6844	2.6860
9	2.1577	11.9763	1.6069	1.8371	5.2277
10	1.0237	29.9370	1.6547	2.3283	3.3920

and high concentration GE images. Whereas the RJMCMC method may resolve each clustered spot into several individual spots as shown in (c).

Finally, our RJMCMC method generates the databases which are shown in Table. 1 and 2. Each table has six columns: index number, position for row, position for column, width for row, width for column and amplitude (intensity) of spots. As we can see in Tables, the amplitudes (intensities) of spots vary from 2.6 to 17.4 and from 5.7 to 503.9 for the Table. 1 and 2 respectively.

5   Conclusions

RJMCMC for 2DGE image analysis has two salient products: restoration and spot finding. 2DGE images suffer from high levels of noise yet RJMCMC extracts the real spots of interest under the assumption of a Gaussian spot shape. This assumption for spot shape implies strong prior information and makes RJMCMC robust against random noise. Another benefit of RJMCMC 2DGE processing

Table 2. Spot information for the second sub-image generated from RJMCMC

num	position (row)	position (col)	width (row)	width (col)	amplitude
1	13.0000	23.0000	2.3559	3.0044	192.5783
2	13.0035	20.0101	1.9893	2.1364	182.8023
3	11.6072	22.8732	1.5061	1.5269	358.6255
4	14.0000	22.0000	3.7394	2.7883	218.3849
5	13.5044	6.5399	2.2925	1.6172	75.2568
6	5.9919	12.9131	2.3382	1.9234	22.7709
7	15.0118	5.0023	1.8144	2.1599	10.4090
8	12.2408	2.1442	2.4446	2.0642	8.5309
9	3.0066	17.4009	1.7251	1.5914	11.7475
10	3.8674	21.2892	1.6920	1.5663	9.8358
11	18.1904	11.8794	1.6128	1.5320	5.7568
12	15.0298	22.8203	1.8769	1.6150	503.9675

is that complicated spots in 2DGE from protein complexes are separated into several small spots. Moreover, the RJMCMC finds some extremely weak spots, based on Gaussian spot shape assumption, which many threshold approaches fail to detect. In addition, RJMCMC does not require that the number of spots be fixed: RJMCMC based on Monte Carlo methods searches for the proper number of spots automatically. However, the radial functions are expressed by rather large matrices, so if there are many spots in the image of interest, RJMCMC for 2DGE can be a time consuming method and possibly impracticable. Also, if the spot shape is very different from a Gaussian shape, RJMCMC in this paper will tend to generate many small Gaussian shaped spots to model the non-Gaussian spot. That is, we note that the proper spot finding may fail for non-Gaussian shaped spots. However, restoration can work even in the case of non-Gaussian spots since the overall restored shape may still be well modelled.

6   Further Work

We present the possibility of RJMCMC to process and analyze 2DGE images with a Gaussian spot shape assumption. However, it is known that the actual shape for 2DGE is non-Gaussian and non-Lorentzian. Therefore, we will incorporate more realistic shapes into the 2DGE image and this will give better Bayesian model for RJMCMC (see spots in Fig. 3). One limitation of RJMCMC for practical use is its computation time. At present, it takes 60 minutes with 10000 iterations on a Pentium CPU at 3.20GHz for the first experimental example. To make RJMCMC more practicable, sub-sampling will be applied. Next, we aim to research labeling of the spots from the RJMCMC output. Last, we will improve more sophisticated algorithms for the background subtraction as shown in Lepski's paper [4].

Acknowledgements

For ChulHun Kang and Tae-Seong Kim, this study was supported by a grant of the Korea Health 21 *R&D* Project, Ministry of Health and Welfare, Republic of Korea (02-PJ3-PG6-EV07-0002).

References

1. Reed, I. S., Yu, X.: Adaptive multiple-band CFAR detection of an optical pattern with unknown spectral distribution, *IEEE Transactions on Acoustics, Speech, and Signal Processing,*. **38**, (1990), 1760–1770
2. Green,P.J.: Reversible Jump Markov Chain Monte Carlo computation and Bayesian model determination, *Biometrika*, (1995), 711-732.
3. Casella,G., Robert,C.P: Rao-Blackwellisation of sampling schemes, *Biometrika*, **83**, (1996), 81-94.
4. Lepski, O. V., Mammen, E., Spokoiny, V. G.: Optimal spatial adaptation to in-homogeneous smoothness: and approach based on kernel estimates with variable bandwith selectors, *The Annuals of statistics,*. **25**, (1997), 929–947.
5. Stephens,M.: Bayesian Methods for Mixtures of Normal Distributions, *PhD thesis*, Magdalen College, Oxford University, (1997)
6. Bergling, H.: Automated 2-D gel electrophoresis spot excision using Ettan Spot Picker in proteomics research , *Life Science News*. Amersham Pharmacia Biotech. (2001)
7. Godsill,S.J.: On the Relationship Between Markov Chain Monte Carlo Methods for Model Uncertainty, *Journal of Computational and Graphical Statistics*, **10**, (2001), 1-19.
8. Dowsey,A.W., Michael,J.D., Yang,G.: The role of bioinformatics in two-dimensional gel electrophoresis, *PROTEOMICS*, **3**, (2002), 1567 - 1596.
9. Pedersen,L.: Analysis of Two-dimensional Electrophoresis images, *PhD thesis*, Technical University of Denmark. (2002)
10. Cutler,P., Heald,G., White,I.R., Ruan,J.: A novel approach to spot detection for two-dimensional gel electrophoresis images using pixel value collection, *PROTEOMICS*, **3**, (2003), 392 - 401.
11. Gorg,A., Weiss,W., Dunn,M.J.: Current two-dimensional electrophoresis technology for proteomics, *PROTEOMICS*, **4**, (2004), 3665 - 3685.
12. Kaczmarek,K., Walczak,B., Jong,S., Vandeginste,B.G.M.: Preprocessing of two-dimensional gel electrophoresis images, *PROTEOMICS*, **4**, (2004), 2377 - 2389.
13. Rohr,K., Cathier,P., Worz,S.: Elastic resistration of gel electrophoresis images Based on Landmarks and Intensities, *International Symposium on Biomedical Imaging (ISBI)*, (2004), 1451-1454.
14. Bettens,E., Scheunders,P., Dyck,D.V., Moens, L., Osta,P.V.: Computer analysis of two-dimensional electrophoresis gels: A new segmentation and modeling algorithm, *Electrophoresis*, **18**, (2005), 792 - 798.
15. Yoon,J., Godsill,S.J., Kupce,E., Freeman,R.: Deterministic and statistical methods for reconstructing multidimensional NMR spectra, *Magnetic Resonance in Chemistry*, **44**, (2006), 197-209.
16. Yoon,J., Godsill,S.J.: Bayesian Inference for Multidimensional NMR image reconstruction, *European Signal Processing Conference (EUSIPCO)*, accepted, (2006).

SimShiftDB: Chemical-Shift-Based Homology Modeling*

Simon W. Ginzinger[1],[**], Thomas Gräupl[2], and Volker Heun[1]

[1] Institut für Informatik, Ludwig-Maximilians-Universität München,
Amalienstr. 17, D-80333 München
{Simon.Ginzinger|Volker.Heun}@bio.ifi.lmu.de
[2] Fachbereich für Computerwissenschaften, Universität Salzburg,
Jakob-Haringer-Str. 2, A-5020 Salzburg
Thomas.Graeupl@sbg.ac.at

Abstract. An important quantity that is measured in NMR spectroscopy is the chemical shift. The interpretation of these data is mostly
done by human experts. We present a method, named *SimShiftDB*, which
identifies structural similarities between a protein of unknown structure
and a database of resolved proteins based on chemical shift data. To
evaluate the performance of our approach, we use a small but very reliable test set and compare our results to those of 123D and TALOS.
The evaluation shows that SimShiftDB outperforms 123D in the majority of cases. For a significant part of the predictions made by TALOS,
our method strongly reduces the error. SimShiftDB also assesses the statistical significance of each similarity identified.

1 Introduction

NMR Spectroscopy is an established method to resolve protein structures on
an atomic level. The NMR structure determination process consists of several
steps, varying in complexity. A quantity that is measured routinely in the beginning is the chemical shift. Chemical shifts are available on a per atom basis
and inherently carry structural information. Chemical shifts in general do not
suffice to calculate the structure of biological macromolecules, such as proteins.
Additional experiments of increased complexity and human expert knowledge
are necessary to obtain the solution.

In an earlier study, the performance of a pairwise chemical shift alignment
algorithm was evaluated. Ginzinger and Fischer [1] were able to show that it
is indeed possible to utilize the information hidden in the chemical shift data
for constructing structurally meaningful alignments. Now we present a method
(called *SimShiftDB*) that searches for similarities between a target protein with
assigned chemical shifts and a database of template proteins for which both

---

* This work was funded by the German Research Foundation (DFG, Bioinformatics
  Initiative).
** Corresponding author.

S. Hochreiter and R. Wagner (Eds.): BIRD 2007, LNBI 4414, pp. 357–370, 2007.

chemical shift data and 3D coordinates are available. For each target-template-pair we calculate a chemical shift alignment. These alignments map a set of residues from the target to a set of residues from the template structure. Therefore, we can build a structural model for the aligned residues from the target based on the coordinates of the associated residues from the template. To give the spectroscopist the possibility to judge over the statistical significance of a certain alignment with shift similarity score S, we calculate the expectation of the number of alignments with score $\geq S$ occurring by chance.

To evaluate the performance of our approach, we compare the backbone angle prediction quality of our method to 123D [2], a threading approach, and to TALOS [3], which tries to calculate backbone angles from the amino acid sequence and associated chemical shift data. We are able to prove empirically that 123D is outperformed significantly by our method. When comparing to TALOS, SimShiftDB performs significantly better for 36% of the target's residues. Our result suggests that both TALOS and SimShiftDB have their strengths and therefore should be used in parallel in the NMR structure determination process.

In the following we describe the template database, the chemical shifts substitution matrices used for scoring chemical shift alignments, the calculation of the expected value and the SimShiftDB algorithm. Afterwards the results on our test set will be presented and discussed.

2   The Template Database

One publicly available database exists which contains chemical shift information for various proteins [4]. However, to date there are only 3750 structure in the *proteins/peptide* - class, which corresponds to about 10% of the structures deposited in PDB. Additionally, there is no standard set of chemical shifts which has to be available for each entry, e.g., one entry may contain only ^{1}H chemical shifts while for a different one just ^{15}N chemical shifts are available. Finally, as pointed out in [5] various errors occur in the data. These problems led us to the conclusion to use a different template database.

Chemical shifts for all structures in the ASTRAL [6] database are calculated using SHIFTX [7]. Chemical shifts predicted with SHIFTX correlate strongly with measured data. It is also shown that the agreement between observed and calculated chemical shifts is an extremely sensitive measure to assess the quality of protein structures. This approach leaves us with a database containing 64,839 proteins with known 3D structure and associated chemical shifts.

3   Substitution Matrices for Shift Data

In order to identify pairs of amino acids with associated chemical shifts which are likely to be structurally equivalent, we derive substitution matrices for chemical shifts using the modus operandi described in [8]. Therefore, a *Standard of Truth* is needed for the calculation of the matrix values. We rely on the DALI database [9] containing a set 188,620 structure-based alignments. Through a sequence-similarity-search [10] we map all sequences being part of a DALI alignment to our template database.

For each amino acid we calculate the minimal and maximal chemical shift of $^{1}H_{\alpha}$, $^{13}C_{\alpha}$, $^{13}C_{\beta}$ and ^{13}C in our template database. Then we divide the range between minimum and maximum into two equal parts. This enables us to classify each chemical shift as either *weak* (situated in the first part of the range) or *strong*.

It is convenient to define a new alphabet on proteins with associated chemical shift sequences, namely Σ^{S}. A letter $\mathcal{A}$ in this alphabet is a tuple (a, s_1, s_2, s_3, s_4), where a is the corresponding amino acid identifier and s_1, s_2, s_3, s_4 are the classifications for the corresponding shifts for $^{1}H_{\alpha}$, $^{13}C_{\alpha}$, $^{13}C_{\beta}$ and ^{13}C, respectively.

We derive the relative frequencies of each of the letters in the template database, denoted by $p_{\mathcal{A}}$. Additionally, we calculate the relative frequencies of all substitution events, denoted by $q_{\mathcal{A},\mathcal{B}}$, which is the relative frequency of letters $\mathcal{A}$ and $\mathcal{B}$ being aligned in the DALI database. To account for the bias of overrepresented folds, we use pseudo counts to give each fold type an equal weight. To do so, each alignment is identified with the fold type associated to the first sequence according to the SCOP [11] classification.

Then, we calculate the well-known log-odds scores [8]

$$o_{\mathcal{A},\mathcal{B}} = \log\left(\frac{q_{\mathcal{A},\mathcal{B}}}{e_{\mathcal{A},\mathcal{B}}}\right), \tag{1}$$

where

$$e_{\mathcal{A},\mathcal{B}} = \begin{cases} 2 * p_{\mathcal{A}} * p_{\mathcal{B}} & \text{if } \mathcal{A} \neq \mathcal{B}, \\ p_{\mathcal{A}}^2 & \text{otherwise.} \end{cases} \tag{2}$$

Finally the log-odds scores are multiplied with a normalization factor η and rounded to the nearest integer. The shift substitution matrix entries $s_{\mathcal{A},\mathcal{B}}$ are then formally defined as

$$s_{\mathcal{A},\mathcal{B}} = \lfloor o_{\mathcal{A},\mathcal{B}} * \eta + 0.5 \rfloor. \tag{3}$$

Here the parameter η is set to 10.

4   E-Values for Chemical Shift Alignments

A shift alignment produced by SimShiftDB is a set of local ungapped alignments which do not overlap. In [12] the authors derive a p-value for multiple ungapped alignments which may be ordered consistently (see page 5874, section *Consistently Ordered Segment Pairs in Sequence Alignments* for details). For this p-value two statistical parameters (λ, κ) have to be calculated. We use the method described in [13] to obtain these parameters. Formally λ is defined as the unique positive solution of the equation

$$\sum_{\mathcal{A},\mathcal{B}\in\Sigma^{S}} p_{\mathcal{A}}^{T} * p_{\mathcal{B}}^{D} * e^{\lambda * s_{\mathcal{A},\mathcal{B}}} = 1, \tag{4}$$

where $p_{\mathcal{A}}^T$ is the probability that the letter $\mathcal{A}$ occurs in the target sequence and $p_{\mathcal{B}}^D$ is the probability that letter $\mathcal{B}$ occurs in the template database.

The parameter κ is calculated as

$$\kappa = e^{\gamma} * \frac{\delta}{(1 - e^{-\lambda * \delta}) * E[S[1]e^{\lambda * S[1]}]}, \tag{5}$$

with

$$\gamma = -2 * \sum_{k=1}^{\infty} \frac{1}{k} (E[e^{\lambda * S[k]} | S[k] < 0] + P(S[k] \geq 0)) \tag{6}$$

and

$$\delta = \gcd\{s_{\mathcal{A},\mathcal{B}} \mid \mathcal{A}, \mathcal{B} \in \Sigma^S\}. \tag{7}$$

Here, $S[k]$ is a random variable representing the sum of the pair scores of an alignment of length k. For further details, we refer to [13].

Using λ, κ as described above, we can normalize the score of an ungapped chemical shift alignment A as follows. Let S_A be the sum of the pairwise scores of the aligned letters from Σ^S. The normalized score S'_A is then defined as

$$S'_A = \lambda * S_A - \ln(n * m * \kappa), \tag{8}$$

where n is the length of the target sequence and m is the length of the template sequence.

Using Equ. (4) from [12], we can calculate the probability that a number of consistently ordered alignments $A_1, \ldots, A_r$ with normalized scores $S'_{A_1}, \ldots, S'_{A_r}$ occurs by chance as

$$P(T', r) = \int_{T' + \ln(r!)}^{\infty} \frac{e^{-t}}{r!(r-2)!} \int_0^{\infty} y^{r-1} e^{-e^{\frac{y-t}{r}}} \, dy \, dt, \tag{9}$$

with

$$T' = \sum_{i=1}^{r} S'_{A_i}. \tag{10}$$

Note that as our alignments are ordered consistently, we start integrating from $T' + \ln(r!)$.

However, one problem remains. $P(T', r)$ does not take into account the database size. Therefore, we additionally calculate the expected number of alignments in our search space with a score not less than the score of the alignment of interest:

$$E(T', r) = P(T', r) * \frac{N}{m}. \tag{11}$$

Here m is the length of the template sequence and N is the number of letters in the database. $E(T', r)$ is the E-value we use to assess the statistical significance of the chemical shift alignments.

5   The Shift Alignment Algorithm

We design a two step algorithm to build a chemical shift alignment for two sequences from the alphabet Σ^S. Initially, a set of local ungapped alignments is constructed. Then we search for a best legal combination of a subset of these alignments.

5.1   Step 1: Calculate Local Alignments

We construct the pair score matrix, containing scores for all pairs of letters $\mathcal{A}$ and $\mathcal{B}$ where $\mathcal{A}$ is a letter of the target and $\mathcal{B}$ is a letter of the template sequence. The score for each pairing is $s_{\mathcal{A},\mathcal{B}}$ as defined in Equ. (3). Then we apply an algorithm [14] which identifies all maximal scoring subsequences (*MSS*) on each diagonal in linear time. An MSS is defined as follows.

Definition 1. *Let $a = (a_1, \ldots, a_n) \in \mathbb{R}^n$ and $a' = (a_i, \ldots, a_j)$ be a subsequence of a. a' is a maximal scoring subsequence if and only if*

(1) All proper subsequences of a' have lower score.
(2) No proper supersequence of a' contained in a satisfies (1).

The MSS can also be interpreted as a local ungapped alignment and will be called a *block* from now on. Note that the algorithm only identifies MSS's with score greater than zero. Additionally, we remove all MSS's of length ≤ 5.

5.2   Step 2: Identify the Best Legal Combination

We now build a DAG (directed acyclic graph) where the blocks correspond to the nodes in the graph. Two blocks may be combined if they are connected by an edge in this graph. Two constraints have to be fulfilled for two blocks (B_1 and B_2) to be connected by an edge from B_1 to B_2:

1. B_1 and B_2 may **not** overlap, neither in the target nor in the template sequence. Additionally, B_1 has to appear before B_2 in the target as well as in the template.
2. Let **d** be the number of residues in the target sequence between the end of B_1 and the beginning of B_2. Let **L** be the last residue from the first block in the template sequence and **F** be the first residue from the second block in the template sequence (see Fig. 1). We require the residues **L** and **F** not to be further apart in the structure than the maximal distance that could be bridged by a polypeptide chain of **d** residues. Here it is assumed that the maximal distance between two C_α atoms in the polypeptide chain is 4.0Å.

We also add an artificial start node to the DAG from which every other node may be reached.

Fig. 1 shows an example of a block matrix with blocks B_1 and B_2 fulfilling constraint 1 and the corresponding check of constraint 2 in the template structure. In this example **d** has to be at least 2, if blocks B_1 and B_2 are to be connected by an edge in the DAG.

In the DAG we then weigh each node with the normalized score (as defined in Equ. (8)) of the corresponding block. Then Procedure 1 identifies the optimal path in the DAG, using $P(T', r)$ as a measure of quality.

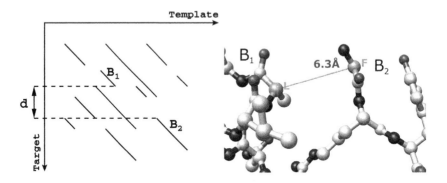

Fig. 1. A block matrix with a gap of length d in the target sequence highlighted and the corresponding gap in the structure of the template sequence

The idea of the algorithm is as follows. Beginning from the artificial start node, we perform a depth first search (*DFS*) in the DAG identifying the lowest scoring path according to $P(T', r)$. However, $P(T', r)$ is not only dependent on the summed score of the blocks but also on r, the length of the path. When reaching a node v in the DFS, it is impossible to determine the overall best successor for this node. However, when the number of allowed successors of v in the path is fixed, the solution may be found. Therefore, we do not save a single best successor for each node, but we keep an array of optimal successors, for each possible number of succeeding blocks, named *succ* in Procedure 1 (succ[v][3], for example, saves the optimal successors of v given that v is first node in a path consisting of three blocks). After the DFS finishes, succ[start] contains the optimal path for each possible path length. Now the highest scoring path in succ[start] corresponds to the combination of blocks achieving the lowest p-value. The running time of procedure 1 is $O(e * (n + m))$ with e being the number of edges in the DAG and n and m being the length of the target and the length of the template, respectively. Note that the DAG is sparsely connected and therefore in practice e is in the order of $(n+m)^2$. In our implementation one database[1] search on an standard laptop (Intel T2500, 2.0 GHz, 1 GB RAM) takes about fifteen minutes. By discarding longer blocks in Step 1 of the algorithm the running time may be strongly decreased. Discarding all blocks with a length less than 10, for example, results in a running time decrease of approximately 80%.

---

[1] 64839 proteins with an average length of 183 residues.

Procedure 1. DFS which fills the array succ

```
/* adj      ... adjacency list of the nodes in the graph
   visited ... array of boolean variables saving the DFS status of each node
   v          ... current node looked at in the graph
   succ   ... two dimensional array saving the optimal successors              */

def  SimShiftDB_DFS(adj,visited,v,succ)
 1: best_succ ← [] /*empty array*/
 2: for w in adj[v] do
 3:    if visited[w] = 0 then
 4:       SimShiftDB_DFS(adj,visited,w,succ)
 5:    end if
 6:    /*merge arrays best_succ and succ[w] favoring higher scoring paths*/
 7:    best_succ ← merge(best_succ, succ[w])
 8: end for
 9: visited[v] ← 1
10: for k in best_succ do
11:    succ[v][k+1] ← best_succ[k]
12: end for
```

6   Results

6.1   Evaluation of the Modeling Performance

To evaluate the performance of our algorithm, we compare our method to 123D, an established threading method, and the standard tool used by spectroscopists working with chemical shifts, namely TALOS. Our target set has to fulfill two constraints:

- The chemical shift data shall be of high quality (not corrupted by errors as noted in [5]).
- Chemical shifts for $^1H_\alpha$, $^{13}C_\alpha$, $^{13}C_\beta$ and ^{13}C have to be available.

As it is often hard to check the reliability of chemical shift data, we use a set of six target structures which were provided by the group of Prof. Dr. Horst Kessler from the Technische Universität München. The data for PH1500-N (unpublished), HAMP (in press), PHS018 (in press), KDP [15] and VAT-N [16] was measured directly by this group. The data for JOSEPHIN [17], which was solved by a different group, was checked for its correctness. As all of these structures were recently resolved, three dimensional data is also available. This way we can reliably check the quality of our predictions. The set-up of our experiment is as follows:

- It is required that all methods give torsion angle prediction for at least 80% of the target protein. For 123D and SimShiftDB, we sort the alignments produced by the quality score of the respective method (alignment score for 123D and e-value for SimShiftDB) and take as many alignments (starting from the best) as necessary such that at least 80% of the residues of the target protein have an assigned residue from a template structure. The assignment is done favoring alignments with better score if a residue from the target structure is assigned multiple times in separate alignments. Concerning the comparison of SimShiftDB to 123D, we additionally discard all alignments with sequence identity $\geq 90\%$ to remove trivial solutions. As TALOS predicts backbone torsion angles for all residues of the target, no additional work is required in this case.

– To evaluate the torsion angle predictions, we build a model for the torsion angles from the target structure (using the torsion angles of the assigned residues from the templates). Then we calculate the torsion angles for our target using STRIDE [18]. Now it is possible to assess the average error in torsion angle prediction by using the STRIDE calculations as a Standard of Truth.

Fig. 2 and Fig. 3 show the percentage of torsion angles per structure where SimShiftDB outperforms 123D and TALOS (for Φ and Ψ angles, respectively). SimShiftDB outperforms 123D in 62% and 69% of all cases and 35% and 36% of all backbone torsion angles predicted by SimShiftDB have a smaller error than those predicted by TALOS. To check that the difference between TALOS and SimShiftDB is not just marginal, we calculate the mean error of both methods for the cases where SimShiftDB outperforms TALOS (see Fig. 4 and 5 for Details). SimShiftDB reduces the error (compared to TALOS) by more than 60%.

6.2   Evaluation of the P-Value Correctness

Two sets, S_1 and S_2, of random chemical shift alignments are constructed as follows. Step 1 of the SimShiftDB algorithm is performed for each target-template-pair. Based on the identified blocks, two DAGs, namely G_1 and G_2, are build. In G_2 nodes which fulfill constraints 1 and 2 (see page 361) are connected, whereas in G_1 constraint 1 has to be fulfilled only. Then ten nodes are drawn from each DAG without replacement. For every node n, we construct a random path in the DAG starting in n until we reach a node with outdegree zero. Each prefix of the path in G_1 (or G_2) yields an alignment, which is added to S_1 (or S_2, respectively).

For each constructed alignment from S_1 or S_2, we calculate the empirical p-value and compare it to the theoretical p-value. The results of this comparison for alignments consisting of one, two or three blocks are shown in Fig. 6 and Fig. 7, for S_1 and S_2, respectively. The empirical p-value is less than the theoretical p-value for both sets, which is also true for alignments consisting of a greater number of blocks (data not shown). Therefore, the theoretical p-value provides a conservative estimate, both in theory and practice.

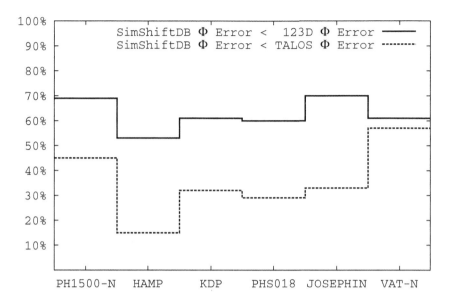

Fig. 2. Percentage of Φ-angle predictions where SimShiftDB outperforms 123D and TALOS

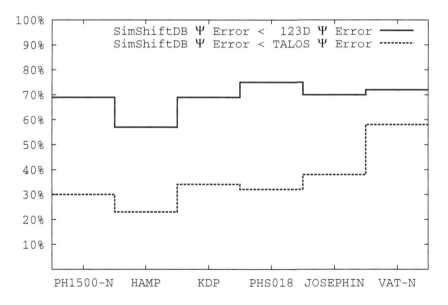

Fig. 3. Percentage of Ψ-angle predictions where SimShiftDB outperforms 123D and TALOS

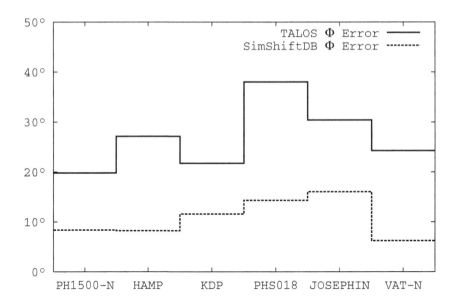

Fig. 4. Φ-angle error compared to STRIDE for those predictions where SimShiftDB outperforms TALOS

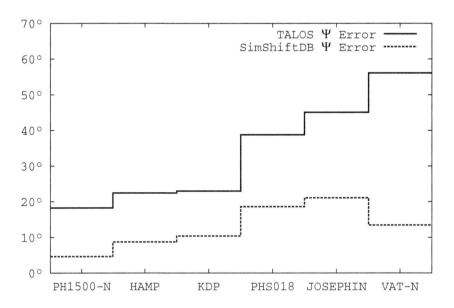

Fig. 5. Ψ-angle error compared to STRIDE for those predictions where SimShiftDB outperforms TALOS

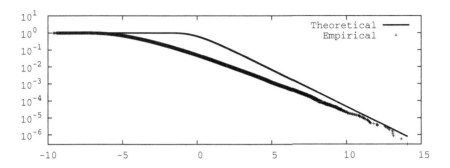

(a) Alignments consisting of one block (3,654,077 data points)

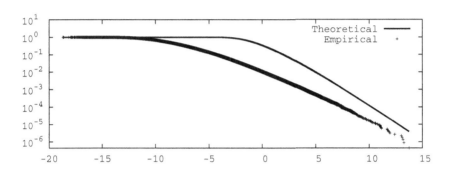

(b) Alignments consisting of two blocks (2,260,265 data points)

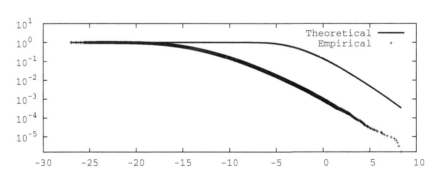

(c) Alignments consisting of three blocks (638,037 data points)

Fig. 6. Empirical versus theoretical p-values for alignments from S_1

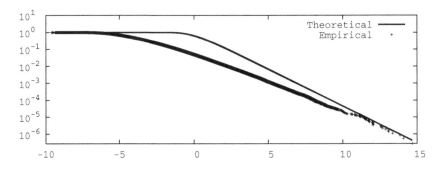

(a) Alignments consisting of one block (3,654,077 data points)

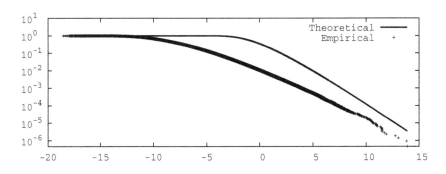

(b) Alignments consisting of two blocks (2,190,528 data points)

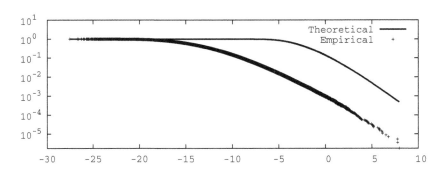

(c) Alignments consisting of three blocks (574,503 data points)

Fig. 7. Empirical versus theoretical p-values for alignments from S_2

7 Discussion

We developed a method which builds models for target proteins of unknown
structure using chemical shift data measured in NMR Spectroscopy. The method
has been evaluated on a small, but very reliable test set. From the results presented in the last section, we draw the following conclusions:

- SimShiftDB strongly outperforms methods which are based on amino acid
  sequence information alone and should therefore be used whenever chemical
  shift data is available.
- When comparing to TALOS, both methods show their strength. However,
  SimShiftDB is able to outperform TALOS in a significant number of cases.
  Therefore we suggest that both TALOS and SimShiftDB should be used
  when trying to solve a protein structure by NMR spectroscopy.

8 Availability

SimShiftDB is available via `http://shifts.bio.ifi.lmu.de`.

Acknowledgments

Thanks go to Joannis Apostolakis, Rolf Backofen and Johannes Fischer for helpful discussions and suggestions. Many thanks also go to Murray Coles for extremely helpful discussions concerning NMR spectroscopy. Simon W. Ginzinger
also wants to thank Manfred J. Sippl and Robert Konrat who first introduced
him to the idea of searching similarities in chemical shift sequences.

References

1. Ginzinger, S.W., Fischer, J.: SimShift: Identifying structural similarities from NMR
   chemical shifts. Bioinformatics **22**(4) (2006) 460–465
2. Alexandrov, N.N., Nussinov, R., Zimmer, R.M.: Fast protein fold recognition via
   sequence to structure alignment and contact capacity potentials. Pac Symp Biocomput (1996) 53–72
3. Cornilescu, G., Delaglio, F., Bax, A.: Protein backbone angle restraints from
   searching a database for chemical shift and sequence homology. J Biomol NMR
   13(3) (1999) 289–302
4. Seavey, B.R., Farr, E.A., Westler, W.M., Markley, J.L.: A relational database for
   sequence-specific protein NMR data. J Biomol NMR **1**(3) (1991) 217–236
5. Zhang, H., Neal, S., Wishart, D.S.: RefDB: a database of uniformly referenced
   protein chemical shifts. J Biomol NMR **25**(3) (2003) 173–195
6. Chandonia, J.M., Hon, G., Walker, N.S., Conte, L.L., Koehl, P., Levitt, M., Brenner, S.E.: The ASTRAL compendium in 2004. Nucleic Acids Res **32**(Database
   issue) (2004) D189–D192
7. Neal, S., Nip, A.M., Zhang, H., Wishart, D.S.: Rapid and accurate calculation of
   protein ^{1}H, ^{13}C and ^{15}N chemical shifts. J Biomol NMR **26**(3) (2003) 215–240

8. Henikoff, S., Henikoff, J.G.: Amino acid substitution matrices from protein blocks. Proc Natl Acad Sci **89**(22) (1992) 10915–10919
9. Holm, L., Sander, C.: Mapping the protein universe. Science **273**(5275) (1996) 595–603
10. Altschul, S.F., Gish, W., Miller, W., Myers, E.W., Lipman, D.J.: Basic local alignment search tool. J Mol Biol **215**(3) (1990) 403–410
11. Murzin, A.G., Brenner, S.E., Hubbard, T., Chothia, C.: SCOP: a structural classification of proteins database for the investigation of sequences and structures. J Mol Biol **247**(4) (1995) 536–540
12. Karlin, S., Altschul, S.F.: Applications and statistics for multiple high-scoring segments in molecular sequences. Proc Natl Acad Sci **90**(12) (1993) 5873–5877
13. Karlin, S., Altschul, S.F.: Methods for assessing the statistical significance of molecular sequence features by using general scoring schemes. Proc Natl Acad Sci **87**(6) (1990) 2264–2268
14. Ruzzo, W.L., Tompa, M.: A linear time algorithm for finding all maximal scoring subsequences. Proc Int Conf Intell Syst Mol Biol (1999) 234–241
15. Haupt, M., Bramkamp, M., Coles, M., Altendorf, K., Kessler, H.: Inter-domain motions of the N-domain of the KdpFABC complex, a P-type ATPase, are not driven by ATP-induced conformational changes. J Mol Biol **342**(5) (2004) 1547–1558
16. Coles, M., Diercks, T., Liermann, J., Groger, A., Rockel, B., Baumeister, W., Koretke, K.K., Lupas, A., Peters, J., Kessler, H.: The solution structure of VAT-N reveals a 'missing link' in the evolution of complex enzymes from a simple $\beta\alpha\beta\beta$ element. Curr Biol **9**(20) (1999) 1158–1168
17. Nicastro, G., Menon, R.P., Masino, L., Knowles, P.P., McDonald, N.Q., Pastore, A.: The solution structure of the josephin domain of ataxin-3: structural determinants for molecular recognition. Proc Natl Acad Sci **102**(30) (2005) 10493–10498
18. Frishman, D., Argos, P.: Knowledge-based protein secondary structure assignment. Proteins **23**(4) (1995) 566–579

Annotation of LC/ESI-MS Mass Signals

Ralf Tautenhahn, Christoph Böttcher, and Steffen Neumann

Leibniz Institute of Plant Biochemistry, Department of Stress and Developmental
Biology, Weinberg 3, 06120 Halle, Germany
{rtautenh|cboettch|sneumann}@ipb-halle.de
http://www.ipb-halle.de

Abstract. Mass spectrometry is the work-horse technology of the emerging field of metabolomics. The identification of mass signals remains the
largest bottleneck for a non-targeted approach: due to the analytical
method, each metabolite in a complex mixture will give rise to a number
of mass signals. In contrast to GC/MS measurements, for soft ionisation
methods such as ESI-MS there are no extensive libraries of reference
spectra or established deconvolution methods. We present a set of annotation methods which aim to group together mass signals measured from
a single metabolite, based on rules for mass differences and peak shape
comparison.

Availability: The software and documentation is available as an R package on http://msbi.ipb-halle.de/

1 Introduction

Metabolomics and especially mass spectrometry have evolved into an important
technology to solve challenges in functional genomics. The ambitious goal is
the unbiased and comprehensive quantification of metabolite concentrations of
organisms, tissues, or cells [Oliver98, Fiehn00].

The combination of chromatographic separation with subsequent mass spec-
trometric detection has emerged as key technology for multiparallel analysis of
low molecular weight compounds in biological systems. Gas chromatography-
mass spectrometry (GC/MS) based techniques are mature and well-established
but restricted to volatile compounds, at least after derivatization. High-
performance liquid chromatography-mass spectrometry (HPLC/MS) facilitates
the analysis of compounds of higher polarity and lower volatility in a much
wider mass range without derivatization. To complement GC/MS based profiling schemes towards a higher coverage of a systems' metabolome LC/MS based
platforms have been developed recently [Roepenack-Lahaye04].

LC/MS techniques require an ionisation of the analytes out of the liquid phase
under atmospherical pressure. This is in sharp contrast to GC/MS, where analytes are ionised and subsequently fragmented in the gas phase by electron
impact (EI). Typical atmospheric pressure ionisation (API) techniques are electrospray ionisation (ESI) and atmospheric pressure chemical ionisation (APCI).

S. Hochreiter and R. Wagner (Eds.): BIRD 2007, LNBI 4414, pp. 371–380, 2007.

Positive-ion API spectra of low molecular weight compounds often comprise simple adduct ions with one or more cations (e.g. $[M + H]^+$, $[M + H + Na]^{2+}$) and cluster ions (e.g. $[2M + H]^+$, $[M + CH_3CN + H]^+$). For sensitive compounds fragment ions (e.g. $[M + H - C_6H_{10}O_5]^+$) can be observed due to a collision induced dissociation in the transfer region of the mass spectrometer. In negative-ion mode ionisation occurs by abstraction of positive ions (e.g. $[M - H]^-$) or adduct formation with anions (e.g. $[M + Cl]^-$).

For the identification of compounds it is a prerequisite to deconvolute a compounds' mass spectra from the GC(LC)/MS raw data. In case of EI huge spectral libraries exist where the extracted spectrum can be searched for. For soft ionisation techniques such as ESI to date such libraries do not exist, making the identification of known compounds and the strucutre elucidation of unknown compounds a serious bottleneck in LC/MS based profiling schemes. Thus, the automatic extraction of a components mass spectrum and elucidation of the chemical relations inbetween these spectra is a prerequisit for high throughput analyses and annotation of LC/MS datasets from metabolomics experiments.

This paper is structured as follows: first the workflow is explained, starting with the machine analysis, signal processing to the annotation with two complementary approaches. The last steps in the workflow are the disambiguation and conflict resolution of the results. In section 3 we evaluate the annotation on a real-world dataset measured from plant seed material. We finish with a conclusion and outlook.

2 Implementation

Our metabolomics pipeline consists of several consecutive processing steps. First the samples are run on our LC/MS platform, followed by a signal processing step which collects raw signals into centroid peak data. Since no identification is available for the mass signals, they have to be aligned based on their mass and retention time, such that the N peaks are recognisable across the M runs, producing a $N \times M$ matrix of intensities.

The annotation procedures operate on individual runs, but also take cross-sample correlation into account. The system is implemented in R and the Bioconductor framework, with some compute-intense tasks being placed in native C-code which is called through an R function.

2.1 Data Acquisition and Preparation

We analysed methanolic seed extracts of Arabidopsis thaliana by capillary high performance liquid chromatography coupled to electrospray ionization quadrupole time-of-flight mass spectrometry (HPLC/ESI-QTOF-MS). In this setup the crude seed extracts are first separated on a modified C18 phase applying an acetonitrile/formic acid-water/formic acid gradient at a flow rate of 5 μL min^{-1}. Eluted compounds are detected with an API QSTAR Pulsar i equipped with an ionspray source in positive ion mode between m/z 100 to 1000 at a mean

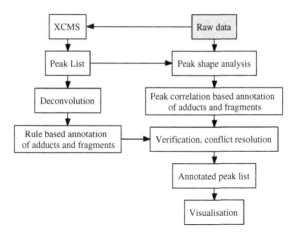

Fig. 1. Workflow of annotation procedure. The "Raw data" acquisition step (upper-right, grey) is carried out on a particular mass spectrometer, the remaining steps are vendor-independent.

resolution of 8000-9000 with an accumulation time of 2 seconds. 16 technical replicates were measured. Raw data files were exported with Analyst QS 1.0 and processed with the XCMS package [Smith06] using conservative parameters (s/n ≥ 3, fwhm=30, bw=10).

2.2 Detection of Known Adducts and Fragments

The observed signal s (measured in m/z) is the result of the molecule mass M modified by the chemical process during the ionisation:

$$s = \frac{nM + a}{z}$$

where $n(n = 1, 2 \ldots)$ is the number of molecules in the ion, a the mass of the adducts (negative for fragments) and z the charge.

For soft ionisation methods such as LC/ESI-MS, different adducts (e.g. $[M + K]^+$, $[M + Na]^+$) and fragments (e.g. $[M - C_3H_9N]^+$, $[M + H - H_2O]^+$) occur. Depending on the molecule having an intrinsic charge, $[M]^+$ may be observed as well. To scan for adducts and fragments with a known mass difference a predefined and user-extensible list (see Tab. 1) is employed.

Theoretically all isotope peaks, adducts and fragments belonging to a single components mass spectrum should have the same retention time. But since peak detection works in the chromatographic domain and independently for all mass signals, they can differ. Especially in the case where chromatographic peaks are broad and noisy, their centroids may be detected in one of the neighbouring scans. To be robust against this effect, a "sliding retention time window" with a user specified width (e.g. $\Delta t = 4s.$, in our case equivalent to two scans) is used.

Table 1. Examples of known adducts and fragments with their mass differences occuring in positive ion mode. The actual difference is calculated considering the charge and the number of molecules M in the observed ion.

Formula	Mass difference in amu
$[M + H]^+$	1.007276
$[M + Na]^+$	22.98977
$[M^+ + Na]^{2+}$	22.98977
$[M + K]^+$	38.963708
$[2M + Na]^+$	22.98977
$[M + H + Na]^{2+}$	23.9976
$[M + H - NH_3]^+$	-16.01872
$[M - C_3H_9N]^+$	-59.073499

For each retention time window all possible combinations

$$M_{ij} = \frac{z_j s_i - a_j}{n_j}$$

of mass signals s_i within the time window and known adducts $A_j = (n_j, z_j, a_j)$, $j = 1 \ldots J$ are calculated. Each group of similar masses M_{ij} (tolerance depends on the machine accuracy) yields one annotation hypothesis. All resulting reasonable adduct/fragment hypotheses are collected for subsequent verification, ambiguity removal and annotation.

Example: Given the mass signals $s_1, s_2, s_3 = (122.99, 138.97, 142.01)$ m/z which are observed in one retention time window. The differences $s_1 - A_{(Na)} = 122.99 - 22.98977 = 100.0002$ and $s_2 - A_{(K)} = 138.97 - 38.963708 = 100.0063$ support the hypothesis that s_1 and s_2 are adducts ($[M + Na]^+$, $[M + K]^+$) of a molecule M with an estimated mass of 100.0033 amu.

2.3   Verification of Annotation Hypotheses

The hypotheses described in the previous section can either be correct and reflect the actual chemical process during ionisation, or they are the result of co-eluting molecules having a known fragment/adduct mass difference by chance. Therefore a verification step is mandatory.

Intensity correlation across samples
Because of the theoretically fixed ratio between molecule and adduct intensities in all observations, a simple verification of the chemical relation for a given pair of mass signals is to calculate the correlation of the integrated peak intensities across all samples in which these peaks have been observed. The neccessary alignment of two or more experiments is also part of the XCMS package.

Intensity correlation in the chromatographic domain
The adducts and fragments of the same molecule have the same intensity ratio also in each individual scan of the LC/MS measurement. Therefore their

extracted ion chromatograms (EIC) are theoretically linearly dependent, and correlation or linear regression ($y = \alpha x + 0$, with subsequent evaluation of the factors α and the residuals) can be used to estimate the similarity between the chromatograms.

Peaks with a low intensity are more subjected to noise influence, their chromatograms can be very flat and/or jagged, resulting in a low correlation.

2.4   Exposing Chemical Relations by Chromatogram Correlation Analysis

Even without any predefined mass differences, valuable chemical hypotheses can be obtained by analysing the chromatogram correlations across all samples. For this purpose EIC - correlations are calculated for *all* pairs of mass signals within each retention time window. This is done for each sample where the examined pair of mass signals was observed. As result of this computation, the distribution of peak shape correlations across the samples can be evaluated.

2.5   Ambiguity Removal and Conflict Resolution

Since all chemical relation hypotheses are searched in a sliding retention time window, all duplicates, subset relations and conflicts between hypotheses have to be eliminated. In a naive approach, this would require to compare all members of all hypotheses against each other. To speed up this procedure, *signatures* are calculated for each chemical relation hypothesis. One hypothesis group H_k consists of several entries in the form (s_i, A_j), which encode the annotation hypothesis. For each entry a signature s_{ij} is calculated as

$$\text{sig}_{ij} = p_1 i + p_2 j \quad (s_i, A_j) \in H_k ,$$

with $p_1, p_2 > max(i, j)$ being prime numbers. Furthermore, a hash value is created for each hypothesis group:

$$\text{hash}(H_k) = \sum_{(s_i, A_j) \in H_k} \text{sig}_{ij} .$$

Using these hash values, hypotheses groups containing the same annotations have the same hash value, and subset relations of hypotheses can be detected efficiently using the signatures.

Furthermore, some efforts are made to resolve chemical conflicts. For example $[M]^+$ and $[M + H]^+$ cannot appear both for the same molecule, because the molecule is supposed to have an intrinsic charge in the first case but not in the second. Therefore, relations of this kind can safely be removed.

3   Results

The evaluation is based on the arabidopsis seed dataset described above. The peak list exported from XCMS contains 1100 mass signals. The allowed mass tolerance for the annotation was $0.005\,m/z$, the retention time window was set to 4 seconds. Larger windows (up to 10 seconds) had little effect.

3.1   Annotation of Known Mass Differences

The rule set used contains 30 known mass differences covering typical adducts with protons, sodium and potassium cations and also 10 fragments with neutral losses. Using only this rule set about 200 signals are annotated as isotope peaks and there are more than 1000 competing annotations for adducts and fragments, especially for retention times with many coeluting substances.

For the verification step the threshold for the correlation of the intensities across the samples is specified as 0.6 and for the chromatogram correlation a minimum of 0.8 in 75 % of the samples is required.

After the verification step and ambiguity removal 10 % of the 1100 mass signals are annotated as isotope peaks and 20 % as adducts and fragments.

Fig. 2 shows an example of rule based annotation. Five signals belonging to the mass spectrum of feruloylcholin could be grouped in the peak list. Chromatogram correlations (Fig. 3) were used only for verification. Some combinations of low intensity signals show only a weak correlation, but the hypotheses is kept as long as the other correlations with these signals are above the threshold.

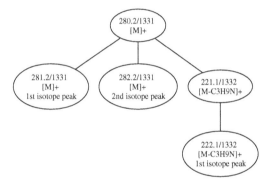

Fig. 2. Graph of mass signals. Nodes are labelled with mass/retention time and a short description, edges indicate an successful verification by correlation across samples and EICs.

3.2   Deconvolution Based on Chromatogram Correlation Analysis

Using only the chromatogram correlations, highly correlated pairs of mass signals were combined to chemical relation hypothesis groups. We used only those pairs with correlation higher than 0.8 in 75 % of the cases (12 out of 16 samples). Manual verification of one hypothesis group confirmed that all grouped signals belong to the mass spectrum of tryptophane (Fig. 4).

3.3   Run-Time

The "Peak shape analysis" step in the workflow (see fig. 1) includes the creation of the extracted ion chromatogram (EIC) for each peak from each file using

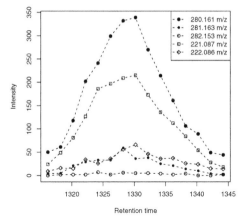
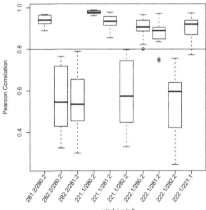

(a) Superimposed EICs for the 5 peaks. The peak shapes are similar, except for the signals with low intensity.

(b) Distribution of pearson correlations for all $\frac{5^2-5}{2}$ pairs of chromatograms shown in (a) across all 16 samples.

Fig. 3. Verification of the annotation shown in Fig. 2. Correlations with the low signal 282.2 m/z are below threshold.

the generic XCMS functions. For each EIC this takes about 0.3 seconds on a standard 2 GHz PC. Depending on the number of peaks, files and the file size this preprocessing step can take up to several hours, which can be reduced by either spreading the computation across a compute cluster (we have good experience using the Sun Grid Engine 6.06) or by using a dedicated and optimized EIC extraction routine. Independent of the calculation method employed, we are currently creating a data warehouse for preprocessed LC/MS data, which includes the EIC for each peak and avoids recomputing EICs for the annotation step altogether.

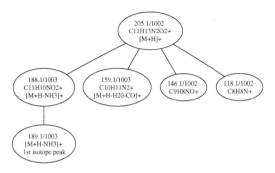

Fig. 4. Graph of extracted signals that belong to the mass spectrum of tryptophane. Nodes are labelled with mass/retention time and a short description, edges indicate a chromatogram correlation above threshold (see 2.4).

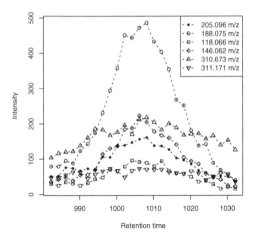

Fig. 5. Superimposed EICs for the 5 peaks shown in fig. 4 and also for the mass signals 310.7 and 311.2 m/z, which are co-eluting, but not chemically related to tryptophane

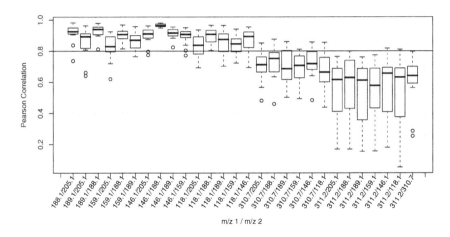

Fig. 6. Distribution of pearson correlations for all pairs of chromatograms shown in fig. 5 across all 16 samples. The mass signals 310.7 and 311.2 m/z are co-eluting, but not chemically related to tryptophane. It can be seen that the chromatogram correlations are significantly lower for *all* combinations with those signals.

Once the EICs are generated, the wall-clock run-time for the annotation (correlation, validation and graphics output) of the sample set of 16 files × 1100 peaks described above is 120 seconds on a standard 2 GHz PC.

4   Conclusion

In metabolomics research the large gap between fingerprint data of unknown mass signals and profiling data for a limited number of metabolites needs to be narrowed down. The peaklists of LC/MS measurements can contain a few thousand peaks, and manual inspection of all of them is simply impossible. The downstream bioinformatics analysis such as hierarchical clustering or self-organising maps is difficult if a large number of observations is caused by artefacts of the analytical process.

The annotation is a valuable addition to the commonly used peaklists, based on both an extensible set of rules and peak shape analysis. Even if no chemical identification is possible, the truly interesting signals become more obvious.

Our annotation framework is also capable of including further sources that are aimed towards metabolite identification, such as the mass decomposition tool IMS [Böcker06] or exact masses from compound libraries such as the metabolic pathways database KEGG [Goto97, Kanehisa06] or KNapSACK [Shinbo06].

Acknowledgements

We thank Dierk Scheel, Edda von Roepenack-Lahaye and Jürgen Schmidt for their valuable discussions. The work is supported under BMBF grant 0312706G.

References

[Böcker06]   Böcker, S., Lipták, Z. and Pervukhin, A.   Decomposing metabolomic isotope patterns. In *WABI 2006 Proc. of Wabi 2006 – 6th Workshop on Algorithms in Bioinformatics*. 2006.

[Fiehn00]   Fiehn, O., Kopka, J., Dörmann, P., Altmann, T., Trethewey, R. and Willmitzer, L. Metabolite profiling for plant functional genomics. *Nature Biotechnology*, vol. 18:115, 2000.

[Goto97]   Goto, S., Bono, H., Ogata, H., Fujibuchi, W., Nishioka, T., Sato, K. and Kanehisa, M. Organizing and computing metabolic pathway data in terms of binary relations. *Pac Symp Biocomput*, pp. 175–186, 1997.

[Kanehisa06]   Kanehisa, M., Goto, S., Hattori, M., Aoki-Kinoshita, K. F., Itoh, M. *et al.* From genomics to chemical genomics: new developments in KEGG. *Nucleic Acids Res*, vol. 34(Database issue):354–357, Jan 2006.

[Oliver98]   Oliver, S., Winson, M., Kell, D. and Baganz, F. Systematic functional analysis of the yeast genome. *Trends Biotechnol*, vol. 16(9):373–378, Sep 1998.

[Roepenack-Lahaye04] Roepenack-Lahaye, E. v., Degenkolb, T., Zerjeski, M., Franz, M., Roth, U. *et al.*   Profiling of Arabidopsis Secondary Metabolites by Capillary Liquid Chromatography Coupled to Electrospray Ionization Quadrupole Time-of-Flight Mass Spectrometry. *Plant Physiology*, vol. 134:548–559, February 2004.

[Shinbo06]   Shinbo, Y., Nakamura, Y., Altaf-Ul-Amin, M., Asahi, H., Kurokawa, K. *et al. Plant Metabolomics*, chap. KNApSAcK: A comprehensive species-metabolite relationship database., pp. 165–181. Biotechnology in Agriculture and Forestry. Springer, 2006.

[Smith06]   Smith, C., Want, E., O'Maille, G., Abagyan, R. and Siuzdak, G. XCMS: Processing mass spectrometry data for metabolite profiling using nonlinear peak alignment, matching and identification. *Analytical Chemistry*, vol. 78(3):779–787, 2006.

Stochastic Protein Folding Simulation in the d-Dimensional HP-Model[*]

K. Steinhöfel[1], A. Skaliotis[1], and A.A. Albrecht[2]

[1] Department of Computer Science
King's College London
Strand, London WC2R 2LS, UK
[2] School of Computer Science
University of Hertfordshire
Hatfield, Herts AL10 9AB, UK

Abstract. We present results from two- and three-dimensional protein folding simulations in the HP-model on selected benchmark problems. The importance of the HP-model for investigating general complexity issues of protein folding has been recently demonstrated by Fu & Wang (LNCS 3142:630–644, 2004) in proving an $exp(O(n^{1-1/d} \cdot \ln n))$ time bound for d-dimensional protein folding simulation of sequences of length n. The time bound is close to the approximation of real folding times of $exp(\lambda \cdot n^{2/3} \pm \chi \cdot n^{1/2}/2)ns$ by Finkelstein & Badretdinov (FOLD DES 2:115–121, 1997), where λ and χ are constants close to unity. We utilise a stochastic local search procedure that is based on logarithmic simulated annealing. We obtain that after $(m/\delta)^{a \cdot D}$ Markov chain transitions the probability to be in a minimum energy conformation is at least $1 - \delta$, where $m \leq b(d) \cdot n$ is the maximum neighbourhood size for a small integer $b(d)$, a is a small constant, and D is the maximum value of the minimum escape height from local minima of the underlying energy landscape. We note that the time bound is sequence-specific, and we conjecture $D < n^{1-1/d}$ as a worst case upper bound. We analyse $D < n^{1-1/d}$ experimentally on selected HP-model benchmark problems.

1 Introduction

In the context of general complexity classifications, protein folding simulations has been proven to be NP-complete for a variety of lattice models [1,2,3], which implies a generic upper bound $exp(O(n))$ for the time complexity of deterministic folding simulations of sequences of length n. One of the most popular lattice models of protein folding is the hydrophobic-hydrophilic (HP) model [4,5,6,7]. In the HP-model, proteins are modelled as chains whose vertices are marked either as H (hydrophobic) or P (hydrophilic); the resulting chain is embedded into some lattice; H nodes are considered to attract each other while P nodes are neutral. An optimal embedding is one that maximizes the number of H-H contacts. The rationale for this objective is that hydrophobic interactions contribute a significant portion of the total energy function. Unlike more sophisticated models of

[*] Research partially supported by EPSRC Grant No. EP/D062012/1.

protein folding, the main goal of the HP-model is to explore broad qualitative questions about protein folding such as whether the dominant interactions are local or global with respect to the chain [6].

In the present paper, we distinguish between proven time bounds of simulations in terms of algorithmic complexity and folding time approximations that are based on a number of (heuristic) assumptions about the underlying kinetics of protein folding. To the best of our knowledge, the first proven time bound for protein folding simulations with an exponent significantly below $O(n)$ is due to Fu and Wang [8]. They proved an $exp(O(n^{1-1/d}) \cdot \ln n)$ algorithm for d-dimensional protein folding simulation to optimum conformations in the HP-model. Linear time algorithms for constant-factor approximations of optimum solutions are reported in [9,10]. Surprisingly, the algorithmic complexity bound by Fu and Wang comes close to the folding time approximation of $exp(\lambda \cdot n^{2/3} \pm \chi \cdot n^{1/2}/2)ns$ by Finkelstein and Badretdinov [11] (asymptotically, within a $\log n$ factor in the exponent). There is an ongoing discussion if folding time approximations that basically rely on the length of sequences only are the appropriate way to estimate the actual folding time (see [12,13,14,15] and the literature therein), and there are instance-specific folding time approximations ranging from a polynomial dependence to an exponential dependence [11,16,17]. In our paper, we present an instance-dependent time bound for protein folding simulations in the HP-model. The simulations are based upon stochastic local search procedures that utilise logarithmic simulated annealing.

Simulated annealing-based search procedures have been discussed and applied in the context of protein folding simulation for about twenty years [18,19,20], shortly after the independent introduction of simulated annealing as a new optimization tool by Kirkpatrick et al. [21] and V. Černy [22] (preprint version in 1982). These applications evolved from earlier methods [23] that were based on Metropolis' algorithm [24]. For on overview on early applications of simulated annealing to protein folding simulations, we refer the reader to [25]. Apart from simulated annealing, search methods like genetic algorithms [26,27], tabu search [28] and ensemble techniques [29,30] have been applied to protein folding simulation; see also [6]. More recent simulated annealing-related application to protein folding simulation can be found in [31,32].

All the above-mentioned applications of simulated annealing to protein folding simulation have in common that they rely on the technique as described in [21]. From an algorithmic point of view, the convergence proof of this simulated annealing procedure is based upon the following observation: At a fixed "temperature" T, the probability distribution over the conformation space approaches in the limit the Boltzmann distribution $e^{-Z(S)/T}/F$, where $Z(S)$ is the value of the objective function for the conformation S and F is a normalisation parameter. If then $T \to 0$, the Boltzmann distribution tends to the distribution over optimum configurations. In practice, however, it is infeasible to perform an infinite number of Markov chain transitions at fixed temperatures, and this explains why this type of simulated annealing with cooling heuristics for T is usually stuck in local minima [33]. But this drawback can be avoided, if the temperature is lowered after each Markov chain transition from step $k - 1$ to k according to $T = T(k) = D/\log(k + 2)$, where D is the maximum value of the minimum escape height from local minima of the underlying conformation space, i.e. the transition probabilities are time-dependent. Under some natural assumptions about the conformation space, the probability distribution over the conformation space tends to the distribution over optimum solutions [34]. Due to the definition of $T(k)$, this optimization method is called logarithmic simulated annealing (LSA). This approach allows to relate the convergence time to structural parameters of the conformations space. Reidys and Stadler [41] utilised this method for a landscape analysis in the wider context of protein folding. In the present paper, we employ the time complexity analysis from [42], where the probability $p = 1 - \delta$ of finding an optimum solution after k Markov chain transitions is related to an exponential function depending on D, δ, and the maximum size m of neighbourhood sets of the conformation space. We expect the method to be adaptable to realistic objective functions and protein sequence geometries, which will be subject of future research.

We focus on the d-dimensional rectangular lattice model, but this restriction is not essential, i.e. extensions to other lattice models, e.g. as described in [43], actually are depending upon the definition of a suitable neighbourhood relation. For the definition of the neighbourhood relation in the conformation space, we employ the pull-move set, which was independently introduced by Lesh et al. [44] and Milostan/Blazewicz et al. [45,46]. We present a time complexity bound of $(m/\delta)^{a \cdot D}$ for protein folding simulation in the HP-model, where m is the maximum neighbourhood size, a is a small constant, and D the above-mentioned maximum escape height from local minima. For this number of steps, the probability of having found an optimum conformation is $p \geq 1 - \delta$. The parameter m depends on the lattice structure, the dimensionality d, and the pull-move set; m can be bounded by $b(d) \cdot n$ for sequences of length n, where $b(d)$ is a small integer depending upon the maximum number of neighbours of individual nodes of sequences. Thus, the time bound depends on the landscape parameter D. One can expect that for protein sequences investigated in [14,15,16,17], the parameter D is relatively small, e.g. even bounded by a constant. Here, we are concerned about the "worst case" behaviour of D, i.e. we try to establish an upper bound of D for arbitrary protein sequences in the HP-model. Based on the time bound by Fu and Wang [8] and our computational experiments, we conjecture $D < n^{1-1/d}$ for arbitrary protein sequences of length n. We present an experimental analysis of $D < n^{1-1/d}$ for $d = 2,3$ on selected benchmark problems defined in [26,44,45,46].

2   LSA for Protein Folding Simulation

For simplicity of presentation, we focus in the present section on the two-dimensional rectangular grid HP-model only. For the three-dimensional case, we refer to [45,46]; the dimensionality affects only the upper bound $b(d) \cdot n$.

384 K. Steinhöfel, A. Skaliotis, and A.A. Albrecht

Anfinsen's thermodynamic hypothesis [48] motivates the attempt to predict protein folding by solving certain optimization problems, but there are two main difficulties with this approach: The precise definition of the energy function that has to be minimised, and the extremely difficult optimization problems arising from the energy functions commonly used in folding simulations [6,49]. In the 2-d rectangular grid HP-model, one can define the minimization problem as follows:

$$\min_{\alpha} E(S, \alpha) \ \text{ for } \ E(S, \alpha) := \xi \cdot HH_c(S, \alpha), \tag{1}$$

where S is a sequence of amino acids containing n elements; $S_i = 1$, if the amino acid on the ith position in the sequence is hydrophobic; $S_i = 0$, if the amino acid on the ith position is polar; α is a vector of $(n - 2)$ grid angles defined by consecutive triples of amino acids in the sequence; HH_c is a function that counts the number of neighbours between amino acids that are not neighbours in the sequence, but they are neighbours on the grid (they are topological neighbours); finally, $\xi < 0$ is a constant lower than zero that defines an influence ratio of hydrophobic contacts on the value of conformational free energy. The distances between neighbouring grid nodes are assumed to be equal to 1. We identify sequences α with conformations of the protein sequence S, and a valid conformation α of the chain S lies along a non-self-intersecting path of the rectangular grid such that adjacent vertices of the chain S occupy adjacent locations. Thus, we define the set of conformations (for each S specifically) by

$$\mathsf{F}_S := \big\{ \, \alpha \text{ is a valid conformation for } S \, \big\}. \tag{2}$$

Since $\mathsf{F} := \mathsf{F}_S$ is defined for a specific S, we denote the objective function by

$$\mathsf{Z}(\alpha) \ := \ \xi \cdot HH_c(S, \alpha). \tag{3}$$

The neighbourhood relation of our stochastic local search procedure is determined by the set of pull moves introduced in [44] for 2-d protein folding simulations in the HP-model (extended to 3-d case in [45,46]). We briefly describe the set of pull moves (for details, we refer the reader to [44]): For simplicity of notation, we identify S_i by i, and we first consider i at step k in location $(x_i(k), y_i(k))$ of the grid. Suppose that a free location L is adjacent to $(x_{i+1}(k), y_{i+1}(k))$ and diagonally adjacent to $(x_i(k), y_i(k))$. The grid vertices $(x_i(k), y_i(k))$, $(x_{i+1}(k), y_{i+1}(k))$, and the free location L constitute three corners of a square; let the fourth corner be the location C. A pull move can be executed, if the location C is either free or equal to $(x_{i-1}(k), y_{i-1}(k))$. If $C = (x_{i-1}(k), y_{i-1}(k))$, the entire local pull move consists of moving the sequence node i to L. If C is free, at first i is moved to L and sequence node $i - 1$ is moved to location C. Then, until a valid conformation is reached, the following actions are performed: Starting with sequence node $j = i - 2$ and potentially down to node 1, we set $(x_j(k + 1), y_j(k + 1)) = (x_{j+2}(k), y_{j+2}(k))$. Similarly, one could consider a pull in the other direction, starting from a free location L adjacent to $(x_{i-1}(k), y_{i-1}(k))$ and diagonally adjacent to $(x_i(k), y_i(k))$. Finally, for technical reasons to make the conformation space reversible, one has to add special moves at the end vertices: Consider two free adjacent locations, where one is adjacent to sequence node n. Then the nodes $n-1$ and n can be moved to these free locations, with subsequent pull moves as before, potentially down to node 1. The same applies to sequence nodes 1 and 2.

Thus, given $\alpha \in \mathsf{F}_S$, for $n-2$ nodes S_i we have at most four possibilities to change the location (potentially with subsequent consequences for the remaining nodes, but $(x_j(k+1), y_j(k+1)) = (x_{j+2}(k), y_{j+2}(k))$ determines a unique action). For the end nodes, the number of changes can be slightly larger, but altogether the number of neighbours is bounded by $b(2) \cdot n$ for a small integer $b(2)$. The same applies to the 3-d case.

Theorem 1. [44] *The set of pull moves is local, reversible, and complete within* F, *i.e. any* $\beta \in \mathsf{F}$ *can be reached from any* $\alpha \in \mathsf{F}$ *by executing pull moves only.*

The set of neighbours of α that can be reached by a single pull move is denoted by N_α, where additionally α is included since the search process can remain in the same configuration. Furthermore, we set

$$N_\alpha := |\mathsf{N}_\alpha| \leq b(d) \cdot n; \tag{4}$$
$$\mathsf{F}_{\min} := \{\alpha : \alpha \in \mathsf{F} \text{ and } Z(\alpha) = \min_{\alpha'} E(S, \alpha')\}. \tag{5}$$

In simulated annealing-based search, the transitions between neighbouring elements are depending on the objective function Z. Given a pair of protein conformations $[\alpha, \alpha']$, we denote by $G[\alpha, \alpha']$ the probability of generating α' from α, and by $A[\alpha, \alpha']$ we denote the probability of accepting α' once it has been generated from α. As in most applications of simulated annealing, we take a uniform generation probability:

$$G[\alpha, \alpha'] := \begin{cases} \frac{1}{N_\alpha}, & \text{if } \alpha' \in \mathsf{N}_\alpha; \\ 0, & \text{otherwise.} \end{cases} \tag{6}$$

The acceptance probabilities $A[\alpha, \alpha']$ are derived from the underlying analogy to thermodynamic systems:

$$A[\alpha, \alpha'] := \begin{cases} 1, & \text{if } Z(\alpha') - Z(\alpha) \leq 0; \\ e^{-\frac{Z(\alpha')-Z(\alpha)}{T}}, & \text{otherwise,} \end{cases} \tag{7}$$

where T is a control parameter having the interpretation of a *temperature* in annealing processes. The probability of performing the transition between α and α' is defined by

$$\mathbf{Pr}\{\alpha \to \alpha'\} = \begin{cases} G[\alpha, \alpha'] \cdot A[\alpha, \alpha'], & \text{if } \alpha' \neq \alpha; \\ 1 - \sum_{\alpha' \neq \alpha} G[\alpha, \alpha'] \cdot A[\alpha, \alpha'], & \text{otherwise.} \end{cases} \tag{8}$$

By definition, the probability $\mathbf{Pr}\{\alpha \to \alpha'\}$ depends on the control parameter T. Let $\mathbf{a}_\alpha(k)$ denote the probability of being in conformation α after k transition steps. The probability $\mathbf{a}_\alpha(k)$ is calculated in accordance with

$$\mathbf{a}_\alpha(k) := \sum_{\beta \in \mathsf{F}} \mathbf{a}_\beta(k-1) \cdot \mathbf{Pr}\{\beta \to \alpha\}. \tag{9}$$

The recursive application of (9) defines a Markov chain of probabilities $\mathbf{a}_\alpha(k)$, where $\alpha \in \mathsf{F}$ and $k = 1, 2, \cdots$. If the parameter $T = T(k)$ is a constant T, the chain is said to be a *homogeneous* Markov chain; otherwise, if $T(k)$ is lowered at each step, the sequence of probability vectors $\mathbf{a}(k)$ is an *inhomogeneous* Markov chain.

In the present paper, we focus on a special type of inhomogeneous Markov chains where the value $T(k)$ changes according to

$$T(k) = \frac{D}{\ln(k+2)}, \; k = 0, 1, ..., \tag{10}$$

where $D > 0$ is independent of k. The choice of $T(k)$ is motivated by convergence properties that have been proved for logarithmic cooling schedules that determine inhomogeneous Markov chains [34]. In order to explain these convergence results, we first need to introduce some parameters characterising local minima of the objective function:

Definition 1. *A conformation $\alpha' \in \mathsf{F}$ is said to be reachable at height h from $\alpha \in \mathsf{F}$, if $\exists \alpha_0, \alpha_1, ..., \alpha_r \in \mathsf{F}$ with $\alpha_0 = \alpha \wedge \alpha_r = \alpha'$ such that $G[\alpha_u, \alpha_{u+1}] > 0, u = 0, 1, ..., (r-1),$ and $\mathsf{Z}(\alpha_u) \leq h$ for all $u = 0, 1, ..., r.$*

We use the notation $H(\alpha \Rightarrow \alpha') \leq h$ for this property. The conformation α is a *local minimum*, if $\alpha \in \mathsf{F}\backslash\mathsf{F}_{\min}$ and $\mathsf{Z}(\alpha') \geq \mathsf{Z}(\alpha)$ for all $\alpha' \in \mathsf{N}_\alpha\backslash\{\alpha\}$.

Definition 2. *Let $\lambda_{\min}$ denote a local minimum, then $D(\lambda_{\min})$ denotes the smallest h such that there exists $\lambda' \in \mathsf{F}$ with $\mathsf{Z}(\lambda') < \mathsf{Z}(\lambda_{\min})$ that is reachable at height $\mathsf{Z}(\lambda_{\min}) + h.$*

Theorem 2. *[34] For $T(k)$ from (10), the convergence $\sum_{\alpha \in \mathsf{F}_{\min}} \mathbf{a}_\alpha(k) \xrightarrow[k \to \infty]{} 1$ of the algorithm defined by (3), ..., (9) is guaranteed if and only if*

*i. $\forall \alpha, \alpha' \in \mathsf{F} \; \exists \alpha_0, \alpha_1, ..., \alpha_r \in \mathsf{F}$ such that $\alpha_0 = \alpha \wedge \alpha_r = \alpha'$
and $G[\alpha_u, \alpha_{u+1}] > 0$ for $u = 0, 1, ..., (r-1);$*
ii. $\forall h : H(\alpha \Rightarrow \alpha') \leq h \iff H(\alpha' \Rightarrow \alpha) \leq h;$
iii. $D \geq \max_{\lambda_{\min}} D(\lambda_{\min}).$

From Theorem 1 and the definition of N_α we immediately conclude that the conditions (i) and (ii) are valid for F. Thus, together with Theorem 2 we obtain:

Corollary 1. *If $D \geq \max_{\lambda_{\min}} D(\lambda_{\min})$, the algorithm defined by (3), ..., (10) and the pull move set from [44] tends to minimum energy conformations in the HP-model.*

The result from [34] has been later been extended with respect to the rate of convergence depending on structural properties of the energy landscape, different types of stochastic processes, and related types of search algorithms; see

[35,36,37,38,39,40] and references therein. The results have in common that long-range convergence properties are considered and/or the impact of parameters such as the size of the neighbourhood relation (degree of adjacent nodes in the landscape) and the definition of generation probabilities are ignored. But these parameters may affect run-time estimations significantly. We employ the time complexity analysis from [42] for protein folding simulation as presented in [50]. The main difference to [42] comes from potentially larger differences of the values of the objective function in single transitions; i.e. we have to consider $\gamma_{\min} := \min_{\alpha \to \alpha'} (Z(\alpha') - Z(\alpha))$ for $(Z(\alpha') - Z(\alpha)) > 0$. We simply take the lower bound $\gamma_{\min} \geq 1$ and obtain

Theorem 3. [42,50] *If* $D \geq \max\limits_{\lambda_{\min}} D(\lambda_{\min})$ *for* F *from (2),* $m := \max\limits_{\alpha} N_\alpha$, *and* $0 < \delta < 1$, *then*

$$k \geq \left(\frac{m}{\delta}\right)^{a \cdot D} \geq \left(\frac{8 \cdot e \cdot (m+1)^3}{\delta}\right)^{(D+1)} \quad implies \qquad (11)$$

$$\sum_{\alpha' \in \mathsf{F}_{\min}} \mathbf{a}_{\alpha'}(k) \geq 1 - \delta. \qquad (12)$$

As can be seen, $a \geq 5$ is sufficient for (11), i.e. a is a small constant.

3 Landscape Analysis on Selected Benchmarks

As mentioned in Section 1 already, the run-time estimation (11) from Theorem 3 is problem-specific, i.e. depends on the parameter D of the landscape induced by an individual protein sequence and the pull-move set. For a problem-independent, worst case upper bound we conjecture $D \leq n^{1-1/d}$, which complies with the result from [8]. A rigorous proof of the conjecture seems to be challenging, and existing techniques for landscape analysis in the context of protein folding [41,51,52,53] do not provide the necessary tools. However, for individual protein sequences, one can proceed as follows: Given a sequence α, the parameter D is estimated in a pre-processing step (landscape analysis), where the maximum increase of the objective function is monitored in-between two successive improvements of the best value obtained so far. This approach usually over-estimates D significantly. Therefore, we are searching for a suitable constant c such that $D' = G_{\mathrm{monit}}/c$ comes close to D, where G_{monit} is the maximum of the monitored increases of the objective function in-between two successive total improvements of the objective function. This estimation D' is then taken (together with the length of α and a choice of δ for the confidence $1 - \delta$) as the setting for the (slightly simplified) run-time estimation according to (11). In our computational experiments on d-dimensional benchmark problems we indeed obtain optimum solutions for smaller values of D than $n^{1-1/d}$.

3.1 Two-Dimensional Simulation

Following the experimental part of [42], we use $(n/\delta)^{D'}$ as a simplified version of (11), i.e. the maximum size m of the neighbourhood set is substituted in (11)

388   K. Steinhöfel, A. Skaliotis, and A.A. Albrecht

by n and the affect of $b(d)$ on the constant a is neglected; actually, based on the computational experiments from [42], we set $a := 1$ and $D' \approx \sqrt{n}/2$ for the case of 2-d protein folding simulation in the HP-model. Furthermore, we compare D' to G_{monit}, i.e. apart from trying to approximate the real D by D', we also try to relate D' to G_{monit}.

Table 1. Selected 2D benchmark problems from [26,44]

name/n	structure	$Z_{\min}$
S36	3P2H2P2H5P7H2P2H4P2H2PH2P	-14
S60	2P3HP8H3P10HPH3P12H4P6HP2HPHP	-35
S64	12HPHPH2P2H2P2H2PH2P2H2P2H2PH2P2H2P2H	
	2PHPHP12H	-42
S85	4H4P12H6P12H3P12H3P12H3PH2P2H2P2H2PHPH	-53
S100	6PHP2H5P3HP5HP2H4P2H2P2HP5HP10HP2HP7H	
	11P7H2PHP3H6PHP2H	-48

The stochastic local search procedure as described in Section 2 was implemented and we analysed the 2-d benchmark problems (cf. [26,44]) presented in Table 1. Unfortunately, information about the exact number of ground states is not available for all of the benchmark problems shown in Table 1. In [44], three ground states are reported for S85, and two ground states for S100.

Table 2. Results for selected 2-d benchmarks; $1 - \delta = 0.51$

name/n	$\sqrt{n}$	G_{monit}	D'	$k = (n/\delta)^{D'}$	k_{av}
S36	6.00	9.25	3.00	$\approx 4.0 \times 10^5$	29,341
S60	≈ 7.74	14.00	3.87	$\approx 1.2 \times 10^8$	30,319
S64	8.00	18.00	4.00	$\approx 2.9 \times 10^8$	259,223
S85	≈ 9.20	21.75	4.60	$\approx 2.0 \times 10^{10}$	13,740,964
S100	10.00	21.50	5.00	$\approx 3.5 \times 10^{11}$	57,195,268

In Table 2 we report results where $Z_{\min}$ was achieved for all five benchmark problems from Table 1. By k_{av} we denote the average number of transitions necessary to achieve $Z_{\min}$ calculated from four successive runs for the same benchmark problem. The same applies to G_{monit}, which is the average from the four runs executed for each of the five benchmark problems. Although by definition D has to be an integer value in the HP-model, we allowed rational values for D'. The simplified version of (11) was calculated for $m = n$ and $\delta = 0.49$, i.e. for a confidence of 51%. As already mentioned, the value of D' was chosen $\approx \sqrt{n}/2$, which was used in (10) for the implementation.

As can be seen, the simplified version of (11) still over-estimates the number of transitions sufficient to achieve $Z_{\min}$ for the selected benchmark problems, which is at least partly due to the setting $m = n$. One can expect $m << n$ for conformations close to ground states. To incorporate improved upper bounds

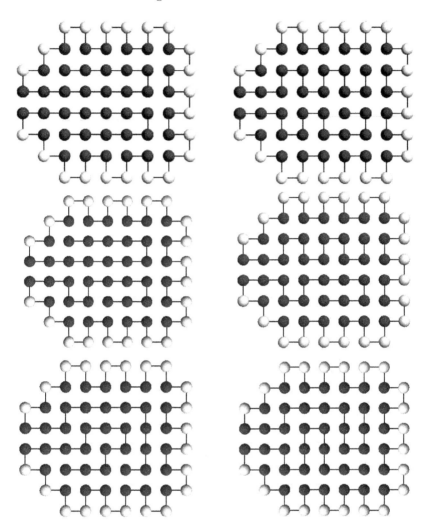

Fig. 1. Six ground states for S64

of m will be subject of future research. For $S64$, we found a local minimum at $Z(\lambda_{\min}) = -41$, and we identified six different ground states with $Z(\alpha') = -42$. The six ground states are shown in Figure 1, and Figure 2 displays ground states for S85 and S100. Based on the data from Table 2, the constant c in $D' = G_{\mathrm{monit}}/c$ ranges from 3.08 to 4.73.

3.2 Three-Dimensional Simulation

In a similar way we proceed for the 3-d benchmark problems described in Table 3, where we also succeeded to calculate optimum solutions with $Z_{\min} = -32$ and $Z_{\min} = -34$, respectively. The benchmark problems correspond to problems

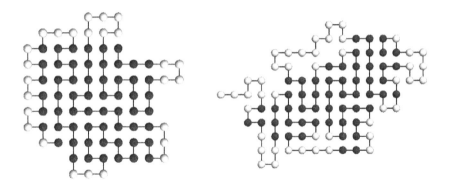

Fig. 2. Ground states for S85 and S100

Table 3. Two 3-d benchmark problems from [47]

name/n	structure	$Z_{\min}$
S48A	HP2H2P4HP3H2P2H2PHP3HPHP2H2P2H3PH8P2H	-32
S48B	4HP2HP5H2PH2P2H2PH6PH2PH3PH2P2H2P2HPH	-34

no. 1 and no. 2 in [46] (Table 1, p. 140), and tabu search with 1000 iterations produced -22 and -20 on these instances. Currently, we execute computational experiments on the remaining eight 3-d instances listed on p. 140 in [46].

We note that in this case we consider the conjecture $D < \sqrt[3]{n^2} = \sqrt[3]{48^2}$. We choose the same constant as before, i.e. we set $D' := \sqrt[3]{n^2}/2$. The parameters are defined in the same way as in Section 3.1. The particular parameter settings and results are shown in Table 4.

Table 4. Results for S48A and S48B from [47]; $1 - \delta = 0.51$

name/n	$\sqrt[3]{n^2}$	D'	$k = (n/\delta)^{D'}$	k_{av}
S48A	≈ 13.20	6.60	$\approx 1.4 \times 10^{13}$	$11,078,395$
S48B	≈ 13.20	6.60	$\approx 1.4 \times 10^{13}$	$8,590,196$

In this case, $k \approx 1.4 \times 10^{13}$ overestimates k_{av} more than 10^6 times (S48A), which might indicate that $D \approx D' < \sqrt[3]{n^2}/c$ for $c > 2$ is more appropriate for the 3-d case, but further information about the 3-d case will be gathered from the ongoing computational experiments on the remaining eight 3-d instances listed in [46]. Optimum conformations for S48A and S48B from Table 3 are shown in Figure 3. Overall, the results encourage us to attempt a formal proof of the conjecture $D \leq n^{1-1/d}$.

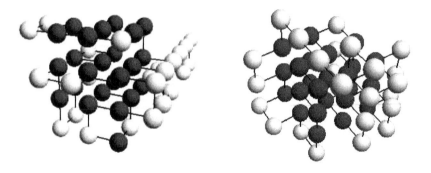

Fig. 3. Ground states for S48A and S48B

4   Concluding Remarks

We analysed the run-time of protein folding simulations in the HP-model, if the underlying algorithm is based on the pull-move set and logarithmic simulated annealing. With our method we obtained optimum solutions on a number of established benchmark problems for 2-d protein folding simulation in the HP-model and two benchmarks for 3-d folding. Future research will be directed towards tight upper bounds of D in terms of the sequence length n, improved upper bounds of the maximum neighbourhood size m, and we intend to extend our computational experiments to more realistic objective functions and folding geometries, along with estimations of D for specific protein sequences.

References

1. Paterson, M., Przytycka, T.: On the complexity of string folding. Discrete Appl. Math. **71** (1996) 217–230.
2. Berger, B., Leighton, T.: Protein folding in the hydrophobic-hydrophilic (HP) model is NP-complete. J. Comput. Biol. **5** (1998) 27–40.
3. Nayak, A., Sinclair, A, Zwick, U.: Spatial codes and the hardness of string folding problems. J. Comput. Biol. **6** (1999) 13–36.
4. Dill, K.A., Bromberg, S., Yue, K., Fiebig, K.M.: D.P. Yee, P.D. Thomas, H.S. Chan, Principles of protein folding - A perspective from simple exact models. Protein Sci. **4** (1995) 561–602.
5. Aichholzer, O., Bremner, D., Demaine, E.D., Meijer, H., Sacristán, V., Soss, M.: Long proteins with unique optimal foldings in the H-P model. Comp. Geom.-Theor. Appl. **25** (2003) 139–159.
6. Greenberg, H.J., Hart, W.E., Lancia, G.: Opportunities for combinatorial optimization in computational biology. INFORMS J. Comput. **16** (2004) 211–231.
7. Schiemann, R., Bachmann, M., Janke, W.: Exact sequence analysis for three-dimensional hydrophobic-polar lattice proteins. J. Chem. Phys. **122** (2005) 114705.
8. Fu, B., Wang, W.: A $2^{O(n^{1-1/d} \cdot \log n)}$ time algorithm for d-dimensional protein folding in the HP-model. In: J. Daz, J. Karhumäki, A. Lepistö, D. Sannella (eds.). Proceedings of the 31st International Colloquium on Automata, Languages and Programming. pp. 630–644. LNCS 3142, Springer-Verlag, Heidelberg, 2004.

9. Hart, W.E., Istrail, S.: Lattice and off-lattice side chain models of protein folding: Linear time structure prediction better than 86% of optimal. J. Comput. Biol. **4** (1997) 241–260
10. Heun, V. Approximate protein folding in the HP side chain model on extended cubic lattices. Discrete Appl. Math. **127** (2003) 163–177.
11. Finkelstein, A.V., Badretdinov, A.Y.: Rate of protein folding near the point of thermodynamic equilibrium between the coil and the most stable chain fold. Fold. Des. **2** (1997) 115–121.
12. Bakk, A., Dommersnes, P.G., Hansen, A., Høye, J.S., Sneppen, K., Jensen, M.H.: Thermodynamics of proteins: Fast folders and sharp transitions. Comput. Phys. Commun. **147** (2002) 307–312.
13. Galzitskaya, O.V., Garbuzynskiy, S.O., Ivankov, D.N., Finkelstein, A.V.: Chain length is the main determinant of the folding rate for proteins with three-state folding kinetics. Proteins **51** (2003) 162–166.
14. Kubelka, J., Hofrichter, J., Eaton, W.A.: The protein folding 'speed limit'. Curr. Opin. Struc. Biol. **14** (2004) 76–88.
15. Zhu, Y., Fu, X., Wang, T., Tamura, A., Takada, S., Saven, J.G.: Guiding the search for a protein's maximum rate of folding. Chem. Phys. **307** (2004) 99–109.
16. Ivankov, D.N., Garbuzynskiy, S.O., Alm, E., Plaxco, K.W., Baker, D., Finkelstein, A.V.: Contact order revisited: Influence of protein size on the folding rate. Protein. Sci. **12** (2003) 2057–2062.
17. Li, M.S., Klimov, D.K., Thirumalai, D.: Thermal denaturation and folding rates of single domain proteins: size matters. Polymer **45** (2004) 573–579.
18. Li, Z., Scheraga, H.A.: Monte Carlo-minimization approach to the multiple-minima problem in protein folding. Proc. Natl. Acad. Sci. USA **84** (1987) 6611–6615.
19. Wilson, S.R., Cui, W., Moskowitz, J.W., Schmidt, K.E.: Conformational analysis of flexible molecules: Location of the global minimum energy conformation by the simulated annealing method. Tetrahedron Lett. **29** (1988) 4373–4376.
20. Kawai, H., Kikuchi, T., Okamoto, Y.: A prediction of tertiary structures of peptide by the Monte Carlo simulated annealing method, Protein Eng. **3** (1989) 85–94.
21. Kirkpatrick, S., Gelatt Jr., C.D., Vecchi, M.P.: Optimization by simulated annealing. Science **220** (1983) 671–680.
22. Černy, V.: A thermodynamical approach to the travelling salesman problem: an efficient simulation algorithm. J. Optimiz. Theory App. **45** (1985) 41–51.
23. Paine, G.H., Scheraga, H.A.: Prediction of the native conformation of a polypeptide by a statistical-mechanical procedure. I. Backbone structure of enkephalin. Biopolymers **24** (1985) 1391–1436.
24. Metropolis, N., Rosenbluth, A.W., Rosenbluth, M.N., Teller, A.H., Teller, E.: Equation of state calculations by fast computing machines. J. Chem. Phys. **21** (1953) 1087–1092.
25. Hansmann, U.H.E., Okamoto, Y.: Monte Carlo simulations in generalized ensemble: Multicanonical algorithm versus simulated tempering. Phys. Rev. E **54** (1996) 5863–5865.
26. Unger, R., Moult, J.: Genetic algorithms for protein folding simulations. J. Mol. Biol. **231** (1993) 75–81.
27. Rabow, A.A., Scheraga, H.A.: Improved genetic algorithm for the protein folding problem by use of a Cartesian combination operator. Protein Sci **5** (1996) 1800–1815.
28. Pardalos, P.M., Liu, X., Xue, G.: Protein conformation of a lattice model using tabu search. J. Global Optim. **11** (1997) 55–68.

29. Hansmann, U.H.E.: Generalized ensemble techniques and protein folding simulations, Comput. Phys. Commun. **147** (2002) 604–607.
30. Eisenmenger, F., Hansmann, U.H.E., Hayryan, S., Hu, C.K.: An enhanced version of SMMP–open-source software package for simulation of proteins. Comput. Phys. Commun. **174** (2006) 422–429.
31. Eastman, P., Grønbech-Jensen, N., Doniach, S.: Simulation of protein folding by reaction path annealing. J. Chem. Phys. **114** (2001) 3823–3841.
32. Klepeis, J.L., Pieja, M.J., Floudas, C.A.: A new class of hybrid global optimization algorithms for peptide structure prediction: integrated hybrids. Comput. Phys. Commun. **151** (2003) 121–140.
33. Aarts, E.H.L.: Local search in combinatorial optimization. John Wiley & Sons, New York, 1998.
34. Hajek, B.: Cooling schedules for optimal annealing. Mathem. Oper. Res. **13** (1988) 311–329.
35. Chiang, T.S., Chow, Y.Y.: A limit theorem for a class of inhomogeneous Markov processes. Ann. Probab. **17** (1989) 1483–1502.
36. Catoni, O.: Rough large deviation estimates for simulated annealing: applications to exponential schedules. Ann. Probab. **20** (1992) 1109–1146.
37. Löwe, M.: Simulated annealing with time dependent energy function via Sobolev inequalities. Stoch. Proc. Appl. **63** (1996) 221–233.
38. Locatelli, M.: Convergence and first hitting time of simulated annealing algorithms for continuous global optimization. Math. Meth. Oper. Res. **54** (2001) 171–199.
39. Johnson, A.W., Jacobson, S.H.: On the convergence of generalized hill climbing algorithms. Discrete Appl. Math. **119** (2002) 37–57.
40. Chen, N., Liu, W., Feng, J.: Sufficient and necessary condition for the convergence of stochastic approximation algorithms. Stat. Probabil. Lett. **76** (2006) 203–210.
41. Reidys, C.M., Stadler, P.F.: Combinatorial landscapes. SIAM Rev. **44** (2002) 3–54.
42. Albrecht, A.A.: A stopping criterion for logarithmic simulated annealing. Computing **78** (2006) 55–79.
43. Backofen, R.: A polynomial time upper bound for the number of contacts in the HP-model on the face-centered-cubic lattice (FCC). J. Discrete Algorithms **2** (2004) 161–206.
44. Lesh, N., Mitzenmacher, M., Whitesides, S.: A complete and effective move set for simplified protein folding. In: Proceedings of the 7th Annual International Conference on Computational Biology. pp. 188–195. ACM Press, New York, 2003.
45. Milostan, M., Lukasiak, P., Dill, K.A., Blazewicz, A.: A tabu search strategy for finding low energy structures of proteins in HP-model. In: Proceedings of the 7th Annual International Conference on Computational Biology. pp. 205–206. ACM Press, New York, 2003.
46. Blazewicz, J., Lukasiak, P., Milostan, M.: Application of tabu search strategy for finding low energy structure of protein. Artif. Intell. Med. **35** (2005) 135–145.
47. Beutler, T.C., Dill, K.A.: A fast conformational search strategy for finding low energy structures of model proteins. Protein Sci. **5** (1996) 2037–2043.
48. Anfinsen, C.B.: Principles that govern the folding of protein chains. Science **181** (1973) 223–230.
49. Neumaier, A.: Molecular modeling of proteins and mathematical prediction of protein structure. SIAM Rev. **39** (1997) 407–460.
50. Albrecht, A.A., Steinhöfel, K.: Run-time estimates for protein folding simulation in the H-P model. In: Online Proceedings of the 9th International Symposium on Artificial Intelligence and Mathematics, Fort Lauderdale, Florida, 2006; http://anytime.cs.umass.edu/aimath06/

51. Eastwood, M.P., Hardin, C., Luthey-Schulten, Z., Wolynes, P.G.: Evaluating protein structure-prediction schemes using energy landscape theory. IBM J. Res. Develop. **45** (2001) 475–497.
52. Wales, D.: Energy landscapes. Cambridge University Press, Cambridge, 2003.
53. Carr, J.M., Trygubenko, S.A., Wales, D.J.: Finding pathways between distant local minima. J. Chem. Phys. **122** (2005) 234903.

Enhancing Protein Disorder Detection by Refined Secondary Structure Prediction

Chung-Tsai Su[1], Tong-Ming Hsu[1], Chien-Yu Chen[2,*], Yu-Yen Ou[3,4], and Yen-Jen Oyang[1]

[1] Department of Computer Science and Information Engineering, National Taiwan University, Taipei, 106, Taiwan, R.O.C.
[2] Department of Bio-industrial Mechatronics Engineering, National Taiwan University, Taipei, 106, Taiwan, R.O.C.
[3] Graduate School of Biotechnology and Bioinformatics, Yuan Ze University, Chung-Li, 320, Taiwan, R.O.C.
[4] Department of Computer Science and Engineering, Yuan Ze University, Chung-Li, 320, Taiwan, R.O.C.
sbb@mars.csie.ntu.edu.tw, cychen@mars.csie.ntu.edu.tw

Abstract. More and more proteins have been observed to display functions through intrinsic disorder. Such structurally flexible regions are shown to play important roles in biological processes and are estimated to be abundant in eukaryotic proteomes. Previous studies largely use evolutionary information and combinations of physicochemical properties of amino acids to detect disordered regions from primary sequences. In our recent work DisPSSMP, it is demonstrated that the accuracy of protein disorder prediction is greatly improved if the disorder propensity of amino acids is considered when generating the condensed PSSM features. This work aims to investigate how the information of secondary structure can be incorporated in DisPSSMP to enhance the predicting power. We propose a new representation of secondary structure information and compare it with three naïve representations that have been discussed or employed in some related works. The experimental results reveal that the refined information from secondary structure prediction is of benefit to this problem.

Keywords: protein disorder prediction; secondary structure; Radial Basis Function Network.

1 Introduction

Intrinsically disordered proteins or protein regions exhibit unstable and changeable three dimensional structures under physiological conditions [1, 2, 3]. Although lacking fixed structures, many unfolded disordered proteins or partial protein regions have been identified to participate in many biological processes and carry out important biological functions [2, 4]. Also, it has been observed that the absence of a rigid structure allows disordered binding regions to interact with several different

---

* Corresponding author.

S. Hochreiter and R. Wagner (Eds.): BIRD 2007, LNBI 4414, pp. 395–409, 2007.
© Springer-Verlag Berlin Heidelberg 2007

targets [5, 6]. Therefore, the automated prediction of disordered regions is a necessary preliminary procedure in high-throughput methods for understanding protein function.

Many studies have demonstrated that the disordered regions can be detected by examining the amino acid sequences [6, 7, 8, 9]. Disordered regions are distinguished from ordered regions by its low sequence complexity [10], amino acid compositional bias [7], high evolutionary tendencies [11], or high flexibility [12]. In our recent work DisPSSMP, we investigated the predicting power of a condensed position specific scoring matrix with respect to physicochemical properties (PSSMP) on the prediction accuracy, where the PSSMP is derived by merging several amino acid columns of a PSSM belonging to a certain property into a single column [13]. Additionally, DisPSSMP decomposes each conventional physicochemical property of amino acids into two disjoint groups which have a propensity for order and disorder respectively. It outperforms the existing packages in predicting protein disorder by employing some new properties that perform better than their parent properties [13].

Several studies have attempted to incorporate the predicted information of secondary structure elements (SSE) in predicting protein disorder [3, 6, 14, 15, 16, 17, 18, 19]. NORSp aims to identify the regions with no sufficient regular secondary structure as disorder, by means of merging predictions of secondary structure from PROFphd, transmembrane helices from PHDhtm, and coiled-coil regions from COILS [14, 15]. The GlobPlot service detects sequence regions of globularity or disorder by calculating a running sum of the propensity for random coils and secondary structures [16]. Meanwhile, some other approaches employed secondary structure information as parts of their features. DISOPRED2 employs the predicted secondary structures from PSIPred to refine its prediction of disordered residues [3, 6]. DISpro combines evolutionary information from PSI-BLAST, secondary structures from SSpro, and relative solvent accessibility from ACCpro for protein disorder prediction [17]. Similarly, VLS2 adopts various features in predicting protein disorder, including amino acid frequencies, spacer frequency, sequence complexity, charge-hydrophobicity ratios, averaged flexibility indices, and averaged PSI-BLAST profiles, as well as the averaged PHD and PSIPred secondary structure predictions [18, 19].

Although employing the predicted information of secondary structure is not new to this problem, it is not clear how much this feature contributes when employed with other important features such as amino acid physicochemical properties. There are also some challenges when retrieving secondary structure information from existing packages of secondary structure prediction. Here we use two examples to illustrate the difficulties. Fig. 1(a) shows the partial results of six famous secondary structure predictors for the protein *platelet membrane glycoprotein IIIa beta subunit* (chain B of PDB structure 1JV2) and Fig. 1(b) for *DNA-directed RNA polymerase II largest subunit* (chain A of 1I3Q). It is observed that the categories of secondary structure predicted from various predictors are sometimes inconsistent and the boundaries they detected are largely unlike, especially for the short segments. Thus, it is of great interest to develop a refined description of SSE that benefits the problem of protein disorder prediction.

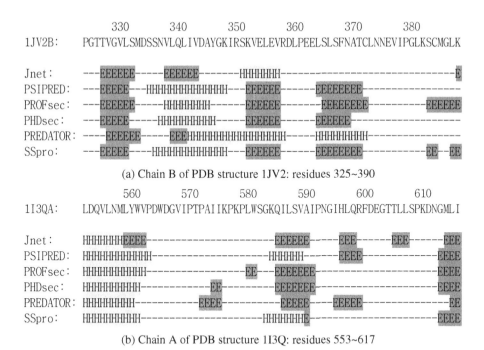

(a) Chain B of PDB structure 1JV2: residues 325~390

(b) Chain A of PDB structure 1I3Q: residues 553~617

Fig. 1. The partial results of several secondary structure predictors on the chain B of PDB structure 1JV2 and chain A of 1I3Q. Helixes are denoted as 'H', strands as 'E', and coils as '-'.

In this study, we propose a new representation to refine the secondary structure information and integrate it with the condensed PSSM features developed in our recent study [13]. The new representation transforms the predicted secondary structure elements (SSE) into a distance-based feature. The proposed idea is compared with three naïve representations that have been discussed or employed in some related works. The experimental results reveal that after the refinement procedure, the influence of potential errors from secondary structure prediction can be effectively reduced.

2   Materials

For training and validation processes, six datasets have been extracted from different databases. The number of chains, ordered/disordered regions, and residues in ordered/disordered regions of each dataset are provided in Table 1. The training data is composed of datasets PDB652, D184, and G200, which are based on the procedures described in the following paragraphs. On the other hand, three independent datasets, which are named R80, U79, and P80, are employed as validation benchmarks as in related studies [20, 21]. These blind testing sets serve as a platform for comparing the performance of different SSE representations.

Table 1. Summary of the datasets employed in this study

Number of :	Training data			Testing data		
	PDB652	D184	G200	R80	U79	P80
Chains	652	184	200	80	79	80
Ordered regions	1281	257	200	151	0	80
Disordered regions	1613	274	0	183	79	0
Residues in ordered regions	190936	55164	37959	29909	0	16568
Residues in disordered regions	49365	27116	0	3649	14462	0
Total residues in the dataset	240301	82280	37959	33558	14462	16568

The first training set PDB652 contains 652 partially disordered proteins from the PDB database [22], each of which contains at least one disordered regions with more than 30 consecutive residues. The second training set D184 is derived from DisProt database [23], a curated database that provides information about proteins that wholly or partially lack a stable three-dimensional structure under putatively native conditions. For the details of the procedures in preparing datasets PDB652 and D184, the readers can refer to our recent work [13]. Different from the set PDB693 in [13], PDB652 excludes the sequences with similarity identity of more than 70% against any protein sequence in the other training sets by running Cd-Hit [24], resulting 652 proteins.

Since PDB652 and D184 contain more than 60% of disordered residues in terminal regions of the proteins, which causes the window-based classifiers to over-predict the terminal residues as disorder, this work collects an additional training set G200 from the PDB database [22] (there are 35579 proteins structures containing 85233 chains in the PDB release of 13-May-2006). After removal of the DNA chains and the protein chains shorter than 80 residues or with disordered regions, only 1847 fully ordered chains remain. Among the completely ordered proteins, 200 of them are randomly selected as the dataset G200. Similarly, we use the same criterion described above to handle the redundancy issue.

There are three independent datasets for evaluation in this study. The first set R80, prepared by Yang et al. [20], contains 80 protein chains from PDB database. The second set U79, provided by Uversky et al. in 2000 [21], contains 79 wholly disordered proteins. Finally, the third testing set P80, organized by PONDR (retrieved in February 2003), includes 80 completely ordered proteins. Like Yang et al. did in their study [20], these testing sets were employed in some recent related studies as a platform in comparison of different approaches in protein disorder prediction. Particularly, the testing sets U79 (wholly disordered proteins) and P80 (entirely globular proteins) examine whether the proposed method is under- or over-predicting protein disorder.

3 Methods

In this section, we first introduce the disorder predictor DisPSSMP, which was proposed in our recent work based on condensed PSSMs considering propensity for order or disorder [13]. Next, four representations of summarizing the local information of secondary structure are presented. Finally, the procedures of constructing the Radial Basis Function Networks (RBFN) classifier and some widely used evaluation measures are described in details.

3.1 Organizing Feature Sets

The disorder predictor DisPSSMP constructs its predicting model based on the condensed position specific scoring matrix with respect to physicochemical properties (PSSMP), which are shown to exhibit better performance on protein disorder prediction than the original PSSM features [13]. The success of DisPSSMP thanks to its invention of considering the disorder propensity of amino acids when searching for an optimized feature combination of PSSMPs. The selected condensed PSSM properties include: *Aliphatic*, *Aromatic$_O$*, *Polar*, and *Small$_D$* [13]. The derived feature set improves the performance of a classifier built with RBFN in comparison with the feature set constructed with PSSMs or PSSMPs that adopt simply the conventional physicochemical properties. The original feature set employed by DisPSSMP is named PSSMP-4 in the rest of this study.

In this study, we aim to investigate how the predicting power of DisPSSMP can be improved when incorporating secondary structure information with PSSMP-4. We propose a new representation named SSE-DIS, and compare it with other representations listed in Table 2, named SSE-BIN, SSE-PRO, and SSE-DEN respectively. Before extracting the features from the results of a secondary structure predictor, a SSE with less than five successive secondary structure residues are removed. We expect the remaining secondary structure segments to provide more reliable information than the original predictions. As summarized in Table 2, SSE-BIN comprises three binary features which correspond to the predicted secondary structure classes (helixes, strands, and coils), respectively. Like DISpro [17], for each residue, only one of the three features is set to 1 according to its SSE class. Another representation, named SSE-PRO, includes three real values which represent the probabilities for each class. Next, SSE-DEN calculates the density of secondary structures within a specific window size. Some pilot experiments based on training data show that the performance of SSE-DEN using various window sizes from 11 to 61 is almost the same. In this regard, a window size of 41 is employed in the experiments reported in this study. The feature SSE-DEN is derived from the concept of NORS (no regular secondary structure) [14]. Finally, the proposed representation SSE-DIS takes the distance of a residue to its nearest secondary structure element. This feature aims to emphasize the locations which are far from the regions consisted of regular secondary structures. The procedures of generating these four representations are exemplified by Fig. 2.

Table 2. The definition of four representations of secondary structure

Name	Description	Number of features*	Parameters associated with the representation
SSE-BIN	Binary values decoding helixes, strands, and coils	3	None
SSE-PRO	Probability values for helixes, strands, and coils	3	None
SSE-DEN	The density of secondary structures	1	A window size for calculating densities. It is set as 41 after some pilot experiments were conducted.
SSE-DIS	The distance from the nearest secondary structure element	1	None

*Each feature is normalized into the interval [0, 1] before they are employed in prediction.

In Fig. 2, column (a) stands for the original protein sequence, and column (b) is the predicted secondary structure element. In this study, we employ Jnet [25] as the secondary structure predictor, which is a neural network secondary structure predictor based on multiple sequence alignment profiles. Next, the predicted information is refined by removal of secondary structure elements with less than five residues, resulting column (c). The refined information is next transformed into features SSE-BIN (d1), SSE-DEN (d3), and SSE-DIS (d4) respectively. At the same time, the feature SSE-PRO (d2) is transformed from the original results of Jnet. The example used in Fig. 2 is from the protein *cys regulon transcriptional activator cysb* (PDB structure 1AL3). For each residue, the feature values falling in a window size of l centered at the given residue are extracted as its feature vector and the experiments of considering different window sizes are shown in the next section.

3.2 Classifier

DisPSSMP adopts the QuickRBF package [26] in constructing RBFN for classification. In this study, we in particular tackle the problem of handling skewed datasets, which stands for the problems with unbalanced numbers of positive (disorder) and negative (order) samples. In the previous implementation of DisPSSMP, equal quantity of residues from ordered and disordered regions was used in constructing the classifier. However, a large volume of ordered regions was lost after randomly removal of unwanted ordered residues. Thus in this study, all of ordered and disordered residues in the training datasets are included to construct the classifier without loss of information. In order to not over-predict residues as ordered, we adopt an alternative function in determining the outputs based on the function values generated by the RBF network. Let the number of the centers in the network be c, the probability distribution functions for the classes order and disorder are represented as follows:

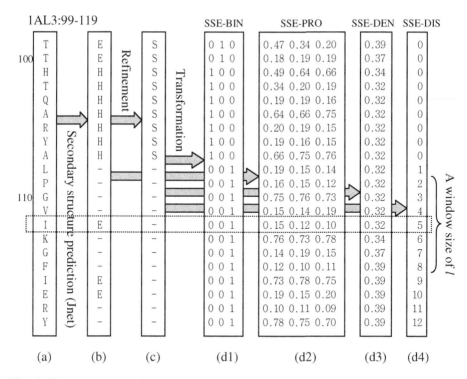

Fig. 2. The procedures of generating different representations (d1~d4) for the predicted secondary structures

$$\begin{bmatrix} \text{Order}(R_i) \\ \text{Disorder}(R_i) \end{bmatrix} = \begin{bmatrix} w_{1,1} & w_{1,2} & \cdots & w_{1,c} \\ w_{2,1} & w_{2,2} & \cdots & w_{2,c} \end{bmatrix} \begin{bmatrix} \phi_1(R_i) \\ \phi_2(R_i) \\ \cdots \\ \phi_c(R_i) \end{bmatrix}, \tag{1}$$

where R_i is the feature vector of a residue in the data sets, $\phi_j(R_i)$ is the j-th kernel function employed, and $w_{1,j}$ and $w_{2,j}$ are the weights of the edges connecting the hidden nodes and output nodes of the RBF network. Since Disorder(R_i) and Order(R_i) are supposed to be in between 0 and 1, they are set to 1 and 0 when the values generated by QuickRBF are larger than 1 or smaller than 0, respectively. Then, the formula for calculating the disorder propensity of residue R_i is represented as follows:

$$\text{Propensity}_D(R_i) = (\text{Disorder}(R_i) - \text{Order}(R_i) + 1)/2. \tag{2}$$

Finally, the alternative classification function Classifier(R_i) is shown in Equation (3). It means that the residue R_i is predicted as disordered if (Propensity$_D(R_i) \geq$ *Threshold*), where the parameter *Threshold* is determined by conducting cross-validation procedure on the training dataset.

$$\text{Classifier}(R_i) = \begin{cases} 1 & \text{if } (\text{Propensity}_D(R_i) \geq Threshold); \\ 0 & \text{otherwise.} \end{cases} \tag{3}$$

3.3   Evaluation Measures

Protein disorder prediction is a binary classification task, and many validation measures have been introduced to this problem [27, 28, 29]. Since the disordered residues of proteins in PDB database are so rare that a skewed dataset is considered here, we employ four measures that are considered proper in this problem together to evaluate the performance of different feature sets. As listed in Table 3, *sensitivity* and *specificity* represent the fraction of disordered and ordered residues correctly identified, respectively. Another commonly used evaluation measure *probability excess*, recommended by CASP6 [28] and Yang *et al.* [20], is employed here, too. Last, the Receiver Operating Characteristic (ROC) curve has been considered as with the most discriminating ability when comparing disorder prediction methods [17, 19, 28]. Indeed, the *area under the ROC curve* (*AUC*) represents the performance of each method fairly. Accordingly, the evaluation in cross validation is also based these measures.

3.4   Constructing Predicting Models with Cross Validation

In order to conduct a five-fold cross validation, all the chains in datasets PDB652, D184, and G200 are randomly split into five subsets of approximately equal sizes. As suggested in DisPSSMP [13], the feature set PSSMP-4 with the window size set as 47 is employed in cross validation to determine the parameter *Threshold* of Equation 3. In general, *sensitivity* increases when *specificity* decreases, and vice versa. Therefore, in this study, *Threshold* is determined by maximizing the probability excess. From Table 4, *probability excess* achieves maximal when *Threshold* is 0.22.

Table 3. The equations of the evaluation measures

Measure	Abbreviation	Equation
Sensitivity	*Sens.*	TP/(TP+FN)
Specificity	*Spec.*	TN/(TN+FP)
Probability excess	*Prob. Excess*	(TP×TN−FP×FN)/((TP+FN)×(TN+FP))
Area under ROC curve	*AUC*	$\sum_{p=1}^{100}[Spec(p/100) - Spec((p-1)/100)] * [Sens(p) + Sens((p-1)/100)]/2$

TP: the number of correctly classified disordered residues; FP: the number of ordered residues incorrectly classified as disordered; TN: the number of correctly classified ordered residues; FN: the number of disordered residues incorrectly classified as ordered; p: the threshold employed in Equation (3) ($p = 1, 2, ..., 100$); $Spec(p/100)$: specificity when the threshold in Equation (3) is $p/100$; $Sens(p/100)$: sensitivity when the threshold in Equation (3) is $p/100$.

Table 4. Cross validation for the training sets with window size of 47 for PSSMP-4

TP	FP	TN	FN	*AUC*	*Threshold*	*Sens.*	*Spec.*	*Prob. Excess*
49672	57826	226233	26809	0.781	0.22	0.650	0.796	0.446

4 Results and Discussions

In this section, we first evaluate how DisPSSMP performs when incorporating four representations of the secondary structure respectively. After that, we show some statistics from the secondary structures predicted by Jnet and discuss how SSE-DIS benefits the protein disorder prediction.

4.1 Results on Testing Data

For the blind testing sets, the comparison of the performance of PSSMP-4 with four representations of secondary structure is performed by a range of the window size l from 0 to 59, while zero means that the feature set comprises only PSSMP-4.

In Fig. 3, the results on the testing set R80 are shown. According to *AUC* in Fig. 3(a), the performance of SSE-DIS is improved when the window size increases. Meanwhile, the performance of SSE-DEN increases slightly when the window size is smaller than 19 and decreases when larger window sizes are considered. On the other hand, the performance of SSE-BIN and SSE-PRO decrease apparently when they are incorporated with PSSMP-4. It is concluded that SSE-DIS performs consistently better than the other representations when different window sizes are considered. Fig. 3(b) shows the comparison based on another measure *probability excess*. From this point of view, the predicting powers of SSE-DIS and SSE-DEN are comparable when the window size is not large. Combining the results of all the testing sets, it reveals that the representations SSE-BIN and SSE-PRO fail to improve the accuracy of DisPSSMP when they are incorporated with the original feature set PSSMP-4.

For wholly ordered or disordered proteins, the comparison is conducted on the testing sets U79 and P80. It can be observed in Fig. 4(a) and (b) that the difference between SSE-DIS and SSE-DEN is more significant in this comparison. Like in Fig. 3(b), Fig. 4(b) shows that SSE-BIN has a better a *probability excess* than PSSMP-4 when l is smaller than 19. It seems that SSE-BIN provides some useful information when the sliding window is small.

4.2 Discussions

We observed that the classifier trained with PSSMP-4+SSE-DIS predicts more disordered residues than the classifier trained from PSSMP-4 alone. Here we use two examples to explain the difference between these two classifiers. The experimental results of the protein *methionyl-tRNA synthetase* (PDB structure 1PG2 in the dataset R80) and the protein *Heat shock transcription factor, N-terminal activation domain*

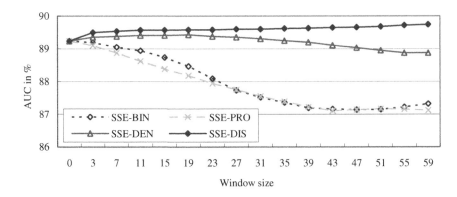

(a) Comparison based the measure *AUC*.

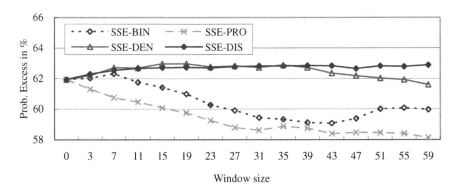

(b) Comparison based on the measure *probability excess*.

Fig. 3. The results on the testing set R80

(u56 from dataset U79) are drawn in Fig. 5(a) and Fig. 5(b), respectively. For 1PG2, there are three disordered regions which are residues of 1-3, 126-184, and 551. The first and third disordered regions are predicted correctly by using PSSMP-4 alone (the blue line) as well as PSSMP-4+SSE-DIS (the pink line). However, for the second disordered region, there are only two residues predicted as disordered by using PSSMP-4, while there are twenty residues predicted as disordered by using PSSMP-4+SSE-DIS. It is shown in the Fig. 5(a) that there are five short segments of secondary structure which are predicted as beta strands by Jnet but have been removed after refinement step. This procedure increases the values of the SSE-DIS near this region and then enlarges the disorder propensity of these residues. Similarly, the accuracy of protein disorder prediction in the first 100 residues of u56, a totally disordered protein, is improved explicitly due to the removal of eight short segments of secondary structure, including seven beta-strands and one helix.

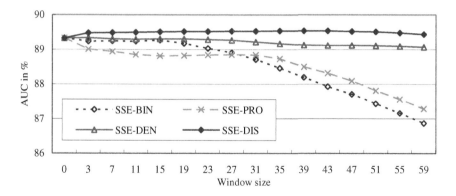

(a) Comparison based on the measure *AUC*.

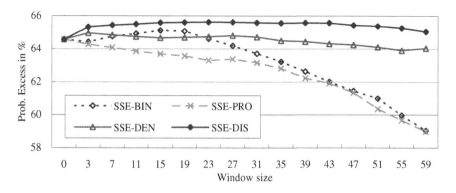

(b) Comparison based on the measure *probability excess*.

Fig. 4. The results on the testing sets U79 and P80

At the end of this section, it is of interest to see the statistics of the predicted secondary structures present in ordered regions and different groups of disordered regions. With all of the datasets used in this study, the ratio of each secondarystructure type in different groups of protein regions is exhibited in Table 5. In Table 5(a), the statistics are calculated from the original results from Jnet, whereas the statistics in Table 5(b) are calculated from the refined results by removing short SSE segments. Here are some observations. First, the ratios of coils in ordered region in Table 5(a) among all datasets are about 50%, which are lower than in short disordered regions (~90%), middle disordered regions (~80%), and long disordered regions (~65%). When comparing Table 5(b) with Table 5(a), it is observed that many beta strands predicted by Jnet are shorter than five successive residues and are not used in constructing the feature sets. Furthermore, it is attractive that the ratios of long disordered regions are more similar to the ratios of ordered regions than the other groups of disordered regions. This phenomenon might correspond to some long

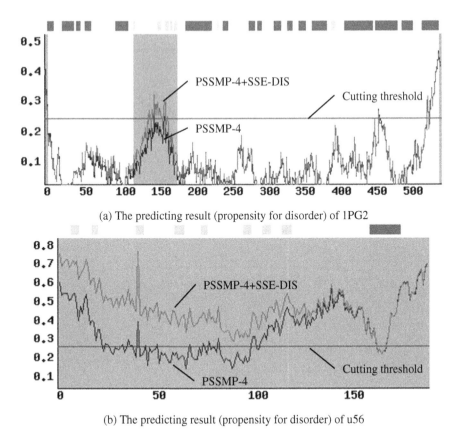

(a) The predicting result (propensity for disorder) of 1PG2

(b) The predicting result (propensity for disorder) of u56

Fig. 5. The comparison of protein disorder prediction with PSSMP-4 and PSSMP-4+SSE-DIS. The figure plots the disorder propensity of a residue (*y-axis*) versus the position of a protein sequence (*x-axis*), where shaded areas are annotated disordered regions (*gray blocks*), the blue line shows the results by using PSSMP-4 alone, the pink line shows the results of using PSSMP-4+SSE-DIS, and the red line presents the cutting threshold of the classification function. The refined secondary structure segments are marked as green blocks (the darker ones), and the segments that contain less than five successive residues and thus have been removed are shown as yellow blocks (the lighter ones).

disordered segments of proteins with specific functions [2, 4]. Since the disorder-to-order transition upon binding occurs in some of long disordered segments of proteins, the segments might comprise secondary structure to stabilize the interfaces or binding domains between a protein and its ligand. This observation needs more investigations and discussions in future studies.

5 Conclusions

In this study we investigate how the predicting power of condensed PSSM features in recognizing protein disorder can be enhanced by secondary structure information. This

Table 5. The statistics of secondary structure in ordered regions and three groups of disordered regions with different lengths

(a) The original results from Jnet

Dataset	Ordered region			Short disordered region(1~4)			Middle disordered region(5~9)			Long disordered region(10~)		
	Helix	Strand	Coil	Helix	Strand	Coil	Helix	Strand	Coil	Helix	Strand	Coil
PDB652	35.2	17.9	46.9	8.7	5.5	85.7	13.8	5.3	80.9	27.7	11.3	61.0
D184	32.0	15.9	52.1	8.2	1.4	90.4	22.0	8.4	69.6	24.5	9.6	65.9
G200	31.3	18.7	49.9									
R80	33.0	17.6	49.5	1.1	5.7	93.1	7.7	5.1	87.2	20.8	9.0	70.2
U79										28.3	9.4	62.3
P80	32.2	17.9	50.0									

(b)The refined results after removal of secondary structure segments with length of less than five successive residues

Dataset	Ordered region			Short disordered region(1~4)			Middle disordered region(5~9)			Long disordered region(10~)		
	Helix	Strand	Coil	Helix	Strand	Coil	Helix	Strand	Coil	Helix	Strand	Coil
PDB652	34.8	14.1	51.1	8.4	3.9	87.7	13.6	3.4	83.0	27.3	8.1	64.6
D184	31.6	12.2	56.2	8.2	0.0	91.8	22.0	6.5	71.5	24.0	6.8	69.2
G200	31.0	14.5	54.4									
R80	32.6	13.7	53.7	0.0	4.6	95.4	7.7	3.8	88.5	20.7	5.8	73.5
U79										27.7	6.5	65.8
P80	31.9	13.3	54.8									

A value in this table is the percentage of a certain type of secondary structure in ordered regions or disordered regions. The empty squares mean that there is no such region in that dataset.

work compares four kinds of representations in depicting secondary structures detected by secondary structure predictors. The results suggest that the proposed representation achieves more coverage of disordered residues without increasing the false positives obviously. The detected disorder information is expected to be useful in protein structure prediction and functional analysis. In the future, there are several directions for further improving the performance of protein disorder prediction. More predicted information from primary sequences can be merged to enhance the predicting power of the classifiers. On the other hand, incorporating more machine learning skills for handling skewed datasets in this problem also deserves further studies.

References

1. Dunker, A.K., Obradovic, Z., Romero, P., Kissinger, C., Villafrance, E.: On the importance of being disordered. PDB Newsletter (1997) 81:3-5
2. Wright, P.E., Dyson, H.J.: Intrinsically unstructured proteins: re-assessing the protein structure-function paradigm. J. Mol. Biol. (1999) 293(2):321-331
3. Ward, J.J., Sodhi, J.S., McGuffin, L.J., Buxton, B.F., Jones, D.T.: Prediction and functional analysis of native disorder in proteins from the three kingdoms of life. J Mol Biol (2004) 337:635-645

4. Fink, A.L.: Natively unfolded proteins. Current Opinion in Structural Biology (2005) 15:35-41
5. Dunker, A.K., Garner, E., Guilliot, S., Romero, P., Albercht, K., Hart, J., Obradovic, Z., Kissinger, C., Villafranca, J.E.: Protein disorder and the evolution of molecular recognition: theory, predictions and observations. Pac Symp Biocomput (1998) 3:473-484
6. Jones, D.T., Ward, J.J.: Prediction of disordered regions in proteins from position specific scoring matrices. Proteins (2003) 53:573-578
7. Romero, P., Obradovic, Z., Kissinger, C., Villafranca, J.E., Dunker, A.K.: Identifying disordered regions in proteins from amino acid sequence. Proc. IEEE Int. Conf. Neural Networks. (1997) 1:90-95
8. Romero, P., Obradovic, Z., Kissinger, C., Villafranca, J.E., Garner, E., Guilliiot, S., Dunker, A.K.: Thousands of proteins likely to have long disordered regions. Pac Symp Biocomput (1998) 3:437-448
9. Obradovic, Z., Peng, K., Vucetic, S., Radivojac, P., Brown, C.J., Dunker, A.K.: Predicting intrinsic disorder from amino acid sequence. Proteins (2003) 53:566-572
10. Wotton, J.C., Federhen, S.: Statistics of local complexity in amino acid sequences and sequence databases. Comput Chem (1993) 17:149-163
11. Brown, C.J., Takayama, S., Campen, A.M., Vise, P., Marshall, T.W., Oldfield, C.J., Williams, C.J., Dunker, A.K.:Evolutionary rate heterogeneity in proteins with long disordered regions. J Mol Evol (July 2002) 55:104-110
12. Vihinen, M., Torkkila, E., Riikonen, P.: Accuracy of protein flexibility predictions. Proteins (1994) 19:141-149
13. Su, C.T., Chen C.Y., Ou, Y.Y.: Protein disorder prediction by condensed PSSM considering propensity for order or disorder. BMC Bioinformatics (2006) 7:319
14. Liu, J., Rost, B.: NORSp: predictions of long regions without regular secondary structure. Nucl. Acids Res. (2003) 31(13):3833-3835
15. Liu, J., Tan, H., Rost, B.: Loopy proteins appear conserved in evolution. J. Mol. Biol. (2002) 322:53-64
16. Linding, R., Russell, R.B., Neduva, V., Gibson, T.J.: GlobPlot: exploring protein sequences for globularity and disorder. Nucl. Acids Res. (2003) 31:3701-3708
17. Cheng, J., Sweredoski, M.J., Baldi, P.: Accurate prediction of protein disordered regions by mining protein structure data. Data Mining and Knowledge Discovery (2005) 11: 213-222
18. Obradovic, Z., Peng, K., Vucetic, S., Radivojac, P., Dunker, A.K.: Exploiting heterogeneous sequence properties improves prediction of protein disorder. Proteins (2005) Suppl 7:176-182
19. Peng, K., Radivojac, P., Vucetic, S., Dunker, A.K., Obradovic, Z.: Length-dependent prediction of protein intrinsic disorder. BMC Bioinformatics (2006) 7:208
20. Yang, Z.R., Thomson, R., McNeil, P., Esnouf, R.M.: RONN: the bio-basis function neural network technique applied to the detection of natively disordered regions in proteins. Bioinformatics Advance Access Published June 9, 2005
21. Uversky, V.N., Gillespie, J.R., Fink, A.L.: Why are "natively unfolded" proteins unstructured under physiologic conditions? Proteins (2000) 41:415-427
22. Berman, H.M., Westbrook, J., Feng, Z., Gilliland, G., Bhat, T.N., Weissig, H., Shindyalov, I.N., Bourne, P.E.: The Protein Data Bank. Nucl. Acids Res. (2000) 28:235-242
23. Vucetic, S., Obradovic, Z., Vacic, V., Radivojac, P., Peng, K., Lakoucheva, L.M., Cortese, M.S., Lawson, J.D., Brown, C.J., Sikes, J.G., Newton, C.D., Dunker, A.K.: DisProt: a database of protein disorder. Bioinformatics (2005) 21:137-140

24. Li, W., Jaroszewski, L., Godzik, A.: Tolerating some redundancy significantly speeds up clustering of large proteins databases. Bioinformatics (2002) 18:77-82
25. Cuff, J.A., Barton, G.J.: Application of enhanced multiple sequence alignment profiles to improve protein secondary structure prediction, Proteins (2000) 40:502-511
26. Ou, Y.Y., Chen, C.Y., Oyang, Y.J.: A Novel Radial Basis Function Network Classifier with Centers Set by Hierarchical Clustering, IJCNN '05. Proceedings. (2005) 3:1383-1388.
27. Melamud, E., Moult, J.: Evaluation of disorder predictions in CASP5. Proteins (2003) 53:561-565
28. Jin, Y., Dunbrack, R.L.: Assessment of disorder predictions in CASP6. Proteins (2005) Early View
29. Ward. J.J., McGuffin, L.J., Bryson, K., Buxton, B.F., Jones, D.T.: The DISOPRED server for the prediction of protein disorder. Bioinformatics (2004) 20:2138-2139

Joining Softassign and Dynamic Programming for the Contact Map Overlap Problem

Brijnesh J. Jain[1] and Michael Lappe[2]

[1] Berlin University of Technology, Faculty IV, Germany
`jbj@cs.tu-berlin.de`
[2] Max Planck Inst. for Molecular Genetics, Berlin, Germany
`lappe@molgen.mpg.de`

Abstract. Comparison of 3-dimensional protein folds is a core problem in molecular biology. The Contact Map Overlap (CMO) scheme provides one of the most common measures for protein structure similarity. Maximizing CMO is, however, NP-hard. To approximately solve CMO, we combine softassign and dynamic programming. Softassign approximately solves the maximum common subgraph (MCS) problem. Dynamic programming converts the MCS solution to a solution of the CMO problem. We present and discuss experiments using proteins with up to 1500 residues. The results indicate that the proposed method is extremely fast compared to other methods, scales well with increasing problem size, and is useful for comparing similar protein structures.

1 Introduction

Solutions to the problem of comparing protein structures are important in structural genomics for (i) protein function determination and drug design, (ii) fold prediction, and (iii) protein clustering [1]. For this, various methods have been devised. The most common are (i) RMSD [12], (ii) Distance Map Similarity [10], and (iii) Contact Map Overlap (CMO) [5,6]. Here, we focus on the CMO scheme.

To structurally compare two proteins using CMO, we first derive contact maps of each protein under consideration. A contact map consists of an ordered set of vertices representing the residues of a protein and of contacts (edges) connecting two vertices if the corresponding residues are geometrically close. The order of the vertices corresponds to the sequence order of the residues. Maximizing the CMO of both proteins aims at finding an order preserving alignment of both contact maps such that the number of common contacts is maximized.

CMO is NP-hard [8] and also hard to approximate [9]. The exponential time complexity initiated ongoing research on devising solutions to the CMO problem. Most methods employ discrete optimization techniques [1,2,3,13,14,17,18] and have been typically applied in a systematic way only to small-sized proteins with up to ≈ 250 residues.

In this paper, we suggest an approach with the following characteristics:

- The major tool of the proposed algorithm is based on nonlinear continuous rather than discrete optimization techniques.

S. Hochreiter and R. Wagner (Eds.): BIRD 2007, LNBI 4414, pp. 410–423, 2007.
© Springer-Verlag Berlin Heidelberg 2007

- The proposed algorithm provides good suboptimal solutions for similar protein structures.
- The proposed algorithm is fast and scales well with problem size. Proteins with about 1500 residues can be aligned in less than 1.5 minutes on a Athlon 64 32000+ processor with 2 GHz. Comparable results have not been reported in the literature so far.

The proposed approach applies the softassign algorithm developed by [7,11] for solving a related problem, the maximum common subgraph (MCS) problem. The MCS problem asks for the maximum number of common contacts without considering the sequence order of the contact maps. To convert the MCS solution of softassign to a feasible solution for CMO, we enhance softassign with a problem-dependent objective function and a dynamic programming post-processing procedure. In experiments using real proteins form the protein database (PDB), we investigated the behavior and performance of the modified softassign approach.

The rest of this paper is organized as follows: Section 2 introduces the CMO problem. In Section 3, we describe the proposed algorithm. Section 4 presents and discusses the results. Finally, Section 5 concludes with a summary of the main results and an outlook for further research.

2   Problem Statement

A *contact map* is an undirected graph $X = (V, E)$ consisting of an ordered set $V = \{1, \ldots n\}$ of vertices and a set $E \subset V^2$ of edges. Vertices represent residues of a folded protein and edges the contacts between the corresponding residues. The vertex set of X is also referred to as $V(X)$ and its edge set as $E(X)$. As usual, the adjacency matrix $\boldsymbol{X} = (x_{ij})$ of X is defined by

$$x_{ij} = \begin{cases} 1 & : & \text{if } (i,j) \in E(X) \\ 0 & : & \text{otherwise} \end{cases}.$$

Our goal is to structurally align two contact maps X and Y. For this, we ask for a partial vertex mapping

$$\phi : V(X) \rightarrow V(Y), \quad i \mapsto i^\phi$$

that maximizes some criterion function subject to certain constraints. In the following we specify both, the criterion function and the constraints.

A *common* or *shared contact* aligned by ϕ is a pair of edges $(i, j) \in E(X)$ and $(r, s) \in E(Y)$ such that $(r, s) = (i^\phi, j^\phi)$. The criterion function to be maximized is of the form

$$f(\phi) = \sum_{i,j \in \mathsf{dom}(\phi)} x_{ij} y_{i^\phi j^\phi}, \tag{1}$$

where $\mathsf{dom}(\phi)$ denotes the domain of the partial mapping ϕ. The terms $x_{ij} y_{i^\phi j^\phi}$ are either one or zero. We have $x_{ij} y_{i^\phi j^\phi} = 1$ if, and only if (i, j) and (i^ϕ, j^ϕ) are common contacts. Hence, $f(\phi)$ counts the number of common contacts aligned by a feasible mapping ϕ.

The first constraint is that the partial mappings are injective. Let $\Pi(X,Y)$ denote the set of all injective partial mapping from $V(X)$ to $V(Y)$. Then the *maximum common subgraph (MCS) problem* is the problem of maximizing $f(\phi)$ subject to $\phi \in \Pi(X,Y)$.

The second constraint demands that a mapping ϕ from $\Pi(X,Y)$ preserves the order of the vertices. A mapping is order-preserving if

$$ i < j \quad \Rightarrow \quad i^\phi < j^\phi $$

for all $i,j \in V(X)$. By $\Pi^+(X,Y) \subseteq \Pi(X,Y)$ we denote the subset of order-preserving mappings. The *maximum contact map overlap problem* is the problem of maximizing $f(\phi)$ subject to $\phi \in \Pi^+(X,Y)$. Thus, the CMO problem can be derived from the MCS problem by additionally imposing the order-preserving constraint.

3   Joining Softassign and Dynamic Programming

Our algorithm proceeds in two stages: In the first stage, we ignore the order-preserving constraint and approximately solve the MCS problem. The second stage converts the solution of the first stage to a feasible solution of CMO.

The main challenge is to find an approximate solution of the MCS problem in the first stage, which is close to a good solution of the CMO problem. To tackle this problem, the algorithm consists of three components:

- **The Softassign Algorithm.** In the first stage, we use softassign [7,11], one of the most powerful method for approximately solving the MCS problem.
- **Dynamic Programming.** The second stage takes the output of softassign as input and converts it to a feasible solution of the CMO problem using a dynamic programming approach. Dynamic programming yields an optimal solution given the output of softassign.
- **Compatibility function.** The compatibility function provides the crucial link between the first and second stage of the algorithm. It aims at reshaping the objective function of the MCS problem such that good local optima correspond to good solutions of the CMO problem.

In the following, we describe each component of the algorithm separately.

3.1   The Softassign Algorithm

Softassign minimizes a continuous quadratic formulation of the MCS problem. Section 3.1.1 presents the quadratic program and in Section 3.1.2, we describe the softassign algorithm. For a detailed presentation of softassign, we refer to [7].

In the following, we assume that X and Y are contact maps consisting of n and m vertices, respectively.

A Continuous Quadratic Program for the MCS Problem. To present a quadratic formulation of the MCS problem, we encode the partial vertex mappings as match matrices. Let $\phi : V(X) \rightarrow V(Y)$ be a mapping from $\Pi(X,Y)$. Then the matrix representation of ϕ is a binary $(n \times m)$-matrix $\boldsymbol{M}_\phi = (m_{ij})$ with

$$m_{ij} = \begin{cases} 1 & : \quad i^\phi = j \\ 0 & : \quad \text{otherwise} \end{cases}$$

We may identify mappings ϕ from $\Pi(X,Y)$ with their associated matrices $\boldsymbol{M}_\phi$. Therefore, we also write $\boldsymbol{M}_\phi \in \Pi(X,Y)$ by abuse of notation.

Using match matrices, the constraint $\phi \in \Pi(X,Y)$ can be rewritten as

$$\sum_{i=1}^{n} m_{ij} \leq 1, \quad \forall j \in V(Y) \tag{2}$$

$$\sum_{j=1}^{m} m_{ij} \leq 1, \quad \forall i \in V(X) \tag{3}$$

$$m_{ij} \in \{0,1\}, \quad \forall i \in V(X), \; \forall j \in V(Y) \tag{4}$$

Let $c_{ijrs} = x_{ir} y_{js}$ denote the *compatibility* of x_{ir} and y_{js}. The quadratic integer formulation of criterion function (1) is

$$E(M) = -\frac{1}{2} \sum_{i=1}^{n} \sum_{j=1}^{m} \sum_{r=1}^{n} \sum_{s=1}^{m} m_{ij} m_{rs} c_{ijrs}. \tag{5}$$

Thus the MCS problem is equivalent to minimizing $E(M)$ subject to the constraints (2)-(4).

We replace constraint (4) by

$$m_{ij} \in [0,1], \quad \forall i \in V(X), \; \forall j \in V(Y). \tag{6}$$

to obtain a continuous version of the quadratic integer program for the MCS problem. Matrices satisfying constraints (2), (3), and (6), are called *doubly stochastic*.

Softassign. Softassign minimizes $E(M)$ subject to the constraints (2), (3), and (6). The core of the algorithm implements a deterministic annealing process with annealing parameter T by the following iteration scheme

$$m_{ij}^{(t+1)} = a_i b_j \exp\left(-\frac{1}{T} \sum_{r=1}^{n} \sum_{s=1}^{m} m_{rs}^{(t)} c_{ijrs}\right), \tag{7}$$

where t denotes the time step. The scaling factors a_i, b_i computed by Sinkhorn's algorithm [16] enforce the constraints of a doubly stochastic matrix.

Algorithm 1 outlines the softassign algorithm (see [7]). To convert the inequality constraints (2) and (3), softassign uses slack variables by adding an extra row and column to the matrix $\boldsymbol{M}$. By $\hat{\boldsymbol{M}} = (\widehat{m}_{ij})$ we denote the augmented $(n+1) \times (m+1)$-matrix. The parameters T_0, T_f, and α describe the annealing schedule. The annealing parameter T is initialized with T_0 and gradually lowered by αT until it reaches the final value T_f. The parameters I_0 and I_1 specify the maximum number of steps of the Assignment and Sinkhorn loop.

Algorithm 1. Softassign

1. **Initialization:** $T \leftarrow T_0$ and $\widehat{m}_{ij} \leftarrow 1 + \varepsilon$
2. **while** $T > T_f$ **do**
   1. **for** $t_0 = 1$ **to** I_0 **do**   (Assignment Loop)
      1. $q_{ij} \leftarrow \partial E / \partial m_{ij}$
      2. $m_{ij} \leftarrow \exp(q_{ij}/T)$
      3. **for** $t_1 = 1$ **to** I_1 **do**   (Sinkhorn Loop)
         1. $r_i \leftarrow \sum_{j=1}^{m+1} \widehat{m}_{ij}$
         2. $\widehat{m}'_{ij} \rightarrow \widehat{m}_{ij}/r_i$
         3. $c_j \leftarrow \sum_{i=1}^{n+1} \widehat{m}'_{ij}$
         4. $\widehat{m}_{ij} \leftarrow \widehat{m}'_{ij}/c_j$
   2. $T \leftarrow \alpha T$

Since $c_{ijrs} = 1$ if, and only if, $(i,r) \in E(X)$ and $(j,s) \in E(Y)$, the complexity of softassign is $O(|E(X)| \cdot |E(Y)|)$.

3.2   Dynamic Programming

We use dynamic programming to convert the output $\boldsymbol{M}$ of softassign to a feasible solution of the CMO problem, which is optimal with regard to $\boldsymbol{M}$.

Generally, let $\boldsymbol{W} = (w_{ij})$ be a $(n \times m)$-matrix with elements $w_{ij} \in [0,1]$. The dynamic programming approach maximizes the weight function

$$\omega(\boldsymbol{P}) = \sum_{i=1}^{n} \sum_{j=1}^{m} w_{ij} p_{ij}$$

subject to $\boldsymbol{P} = (p_{ij}) \in \Pi^+(X, Y)$.

The main idea of dynamic programming is to use optimal solutions of subproblems to find an optimal solution of the overall problem. To describe the subproblems, we need the notion of *induced subgraph*. Let Z be a graph and let $k \leq |V(Z)|$. By Z_k we denote the subgraph of Z induced by the first k vertices from Z. The vertex and edge set of Z_k are given by

$$V(Z_k) = \{1, \ldots, k\} \subseteq V(Z) \quad \text{and} \quad E(Z_k) = E(Z) \cap V(Z_k)^2.$$

Let $k \leq n$ and $l \leq m$. The (k,l)-th subproblem maximizes

$$\omega(\boldsymbol{P}) = \sum_{i=1}^{k} \sum_{j=1}^{l} w_{ij} p_{ij}$$

subject to $\boldsymbol{P} \in \Pi^+(X_k, Y_l)$.

The dynamic programming algorithm consists of two stages. In the first stage, it converts the matrix W to a score matrix $S = (s_{kl})$. The scores s_{kl} correspond to optimal solutions of the (k, l)-th subproblem with respect to X_k and Y_l. In the second stage, the algorithm selects in a backward process correspondences with highest score such that all constraints of the CMO problem are satisfied.

The following algorithm outlines the procedure to construct a scoring matrix S. The vector s^* records the highest score for each column.

Algorithm 2. Scoring

1. **Initialization:** $S = 0$ and $s^* = 0$
2. **for** $i = 1$ **to** n **do**
    1. **for** $j = 1$ **to** m **do**
        1. $s_{ij} \leftarrow w_{ij} + \max\{s_1^*, \ldots, s_{j-1}^*\}$
    2. **for** $j = 1$ **to** m **do**
        1. $s_j^* \leftarrow \max\{s_j^*, s_{ij}\}$

The complexity of Algorithm 2 is $O(nm)$. The matrix S has the property that $s_{ij} \leq s_{kl}$ for all $i < k$ and $j < l$. We exploit this property in a backward process to construct a feasible solution of the CMO problem.

Algorithm 3. Recover feasible Solution

1. **Initialization:** $P = 0$
2. $k \leftarrow m$
3. **for** $i = n$ **to** 1 **do**
    1. $k' \leftarrow \arg\max_{j \leq k}\{s_{ij}\}$
    2. $p_{ik'} \leftarrow 1$
    3. $k \leftarrow k' - 1$

Without using a more sophisticated data structure the complexity of Algorithm 3 is $O(nm)$.

3.3 Compatibility Function

In its original formulation, softassign minimizes the objective function (5), where the compatibilities are of the form $c_{ijrs} = x_{ir}y_{js}$. This choice of compatibilities ignores the order-preserving constraint for each partial mapping $\{i \mapsto j, r \mapsto s\}$ and may therefore result in a poor final solution for CMO.

Figure 1 illustrates some of the problems using the original compatibilities. Suppose that the MCS problem is the problem of maximizing an objective $f_{MCS}(M)$ subject to $M \in \Pi(X, Y)$. For example, $f_{MCS}(M) = -E(M)$, where $E(M)$ is the objective function (5). Similarly, the COM problem is the problem of maximizing some objective function $f_{CMO}(P)$ subject to $P \in \Pi^+(X, Y)$. Figure 1 shows that f_{MCS} has three maxima M_1, M_2, and M_3. Applying dynamic programming converts the local maxima M_i to feasible solutions P_i of the CMO problem. The following undesirable situations can occur:

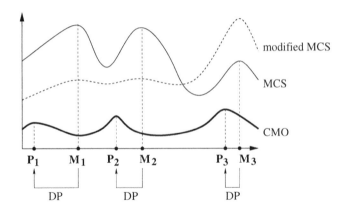

Fig. 1. Idealized example for illustrating the effects of compatibility values. The horizontal axis represents the continuous search space. The vertical axis shows the values of (fictive) objective functions to be maximized for the CMO, original MCS, and modified MCS problem. The original MCS uses $c_{ijrs} = x_{ir}y_{js}$ as compatibility values. The modified MCS uses more appropriate compatibility values for the CMO problem. Dynamic programming (DP) transforms the local maxima M_i of the original and modified MCS problem to local maxima P_i of the CMO problem.

- $f_{MCS}(\boldsymbol{M}_1) = f_{MCS}(\boldsymbol{M}_2)$, but $f_{CMO}(\boldsymbol{P}_1) < f_{CMO}(\boldsymbol{P}_2)$.
- $f_{MCS}(\boldsymbol{M}_3) < f_{MCS}(\boldsymbol{M}_i)$, but $f_{CMO}(\boldsymbol{P}_3) > f_{CMO}(\boldsymbol{P}_i)$ for $i = 1, 2$.

This observation suggests to define a compatibility function that reshapes the objective function of MCS in a similar way as shown in Figure 1.

The second issue to take into account when constructing a suitable compatibility function is time complexity. The time complexity of softassign using modified compatibility values is $O(|C|)$, where $C = \{(i, j, r, s) : c_{ijrs} \neq 0\}$. Thus, to keep the time complexity low the number of nonzero values c_{ijrs} should be small. Therefore, we avoid penalties $c_{ijrs} < 0$ for partial mappings $\{i \mapsto j, r \mapsto s\}$ that violate the constraints.

We suggest the following modified compatibility function:

$$
c_{ijrs} = \begin{cases} 1/(1 + \alpha|r_{ir} - r_{js}|) & : \quad (i, r) \sim (j, s) \\ \beta & : \quad r_{ir} = r_{js} = 1 \ , \\ 0 & : \quad \text{otherwise} \end{cases}
$$

where α and β are problem dependent parameters. The quantity $r_{kl} = |k - l|$ is the *range* of edge (k, l). The notion $(i, r) \sim (j, s)$ means that the partial mapping $\phi = \{i \mapsto j, r \mapsto s\}$ satisfies both constraints of the CMO problem, that is $\phi \in \Pi^+(X, Y)$.

The modified compatibility function has the following characteristics:

- Since contact maps are sparse, most compatibility values are zero keeping the time complexity of softassign low.
- Only feasible partial mappings $\phi = \{i \mapsto j, r \mapsto s\}$ and their direct neighbors are positively weighted.

– Common contacts with similar range are considered to be more compatible to one another than common contacts with different range.

3.4   The SADP Algorithm: Assembling the Components

The SADP (Softassign and Dynamic Programming) procedure to approximately solve the CMO problem operates as follows:

Algorithm 4. CMO Algorithm

1. Define compatibility function
2. Algorithm 1: $M = \text{softassign}(X, Y)$
3. Algorithm 2: $S = \text{getScoreMatrix}(M)$
4. Algorithm 3: $P = \text{getSolution}(S)$

Note that for solving the CMO problem, it is not necessary to exactly meet the constraints of a doubly stochastic matrix in Step 2. The task of softassign is merely to return a useful weighted matrix that yields a good scoring in Step 3. Therefore, we can terminate the Assignment and Sinkhorn loop after a maximum number of iteration steps. Additionally, we terminate a loop prematurely when the change of the previous and current match matrix is below a given threshold.

4   Experiments

To assess the performance, we applied SADP to three test problems. The algorithm was implemented in Java using JDK 1.2. Whenever possible, we compared SADP with a C implementation of the CAP algorithm proposed by Caprara and Lancia [1].[1] All experiments were run on a Linux PC with Athlon 64 32000+ processor (2.0 GHz CPU).

4.1   Sokol Test

In our first test series, we compared the performance of SADP against the following algorithms implemented in C:

Acronym	Reference	Platform
CAP	Caprara & Lancia [1]	Linux PC, Athlon 64 3200+, 2.0 GHz CPU
STR	Strickland et al. [17]	SGI workstation, 200 MHz CPU
XIE	Xie & Sahinidis [18]	Dell workstation, 3.0 GHz CPU

As test instances, we used the Sokol test set consisting of 11 alignment pairs of small-size proteins compiled by [17].

---

[1] The implementation of CAP has been provided by Lancia. In all experiments, we used the default parameter setting.

Table 1. Results of the Sokol test set. The first column shows the proteins to be structurally aligned. Columns 2-5 show the number of vertices and edges of the contact maps. Column 6 with identifier f_{SADP} shows the number of shared contacts found by SADP. Column 7 with identifier f_* shows the optimal solution found by CAP, STR, and XIE. Columns with identifiers of the form t_{ALG} refer to the computation time required by the respective algorithm $\text{ALG} \in \{\text{CAP}, \text{SADP}, \text{STR}, \text{XIE}\}$. Computation times are measured in seconds. Entries with value 0 (0.00) refer to computation times of less than 1 (0.01) second.

X-Y	$|V(X)|$	$|E(X)|$	$|V(Y)|$	$|E(Y)|$	f_{SADP}	f_*	t_{SADP}	t_{CAP}	t_{STR}	t_{XIE}
3ebx-1era	48	52	48	49	31	31	0.08	1.88	236	0.72
1bpi-2knt	51	58	44	42	27	29	0.06	1.94	182	0.46
1knt-1bpi	43	43	51	58	28	30	0.06	1.47	110	0.38
6ebx-1era	35	34	48	49	19	20	0.04	1.11	101	0.73
2knt-5pti	44	42	47	52	25	28	0.05	1.08	95	0.32
1bpi-1knt	51	58	43	43	28	30	0.05	1.53	19	0.24
1bpi-5pti	51	58	47	52	41	42	0.05	0.88	30	0.29
1knt-2knt	43	43	44	42	39	39	0.03	0.00	0	0.15
1knt-5pti	43	43	47	52	26	28	0.05	1.16	46	0.57
1vii-1cph	20	19	12	9	6	6	0.00	0.02	0	0.00
3ebx-6ebx	48	52	35	34	27	28	0.04	0.18	6	0.12
Total:					297	311	0.51	11.24	825	3.98

Table 1 summarizes the results. The results for STR and XIE are taken from [18]. The CAP, STR, and XIE algorithm returned an optimal solution f_* in all 11 trials giving a total of 311 common contacts. With 297 shared contacts, the solution quality of SADP is in average about 95.5% to optimal.

A fair comparison of computation time is difficult, because the other three algorithms were implemented in C and two of them (STR, XIE) were tested on different platforms. The results indicate that SADP has the best time performance. This indication will be approved for large-scale problems.

4.2 Experiment with Large-Scaled Data

The aim of our second test series is to evaluate the behavior and performance of SADP for large-scaled data. A comparison of SADP against the other three algorithms mentioned in the Sokol test is not possible because of time constraints. For this reason, we designed a semi-synthetical evaluation procedure.

We randomly selected 21 proteins from PDB having 800 up to 1500 residues. The corresponding contact maps have the following characteristics:

Vertices				Edges			
max	min	mean	std	max	min	mean	std
1488	801	959.4	172.2	5273	2717	3287	604.1

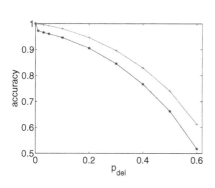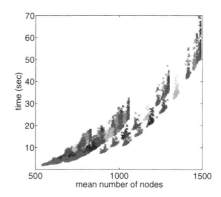

Fig. 2. Left: Average accuracy of SADP as a function of p_{del}. The upper curve shows the average accuracy normalized by the lower bound $|E(Y_2)|$ and the lower curve is the average accuracy normalized by $|E(Y)|$. **Right:** Computation time in *sec* as a function of the mean $(|V(X)| + |V(Y)|)/2$ for each trial (X, Y).

Let $\mathcal{X}$ denote the set of all 21 selected contact maps. We structurally aligned pairs (X, Y) of contact maps, where X is a contact map from $\mathcal{X}$ and Y is a corrupted version of X. To create Y, we proceeded as follows: First, we randomly deleted each vertex of X with probability p_{del}. Let Y_1 denote the resulting contact map. Next, we randomly deleted each edge from Y_1 with the same probability p_{del} and to obtain contact map Y_2. In a last step, we randomly connected each vertex with 10% probability to a randomly chosen vertex by an edge to obtain the final corrupted version Y. For each contact map $X \in \mathcal{X}$ and each parameter $p_{del} \in \{0.01, 0.03, 0.05, 0.1, 0.2, 0.3, 0.4, 0.5\}$, we generated 100 corrupted copies Y giving a total of 20,000 trials. This setting allows us to assess the solution quality of SADP: Given an alignment pair (X, Y), we find that $|E(Y_2)|$ is a lower bound of the maximum number of shared contacts, because Y_2 is the contact map after randomly removing edges from X. Hence, Y_2 and X have $|E(Y_2)|$ edges in common. In the final step we randomly add edges to Y_2 to obtain Y. Therefore X and Y share at least $|E(Y_2)|$ common edges.

Figure 2 shows the accuracy and computation time of SADP. From the left plot of Figure 2 we see that the accuracy of softassign degrades with increasing level of corruption. This observation is in accordance with [?] for the maximum common subgraph problem. First attempts to improve this deficiency apply kernelized comptatibiltiy functions [?]. Thus, in its current form, SADP is suitable for problems, where detection of mid to high structural similarity between two proteins is relevant.

The right plot of Figure 2 shows that the computation time of SADP is less than 1.5 minutes for aligning the largest contacts maps of order ≈ 1500. Aligning contact maps of order ≈ 800 took about 10-15 seconds. In contrast, we aborted CAP applied on two contact maps of order ≈ 800 after 5 days without obtaining a result.

Table 2. Protein domains of the Skolnick test set. Columns with identifier ID refer to the index assigned to the domains; columns with identifier PDB refer to the PDB code for the protein containing the domain; and columns with identifier CID refer to the chain index of a protein. If a protein consists of a single chain, the corresponding entry in the CID column is left empty. Note that the IDs differ from those used in [14].

ID	PDB	CID	ID	PDB	CID	ID	PDB	CID	ID	PDB	CID
1	1b00	A	11	4tmy	B	21	2b3i	A	31	1tri	
2	1dbw	A	12	1rn1	A	22	2pcy		32	3ypi	A
3	1nat		13	1rn1	B	23	2plt		33	8tim	A
4	1ntr		14	1rn1	C	24	1amk		34	1ydv	A
5	1qmp	A	15	1baw	A	25	1aw2	A	35	1b71	A
6	1qmp	B	16	1byo	A	26	1b9b	A	36	1bcf	A
7	1qmp	C	17	1byo	B	27	1btm	A	37	1dps	A
8	1qmp	D	18	1kdi		28	1hti	A	38	1fha	
9	3chy		19	1nin		29	1tmh	A	39	1ier	
10	4tmy	A	20	1pla		30	1tre	A	40	1rcd	

4.3   Skolnick Clustering Test

The aim of the Skolnick clustering test originally suggested by Skolnick and described in [14] is to classify 40 proteins into four families according to their cluster membership. The protein domains are shown in Table 2. In the following, we use their assigned indexes to refer to the respective domains.

Table 3 describes the protein domains and their families. Its fourth column with identifier *Seq. Sim.* indicates that sequence alignment fails for clustering the protein domains according to their family membership. This motivates structural alignment for solving the Skolnick clustering test.

The contact maps of the domains were provided by [18] and differ slightly from the ones compiled by [14]. To cluster the Skolnick data, we used the similarity measure [18]

$$\sigma(X, Y) = \frac{2f_{\text{SADP}}(X, Y)}{|E(X)| + |E(Y)|},$$

where $f_{\text{SADP}}(X, Y)$ denotes the number of shared contacts found by SADP.

Table 3. Protein domains of the Skolnick test set and their categories as taken from [1]. Shown are the characteristics of the four families, the mean number of residues, the range of similarity obtained by sequence alignment and the identifiers of the protein domains.

Family	Style	Residues	Seq. Sim.	Proteins
1	alpha-beta	124	15-30%	1-14
2	beta	99	35-90%	15-23
3	alpha-beta	250	30-90%	24-34
4		170	7-70%	35-40

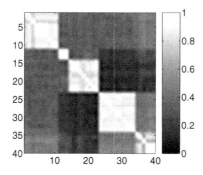

Fig. 3. Pairwise similarity matrix of the Skolnick clustering test. The color bars show the gray-scaled shading for similarities in the range $[0, 1]$. Brighter shadings indicate larger similarities and vice versa. We can identify five clusters.

Table 4. Descriptions of the clusters of proteins from the the Skolnick test. Fold, family, and superfamily are according to SCOP.

Cluster	Domains	Fold	Superfamily	Family
1	1-8, 12-14	Flavodin-like	Che Y-like	Che Y-related
2	9-11	Microbial ribonucleases	Microbial ribonucleases	Fungi ribonucleases
3	15-23	Cuperdoxin-like	Cuperdoxins	Plastocyanim-like Plastoazurin-like
4	24-34	TIM-beta alpha-barrel	Triosephosphate isomerase (TIM)	Triosephosphate isomerase (TIM)
5	35-40	Ferritin-like	Ferritin-like	Ferritin

For 780 pairwise structural alignments, SADP required about 12 minutes. Figure 3 summarizes the results. The results indicate an agreement with the SCOP categories as shown in Table 4. Hence, SADP has sufficient discriminative power to solve the Skolnick clustering test within a relatively short period of time.

5   Conclusion

We proposed a continuous solution to the CMO problem combining softassign with dynamic programming linked together via a problem dependent compatibility function. Experiments show that the proposed method is extremely fast and well suited for similar protein structures, and therefore useful for solving clustering tasks.

Since we have not thoroughly tested alternative compatibility functions, one future research direction aims at constructing compatibility functions that improve the performance of SADP, in particular for dissimilar pairs of contact maps.

References

1. A. Caprara and G. Lancia. Structural alignment of large-size proteins via Lagrangianrelaxation. In *RECOMB,* pages 100–108, 2002.
2. A. Caprara, R. Carr, S. Istrail, G. Lancia, and Brian Walenz. 1001 optimal pdb structure alignments: Integer programming methods for finding the maximum contact map overlap. *Journal of Computational Biology,* 11(1):27-52, 2004.
3. R. Carr, W. Hart, N. Krasnogor, J. Hirst, E. K. Burke, and J. Smith. Alignment of protein structures with a memetic evolutionary algorithm. In *GECCO 02: Proceedings of the Genetic and Evolutionary Computation Conference,* pages 1027-1034, San Francisco, CA, USA, 2002.Morgan Kaufmann Publishers Inc.
4. A. M. Finch, R. C. Wilson, and E. R.Hancock. An energy function and continuous edit process for graph matching. *Neural Computation,* 10(7):1873-1894, 1998.
5. A. Godzik, A. Kolinski, and J. Skolnick. Topology fingerprint approach to the inverse protein folding problem. *Journal of Molecular Biology,* 5(1):227-238, 1992.
6. A. Godzik and J. Skolnick. Flexible algorithm for direct multiple alignment of protein structures and sequences. *Computer Applications in the Biosciences,* 10(6):587-596, 1994.
7. S. Gold and A.Rangarajan. A graduated assignment algorithm for graph matching. *IEEE Transactions on Pattern Analysis and Machine Intelligence,* 18(4):377-388, 1996.
8. D. Goldman. *Algorithmic Aspects of Protein Folding and Protein Structure Similarity.* PhD thesis, University of California, Berkerly, 2000.
9. D. Goldman, S. Istrail, and C. H. Papadimitriou. Algorithmic aspects of protein structure similarity. In *40th Annual Symposium on Foundations of Computer Science,* pages 512-521. IEEE Computer Society, 1999.
10. L. Holm and C. Sander. Protein structure comparison by alignment of distance matrices. *Journal of Molecular Biology,* 233:123-138, 1993.
11. S. Ishii and M. Sato. Doubly constrained network for combinatorial optimization. *Neurocomputing,* 43:239-257, 2002.
12. W. Kabash. A solution for the best rotation to relate two sets of vectors. *Acta Crystallographica,* A32(5):922-923, 1978.
13. N. Krasnogor. Self generating metaheuristics in bioinformatics: The proteins structure comparison case. *Genetic Programming and Evolvable Machines,* 5(2):181-201, 2004.
14. G. Lancia, R. Carr, B. Walenz, and S. Istrail. 101 optimal pdb structure alignments: a branch-and-cut algorithm for the maximum contact map overlap problem. In *RECOMB 01: Proceedings of the fifth annual international conference on Computational biology,* pages 193-202, NewYork, NY, USA, 2001. ACM Press.
15. M.A. Lozano and F. Escolano. A significant improvement of softassign with diffusion kernels. In A. Fred, Terry Caelli, R. P. W. Duin, A. Campilho, and D. de Ridder, editors, *Structural, Syntactic, and Statistical Pattern Recognition: Joint IAPR International Workshops, SSPR 2004 and SPR 2004,* volume 3138 of *Lecture Notes in Computer Science,* pages 76-84, 2004.
16. R. Sinkhorn. A relationship between arbitrary positive matrices and doubly stochastic matrices. *Annals of Mathematical Statistics,* 35(2):876-879, 1964.

17. D. M. Strickl and, E. Barnes, and J. S. Sokol. Optimal protein structure alignment using maximum cliques. *Operations Research,* 53(3):389-402, 2005.
18. W. Xie and N. V. Sahinidis. A branch-and-reduce algorithm for the contact map overlap problem. In A. Apostolico, C. Guerra, S. Istrail, P. Pevzner, and M. Waterman, *editors, RECOMB,* volume 3909 of *Lecture Notes in Computer Science,* pages 516-529. Springer Berlin / Heidelberg, 2006.

An Evaluation of Text Retrieval Methods for Similarity Search of Multi-dimensional NMR-Spectra

Alexander Hinneburg[1], Andrea Porzel[2], and Karina Wolfram[2]

[1] Institute of Computer Science
Martin-Luther-University of Halle-Wittenberg, Germany
`hinneburg@informatik.uni-halle.de`
[2] Leibniz Institute of Plant Biochemistry (IPB), Germany
{`aporzel,kwolfram`}`@ipb-halle.de`

Abstract. Searching and mining nuclear magnetic resonance (NMR)-spectra of naturally occurring substances is an important task to investigate new potentially useful chemical compounds. Multi-dimensional NMR-spectra are relational objects like documents, but consists of continuous multi-dimensional points called peaks instead of words. We develop several mappings from continuous NMR-spectra to discrete text-like data. With the help of those mappings any text retrieval method can be applied. We evaluate the performance of two retrieval methods, namely the standard vector space model and probabilistic latent semantic indexing (PLSI). PLSI learns hidden topics in the data, which is in case of 2D-NMR data interesting in its owns rights. Additionally, we develop and evaluate a simple direct similarity function, which can detect duplicates of NMR-spectra. Our experiments show that the vector space model as well as PLSI, which are both designed for text data created by humans, can effectively handle the mapped NMR-data originating from natural products. Additionally, PLSI is able to find meaningful "topics" in the NMR-data.

1 Introduction

Nuclear magnetic resonance (NMR)-spectra are an important fingerprinting method to investigate the chemical structure of organic compounds from plants or other tissues. Two-dimensional-NMR spectroscopy is able to capture the influences of two different atom types at the same time (e.g. ^{1}H, hydrogen and ^{13}C carbon). The result of an 2D-NMR experiment can be seen as an intensity function measured over two variables[1]. Regions of high intensity are called peaks, which contain the real information about the underlying molecular structure. The usual visualizations of 2D-NMR spectra are contour plots as shown in figure 1. An ideal peak would register as a small dot, however, due to the limited resolution available (dependent on the strength of the magnetic field)

---

[1] The measurements are in parts per million (ppm).

S. Hochreiter and R. Wagner (Eds.): BIRD 2007, LNBI 4414, pp. 424–438, 2007.

multiple peaks may appear as a single merged object with non-convex shape. In the literature peaks are noted by their two-dimensional positions without any information about the shapes of the peaks. Content-based similarity search of 2D-NMR spectra would be a valuable tool for structure investigation by comparing spectra of unknown compounds with a set of spectra, for which the structures are known. While the principle is already in use for 1D-NMR spectra [7,1,11,6,2], to the best of our knowledge, no effective similarity search method is known for 2D-NMR-spectra.

Simplified, a 2D-NMR spectrum is a set of two-dimensional points. There is an analogy to text retrieval, where documents are usually represented as sets of words. Latent space models [5,9,3] were successfully used to model documents and thus improved the quality of text retrieval. Recently, a diversity of text mining approaches for different problems [4,12,8] have been proposed, which make use of probabilistic latent space models. The goal of this work is to show by example how to apply text retrieval and mining methods to biological data originating from experiments.

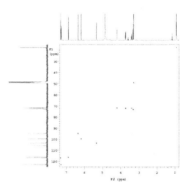

Fig. 1. 2D-NMR spectrum of quercetrin. The plots at the axes are the corresponding 1D-NMR spectra.

The contribution of this paper are methods to map 2D-NMR spectra to discrete text-like data, which can be analyzed and searched by any text retrieval method. We evaluate on real data the performance of two text retrieval methods, namely the standard vector space model [10] and PLSI [5] in combination with our mapping methods for 2D-NMR spectra. Additionally, we propose a simple similarity function, which operates directly on the peaks of the spectra and serves as bottom line benchmark in the experimental evaluation. Our results indicate at a larger scope that text retrieval and mining methods, designed for text data created by humans, in combination with appropriate mapping functions may yield the potential to be also successful for experimental data from naturally occurring objects. In this paper we consider exemplarily ^{1}H, ^{13}C one-bond heteronuclear shift correlation 2D-NMR spectra.

The paper is structured as follows: in section 2, we propose a simple similarity function as bottom line benchmark and define fuzzy duplicates. In section 3, we propose the mapping functions for 2D-NMR spectra while in section 4, we introduce briefly the used text retrieval methods. In section 5, we describe our experimental evaluation and section 6 concludes the paper.

2   Directly Computing Similarity

In this section, we introduce a method to directly compute similarity between pairs of spectra. This method will be used in the experiments as a bottom line benchmark. We also propose on the basis of direct similarity a definition of fuzzy duplicates.

First we give a formal definition of 2D-NMR spectra. A two-dimensional NMR-spectrum of an organic compound captures many structural character-istics like rings and chains. Most important are the positions of the peaks. As the shape of a peak and its height (intensity) strongly varies over different ex-periments with the same compound, the representation of a spectrum includes the peak positions only.

Definition 1. *A **2D NMR-spectrum** A is defined as a set of points* $\{x_1, \ldots, x_n\} \subset \mathbb{R}^2$. *The* $|\cdot|$ *function denotes the size of the spectrum* $|A| = n$.

The size of a spectrum is typically between 4 and 30 for small molecules found in naturally occurring products.

As a peak in a spectrum has two numeric attributes, which can vary con-tinuously, we formalize the notion of matching peaks. A simple but effective approach is to require that a peak matches other peaks only within a certain spatial neighborhood. The neighborhood is defined by the ranges α and β.

Definition 2. *A peak x from spectrum A **matches** a peak y from spectrum B, iff* $|x.c - y.c| < \alpha$ *and* $|x.h - y.h| < \beta$, *where .c and .h denote the NMR measurements for carbon and hydrogen respectively.*

Note that a single peak of a spectrum can match several peaks from another spectrum. Given two spectra A and B, the subset of peaks from A which find matching partners in B is denoted as $matches(A, B) = \{x \colon x \in A, \exists y \in B \colon x \text{ matches } y\}$. The function $matches$ is not symmetric, but helps to define a symmetric similarity measure.

Definition 3. *Let be A and B two given spectra and* $A' = matches(A, B)$ *and* $B' = matches(B, A)$, *so the **similarity** is defined as*

$$sim(A, B) = \frac{|A'| + |B'|}{|A| + |B|}$$

The measure is close to one if most peaks of both spectra are matching peaks. Otherwise the similarity drops towards zero.

An important application of similarity search is the detection of duplicates to increase the data quality of a collection of 2D-NMR-spectra. Clearly a naive definition of duplicates does not work, like two duplicate spectra A and B need to have the same size and the peaks at the same positions. The reason is that the spectra are measured experimentally and so the peak positions differ even if the same probe is analyzed twice. So flexibility should be allowed for the peak positions. Another problem appears when two spectra of the same substance are measured with different resolutions. In case a spectrum is measured with low resolution it may happen that neighboring peaks are merged to a single one. A restriction to an one-to-one relationship between matching peaks can not handle such cases.

We propose a definition of fuzzy duplicates based on the direct similarity measure, which can deal with both of the mentioned problems.

Definition 4. *A pair of 2D-NMR-spectra A and B are* **fuzzy duplicates**, *iff* $sim(A, B) = 1$.

By that definition it is only required that every peak of a spectrum finds at least one matching peak in the other spectrum.

The matching criterion checks only local properties of the spectra. Therefore the direct similarity function can not account for typical chemical substructures described by typical constellations of multiple peaks. For that it would be necessary to check if several peaks are present at the same time. To capture those more abstract properties more sophisticated methods are needed as shown in the next sections.

3   Mapping of NMR Spectra

In this section we propose different methods to map the peaks of an NMR-spectrum from the continuous space of measurements to a discrete space of words. With the help of such a mapping, methods for text retrieval like PLSI can be directly applied. However, the quality of the similarity search depend on how the peaks are mapped to discrete words. A preliminary study of the proposed mappings appeared as poster in [13].

Like a 2D-NMR spectrum consists of a set of peaks, a document consists of many words, which typically are modeled as a set. So assuming a 2D-NMR spectrum can be transformed into a text-like object by mapping the continuous 2D peaks to discrete variables, a variety of text retrieval models can be applied. However, it is an open question, whether models designed for quite different data, namely texts created by humans, are effective on data which comes for naturally occurring compounds and thus do not include human design patterns. For 2D-NMR spectra similarity search it is not clear, what is the best way to map the peaks of a spectrum to discrete words. We develop methods for this task in the next section. That will enable us to tackle the question, whether methods like the vector space model or PLSI, which is designed for text data, remains effective for experimental data from natural products.

3.1   Grid-Based Mapping

We introduce a simple grid-based method, on which we will build more sophisticated methods. A simple grid-based method is to partition each of the both axis of the two-dimensional peak space into intervals of same size. Thus, an equidistant grid is induced in the two-dimensional peak space and a peak is mapped to exactly one grid cell it belongs to. When a grid cell is identified by a discrete integer vector consisting of the cells coordinates the mapping of a peak $x \in \mathbb{R}^2$ is formalized as

$$g(x) = (g_c(x.c), g_h(x.h)) \text{ with } g_c(x.c) = \left\lfloor \frac{x.c}{w_c} \right\rfloor, \ g_h(x.h) = \left\lfloor \frac{x.h}{w_h} \right\rfloor$$

The quantities w_c and w_h are the extensions of a cell in the respective dimensions, which are parameters of the mapping. The grid is centered at the origin of the peak space. The cells of the grid act as words. The vocabulary generated by the mapped peaks consists of those grid cells which contain at least one peak. Empty grid cells are not included in the vocabulary. A word consists of a two-dimensional discrete integer vector.

Unfortunately the grid-based mapping has two disadvantages. First, close peaks may be mapped to different grid cells. This may lead to poor matching of related peaks in the discrete word space. Second, peaks of new query spectra are ignored when they are mapped to grid cells not included in the vocabulary. So some information from the query is not used for the similarity search which may weaken the performance.

3.2   Redundant Mappings

We propose three mappings which introduce certain redundancies by mapping a single peak to a set of grid cells. The redundancy in the new mappings shall compensate for the drawbacks of the simple grid-based mapping.

Shifted Grids. The first disadvantage of the simple grid-based method is that peaks which are very close in the peak space may be mapped to different grid cells, because a cell border is between them. So proximity of peaks does not guaranty that they are mapped to the same discrete cell.

Instead of mapping a peak to a single grid cell, we propose to map it to a set of overlapping grid cells. This is achieved by several shifted grids of the same granularity. In addition to the base grid some grids are shifted into the three directions $(1,0)(0,1)(1,1)$. An illustration of the idea is sketched in figure 2. In figure 2, one grid is shifted in each of the directions by half of the extent of a cell. In general, there may be $k-1$ grids shifted by fractions of $1/k, 2/k, \ldots, k-1/k$ of the extent of a cell in each direction respectively. For the mapping of the peaks to words which consist of cells from the different grids, two additional dimensions are needed to distinguish (a) the $k-1$ grids in each direction and (b) the directions themselves. The third coordinate represents the fraction by which a

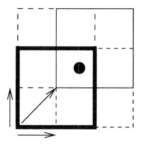

Fig. 2. The four grids are marked as follows: base grid is bold, $(1,0)$, $(0,1)$ are dashed and $(1,1)$ is normal

cell is shifted and the fourth one represents the directions by the following coding: value 0 is (0,0), 1 is (1,0), 2 is (0,1) and 3 is (1,1). So each peak is mapped to a finite set of four-dimensional integer vectors. The mapping of a peak $x \in \mathbb{R}^2$ is

$$s(x) = \{(g_c(x.c), g_h(x.h), 0, 0)\} \cup$$
$$\bigcup_{i=1}^{k-1} \Big\{ (g_c(x.c + {}^i\!/k \cdot w_c), g_h(x.h), i, 1),$$
$$(g_c(x.c), g_h(x.h + {}^i\!/k \cdot w_h), i, 2),$$
$$(g_c(x.c + {}^i\!/k \cdot w_c), g_h(x.h + {}^i\!/k \cdot w_h), i, 3) \Big\}$$

Thus, a single peak is mapped to $3(k-1)+1$ words. A nice property of the mapping is that there exists at least one grid cell for every pair of matching peaks both peaks are mapped to.

Different Resolutions. The second disadvantage of the simple grid-based mapping comes from the fact that empty grid cells (not occupied by at least one peak from the set of training spectra) do not contribute to the representation to be learned for similarity search. So peaks of new query spectra mapped to those empty cells are ignored. That effect can be diminished by making the grid cells larger. However, this is counterproductive for the precision of the similarity search due to the coarser resolution. Thus, there are two contradicting goals, namely (a) to have a fine resolution to handle subtle aspects in the data and (b) to cover at the same time the whole peak space by a coarse resolution grid so that no peaks of a new query spectrum have to be ignored.

Instead of finding a tradeoff for a single grid, both goals can be served by combining simple grids with different resolutions. Given l different resolutions $\{(w_c^{(1)}, w_h^{(1)}), \ldots, (w_c^{(l)}, w_h^{(l)})\}$ a peak is mapped to l grid cells of different sizes. In order to distinguish between the different grids an additional discrete dimension is needed. So the mapping function is

$$r(x) = \bigcup_{i=1}^{l} \{(g_c^{(i)}(x), g_h^{(i)}(x), i)\}$$

with $g_c^{(i)}$ and $g_h^{(i)}$ use $w_c^{(i)}$ and $w_h^{(i)}$ respectively. Note that a hierarchical, quad-tree like partitioning is a special case of the proposed mapping function with $w_c^{(i)} = 2^{i-1}w_c$ and $w_h^{(i)} = 2^{i-1}w_h$.

Combining shifted Grids with different Resolutions. Both methods are designed to compensate for different drawbacks of the simple grid mapping. So it is natural to combine both mappings. The parameters of such a mapping are the number of shifts k, the number of different grid cell sizes l and the actual sizes $\{(w_c^{(1)}, w_h^{(1)}), \dots, (w_c^{(l)}, w_h^{(l)})\}$. Beside the two coordinates for the grid cells, additional discrete dimensions are needed for the shift, the direction and the grid resolution. Using the the definitions from above the mapping function of the combined mapping of a peak is

$$c(x) = \bigcup_{i=1}^{l} \{(g_c^{(i)}(x.c), g_h^{(i)}(x.h), 0, 0, i)\} \cup$$

$$\bigcup_{j=1}^{k-1} \Big\{ (g_c^{(i)}(x.c + j/k \cdot w_c^{(i)}), g_h^{(i)}(x.h), j, 1, i),$$

$$(g_c^{(i)}(x.c), g_h^{(i)}(x.h + j/k \cdot w_h^{(i)}), j, 2, i),$$

$$(g_c^{(i)}(x.c + j/k \cdot w_c^{(i)}), g_h^{(i)}(x.h + j/k \cdot w_h^{(i)}), j, 3, i) \Big\}$$

Thus a single peak is mapped to $l(3(k-1)+1)$ words. In the next section, we give an overview of the text retrieval methods used.

4   Models for Text Retrieval

We briefly introduce the essentials of the vector space model and PLSI to make the paper self contained. In the vector space model, documents are represented by vectors which have as many dimensions as there are words in the used vocabulary of the document collection. Each component of a document vector reflects the importance of the corresponding word for the document. The typical quantity used is the raw term frequency (tf) of that word for the document, say the number of occurrences of that word in a document d. In order to improve the retrieval quality, those vectors are reweighted by multipling tf with the inverse document frequency (ifd) of a word. The inverse document frequency measure is large, if a word is included in only a small percentage of the documents in the collection. Formally, we denote the set of documents by $D = \{d_1, \dots, d_J\}$ and the vocabulary by $W = \{w_1, \dots, w_I\}$. The term frequency of a word $w \in W$ in a document $d \in D$ is denoted as $n(d, w)$ and the reweighed quantity is $\hat{n}(d, w) = n(d, w) \cdot idf(w)$. The similarity between a query document q and a document d from the collection is

$$S(d, q) = \frac{\sum_{w \in W} \hat{n}(d, w) \cdot \hat{n}(q, w)}{\sqrt{\sum_{w \in W} \hat{n}(d, w)^2} \cdot \sqrt{\sum_{w \in W} \hat{n}(q, w)^2}}$$

This can be interpreted as the cosine of the angles between the two vectors.

Probabilistic latent semantic indexing (PLSI) introduced in [5] extends the vector space model by learning topics hidden in the data. The training data consists of a set of document-word pairs $(d^{(i)}, w^{(i)})_{i=1,...,N}$ with $w^{(i)} \in W$ and $d^{(i)} \in D$. The joint probability of such a pair is modeled according to the employed aspect model as $P(d, w) = \sum_{z \in Z} P(z) \cdot P(w|z) \cdot P(d|z)$. The z are hidden variables, which can take K different discrete values $z \in Z = \{z_1, ..., z_K\}$. In the context of text retrieval z is interpreted as an indicator for a topic. Two assumptions are made by the aspect model. First, it assumes pairs (d, w) to be statistically independent. Second, conditional independence between w and d is assumed for a given value for z.

The probabilities necessary for the joint probability $P(d, w)$, namely $P(z)$, $P(w|z)$ and $P(d|z)$, are derived by an expectation maximization (EM) learning procedure. The idea is to find values for unknown probabilities, which maximize the complete data likelihood

$$P(S, z) = \prod_{(d^{(i)}, w^{(i)}) \in S} [P(z) \cdot P(w^{(i)}|z) \cdot P(d^{(i)}|z)]$$

$$= \prod_{d \in D} \prod_{w \in W} [P(z) \cdot P(w|z) \cdot P(d|z)]^{n(d,w)}$$

with $S = \{(d^{(i)}, w^{i)})_{i=1,...,N}\}$ is the set of all document-word pairs in the training set. In the E-step the posteriors for z are computed.

$$P(z|d, w) = \frac{P(z) \cdot P(w|z) \cdot P(d|z)}{\sum_{z' \in Z} P(z') \cdot P(w|z') \cdot P(d|z')}$$

The subsequent M-step maximizes the expectation of the complete data likelihood respectively to the model parameters, namely $P(z)$, $P(w|z)$ and $P(d|z)$.

$$P(d|z) = \frac{\sum_{w \in W} P(z|d, w) \cdot n(d, w)}{\sum_{w \in W} \sum_{d' \in D} P(z|d', w) \cdot n(d', w)}$$

$$P(w|z) = \frac{\sum_{d \in D} P(z|d, w) \cdot n(d, w)}{\sum_{w' \in W} \sum_{d \in D} P(z|d, w') \cdot n(d, w')}$$

$$P(z) = \frac{\sum_{w \in W} \sum_{d \in D} P(z|d, w) \cdot n(d, w)}{\sum_{w \in W} \sum_{d \in D} n(d, w)}$$

The EM algorithm starts with random values for the model parameters and converges by alternating E- and M-step to a local maximum of the likelihood.

There are serval ways possible to answer similarity queries using the trained aspect model. Because of its simplicity, we adopt the PLSI-U variant from [5]. The idea is to extend the cosine similarity measure from the tf-idf vector space model. The extension by Hofmann treats the learned multinomials $P(w|d)$ as term frequencies (tf). Note that $P(w|d) = P(d, w)/P(d)$ with $P(d) = \sum_{w' in W} n(d, w')/N$. The multinomials $P(w|d)$ are smoothed variants of the original term frequencies $\tilde{P}(w|d) = n(d,w)/(\sum_{w'\in W} n(d,w'))$. The proposed tf-weights are linear combinations of the multinomials $P(w|d)$ and $\tilde{P}(w|d)$. Thus, the new tf-idf weights used for the documents within the similarity calculation are

$$\hat{n}(d,w) = (\lambda \cdot P(w|d) + (1-\lambda) \cdot \tilde{P}(w|d)) \cdot idf(w)$$

with $\lambda \in [0,1]$. Hofmann suggests in [5] to set $\lambda = 0.5$. The tf-idf weights for the query are determined as in the standard vector space model. The smoothed tf-weight for a word which actually does not appear in a document may be still non-zero if the word belongs to a topic which is covered by the particular document. In that way a more abstract similarity search becomes possible.

5   Evaluation and Results

In this section we present the results for duplicate detection, a comparison of the effectiveness of the mappings for similarity search, and mining aspects of 2D-NMR-data.

5.1   2D-NMR-Data

The substances included in the database are mostly secondary metabolites of plants and fungi. They cover a representative area of naturally occurring compounds and originate either from experiments or from simulations[2] based on the known structure of the compound. The database includes about 587 spectra, each has about 3 to 35 peaks. The total number of peaks is 7029. Ten small groups of chemically similar compounds are included in the database for controlled experiments. The groups with the number of spectra and number of peaks are listed in table 1 left. The peak space with all peaks in the database is shown in figure 3 right. Two groups, steroids and flavonoids, are selected as examples and shown with their peak distribution within figure 3 right.

Natural steroids occur in animals, plants and fungi. They are vitamins, hormones or cardioactive poisons like digitalis or oleander. The steroids in the database ar mostly hormones like androgens and estrogens. Flavonoids are aromatic substances (rings). Some flavonoids decrease vascular permeability or possess antioxidant activity which can have an anticarcinogenic effect.

5.2   Detection of Duplicates

We used the direct similarity function introduced in section 2 to detect duplicates in the database. With a setting of $a = 3ppm$ and $b = 0.3ppm$, which are reasonable tolerances, 54 of 171991 possible pairs are reported as fuzzy duplicates. An inspection by hand revealed that 30 pairs are just very similar spectra, but 24 are candidates for real duplicates. Many of the found pairs come from the groups shown in table 1. Some pairs consist of an experimental and a simulated

---

[2] ACD/2D NMR predictor, version 7.08, http://www.acdlabs.com/

Table 1. Groups with number of spectra and range of peaks

Group	#Spectra	#Peaks
Pregnans	11	17–26
Anthrquinones	8	3–6
Aconitanes	8	22–26
Triterpenes	17	24–31
Flavonoids	18	5–8
Isoflavonoids	16	5–7
Aflatoxins	8	8–10
Steroids	12	16–23
Cardenolides	15	18–25
Coumarins	19	3–8

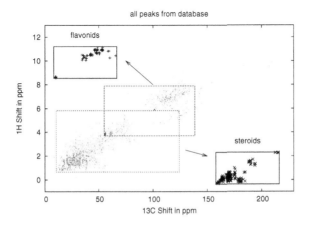

Fig. 3. Distribution of the peaks of all spectra with the distribution within the groups of flavonoids and steroids

spectrum of the same substance which confirms the usefulness of the definition. There was also a surprise, namely the pair Thalictrifoline/Cavidine. Both structures differ only in the stereochemical orientation of one methyl group. Evidently, in this case the commercial software package used for the simulation is not able to reflect the different stereochemistry in calculated spectra. Duplicate search could be also done using tf-idf vectors and cosine similarity. The problem with that approach is to find a general minimum cutoff value for the similarity. Since it is difficult to interpret cosine similarity at the peak level, our simple definition or fuzzy duplicates is preferable for users for biochemistry. In the future, fuzzy duplicates will be used to improve the quality of collections of 2D-NMR spectra.

5.3   Performance Evaluation

The different methods for similarity search of 2D-NMR-spectra are compared using recall-precision curves. The search quality is high, when both – recall and precision – are high. So the upper curves are the best.

First, a series of experiments is conducted using our proposed mapping functions in combination with the vector space model. Each spectrum from the ten groups is used as a query while the rest of the respective group should be found as answers. The plots in figure 4 and 5 show averages over all queries. The results for the simple grid-based mapping are shown in figure 4a. The sizes of the grid cells are varied over $w_c = 4, 6, 8, 10$ and $w_h = 0.4, 0.6, 0.8, 1.0$ respectively. Small sizes give the best results.

The use of shifted grids improves the performance substantially over simple grids, as shown in figure 4b,c. The plots show the experiments for $k = 2, 3$. The results for $k = 2$ and $k = 3$ are almost identical. However, the vocabulary for $k = 2$ is much smaller. In practise, the smaller model with $k = 2$ shifts is favored.

Also the mapping based on grids with different grid cell sizes are assessed. Due to lack of space, only the results from combinations of $w_c^{(1)} = 4, w_h^{(1)} = 0.4$ with other sizes are reported, because those performed best among all combinations. Figure 4d shows that also the mapping based on different grid cell sizes outperforms the simple grid-based mapping. But the improvement is not as much as for shifted grids. The set of resolutions $\{(w_c^{(1)} = 4, w_h^{(1)} = 0.4), (w_c^{(2)} = 10, w_h^{(2)} = 1.0)\}$ performs best.

Also, experiments are performed with the combination of the previous two mappings, namely a combination of shifted grids with those of different resolutions. The performance results are shown in figure 4e which indicates that the best combination, namely the resolution set $\{(w_c^{(1)} = 4, w_h^{(1)} = 0.4), (w_c^{(2)} = 10, w_h^{(2)} = 1.0)\}$ with $k = 2$ shifts, outperforms both previous mappings. This is more clearly seen in figure 4f which compares the best performing settings from the above experiments.

Next, a series of similar experiments is conducted using our proposed mapping functions in combination with PLSI. Random initialization is used for the EM training algorithm described in section 4. All curves are averages from cross validation over all groups. As PLSI is trained on the data beforehand, we used cross validation where the current query is not included in the taining data. As the groups are very small, the leave-one-out cross validation scheme is employed. The results for PLSI are shown in figure 5a-f. PLSI requires to chose the number of hidden aspects. For the experiments reported so far, the PLSI model is used with 20 hidden aspects. Also different numbers of aspects are tested using the best combination of mappings. Figure 5g shows that the performance with 10 aspects drops a bit The increase in the numbers of aspects from 20 to 32 is only marginally reflected in increase of search performance. So 20 is a reasonable number of aspects for the given data.

In summery, the experiments with both text retrieval methods show, that the mappings based on shifted grids and those with different resolutions perform significantly better than the simple grid-based mapping. In both cases, the combination of shifted grids and grids with different resolutions is even better than the individual mappings. The comparison between PLSI and the vector space model (figure 5h) shows that both have similar performance for small recall but for large recall PLSI has a better precision.

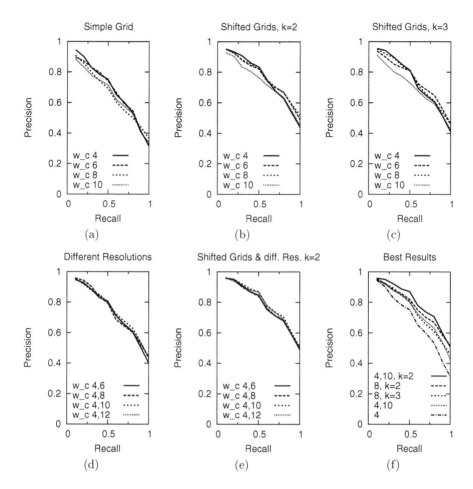

Fig. 4. Average recall-precision curves using the vector space model

Last, the direct similarity function is tested (figure 5i). The size of the matching neighborhood is varied over $\alpha = 4, 6, 8, 10$ and $\beta = 0.4, 0.6, 0.8, 1.0$ respectively. The search quality is quite low. In fact on average, it fails to deliver a spectrum from the answer set in the top ranks which is indicated by the hill-like shape of the curves.

In conclusion, the results prove experimentally that the vector space model as well as the PLSI model, which are designed for text retrieval, are indeed effective for similarity search of 2D-NMR spectra from naturally occurring products.

5.4   Analysis of the Latent Aspects

We analyzed the latent aspects learned by the PLSI model using the mapping based on the combination of shifted grids with different resolutions. The grid cells (words) with high probability for a given aspect are plotted together to

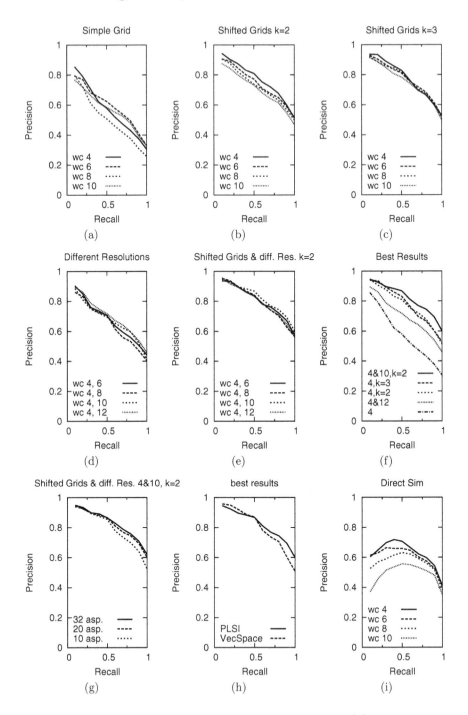

Fig. 5. Average recall-precision curves from leave-one-out cross validation experiments with the PLSI model (a-g), best results of PLSI and vector space model (h) and results for the direct similarity (i)

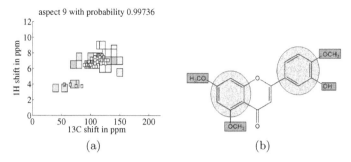

(a)   (b)

Fig. 6. (a) Main aspect of the flavonoid group which includes the region of aromatic rings (upper right cluster) and the region for oxygen substituents (lower left cluster). The gray shades indicate the strength of the association between grid cell and aspect. (b) An example of a flavonoid (3'-Hydroxy-5,7,4'-trimethoxyflavone) where the aromatic rings and the oxygen substituents (methoxy groups in this case) are marked.

describe the aspects meaning. Some aspects specialized on certain regions in the peak space which are typical for distinct molecule fragments like aromatic rings or alkane skeletons. However, also more subtle details of the data are captured by the aspect model. For example, the main aspect for the group of flavonoids specializes not only on the region for aromatic rings which are the main part of flavonoids. It also includes a smaller region which indicates oxygen substitution. A closer inspection of the database revealed that indeed many of the included flavonoids do have several oxygen substitutes. The main aspect for flavonoids with the respective peak distribution of the flavonoid group is shown in figure 6a. We believe a detailed analysis of the aspects found by the model may help to investigate unknown structures of new substances when their NMR-spectra are included in the training set.

6   Conclusion

We proposed redundant mappings from continuous 2D-NMR spectra to discrete text-like data which can be processed by any text retrieval method. We demonstrated experimentally the effectiveness of the our mappings in combination with the vector space model and PLSI. Further analysis revealed that the aspects found by PLSI are chemically relevant. In future research we will study more recent text models like LDA [3] in combination with our mapping methods.

References

1. A. Tsipouras, J. Ondeyka, C. Dufresne et al. Using similarity searches over databases of estimated c-13 nmr spectra for structure identification of natural products. *Analytica Chimica Acta*, 316:161–171, 1995.
2. A. S. Barros and D. N. Rutledge. Segmented principal component transform-principal component analysis. *Chemometrics & Intelligent Laboratory Systems*, 78:125–137, 2005.

3. D. M. Blei, A. Y. Ng, and M. I. Jordan. Latent dirichlet allocation. *Journal of Machine Learning Research*, 3:993–1022, 2003.
4. L. Cai and T. Hofmann. Text categorization by boosting automatically extracted concepts. In *SIGIR '03*, 2003.
5. T. Hofmann. Probabilistic latent semantic indexing. In *SIGIR '99*, 1999.
6. P. Krishnan, N. J. Kruger, and R. G. Ratcliffe. Metabolite fingerprinting and profiling in plants using nmr. *Journal of Experimental Botany*, 56:255–265, 2005.
7. M. Farkas, J. Bendl, D. H. Welti et al. Similarity search for a h-1 nmr spectroscopic data base. *Analytica Chimica Acta*, 206:173–187, 1988.
8. Q. Mei and C. Zhai. Discovering evolutionary theme patterns from text: an exploration of temporal text mining. In *KDD '05*.
9. A. Popescul, L. H. Ungar, D. M. Pennock, and S. Lawrence. Probabilistic models for unified collaborative and content-based recommendation in sparse-data environments. In *UAI '2001*.
10. G. Salton, A. Wong, and C. S. Yang. A vector space model for automatic indexing. *Commun. ACM*, 18(11):613–620, 1975.
11. C. Steinbeck, S. Krause, and S. Kuhn. Nmrshiftdb-constructing a free chemical information system with open-source components. *J. chem. inf. & comp. sci.*, 43:1733 –1739, 2003.
12. M. Steyvers, P. Smyth, M. Rosen-Zvi, and T. Griffiths. Probabilistic author-topic models for information discovery. In *KDD '04*.
13. K. Wolfram, A. Porzel, and A. Hinneburg. Similarity search for multi-dimensional nmr-spectra of natural products. In *PKDD '06*, 2006.

Ontology-Based MEDLINE Document Classification

Fabrice Camous[1], Stephen Blott[1], and Alan F. Smeaton[2]

[1] School of Computing
fcamous@computing.dcu.ie,
sblott@computing.dcu.ie
[2] Centre for Digital Video Processing & Adaptive Information Cluster,
Dublin City University,
Glasnevin, Dublin 9, Ireland
asmeaton@computing.dcu.ie

Abstract. An increasing and overwhelming amount of biomedical information is available in the research literature mainly in the form of free-text. Biologists need tools that automate their information search and deal with the high volume and ambiguity of free-text. Ontologies can help automatic information processing by providing standard concepts and information about the relationships between concepts. The Medical Subject Headings (MeSH) ontology is already available and used by MEDLINE indexers to annotate the conceptual content of biomedical articles. This paper presents a domain-independent method that uses the MeSH ontology inter-concept relationships to extend the existing MeSH-based representation of MEDLINE documents. The extension method is evaluated within a document triage task organized by the Genomics track of the 2005 Text REtrieval Conference (TREC). Our method for extending the representation of documents leads to an improvement of 17% over a non-extended baseline in terms of normalized utility, the metric defined for the task. The SVM^{light} software is used to classify documents.

1   Introduction

In the current era of fast sequencing of entire genomes, genomic information is becoming extremely difficult to search and process. Biologists spend a considerable part of their time searching the research literature. An important task for genomic database curators is to locate experimental evidence in the literature to annotate gene records.

The increasing and overwhelming volume to search through, coupled with the ambiguity [23] of unstructured information found in free-text make manual information processing prohibitively expensive. Biomedical ontologies, when available, can help disambiguate the information expressed with natural languages. They offer standard terms that only relate to specific concepts and therefore reduce the impact of synonyms and polysems. They also contain information about the relationships between concepts and this information can be used to express semantic similarities between concepts.

S. Hochreiter and R. Wagner (Eds.): BIRD 2007, LNBI 4414, pp. 439–452, 2007.

In the biomedical literature, MeSH[1] is used to annotate the conceptual content of the MEDLINE database records. The MeSH ontology is organized into a semantic network that includes several hierarchies[2]. A semantic network is broadly a set of nodes (representing the concepts) and a set of edges, or links (representing the relationships between the concepts).

We hypothesize that the semantic information contained in the MeSH hierarchies can benefit the representation of documents. Our approach extends initial MeSH-based document representations with additional concepts that are semantically close to the initial representation. The semantic similarity between concepts is derived from the distance that separates them in the MeSH hierarchies. The document representation extension method described in this paper is domain-independent and can be applied to other ontologies.

Our approach is evaluated in the context of a binary classification or triage of documents. The triage corresponds to a stage in the information retrieval process where the possibly relevant documents are selected from the mass of non-relevant documents before being thoroughly examined later on. In particular, our method is assessed with a document triage task organized by the Genomics track of the 2005 Text REtrieval Conference (TREC)[3]. The SVM^{light} software [13] is used to classify documents.

The paper is organized as follows. Section 2 describes our ontology-based document representation extension method. Section 3 presents the evaluation framework of our approach, including related work and the results we have obtained. Finally, Section 4 concludes with future research directions.

2   Methodology

This section describes our ontology-based document representation extension. The extension method can be applied to any ontology-based document representation. In this paper, however, we focus on MEDLINE records and the MeSH fields they contain. Some background information about the MEDLINE database and the MeSH ontology is given in Section 2.1. Our method includes comparing concepts semantically with the MeSH hierarchies. Therefore, some background about network-based semantic measures is also available in Section 2.2.

2.1   MEDLINE and MeSH

MEDLINE, the U.S. National Library of Medicine (NLM)[4] bio-medical abstract repository, contains approximately 14 million reference articles from around 4,800 journals. Approximately 400,000 new records are added to it each year. Despite the growing availability of full-text articles on the Web, MEDLINE remains in practice a central point of access to bio-medical research [9,10].

---

[1] http://www.nlm.nih.gov/mesh/meshhome.html
[2] The term *hierarchy* is commonly used to refer to the main parts of the MeSH ontology, which are in fact single-rooted directed acyclic graphs.
[3] http://trec.nist.gov/
[4] http://www.nlm.nih.gov/

The MEDLINE record fields include text-based fields, the title and abstract fields, and ontology-based fields, the MeSH fields. Most MEDLINE records contain 10-12 MeSH fields. MeSH is a biomedical controlled vocabulary produced by the U.S. NLM and used since 1960. MeSH 2004 includes 22,430 descriptors, 83 qualifiers, and 141,455 supplementary concepts. Descriptors are the main elements of the vocabulary. Qualifiers are assigned to descriptors inside the MeSH fields to express a special aspect of the concept. For example, the descriptor *Alcoholism* can be associated with the qualifier *genetics* or the qualifier *psychology*. Each qualifier provides a different context for the concept *Alcoholism*. Descriptors and qualifiers are both organized in several hierarchies. However, the qualifier hierarchies are much smaller and shallower than the descriptor hierarchies. Figure 1 shows a simplified representation of two MeSH descriptor hierarchies, *Diseases* and *Chemical and Drugs*, combined in a unique hierarchy with the addition of an artificial *mesh* root node. Note that only the top concept of the *Chemical and Drugs* hierarchy is represented.

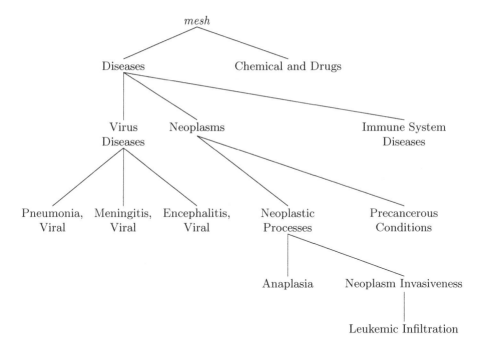

Fig. 1. A partial representation of two descriptor hierarchies combined in one hierarchy

The relationships between descriptor or qualifier nodes in the MeSH hierarchies are of the *broader-than/narrower-than* type [2]. The *narrower-than* relationship is close to the hypernymy (*is-a*) relationship, but it can also include a meronymy (*part-of*) relationship. Inversely, the *broader-than* relationship is close to the hyponymy (*has-instance*) relationship, and can also include a holonymy (*has-a*) relationship. In MeSH, one concept is narrower than another if the documents it retrieves in a search are contained in the set of documents retrieved by the broader concept.

An example of MEDLINE record, describing a full-text article, is shown in Figure 2. It includes textual fields, such as title and abstract, as well as MeSH fields. Beside the textual fields, MeSH fields are a useful source of structured and standardized information. Unlike the free-text content of the title/abstract fields, the MeSH fields unambiguously associate a single term to a single concept. In addition, MeSH concepts are assigned to the records after the examination of the entire research article by human indexers. Consequently, they can complement or add to the information contained in the title and abstract. Finally, previous work by [7] showed that indexing consistency was high amongst NLM human indexers: the authors reported a mean agreement in index concepts between 33%-74% for different experts.

A MEDLINE MeSH field is a combination of a MeSH descriptor with zero or more MeSH qualifiers. In Figure 2, *Centrioles/*ultrastructure* is the combination of descriptor *Centrioles* with qualifier *ultrastructure*. MeSH fields can describe major themes of the article (a concept that is central to the article), or minor themes (a peripheral concept). A star is used to distinguish the major themes from the minor ones. Therefore, the association *Centrioles/*ultrastructure* is a major theme of the MEDLINE record of Figure 2, along with *Organelles/*ultra structure* and *Steroids/*analysis*. In contrast, *Cilia/ultrastructure* and *Respiratory Mucosa/cytology* are minor themes in this record.

2.2   Network-Based Semantic Measures

Network-based measures are usually classified into two groups: edge-based measures and information-based measures [2]. Edge-based measures only use hierarchy information whereas information-based measures integrate concept frequency information from a corpus.

Network-based measures can also be classified into the two following categories:

1. measures that consider all the hierarchy relationships to correspond to the same semantic distance (simple edge weighting), and
2. measures based on the assumption that hierarchy relationships are associated with various semantic distances (complex edge weighting).

In this paper, we use a measure from the first category, the edge count method, to compare two concepts semantically. The edge count method considers that all edges between concepts in the hierarchy correspond to the same semantic distance. For example, in Figure 1, the edge connecting *Diseases* to *Neoplasms* is assumed to represent the same semantic distance as the edge connecting *Neoplasm Invasiveness* to *Leukemic Infiltration*.

The assumption behind the edge count approach is that the semantic distance between two concepts is proportional to the number of edges that separate them in the hierarchy. For example, the edge count between *Pneumonia, Viral* and *Meningitis, Viral* in Figure 1 is 2, and the edge count between *Pneumonia, Viral*

```
PMID - 10605436
TI     - Concerning the localization of steroids in centrioles
         and basal bodies by immunofluorescence.
AB     - Specific steroid antibodies, by the
         immunofluorescence technique,
         regularly reveal fluorescent centrioles
         and cilia-bearing basal bodies in ...
AU     - Nenci I
AU     - Marchetti E
MH     - Animals
MH     - Centrioles/*ultrastructure
MH     - Cilia/ultrastructure
MH     - Female
MH     - Fluorescent Antibody Technique
MH     - Human
MH     - Lymphocytes/*cytology
MH     - Male
MH     - Organelles/*ultrastructure
MH     - Rats
MH     - Rats, Sprague-Dawley
MH     - Respiratory Mucosa/cytology
MH     - Steroids/*analysis
MH     - Trachea
```

Fig. 2. A MEDLINE record example (PMID: PubMed ID, TI: title, AB: abstract, AU: author, MH: MeSH field)

and *Neoplastic Processes* is 4. Therefore, according to edge count, *Meningitis, Viral* is closer semantically to *Pneumonia, Viral* than *Neoplastic Processes* is.

In some hierarchies, there is more than one possible path between two concepts. In the MeSH ontology, concepts can have several parents and children (1.8 children per descriptor on average). A widely-used solution is to consider only the shortest path between two concepts when calculating the semantic distance. [16] uses this approach with an inter-concept measure defined as the shortest path p in the set of possible paths P between two concepts c_1 and c_2 in terms of edge count:

$$dist_{rada1}(c_1, c_2) = \min_{p \in P} (edge\_count(c_1, c_2)) \qquad (1)$$

where $edge\_count$ is the number of edges separating c_1 from c_2 on a path $p \in P$.

Concepts are often clustered together to describe a document or a query. In order to compare two sets of concepts, the inter-concept semantic comparisons need to be combined. This combination results in an inter-document measure. A simple approach is to derive the semantic similarity between two sets of concepts from the combination of all the possible inter-concept comparisons from the two sets. [16] defines a distance measure between two sets of concepts A and B containing m and n concepts respectively as the average of all the $m \times n$ inter-concept distance measures:

$$dist_{rada2}(A, B) = \frac{1}{m \times n} \times \sum_{c_i \in A, c_j \in B} dist_{rada1}(c_i, c_j) \qquad (2)$$

where $dist_{rada1}$ is defined by Equation 1.

2.3   Our Method

MeSH-based document representations are expressed with vectors containing 22,513 elements (22,430 descriptors and 83 qualifiers). Most MEDLINE documents contain 10-12 MeSH fields. Each field contain one descriptor associated optionally to one or more qualifiers. This means that originally the MeSH-based document vectors contain essentially zero values.

Our main idea is that the MeSH hierarchy can be used to add new MeSH concepts to the original document MeSH content. Descriptors and qualifiers have locations in the MeSH hierarchies from which an evaluation of their semantic similarity can be derived. For example, if *Neoplastic Processes* is found in a MeSH field, it can be assumed that the document is also about *Neoplasms* or *Anaplasia*, to some degree (see Figure 1). The challenge is to measure how close semantically the added concepts are to the initial representation, which can contain several concepts.

We use a weighting scheme to distinguish the original MeSH concepts from the ones derived from the extension process. In our experiment, the original MeSH concepts receive a weight w_o of 1, whereas derived MeSH concepts receive a weight w_d, where $0 \leq w_d < 1$, depending on how semantically close they are to the original MeSH representation.

First, we process the MeSH fields of MEDLINE documents in the following way. Descriptors and qualifiers are chosen as the minimal units of information or features. Associations between descriptors and qualifiers found in the MeSH fields (see Section 2.1) are dropped and a qualifier appearing in several associations is considered to appear only once in the document. MeSH field distinctions between major and minor themes (see again Section 2.1) are also ignored. Keeping descriptors and qualifiers as minimal information units allows us to keep track of the concepts they represent and their locations in the MeSH hierarchies. The initial representation obtained is binary. Table 1 shows the binary representation of the MeSH content of the document appearing in Figure 2. Note that *ultrastructure* and *cytology* are only represented once although they appear in three and two MeSH fields, respectively. Moreover, some concepts, such as *Human*, *Female*, and *Male* are not represented at all as they belong to a list of common MeSH concepts, or stopwords, called the Check Tags (CTs). The full list of CTs for MeSH 2004 is given in Table 2. All other concepts not occurring in the document are given a zero weight (not shown in Table 1).

Second, we extend the representations by adding MeSH concepts not in the document initially with a weight w_d, which is derived from the semantic distance

Table 1. Example of binary representation

concept	weight
Animals	1
Centrioles	1
ultrastructure	1
Cilia	1
Fluorescent Antibody Technique	1
Lymphocytes	1
cytology	1
Organelles	1
Rats	1
Rats, Sprague-Dawley	1
Respiratory Mucosa	1
Steroids	1
analysis	1
Trachea	1

Table 2. List of Check Tags (CTs) for MeSH 2004

CTs list
Comparative Study
English Abstract
Female
Human
In Vitro
Male
Support, Non-U.S. Gov't
Support, U.S. Gov't, Non-P.H.S.
Support, U.S. Gov't, P.H.S.

between the added concept and the initial representation. From Equation 2, we obtain a semantic distance between a concept c_i and a document D containing initially m concepts:

$$dist_{ext}(c_i, D) = \frac{1}{m} \sum_{j=1}^{m} dist_{radal}(c_i, c_j)$$

where $dist_{radal}$ is defined by Equation 1. The distance is then turned into a weight w_d with the following formula:

$$w_d(c_i, D) = 1 - (dist_{ext}(c_i, D) / max\_dist)$$

where max_dist is the maximum edge count between two concepts in the hierarchy.

As in Figure 1, an artificial *mesh* root node is placed at the top of all the MeSH hierarchies (see Figure 3) in order to be able to compare all MeSH concepts with

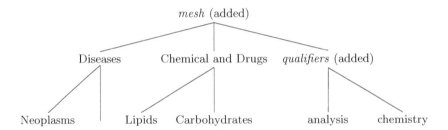

Fig. 3. *mesh* and *qualifiers* nodes add-up (only a few child nodes depicted for clarity)

each other. Additionally, a *qualifier* node is placed between the *mesh* node and all the root nodes of the qualifier hierarchies. This node is created to compensate for the shallowness and the small size of the qualifier hierarchies. It also implicitly increases the distinction between descriptors and qualifiers by increasing the network distance between any descriptor and any qualifier by one edge. In the combined hierarchy obtained, the maximum distance between two concepts is 23 edges.

3   Evaluation

Our representation extension method is evaluated in the context of document binary classification or triage. Our evaluation framework uses a MEDLINE triage task organized by the Genomics track of the 2005 Text REtrieval Conference (TREC). This task is described in Section 3.1. Related work is reviewed in Section 3.2. The document classification is done with the SVM^{light} software [13] (Section 3.3), and the results are presented in Section 3.4.

3.1   TREC 2005 Genomics Track GO Triage Task

The Text REtrieval Conference (TREC) includes a Genomics track since 2003. The TREC guidelines and common evaluation procedures allow research groups from all over the world to evaluate their progress in developing and enhancing information retrieval systems.

One of the tasks of the 2004 and 2005 Genomics track was a biomedical document triage task for gene annotation with the Gene Ontology (GO) [8]. GO is used by several model organism databases curators in order to standardize the description of genes and gene products. The triage task simulated one of the activities of the curators of the Mouse Genome Informatics (MGI) group [5]. MGI curators manually select biomedical documents that are likely to give experimental evidence for the annotation of a gene with one or more GO terms.

For both 2004 and 2005 GO triage tasks, the same subset of three journals from 2002 and 2003 was used. The subset contained documents that had been selected or not for providing evidence supporting GO annotation. The 5837 documents from 2002 were chosen as training documents, and the 6043 from 2003 as test documents. In 2004, they contained 375 and 420 positive examples (documents labeled relevant), respectively. In 2005, the number of positive examples for the training and test documents was updated to 462 and 518, respectively.

The triage task was evaluated with four metrics: precision, recall, F-score, and a normalized utility measure U_{norm}, defined by the following formulae:

$$precision = \frac{TP}{TP + FP}$$

$$recall = \frac{TP}{TP + FN}$$

where TP, FP, and FN stand for true positives, false positives, and false negatives, respectively.

$$F - Score = \frac{2 \times Precision \times Recall}{Precison + Recall}$$

$$U_{norm} = U_{raw}/U_{max}$$

where U_{max} is the best possible score, and U_{raw} is calculated with:

$$U_{raw} = (u_r \times TP) + (u_{nr} \times FP) \tag{3}$$

where u_r is the relative utility of a relevant document and u_{nr} the relative utility of a non-relevant document. With u_{nr} set at -1 and u_r assumed positive, u_r was determined by preferred values for U_{norm} in 4 boundary cases: completely perfect prediction, all documents judged positive (triage everything), all documents judged negative (triage nothing), and completely imperfect prediction. With different numbers of positive examples for classification between 2004 and 2005, $u_r = 20$ and $u_r = 11$ were chosen in 2004 and 2005, respectively.

Our experiment used the updated number of positive examples of 2005 for the classification, and the values $u_r = 11$ and $u_{nr} = -1$. A detailed description of TREC 2004 and 2005 Genomics track tasks can be found in [9,10].

3.2 Related Work

Most approaches used in the 2004 and 2005 GO triage task extract features from various text fields along with the MeSH fields [6,3]. In contrast, Seki et al. [20] and Lee et al. [14] describes experiments with MeSH-only document representations.

Seki et al. [20] describes a process that trains a naïve Bayes classifier using the MeSH features to represent documents. A gene filter is also applied based on gene names found in the text with the YAGI[5] (Yet Another Gene Identifier) software [21]. The results shown in Table 3 include the gene filter. Seki et al. does not specify how the MeSH fields were processed, i.e. whether, for example, groups of words corresponding to descriptors were split.

---

[5] http://www.cs.wisc.edu/~bsettles/yagi

Lee et al. [14] used MeSH for document representation and compared this approach to other representations such as abstract-based and caption-based representations. The semantic network of the UMLS[6] (Unified Medical Language System) was also used to map terms from MeSH, free-text, captions to UMLS semantic types, which are broad medical concepts. An SVM classifier was used and cross-validation results on the task training documents showed the MeSH representation to give the best performance with 0.4968 in normalized utility. The second best representation was obtained by using the UMLS semantic types of the MeSH terms (0.4403). The results in Table 3 correspond to a combination of three representation: MeSH, captions, and UMLS semantic types associated with MeSH.

The best results for the 2004 and 2005 triage tasks used methods that extract features from other fields than MeSH (title, abstract, full-text) and use domain-dependant techniques (term extraction, gene filtering). Dayanik et al. [4] obtains the best normalized utility (0.6512) of the 2004 GO triage task. Documents are represented with features extracted from the title, abstract and MeSH fields of the MEDLINE document format. Niu et al. [15] achieves the best normalized utility (0.5870) of the 2005 GO triage task using the SVM^{light} software [13]. Features are first extracted from full-text articles. The extraction is followed by porter stemming, stopwording, and a domain-specific term extraction method that is using corpus comparison.

Table 3. Results of classification of test documents for related work

Participant	Precision	Recall	F-score	Normalized Utility
Seki et al. (2004)	0.1118	0.7214	0.1935	0.4348
Dayanik et al. (2004)	0.1579	0.8881	0.2681	0.6512
Lee et al. (2005)	0.1873	0.8803	0.3089	0.5332
Niu et al. (2005)	0.2122	0.8861	0.3424	0.5870

3.3   Text Categorization and SVM^{light}

A discussion on text categorization and machine learning techniques is beyond the scope of this paper and is found elsewhere [19]. Our experiment focuses on evaluating an ontology-based document representation extension method and does not aim at comparing several text categorization techniques.

We use the SVM^{light} software which is an implementation of the Support Vector Machine method by Joachims [13]. The SVM learner defines a decision surface between the positive and negative examples of the training set. SVM training leads to a quadratic optimization problem and learning from large training sets can quickly become computationally expensive. The SVM^{light} implementation allows for large-scale SVM learning at lower computational cost.

We use the default settings for the learning module (svm_learn) and the classification module (svm_classify) of SVM^{light}. Default settings include the use of

---

[6] http://umlsinfo.nlm.nih.gov/

a linear kernel. Moreover, parameter C, allowing a trade-off between the training error and the classifier complexity, is determined automatically. The only modification is the setting of parameter j (cost-factor by which training errors on positive examples out-weight errors on negative examples, the default being 1) to 11, similarly to Subramaniam et al. [22]. The j parameter allows us to tune the classifier to the difference between u_r (= 11), the relative utility of a true positive, and u_{nr} (= -1), the relative utility of a false positive (see Equation 3, Section 3.1).

Our experiment evaluates the effect of hierarchy information on the automatic parameter tuning of SVM^{light} during the learning of a classifier. The automatic tuning was shown to perform well for text categorization [12].

3.4   Results

In order to limit the size of and the noise contained in the extended representations, we use a threshold t for the weight w_d of each candidate concept to extension. We experiment with different threshold values with the training set using 4-fold cross validation. The value $t = 0.6975$ is found to give the best normalized utility on average. The documents from the training and test sets are then extended with threshold $t = 0.6975$. SVM^{light} learns a classifier with the entire training set and the classifier obtained is used on the test set. Table 4 shows the results in terms of Precision, Recall, F-Score and normalized utility for the classification of the test documents. The results are compared to the simple use of the initial un-extended MeSH representation, and to the best result from the 2005 GO triage task in terms of normalized utility [15].

Table 4. Result of extension for the classification of test documents with $t = 0.6975$

	Precision	Recall	F-Score	U_{norm}
$t = 1$	0.2980	0.6139	0.4013	0.4824
$t = 0.6975$	0.2041	0.8745	0.3309	**0.5644** (+17%)
Niu et al. (2005)	0.2122	0.8861	0.3424	0.5870

The extension of MeSH-based document representations with new concepts leads to a drop in precision but an increase in recall. A value of 1 for the threshold corresponds to no extension. The normalized utility increases up to 0.5644 (+17% compared to our simple non-extended binary representation) for a threshold of 0.6975. Because of the importance of recall in this particular triage task and therefore in the utility function, our document representation extension has a positive impact on the normalized utility.

A normalized utility value of 0.5644 is not far from the highest score for the 2005 GO triage task (0.5870) and is positioned inside the top ten runs from a total of 47. The best run of the track used a domain-specific technique, and the free-text fields of MEDLINE records [15] (see Section 3.2). This suggests that the results of our domain-independent method could be further improved by the use of other fields than MeSH, and by domain-specific knowledge.

Most importantly, the improvements over non-extended MeSH-only document representations show that the hierarchical structure of the MeSH ontology can be used with little modification (*mesh* and *qualifier* nodes add-up) to build extended document representations that are beneficial to the GO triage task. Finally, our extension is based on a simple edge count method to calculate the semantic distance between concepts. More refined semantic measures integrating more information from the hierarchy and the domain could yield better results.

In our experiments, classifiers were learnt with the SVM^{light} software with default settings that include to the automatic tuning of C (see Section 3.3). In future work, we will manually tune C on the training set, and compare the manual tuning to our representation extension method. The default settings of SVM^{light} use a linear kernel function to separate the training data. In future work we will experiment with other kernel functions available with the software, such as polynomial, radial basis function, and sigmoid tanh[7].

4   Conclusion and Future Work

With the growing availability of biomedical information mainly in the form of free-text biologists need tools to process information automatically. In the early stage of information retrieval it is useful to select the probably relevant information from the mass of non-relevant information. Such a selection can be done with binary classification (also called triage) techniques.

Free-text is an ambiguous information representation. It can contain synonyms and polysems that require external knowledge for disambiguation. An alternative is to use ontologies to represent information. Ontologies provide standard terms for naming concepts and explicitly define relationships between concepts.

In this paper we proposed a method that extend ontology-based representations of biomedical documents. The method used in our experiment included the use MeSH for biomedical document representation. The initial MeSH-only representations were then extended with concepts that were semantically similar within the MeSH hierarchy. A simple edge count distance measure was used to evaluate semantic proximity between concepts.

Our method was evaluated on a document triage task that consisted in selecting documents containing experimental evidence for the annotation of genes in a model organism database. The triage task was organized by the Genomics track of the 2005 Text REtrieval Conference (TREC). Our document representation extension method led to an increase of 17% of normalized utility, the metric defined for the triage task. The normalized utility obtained, 0.5644, positions our method amongst the top ten runs out of 47 for the 2005 task, without the use of domain-specific techniques and by relying only on the MeSH document representation. This suggests that our results could be improved by integrating other MEDLINE fields and using domain-specific knowledge.

---

[7] http://svmlight.joachims.org/

In future work we will evaluate our domain-independent method with other ontologies and in other contexts. The Gene Ontology is an example of a taxonomy that provides an ontology-based description of gene records in model organism databases. Ontology-based representation extension can also be applied to document ad hoc retrieval and document clustering.

We also want to experiment with other measures in order to evaluate inter-concept semantic similarity. The edge count used in this paper assumes that the hierarchy edges correspond to the same semantic distances. Other measures are based on the assumption that edge distances vary in the the hierarchy. They calculate this variation with hierarchy information [11], or corpus information [17,18].

Acknowledgments

This research was supported by Enterprise Ireland under the Basic Research Grants Scheme project number SC-2003-0047-Y and by the European Commission under contract FP6-027026-K-SPACE.

References

1. F. Azuaje, H. Wang and O. Bodenreider (2005): Ontology-driven similarity approaches to supporting gene functional assessment. In proceedings of the ISMB'2005 SIG meeting on Bio-ontologies 2005, p9-10, 2005.
2. A. Budanitsky (1999): Lexical Semantic Relatedness and its Application in Natural Language Processing. Technical Report CSRG-390, Department of Computer Science, University of Toronto, August 1999.
3. A.M. Cohen, J. Yang and W.R. Hersh (2005). A comparison of techniques for classification and ad hoc retrieval of biomedical documents. Proceedings of the Fourteenth Text Retrieval Conference (TREC 2005), November, Gaithersburg, Maryland.
4. A. Dayanik, D. Fradkin, A. Genkin, P. Kantor, D. D. Lewis, D. Madigan and V. Menkov (2004): DIMACS at the TREC 2004 Genomics Track. Proceedings of the Thirteenth Text Retrieval Conference (TREC 2004), November, Gaithersburg, Maryland.
5. J. T. Eppig, C. J. Bult, J. A. Kadin, J. E. Richardson and J. A. Blake (2005): The Mouse Genome Database (MGD): from genes to mice - a community resource for mouse biology. Nucleic Acids Res 2005; 33: D471-D475.
6. S. Fujita (2004): Revisiting again document length hypotheses - TREC 2004 Genomics Track experiments at Patolis. Proceedings of the Thirteenth Text Retrieval Conference (TREC 2004), November, Gaithersburg, Maryland.
7. M. E. Funk, C. A. Reid and L. S. McGoogan (1983): Indexing consistency in MEDLINE. Bull Med Libr Assoc., Vol. 71(2), p176-183, 1983.
8. Gene Ontology Consortium (2004): The Gene Ontology (GO) database and informatics resource. Nucleic Acids Research, Vol. 32, 2004.
9. W. R. Hersh, R. T. Bhuptiraju, L. Ross, P. Johnson, A. M. Cohen and D. F. Kraemer (2004): TREC 2004 Genomics Track Overview. Proceedings of the Thirteenth Text Retrieval Conference (TREC 2004), November, Gaithersburg, Maryland, 2004.

10. W. R. Hersh, A. M. Cohen, J. Yang, R. T. Bhuptiraju, P. M. Roberts and M. A. Hearst (2005): TREC 2005 Genomics Track Overview. Proceedings of the Fourteenth Text Retrieval Conference (TREC 2005), November, Gaithersburg, Maryland, 2005.
11. J. J. Jiang and D. W. Conrath (1997): Semantic Similarity Based on Corpus Statistics and Lexical Taxonomy. Proceedings of ROCLING X, Taiwan, 1997.
12. T. Joachims (1998). Text Categorization with Support Vector Machines: Learning with Many Relevant Features. Proceedings of the European Conference on Machine Learning, Springer, 1998.
13. T. Joachims (1999): Making large-Scale SVM Learning Practical. Advances in Kernel Methods - Support Vector Learning, B. Schlkopf and C. Burges and A. Smola (ed.), MIT-Press, 1999.
14. C. Lee, W.-J. Hou and H.-H. Chen (2005): Identifying Relevant Full-Text Articles for Database Curation. Proceedings of the Fourteenth Text Retrieval Conference (TREC 2005), November, Gaithersburg, Maryland, 2005.
15. J. Niu, L. Sun, L. Lou, F. Deng, C. Lin, H. Zheng and X. Huang (2005): WIM at TREC 2005. Proceedings of the Fourteenth Text Retrieval Conference (TREC 2005), November, Gaithersburg, Maryland, 2005.
16. R. Rada, H. Mili, E. Bicknell and M. Blettner (1989): Development and application of a metric on semantic nets. In IEEE Transaction on Systems, Man, and Cybernetics, Vol 19(1), p17-30, 1989.
17. P. Resnik (1995): Using Information Content to Evaluate Semantic Similarity in a Taxonomy. Proceedings of the 14th International Joint Conference on Artificial Intelligence, p448-453, 1995.
18. R. Richardson and A. F. Smeaton (1995): Using Wordnet in a Knowledge-Based Approach to Information Retrieval. Working paper CA-0395, School of Computer Applications, Dublin City University, Dublin, 1995.
19. F. Sebastiani (2002): Machine learning in automated text categorization. ACM Comput. Surv., Vol. 34(1), 2002.
20. K. Seki, J. C. Costello, V. R. Singan and J. Mostafa (2004): TREC 2004 Genomics Track experiments at IUB. Proceedings of the Thirteenth Text Retrieval Conference (TREC 2004), November, Gaithersburg, Maryland, 2004.
21. B. Settles (2004). Biomedical Named Entity Recognition Using Conditional Random Fields and Rich Feature Sets. Proceedings of the COLING 2004 International Joint Workshop on Natural Language Processing in Biomedicine and its Applications (NLPBA)", Geneva, Switzerland.
22. L. V. Subramaniam, S. Mukherjea and D. Punjani (2005): Biomedical Document Triage: Automatic Classification Exploiting Category Specific Knowledge. Proceedings of the Fourteenth Text Retrieval Conference (TREC 2005), November, Gaithersburg, Maryland, 2005.
23. M. Sussna (1993): Word sense disambiguation for free-text indexing using a massive semantic network. Proceedings of the second international conference on Information and knowledge management, p67-74, Washington, D.C., United States, 1993.

Integrating Mutations Data of the TP53 Human Gene in the Bioinformatics Network Environment

Domenico Marra and Paolo Romano

Bioinformatics and Structural Proteomics, National Cancer Research Institute,
Largo Rosanna Benzi 10, I-16132 Genova, Italy
{paolo.romano, domenico.marra}@istge.it

Abstract. We present in this paper some new network tools that can improve the accessibility of information on mutations of the TP53 human gene with the aims of allowing for the integration of this data in the growing bioinformatics network environment and of demonstrating a possible methodology for biological data integration. We implemented the IARC TP53 Mutations Database and related subsets in an SRS site, set up some Web Services allowing for a software oriented, programmatic access to this data, created some demo workflows that illustrate how to interact with Web Services and implemented these workflows in the biowep (Workflow Enactment Portal for Bioinformatics) system. In conclusion, we discuss a new flexible and adaptable methodology for data integration in the biomedical research application domain.

Keywords: data integration, molecular biology databases, bioinformatics, Web Services, XML, workflow management systems, workflow enactment portal.

1 Introduction

1.1 Integration of Biological Information

The huge amount of biological information that is currently available on the Internet resides in data sources that have heterogeneous DBMS, data structures, query methods and information. Otherwise, data integration is needed to achieve a wider view of all available information, to automatically carry out analysis involving more databases and software and to perform large scale analysis. Only a tight integration of data and analysis tools can lead to a real data mining.

Integration of data and processes needs stability: sound knowledge of the domain, well defined information and data, leading to standardization of schemas and formats, clear definition of goals. Instead, integration fears heterogeneous data and systems, uncertain domain knowledge, highly specialized and quickly evolving information, lacking of predefined, clear goals, originality of procedures and processes.

This essentially is the situation of data integration in biology, where a pre-analysis and reorganization of the data is very difficult, because data and related knowledge change very quickly, complexity of information makes it difficult to design data models which can be valid for different domains and over time, goals and needs of researchers evolve very quickly according to new theories and discoveries, thus leading

S. Hochreiter and R. Wagner (Eds.): BIRD 2007, LNBI 4414, pp. 453–463, 2007.

to new procedures and processes. So, traditional tools, like data warehouses and local integration tools (e.g., SRS), pose many problems deriving from the size of sources, the need for frequent updating and for coping with frequently modified data structures.

Therefore, integrating biological information in a distributed, heterogeneous environment requires new flexible, expandable and adaptable technologies and tools that allow for moving from an interactive to an automated approach to data analysis. In literature, proposals in this direction can already be found [1,2]. In this context, the following methodology could be devised:

- XML schemas can be used for the creation of the models of the information,
- XML based languages can be adopted for data representation and exchange,
- Web Services can be made available for the interoperability of software
- computerised workflows can be created and maintained for the definition and execution of analysis processes
- Workflow enactment portals can be installed for the widespread utilization of automated processes

Common vocabularies, when existing, can significantly improve interoperability. Ontologies, especially, when available and widespread used can add semantic metadata to the resources, improve data accessibility, support integrated searches. Unfortunately, the available biomedical ontologies are currently not so widely used and the vast majority of data sources do not use them.

The above methodology can partially cope with the problem of format changes in data sources since it limits interoperability issues to the interface level. This implies that changes, even deep ones, can occur at the data management level without influencing how external applications can interact with the system. Of course, care must be taken by databases' curators to ensure that the interface is still working according to advertised API (Application Programming Interfaces) and these are based on consensus data models.

In this paper, we approach the problem of making information on the mutations of the TP53 human gene as widely available as possible, by implementing some tools through the above approach.

1.2   The TP53 Human Gene and Its Functions

The importance of the function of the p53 protein (reference sequence is SwissProt P04637) in the regulation of the apoptosis process in the cell is well known and recognized. This protein can be considered as a tumour suppressor, in the sense that it prevents the occurrence of tumours by promoting apoptosis in pre-cancer, altered cells. Many studies confirmed that modifications in the regulation of the apoptosis process are positively correlated to the presence of mutations in the TP53 gene (located on the short (p) arm of chromosome 17 at position 13.1, reference sequence is EMBL X02469), thus suggesting that it can, at least partially, be the cause for missing functionality; moreover, the presence of mutations in the TP53 gene has been observed in about half of all cancers, in the 80% of all colon cancer tumours, in the 50% of lung cancer tumours, and in the 40% of breast cancer tumours.

The correlation between mutations of the TP53 gene and the tumour regulation process has been described in many papers. Recently, Human Mutation devoted a special issue on TP53 gene mutations and cancer [3]. The tumour protein p53, expressed by the TP53 gene, can bind directly to DNA. Every time a mutation (damage) occurs in the cell's DNA, the p53 protein becomes involved in determining whether the DNA will be repaired or not. In the former case, p53 activates other genes to fix the damage, while, in the latter case, p53 prevents the cell from dividing and signals it to undergo apoptosis. This process prevents cells with mutated or damaged DNA from dividing, which in turn avoids the development of tumours. Mutations in the TP53 gene usually prevent tumour protein p53 from binding to DNA correctly, lowering the protein effectiveness in regulating cell growth and division. As a result, DNA damage accumulates and cells may divide in an uncontrolled way, leading to a cancerous tumour.

In literature, many TP53 mutations have been described and mapped onto the normal gene. Consequently, databases collecting these information have been created in the past years and made available to the scientific community through the Internet.

The TP53 Mutation Database [4] of the International Agency for the Research on Cancer (IARC) currently is the biggest and most detailed database. It includes somatic mutations (mutation type with references, mutation prevalence, mutation and prognosis), germline mutations (both data and references), polymorphisms, mutant functions and cell line status.

Release 10 includes 21,587 somatic mutations whose description has been derived from 1,839 papers which are included in the Medline database. Information on somatic mutations includes data on the mutation, the sample, the patient and his/her life style. Reference vocabularies and standardized annotations are used extensively for the description of the mutation, tumour site, type and origin and for literature references. Examples of the former are ICD-O (International Classification of Diseases – Oncology) and SMOMED nomenclatures.

This database is available through the IARC web site. Queries can only be executed on-line and imply a human interaction. Moreover, there are datasets (e.g., mutation and prognosis, germline mutations, polymorphisms and cell lines status) that are not searchable by online queries. Results can be downloaded in a tab delimited format, but partial downloads, e.g., subsets of all information available for a given mutation, are not allowed. Main accessibility limitations are the lacking of links to other bioinformatics databases and tools and the user interface, an interactive query form accessible through a standard web page, that makes it difficult to access data through software programs.

In the perspective of an integrated bioinformatics network environment, interoperability between services should be stressed as much as possible, both by setting up links to external resources and by implementing machine-oriented interfaces, that is network interfaces devoted to retrieval of information by software programs, instead of interactive interfaces devoted to researchers.

SRS (Sequence Retrieval System, also known as Sequence Retrieval Software) is a well know search engine for biomedical databanks [5]. Among its most original and useful features are the possibility of querying many databases together and of integrating data retrieval and analysis in the same tool. Moreover, SRS offers XML based input/output, since it is able to manage XML structured databases and can output results by using predefined Document Type Definitions (DTDs). An SRS implementation of the IARC TP53 database had already been set up at the European Bioinformatics Institute – EBI, in the sphere of a mutation database access integration project, whose goal was to find common data structures in mutation databases and to create a unified access tool. This project led to the setting up of SRS libraries referring to about 50 gene mutation databases, two of which related to the TP53 gene: the IARC TP53 database and the UMD-p53 mutation database [6]. A special database linking those records in the two libraries that relate to the same mutation was also set up with the aim of integrating information about the same mutation available in different databases and offering a unique, integrated list of all available mutations. The EBI implementation refers to Release 3 of the IARC database (available before 2000, it contains 10,411 mutations) and it includes somatic mutations data only.

In order to improve automated, software driven access to the IARC TP53 database, we have designed and implemented a new SRS based service which is focused on the TP53 somatic mutations data, but it also includes all supplementary datasets provided by IARC. Integrating these different datasets into SRS and adding links to relevant bioinformatics databases would present a real advantage over currently available systems from IARC and EBI. Furthermore, we implemented Web Services that allows for a programmatic access to the databases, created some demo workflows for demonstrating how Web Services can be easily accessed and made these workflows available in biowep, our Workflow Enactment Portal for Bioinformatics.

2   Methods

2.1   The Implementation of the IARC TP53 Mutation Database in SRS

IARC TP53 mutation database (Release 10) and all related data subsets have been downloaded[1] and reformatted for easier indexing and searching through SRS. Reformatting was carried out by using a mixed manual and automatic procedure, so that it can be replicated for future releases of the database with a minimal effort, although heavy changes in the data structure and contents of the database would strongly limit the reproducibility of procedures. It also involved a few changes and additions in the database fields and their contents (e.g., the field ExonIntron was split into two new fields, Exon and Intron; the field Tumor origin now includes the new Metastasis localization subfield; new unique record identifiers were added when missing). All datasets have then been implemented in a purpose SRS site as separate libraries. Data fields have been defined in agreement with controlled vocabularies that were used for data input at IARC, so that SRS indexes are based on them.

Links between libraries have been created whenever possible. HTML links can be used while navigating results pages and include links to Medline at NCBI through Pubmed IDs. SRS links have been defined so that they can be used for the creation of data views incorporating information from more databases. They also allow to execute queries including search conditions restricted to linked record subsets only.

---

[1] At the time of writing the final version of the paper, the Release 11 of the database had already been released, but we did not yet updated the SRS site. We are now upgrading the system and, at the time of the publication, the new release will probably be available on-line.

Data views have been created for displaying both complete and focused data sets, the latter starting from mutation, sample or patient specific data.

XML based data interchange between software tools is facilitated by the definition of an ad hoc Document Type Definition (DTD). A purpose DTD, has been designed for somatic mutations data. It includes three main groups of elements:

- MutationData: including type, location, effects,
- SampleData: including tumour origin, topography, sample source,
- PersonInfo: including sex, age, ethnics, clinical status, life styles, expositions.

Loaders, views and XMLPrintMetaphors (these are all SRS objects) have been defined, so that SRS can create XML formatted outputs based on the above DTD. Outputs can of course be downloaded.

All IARC TP53 Database datasets have been made available on-line and can easily and effectively be queried through SRS query forms. Due to the careful definition of data fields, terms included in the controlled vocabularies that were used during data input at IARC can also be used from within the SRS extended query form, thus allowing for a data-driven search.

2.2 Implementation of Web Services

Original Web Services have been implemented, allowing for the retrieval of information from TP53 databases through a programmatic interface. Web Services are also viable for an efficient inclusion in pipelines and workflows that are implemented by using some workflow management systems. We implemented Web Services allowing for the execution of a search by interesting properties towards the IARC TP53 Mutation Database.

Integration of these data with other sources was planned by using either unique IDs or common terms. In order to do this, two types of services were implemented, searching either for a specific feature and returning IDs or searching for an ID and returning full records.

Web Services were implemented by using Soaplab, a SOAP-based Analysis Web Service providing a programmatic access to applications on or through remote computers [7]. Soaplab allows for the implementation of Web Services offering access to local command-line applications, like the EMBOSS software, and to the contents of ordinary web pages. The only requirements of Soaplab are the Apache Tomcat servlet engine with the Axis SOAP toolkit, Java and, optionally, perl and mySQL. Once the server has been installed, new Web Services are added by defining simple definitions of the related execution commands (see code below).

```
# ACD definition TP53 Mutations Database Web Services
# Search db by codon number and get IDs
appl: getIdsByCodonNumber [
    documentation: "Get ids by codon number"
    groups: "O2I"
    nonemboss: "Y"
    comment: "launcher get"
    supplier: http://srs.o2i.it/srs71bin/cgi-bin/wgetz
    comment: "method [{$libs}-Codon_number:'$codon'] -lv
```

```
2000000 -ascii"
  ]
string: libs [ parameter: "Y" ]
string: codon [ parameter: "Y" ]
outfile: result [ ]

# Search by mutation IDs and get all related Pubmed IDs
appl: getP53PubMedIdByIds [
  documentation: "Get PubMed id by TP53 mutation ids"
  groups: "O2I"
  nonemboss: "Y"
  comment: "launcher get"
  supplier: http://srs.o2i.it/srs71bin/cgi-bin/wgetz
  comment: "method [TP53_iarc:'$id'] -f pen -lv 2000000
-ascii"
  ]
string: id [ parameter: "Y" ]
outfile: result [ ]
```

Definitions are written in the AJAX Command Definition (ACD) language, specially designed for EMBOSS software suite and reused in the Soaplab context, and are then converted into XML before they can be used. In our case, all Web Services have been defined as remote calls to the SRS site. Essential differences between Web Services definitions consist in the input and output parameters and in the specification of the URL address where these parameters must be inserted.

2.3 Creation of Demo Workflows

Web Services are especially useful for the automation of data analysis processes and these can be created and executed by using workflow managements systems. We developed some demo workflows with the aim of providing the researcher with simple tools. We adopted the following criteria:

- use of a limited number of query options (inputs to the Web Services), usually not more than two,
- use of clear, simple parameters/values pairs: exon, intron and codon, all given as integer numbers, and similar parameters were used as inputs,
- production of simple results/outputs.

Workflows have been designed to access to and to retrieve data from the IARC TP53 Mutations Database through our Web Services. They can also be viewed as an example of data integration that can be achieved through workflow management systems. We used the Taverna Workbench [8], a workflow management systems that was developed within the myGrid initiative [9], and workflows are available on-line, in the XML Simple Conceptual Unified Flow Language (XScufl) format, for use by interested users working with this tool.

Eight workflows have been developed: they can be divided into two groups on the basis of the type of query and of its output. The first group, comprising six workflows, can be described as a set of queries that interrogates the TP53 Somatic Mutations Database; according to input parameters, the output is one or more sets of somatic mutation entries in the flat format: each information is on a separate line of text, identified by a label which is followed by the data.

The output doesn't offer any new information, it just includes some records of the mutations database. The workflows can then be used for a basic level of analysis, regarding distributions of one or more somatic mutations in the database.

The second group, including two workflows, includes searches that involve two or more databases that are queried either serially or in parallel; the output consists in one or more sets of entries from searched databases that can be integrated. A certain level of data mining is here possible, since the retrieval of data from more databases offer the researcher a unseen, unknown view of available information.

The support site includes a short description of some of the workflows, together with a link to the workflow itself, expressed in the XScufl format, both in XML and as a text file.

2.4 Workflow Enactment by End Users

Workflow Management Systems can support the creation and deployment of complex procedures, but they assume that researchers know which bioinformatics resources can be reached through a programmatic interface and that they are skilled in programming and building workflows. Therefore, they are not viable to the majority of unskilled researchers.

We designed biowep, a Workflow Enactment Portal for Bioinformatics, that is a web based client application allowing for the selection and execution of a set of predefined workflows by unskilled end users [10]. The system is freely available on-line. Contrary to the other workflow enacters that run workflow on the user's computer, in biowep the enactment of workflows is carried out on the server. Workflows can be selected by the user from a list or searched through the description of their main components. Actually, these are first annotated on the basis of their task and data I/O and this annotation is inserted into a database that can be queried by end users. Results of workflows' executions can be saved and later reused or reanalyzed.

Biowep includes a detailed description of all workflows and a useful description of input data that helps end user in launching the workflows.

3 Results

The IARC TP53 Somatic Mutations Database, including somatic mutations data and literature references, mutation prevalence, mutation and prognosis, germline mutations and references, cell line status and mutation functions, is available on-line in a purpose SRS site and can be easily and effectively queried by using standard SRS query forms. Due to the careful definition of data fields, terms included in the controlled vocabularies that were used during data input at IARC can also be used from within the SRS extended query form, thus allowing for a data-driven search. This query form includes an input field for each database field. Input fields are formatted according to the type of the corresponding database field. They can therefore be based on checkboxes, numeric ranges, text fields and multi select boxes having the items of the input controlled vocabularies as reference lists from which to make the selection. This allows to achieve the same functionalities of the IARC advanced search form.

SRS links allow to retrieve data from one database by also imposing restrictions on the other. E.g., retrieve all papers describing a defined mutation type or all mutations described by papers appeared in a given journal. HTML links from the literature references database to PubMed site at the National Center for Biotechnology Information (NCBI) allow a direct access to available information, either abstract or full text.

Purpose views (MutationData, SampleData and PersonInfo) can be used to focus attention on related information of a datasets, hiding not pertinent information, thus achieving a more compact output. By using SRS views manager, personal data views can also be created. Finally, users, after a fruitful query, can opt for displaying and/or downloading results as either a text file, a generic XML file (an XML file based on the database description in the SRS configuration files) or a specific XML file (an XML file based on the newly defined DTD).

A set of original Web Services have been designed and implemented by using Soaplab and are now available on-line, both through their descriptions expressed in the Web Services Description Language (WSDL) and the Soaplab server. A list of Web Services that are currently available is presented in the following table 1.

Table 1. List of Web Services related to IARC TP53 Mutation Database and implemented by using Soaplab tool. In column 'Input', lib stays for the name of the library, the term by which databases are referenced in SRS. See http://www.o2i.it/webservices/ for a sample list of available Web Services.

Web Service Name	Input	Output
getP53MutationsByProperty	lib, text	Full record
getP53MutationsByIds	Id	Full record
getP53MutationIdsByType	lib, mutation type	id(s)
getP53MutationIdsByEffect	lib, effect	id(s)
getP53MutationIdsByExon	lib, exon number	id(s)
getP53MutationIdsByIntron	lib, intron number	id(s)
getP53MutationIdsByCodonNumber	lib, codon number	id(s)
getP53MutationIdsByCpgSite	lib, cpg site (true/false)	id(s)
getP53MutationIdsBySpliceSite	lib, splice site (true/false)	id(s)
getP53MutationIdsByTumorOrigin	lib, origin (primary, secondary, ..)	id(s)

This table lists all Web Services that have been created with the aim of allowing data retrieval from IARC TP53 Mutation Database. Here id refers to the Mutation_ID defined by IARC, while lib refers to the name of involved database. This currently can only be 'TP53_iarc' but it has been parameterized in view of future extensions of the system, allowing to search on other databases as well.

Furthermore, workflows have been designed to access to and to retrieve data from the IARC TP53 Mutations Database Web Services. They can also be viewed as an example of how data integration can be achieved through workflow management systems.

Table 2. Short description of two demo workflows related to the IARC TP53 Mutation Database. These workflows can be enacted through the biowep workflow enactment portal, where they are described in details. Support for launching them is also given in the site.

Workflow name and description	Inputs	Outputs
GetTP53MutationsByIntronAndEffect		
Retrieves all TP53 mutations occurring in the specified intron and having the specified effect.	**intron:** intron where mutation occurs (integer number) **effect:** mutation effect (string of characters, can be "fs", "missense", "na", "nonsense", "other", "silent", "splice")	**ids:** list of mutations' ids **mutations:** full mutations' data in a flat format (text only, one line per information, data preceeded by a label)
GetTP53MutationsByIntronAndEffect2		
Retrieves all TP53 mutations that: • occurs in the specified intron OR have the specified effect • occurs in the specified intron AND have the specified effect • occurs in the specified intron BUT DOES NOT have the specified effect	**intron:** intron where mutation occurs (integer number) **effect:** mutation effect (string of characters, can be "fs", "missense", "na", "nonsense", "other", "silent", "splice")	**mutation_in_or:** full records of mutations retrieved by applying a logical OR to results obtained by single queries **mutation_in_but_not:** full records of mutations retrieved by removing results of the second query from results of the first query **mutation_in_and:** full records of mutations retrieved by applying a logical AND to results obtained by single queries

XScufl formatted versions of these workflows have been created and are available online for use by interested users working with Taverna Workbench. In table 2, some workflows are briefly presented.

Workflows are available on-line through the biowep workflow enactment portal and can be freely executed by interested end-users.

4   Conclusions

We believe automation of data analysis and retrieval processes will offer bioinformatics the possibility of implementing a really machine-oriented, distributed analysis environment. This environment will give researchers the possibility of improving efficiency of their procedures and will allow for the implementation of data integration tools able to significantly improve biological data mining. At the same time, it can protect intellectual property rights for data, algorithms and source code, that would not be copied and would remain on the owners' information system, for those data owners and curators that will adopt such data distribution technologies.

We have presented in this paper a methodology for the implementation of new flexible tools based on the extensive use of XML languages, Web Services, workflows and a portal for workflow enactment. This methodology, that can be improved by the widespread adoption of semantic tools, like ontologies, has been applied to the problem of the integration of information on the mutations of the TP53 human gene in the bioinformatics environment. Starting from an original SRS – based implementation of the IARC TP53 Somatic Mutations Database, we implemented Web Services and created and made available through the biowep workflow enactment system some demo workflows.

It is our opinion that the further development and implementation of Web Services allowing the access to and retrieval from an exhaustive set of molecular biology and biomedical databases being carried out by many research centres and network service providers in the biological and medical domains and the creation of effective and useful workflows by interested scientists through widely distributed workflows management systems such as those presented in this paper will significantly improve automation of in-silico analysis.

We welcome collaborations with all researchers that would be willing to contribute to our effort by submitting databases to our SRS implementation and developing both new Web Services and workflows for their scientific domains of knowledge and special research interests.

Acknowledgments. This work has partially been carried out with the contribution of the Italian Ministry for Education, University and scientific and technical Research (MIUR), Strategic Project "Oncology over Internet". We wish to thank Gilberto Fronza, Alberto Inga and Magali Olivier for their useful suggestions.

References

[1] Jamison, D.C.: Open Bioinformatics (editorial). Bioinformatics 19 (2003) 679-680
[2] Stein, L.: Creating a bioinformatics nation. Nature 417 (2002) 119-120
[3] Soussi, T., Béroud, C. (eds): Special Issue: Focus on p53 and Cancer, Hum. Mutat. 21 (2003) 173-330
[4] Olivier, M., Eeles, R., Hollstein, M., Khan, M. A,, Harris, C. C., Hainaut, P.: The IARC TP53 Database: new online mutation analysis and recommendations to users. Hum. Mutat. 19 (2002) 607-614
[5] Etzold, T., Ulyanov, A., Argos, P.: SRS: information retrieval system for molecular biology data banks. Meth. Enzymol. 266 (1996) 114-128
[6] Béroud, C., Soussi, T.: The UMD-p53 database: new mutations and analysis tools, Special Issue: Focus on p53 and Cancer, Hum. Mutat. 21 (2003) 176-181
[7] Senger, M., Rice, P., Oinn, T.: Soaplab - a unified Sesame door to analysis tools. In: Cox S. J., editor. Proceedings of the UK e-Science All Hands Meeting 2003; September 2–4; Nottingham, UK. (2003) 509–513

[8] Oinn, T., Addis, M. J., Ferris, J., Marvin, D. J., Senger, M., Carver, T., Greenwood, M., Glover, K., Pocock, M. R., Wipat, A., Li, P.: Taverna: a tool for the composition and enactment of bioinformatics workflows. Bioinformatics 20 (2004) 3045-3054

[9] Stevens, R., Robinson, A., Goble, C.: myGrid: personalised bioinformatics on the information grid. Bioinformatics 19 (2003) i302-i304

[10] Romano P, Bartocci E, Bertolini G, De Paoli F, Marra D, Mauri G, Merelli E, Milanesi L, Biowep: a workflow enactment portal for bioinformatics applications, BMC Bioinformatics 2007, 8 (Suppl 1):S19

Web Sites

IARC TP53 Somatic Mutations Database: http://www.iarc.fr/p53/

IARC TP53 Somatic Mutations Database on SRS (it includes support pages for downloading data, implementing the databases, download and enact workflows, etc...): http://srs.o2i.it/

AJAX Command Definition: http://emboss.sourceforge.net/developers/acd/

Soaplab: http://www.ebi.ac.uk/soaplab/

TP53 related demo workflows: http://www.o2i.it/workflows/

Description of sample TP53 related Web Services: http://www.o2i.it/webservices/

TP53 Soaplab based Web Services: http://www.o2i.it:8080/axis/services

biowep: http://bioinformatics.istge.it/biowep/

Efficient and Scalable Indexing Techniques for Biological Sequence Data

Mihail Halachev, Nematollaah Shiri, and Anand Thamildurai

Dept. of Computer Science and Software Engineering
Concordia University, Montreal, Canada
{m_halach,shiri,a_thamil}@cse.concordia.ca

Abstract. We investigate indexing techniques for sequence data, crucial in a wide variety of applications, where efficient, scalable, and versatile search algorithms are required. Recent research has focused on suffix trees (ST) and suffix arrays (SA) as desirable index representations. Existing solutions for very long sequences however provide either efficient index construction or efficient search, but not both. We propose a new ST representation, STTD64, which has reasonable construction time and storage requirement, and is efficient in search. We have implemented the construction and search algorithms for the proposed technique and conducted numerous experiments to evaluate its performance on various types of real sequence data. Our results show that while the construction time for STTD64 is comparable with current ST based techniques, it outperforms them in search. Compared to ESA, the best known SA technique, STTD64 exhibits slower construction time, but has similar space requirement and comparable search time. Unlike ESA, which is memory based, STTD64 is scalable and can handle very long sequences.

Keywords: Biological Databases, Sequences, Indexing, Suffix Trees.

1 Introduction

Advances in technology have lead to generation of enormous data sets that have to be efficiently stored, processed, and queried. Traditional database technology is inadequate for some of these data sets. One example of such data type is sequence data, represented as strings over some alphabet. For instance, a DNA sequence is a string over the nucleotide alphabet $\Sigma = \{A,C,G,T\}$, and a protein sequence is a string over the amino acid alphabet of size 23. The size of GenBank [11] and its collaborating DNA and protein databases which contain data coded as long strings has reached 100 Giga bases [27], and the size of GenBank alone doubles every 17 months. Other examples of sequence data include natural text, financial time series, web access patterns, data streams generated by micro sensors, etc.

An important question is: What is an appropriate model for indexing sequence data? Two main issues that need to be addressed are the required storage space for the model and its construction time. Further, searching in molecular sequence data is central to computational molecular biology, and hence a desired model should provide fast access and efficient support for various types of search. Examples of such applications include local and global alignment, searching for motifs, sequence-tagged sites

S. Hochreiter and R. Wagner (Eds.): BIRD 2007, LNBI 4414, pp. 464–479, 2007.

(STSs), expressed-sequence-tags (ESTs), finding exact and approximate repeats and palindromes, recognizing DNA contamination, DNA sequence assembly, etc. A core part of the solution in these applications, given a query pattern, is to find substrings in the database which are same or similar to the query pattern. While string searching is an old topic and has been studied extensively, the enormous size of ever-growing biological sequence databases, and the variety and volume of queries to be supported in many real-life applications pose new research challenges.

There is a growing need for more scalable indexing structures and techniques to support efficient and versatile search in biological sequence data. In this context, the contributions of this paper are two. First, we propose a novel suffix tree representation (STTD64) for indexing large sequence data using typical desktop computers, aimed at supporting a variety of search operations on sequence data. We also develop construction and exact match search algorithms. The proposed technique is scalable; it can handle the entire human genome (of size 2.8 GB) on typical desktops. Second, through extensive experiments with real-life data, we study and compare the performance of our technique with three major indexing techniques: one based on suffix arrays, and the other two based on suffix trees. We report the experimental results on construction time, storage requirement, and exact match search performance. The results suggest that STTD64 is a desired technique for sequence databases larger than 250MB, which are common in bioinformatics applications.

The rest of the paper is organized as follows. Section 2 reviews research on indexing techniques for sequence data. In section 3, we present the considered existing representations and introduce STTD64, our proposed ST representation. Section 4 describes the corresponding index construction algorithms. Section 5 discusses the exact match search problem which uses the indexes. The experimental results are presented and discussed in Section 6. We conclude the paper in Section 7.

2 Background and Related Work

There are quite a few index structures on sequence data, including PATRICIA trees [23], inverted files [32], prefix index [16], String B-Tree [10], q-grams [5], [25], [26], suffix trees (ST) [31], and suffix arrays (SA) [20]. An important characteristic of biological sequence data that distinguishes it from other types of sequence data is that it does not have a structure, i.e., it cannot be meaningfully broken into words/parts. This explains why inverted files, prefix index, and string B-Tree are not suitable for biological sequence data. While q-grams support fast exact match search, it has been shown that they are not efficient for low similarity searches [24]. Thus, there is an increasing interest on ST and SA as desirable index structures to support a wide range of applications on biological sequence data.

2.1 Suffix Trees and Suffix Arrays

Given an input sequence D of size n, a suffix tree (ST) is a rooted directed tree with n leaves, labeled 1 to n. It supports a wide range of applications efficiently [13]. As a result of their versatility and efficiency, STs are incorporated in many applications, including MUMmer [7], REPuter [19], and SMaRTFinder [22].

There are a number of ST representations [4], [12], [14], [17], and [30]. The most space efficient ST representation is the one proposed by Giegerich et al. [12], to which we refer as STTD32 (for Suffix Tree, Top Down, 32 bits). In the worst case, it requires 12 bytes per character in the sequence D, and 8.5 bytes per character in D on average. Thus, indexing large sequences requires either large RAM or efficient disk based ST construction algorithms. In related literature, known are several in-memory ST construction algorithms [21], [30], and [31] which, for finite alphabets, build the ST in time linear to the size of D. The drawback of these techniques is their limitation on the size of the input sequence, i.e., they are not scalable to the sequences in the bioinformatics domain. Some recent studies [15], [29], focused on disk-based ST construction algorithms. Tian et al. [29] proposed a Top-Down, Disk-based (TDD) ST indexing technique, which uses a variant of STTD32 representation, and employs partitioning and buffering techniques which make good use of available memory and cache. Through TDD, they were able for the first time to construct the ST for the entire human genome in 30 hours, on a regular desktop computer. Despite its worst case time complexity $O(n^2)$, TDD outperforms the previously known construction techniques, due to better utilization of RAM and cache memory.

A *basic* SA [20] is essentially a list of pointers to the lexicographically sorted suffixes in text D. It is a space efficient alternative to ST requiring only $4n$ bytes. To improve the search performance, an additional table called *longest common prefix* (*lcp*) table, of size n, is proposed [20]. In a recent work, Abouelhoda et al. [1] proposed the *enhanced* suffix array (ESA), which is a collection of tables, including the basic SA and a *lcp* table similar to the one in [20]. ESA has the same search complexity as ST based solutions [1], achieved at the cost of increased space for storing the additional tables. ESA is a memory based technique and has a theoretical input sequence size of up to 400 million symbols on 32-bit architectures, as mentioned in [18]. For longer sequences, a 64-bit implementation of ESA is available, however it runs on large server class machines [18]. For such sequences, there are disk-based SA construction algorithms [6], most notably DC3 [8], using which a basic SA was constructed for the human genome in 10 hours on a computer with two 2GHz processors, 1GB RAM, and 4x80 GB hard disks.

2.2 Considered Indexing Techniques

Each indexing technique consists of an index representation, a construction and search algorithms. The four indexing techniques on which we focus and evaluate are as follows. The first technique, STTD32, uses the most space efficient ST representation, proposed in [12]. We have extended the work in [12] by implementing disk based construction and search algorithms. The second technique we consider is TDD, for its superior construction performance [29], compared to existing ST techniques. It uses a variant of the STTD32 ST representation, discussed further in Section 3. The third technique is based on our novel ST representation, called STTD64 (for Suffix Tree, Top Down, 64 bits). We have developed and implemented the corresponding disk based construction and search algorithms. All these three ST based techniques have the same complexity for exact match search (EMS) $O(m + z)$, where m is the query size and z is the number of exact occurrences of the query in the database sequence. These techniques differ in the structure and size of the index, introducing certain practical limits and affecting index construction and search performances, which we investigate experimentally.

The fourth technique we consider in our experiments is ESA (Enhanced SA). ESA is a memory based technique and provides search times comparable to the ST based techniques, for the sequences it can handle. We do not include the disk based DC3 technique in our experiments for two reasons. First, although superior to the previously known disk based SA techniques [8], DC3 does not perform well on a single processor machine and in such conditions is inferior to TDD in terms of construction time, as shown in [29]. Second, DC3 generates only a *basic* SA. The search complexity for EMS algorithm based on basic SA is $O(m\log n + z)$, where n is typically large in the biological sequence domain. The disparity in the search complexity between ST based and basic SA based solution becomes even wider for more involved search tasks. Other disk based SA techniques either lack scalability or exhibit slower construction time, and hence we do not consider them.

In the following sections we discuss the index representations, construction and search algorithms employed in each of the four alternative techniques.

3   Index Representations

In this section, we review the index representations considered in this work. We start with STTD32, proposed in [12]. We will then consider TDD [29]. Next, we introduce our STTD64 representation which overcomes the limitations of STTD32 and TDD. Last, we review the ESA [1] index representation. We illustrate our discussion by considering an input sequence D = AGAGAGCTT$.

3.1   STTD32 Index Representation

A high-level, graphical illustration of the ST for D is shown in Figure 1 in which the numbers in squares enumerate the tree nodes. Each edge is labeled with the corresponding characters from D. The number below each leaf node s gives the starting position in D at which the suffix indicated by the edge labels on the path from the root to s can be found. Figure 2 shows the actual STTD32 representation of ST. We next review some concepts and definitions for STTD32, taken from [12].

For a leaf node s in a suffix tree, the leaf set of s, denoted $l(s)$, contains the position i in text D at which we can find the string denoted by the edge labels from the root to s. For example, for leaf node 15 in Figure 1, we have that $l(15)=\{3\}$. For a branching node u in a ST, the leaf set of u is defined based on the leaf sets of the children of u, i.e., $l(u) =\{l(s)|\ s\ is\ a\ leaf\ node\ in\ the\ subtree\ rooted\ at\ u\}$. For instance, for node 8 in Figure 1, its leaf set would be $l(8)=\{l(14),\ l(15)\}=\{1,3\}$. There is a total order $\prec$ defined on the nodes in the tree, as follows. For any pair of nodes v and w which are children of the same node u, we define $v \prec w$ if and only if $min\ l(v) < min\ l(w)$. For example, we have that node 6 $\prec$ node 8, since $min\ l(6) = 0 < min\ l(8) =1$. For a node v, its *left pointer*, denoted $lp(v)$, is defined as $min\ l(v)$ plus the number of characters on the path from the root to the parent of v. For example, $lp(8) = 1+1=2$. A branching node u in the ST occupies two adjacent STTD32 elements. The first element contains the lp value of u and two additional bits, called the *rightmost bit* and the *leaf bit*. If the rightmost bit is 1, it indicates that u is the last (w.r.t. $\prec$) node at this level. This is indicated in Figure 2 by the superscript R in the corresponding elements. For every branching node, the leaf bit is always 0. In the second STTD32 element, allocated for a branching node u, we store the *firstchild* pointer. It points to the position in STTD32 at which the first child node of u (w.r.t. $\prec$) is stored. For example, in Figure 2, the branching node 2 is stored in the first two elements in STTD32, i.e., STTD32[0] and STTD32[1]. Its first child is a branching node 6 (see Figure 1). Hence, the value in the second element allocated for node 2 (i.e., STTD32[1] = 8) points to position STTD32[8], which is the first of the two STTD32 elements that store node 6. The arrows above the STTD32 in Figure 2 emphasize these pointers for illustration only. A leaf node in ST occupies a single element in STTD32, in which we store the same information as in the first element allocated for a branching node: the *lp* value, the leaf bit (always set to 1), and the rightmost bit. The leaf nodes in Figure 2 are shown in grey. For the STTD32 in this figure, it records branching node 2 in the first two STTD32 elements, branching node 3 is recorded in the next two elements, leaf node 1 is recorded in STTD32[4], etc.

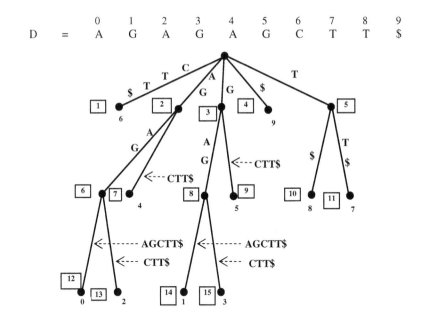

Fig. 1. ST for Sequence AGAGAGCTT$

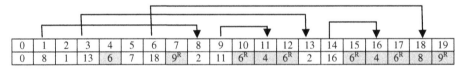

Fig. 2. STTD32 Representation of ST for AGAGAGCTT$

However, STTD32 representation has the following limitation. The lp value has a range between 0 and $|D| = n$. Since 2 bits are reserved for storing the *rightmost bit* and the *leaf bit*, 30 bits are only available in STTD32 for storing the lp value. This limits the size $|D|$ of the data which STTD32 can handle to about 1 billion characters. Both TDD and STTD64 overcome this limitation, as described below.

3.2 TDD Representation

The TDD representation overcomes the 1 GB limitation as follows. Introduced are two additional bitmap arrays, one for storing the leaf bits, and the other for storing the rightmost bits. In this way, for each ST node, all 32 bits are available for recording its lp value. However, this approach introduces some problems. The two additional bitmap arrays are kept fully in memory, which significantly reduces the size of available memory. As a result, the construction and search algorithms perform additional I/O operations, leading to decreased performance. In order to provide an efficient and scalable indexing technique, we propose the STTD64 representation.

3.3 STTD64 Representation

Since STTD64 (64 bit) is inspired by its 32 bit counterpart, the two indexing schemes share some common properties. First, in STTD64 the ST nodes are evaluated and recorded in the same top-down manner as in STTD32. Second, the pointer value associated with each branch node u in STTD64 points to the first child of u. Finally, in both ST representations, we store the leaf and the rightmost bits for every ST node.

These two indexing schemes differ in several ways. First, every node in STTD64 occupies a single array element, regardless of being a branch or a leaf node. Second, each node in STTD32 is 4 bytes, while it is 8 bytes in STTD64. Figure 3 (a) and (b) depict the structures of branch and leaf nodes in STTD64, respectively.

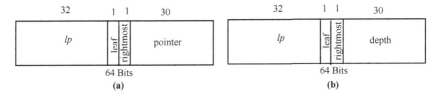

Fig. 3. (a) Branch Node in STTD64 (b) Leaf Node in STTD64

For both types of nodes, the first 32 bits store the lp value, thus overcoming the 1 GB limit. Bit 33 records the leaf bit value and bit 34 records the rightmost bit value. The last 30 bits available for a branch node are used by STTD64 to store the pointer to the first child. The third, and most important difference between STTD32 and STTD64, is that in the latter we use the last 30 bits available for a leaf node to store its *depth*. The depth of a leaf node s is defined as the number of characters on the path from the root of the ST to the parent of s. For example, for the leaf node 14 in Figure 1, the depth is 3 = |GAG|. In Section 5, we discuss the theoretical advantages of recording the depth values. Practical implications are discussed in Section 6.

For the input sequence D, Figure 4 shows the corresponding STTD64 representation. The elements in STTD64 (enumerated in the first row in Figure 4) correspond to the following nodes shown in Figure 1: 2, 3, 1, 5, 4, 6, 7, 12, 13, 8, 9, 14, 15, 11, and 10. The 32 bit lp values are shown in the middle row, and the *pointer/depth* values are shown in the last row. The same convention as for the STTD32 is used to mark a leaf node, a rightmost node, and branch node pointers.

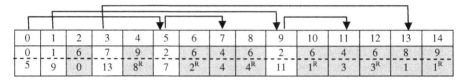

Fig. 4. STTD64 Representation of ST

The improvement of STTD64 over STTD32 comes at a price: increased storage space. For a given sequence D with n characters, in the worst case STTD64 requires 16 bytes per input character. Considering various types of input data, our experiments results reported in Section 6 indicate that STTD64 on average requires about 13 bytes per character in D, thus making it comparable with ESA.

3.4 ESA Representation

An ESA index is a collection of 10 tables (including the basic SA and lcp table) containing various information about the sequence. The average index size observed in our experiments is 13 bytes per character in D. See [1] for details on these tables.

4 Construction Algorithms

The disk based TDD and STTD32 construction algorithms extend the memory based *wotdeager* algorithm [12], by introducing partitioning and buffering techniques. Our STTD64 algorithm is an extension of STTD32. Since the three construction algorithms share a significant overlap, we illustrate them using Figure 5. Both TDD and STTD32 construction algorithms execute Phase 1 and Phase 2. STTD64 performs an additional step, Phase 3, where the depth values are recorded. This additional phase is sequential and requires $O(n)$ time. Thus, all construction algorithms have a worst case time complexity of $O(n^2)$.

5 Exact Match Application

While our focus is on efficient ST representation, evaluation of such proposals should be complemented by a study of performance gains observed for various search applications which use the index as an entry point to the underlying data.

ST Construction Algorithms (*text D, prefixlen*)

Phase1: [Partitioning]
  Scan *D* and based on the first *prefixlen* symbols (default = 1)
  of each suffix, distribute all suffixes to their proper *Partition*.
Phase2: [Construction]
  1. For each *Partition* do the following:
  2. Populate *Suffixes* from current partition
  3. Sort *Suffixes* on the first symbol using *Temp*
  4. while there are more *Suffixes* to be processed
  5.   Output leaf nodes to the *Tree* and
       if a branching node is encountered, process it as follows:
  6.   Push the remaining *Suffixes* onto the Stack.
  7.   For the current branching node, compute the Longest Common
       Prefix (LCP) of all the *Suffixes* in this range by checking *D*
  8.   Sort *Suffixes* on the first symbol using *Temp*
       end while
  9. while Stack is not empty
  10. Pop the unevaluated *Suffixes* from Stack
  11. Go to step 3.
       end while
end TDD/STTD32
Phase 3: [Depth Filling]
  12. Traverse the *Tree* by following the branch node pointers and
      compute and record the *depth* value at each leaf node
end STTD64

Fig. 5. Construction Algorithms for TDD/STTD32/STTD64

In this paper we restrict the study of search performance to the exact match search (EMS) problem, i.e., finding all the positions in *D* which match exactly a query *P*. The restriction to EMS is justified as follows. From a practical point of view, EMS is at the core of numerous exact and approximate search applications. For example, in computational biology, it has several standalone applications, e.g., finding different types of repeats and palindromes, searching for motifs, etc. Also, EMS is seen as a basic step in more complex applications as in BLAST [2, 3], a popular similarity search tool for biological sequences. While our focus in this paper is to assess the performance of our technique using EMS, our ultimate goal, as mentioned earlier, is to provide a single solution to support this and other search operations efficiently.

For TDD and ESA, we use the EMS algorithms provided as part of these software packages. The EMS algorithms for STTD32 and STTD64 proceed as follows. In phase 1, starting from the root, EMS traverses to lower levels of the ST by subsequently matching characters in *P* to the labels of the ST branches. Upon successful matching of all *P* characters, the last ST node found is called *answer root*, and the subtree rooted at that node is called *answer subtree*. For example, for our sequence *D* and query *P* = 'A', the answer root found is node 2, and leaf nodes 12, 13, and 7 represent the set of all answers to *P* (see Figure 1). In phase 2, each leaf node *s* in the answer subtree is visited, and each starting position of *P* in *D*, $occ(s)$, is computed as $occ(s) = lp(s) - depth(s)$, where $lp(s)$ is the *lp* value stored in the ST for *s*, and $depth(s)$ is the number of characters on the path from the root to the parent of *s*.

In the case of STTD32 representation, $depth(s)$ is determined by explicitly traversing the ST from the answer root to *s*. In our previous work [14], we showed that such traversals result in fair amount of jumps between different nodes in the ST.

472   M. Halachev, N. Shiri, and A. Thamildurai

For longer sequences, for which the answer subtrees do not fit in memory, these traversals will lead to a considerable number of ST buffer misses and excessive disk I/Os. In a separate study, we performed numerous experiments to determine the most appropriate ST buffer size. We found that the best performance for STTD32 EMS is achieved using a buffer size that is half the size of the cache, i.e., 1MB in our computing environment and this buffer size is used in our experiments.

For the proposed STTD64 representation, there is no need to traverse the ST in order to compute $depth(s)$; these values are available in the last 30 bits in a leaf node. Thus, once the answer root is found, in phase 2 EMS performs a sequential read of the leaf nodes in the answer subtree. For each such node s, the corresponding $occ(s)$ is computed by subtracting the $depth(s)$ from the $lp(s)$ value.

To illustrate the EMS algorithm using STTD64, consider our running example of D and query P = 'A'. In phase 1, EMS finds the answer root to be node 2 (Figure 1), recorded at array position STTD64 [0] (Figure 4). In phase 2, EMS reads all ST nodes within the answer subtree boundaries and finds the leaf nodes 7, 12, and 13, recorded at STTD64[6], STTD64[7], and STTD64[8], respectively. Using these leaf nodes we get all the positions in D where P occurs: STTD64[6].lp – STTD64[6].$depth$=6 – 2=4; STTD64[7].lp – STTD64[7].$depth$ = 4 – 4 = 0; and STTD64[8].lp – STTD64[8].$depth$ = 6 – 4 = 2.

The major advantage of having the $depth$ values in STTD64 is the increased efficiency gained by essentially making the disk accesses in phase 2 of EMS focused and sequential, which otherwise (i.e. for STTD32) are dispersed and random. It can be proved that the total number of disk reads performed by STTD64 EMS algorithm is $O(|\Sigma|*m + z)$, where $|\Sigma|$ is the alphabet size. This includes $|\Sigma|*m$ random reads and z sequential (bulk) reads, where z is much larger than $|\Sigma|*m$.

6   Experiments and Results

In this section, we describe the experiments carried to evaluate performance of the indexing techniques, and compare them based on their construction times, storage requirements, and exact match search times. We used a standard desktop computer with Intel Pentium 4 @ 3 GHz, 2GB RAM, 300GB HDD, and 2MB L2 cache, running Linux kernel 2.6. All I/O operations are un-buffered at the OS level. All times reported are in seconds and storage requirements reported are in MB.

Vmatch [18] is a commercial software tool (free to academia) that implements ESA. For TDD, we used the C++ source code, available online at the authors' site [29]. We have implemented the STTD32 and STTD64 construction and search algorithms in C.

To study scalability of these indexing techniques, we classify our experiments based on the input sequence size into: (1) short sequences of up to 250MB, (2) medium size sequences of up to 1GB, and (3) long sequences with more than 1GB symbols. Let us call them as type 1, type 2, and type 3, respectively. The basis for this classification is the restriction on the size of the input sequence that could be handled by these techniques. To study the effect of the alphabet size, we have considered, as done in [1, 29], different types of data (DNA, protein, as well as natural texts).

6.1   Type 1 Sequences (Up to 250MB)

The 250MB boundary was considered for being the practical limit for ESA on our desktop. We consider the following sequences: (1) all 24 human chromosomes (DNA data, [27]); (2) the entire SwissProt database (protein data, [9]); and (3) natural texts from the Gutenberg project [28]. Figure 7 shows the sizes of the DNA sequences, whose alphabet size is 4, as in [1, 15, 29]. For SwissProt database, we removed header lines, new line symbols, and blanks. The resulting sequence, called *sprot*, is about 75 million characters over an alphabet of size 23. The natural text sample was cleaned by removing non-printable symbols. The resulting sequence, called *guten80*, contains the first 80 million characters from [28] over an alphabet of 95 symbols.

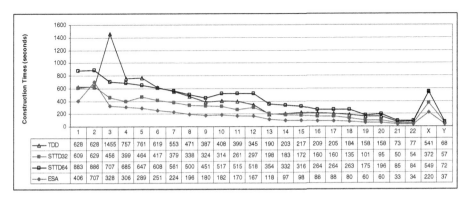

	1	2	3	4	5	6	7	8	9	10	11	12	13	14	15	16	17	18	19	20	21	22	X	Y
TDD	628	628	1455	757	761	619	553	471	387	408	399	345	190	203	217	209	205	184	158	158	73	77	541	68
STTD32	609	629	456	399	464	417	379	338	324	314	261	297	198	183	172	160	160	135	101	95	50	54	372	57
STTD64	883	886	707	685	647	608	561	500	451	517	515	518	354	332	316	264	264	263	175	196	85	84	549	72
ESA	406	707	328	306	289	251	224	196	180	182	170	167	118	97	98	88	88	80	60	60	33	34	220	37

Fig. 6. Construction Times for Human Chromosomes

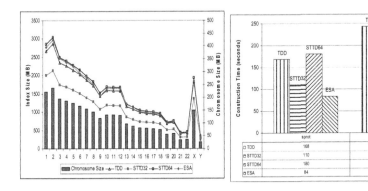

Fig. 7. Human Chromosome & Index Sizes   **Fig. 8.** Construction time for *sprot* and *guten80*

The index construction algorithms applicable for type 1 sequences are ESA, TDD, STTD32, and our proposed STTD64. Figure 6 shows the construction time for DNA data and Figure 8 shows this time for protein and natural text. It can be seen that ESA has the fastest index construction time for type 1 sequences. This is explained by the fact that ESA is a memory based technique, which uses no buffering and hence has no buffer overhead. STTD32 is faster compared to TDD due to better utilization of available memory (STTD32 does not use bitmaps). On the other hand, TDD has a slight advantage over STTD64 in construction time, due to its smaller index size.

We also looked at storage requirements of the various indexing techniques we considered. For DNA data ($|\Sigma| = 4$), STTD32 requires $9.5n$ bytes per character on average, while TDD and STTD64 require on average $12.7n$ and $13.4n$ respectively (Figure 7). We noted that for the ST based approaches, a larger alphabet size always results in reduced storage requirements due to larger branching factor, and hence smaller number of branching nodes. For *sprot* with $|\Sigma| = 23$, the storage using STTD32, TDD, and STTD64 are $8.5n$, $11.8n$, and $12.5n$, and for *guten80* ($|\Sigma| = 95$) they are $8.1n$, $11.4n$, $12.1n$, respectively (Figure 9). The ESA storage requirements we measured for DNA data, *sprot*, and *guten80* are $13.1n$, $13.3n$, and $12.2n$, respectively. It should be noted that ESA needs only three of its tables (of total size $5n$) for exact match search. However, to support a variety of applications, it needs its other tables, resulting in increased memory requirement.

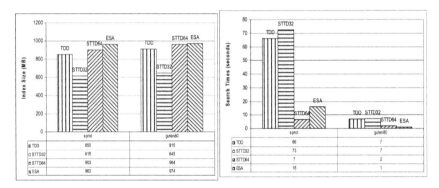

Fig. 9. Index Sizes for *sprot* and *guten80*   **Fig. 10.** Search Times for *sprot* and *guten80*

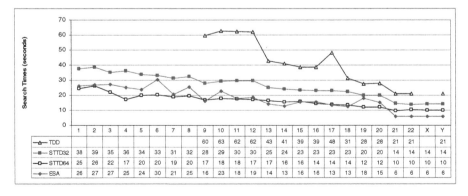

Fig. 11. Search Times for Human Chromosomes

We now discuss the conducted search experiments. The query set for the individual chromosomes is generated as follows. First, we randomly sample each of the 24 chromosomes to obtain 25 substrings of size 7 characters, 25 of size 11, and 25 of size 15. These sizes are the same as query words used in BLAST. In addition to these 75 words per chromosome, we also include in the query set their reverse (as done in [1]), in order to simulate the cases for which there is no exact match of query in D. Thus our query set contains a total of 3600 (150 queries per chromosome) words of sizes 7, 11, and 15 characters each. The query set for *sprot* is obtained in similar way and it contains 500 query words of size 2 characters and 500 words of size 3 characters. The query word sizes are the same as used in BLAST. The query set for *guten80* contains the 100 most frequent words found in the Gutenberg collection, and each word is of size greater than 3 characters. In all our experiments, the search time reported includes the time spent to compute all the positions in the database where exact matches occur, however their output is suppressed.

The search performances of STTD64 and ESA are comparable (Figures 10 and 11), which is an important result to note since the former is a disk based technique, while the latter is memory based. This result validates our idea of storing and using the *depth* value in STTD64, which leads to mostly sequential, bulk reads from the disk during the search. The benefit of the *depth* value is further illuminated when comparing STTD64 to STTD32, where the former is 1.5 times faster on average than the latter for DNA data. The superiority of STTD64 and ESA over STTD32 and TDD becomes even clearer when considering protein data and natural texts, on which we observe an order of magnitude speedup (Figure 10).

While for protein and natural text data, TDD has a performance comparable to STTD32 (Figure 10), for DNA data (Figure 11), we found that TDD to be the slowest of all. Further, TDD could not be used to search in larger chromosomes (e.g., chr 1 to 8 and chr X). This is because the EMS algorithm provided for TDD is memory based, and for larger chromosomes the available 2GB memory is insufficient.

6.2 Type 2 Sequences (Up to 1 GB)

Type 2 sequences are of size between 250 million and 1 billion symbols. The upper bound is the theoretical limit of STTD32 representation, as described in Section 3.1. In this set of experiments, we consider the entire TrEMBL database (protein data, [9]), and the entire Gutenberg collection. We preprocessed the data as explained above to obtain the *trembl* sequence of about 840 million symbols, and the *guten* sequence of about 400 million symbols. The techniques that are applicable for this input size range are STTD32, STTD64, and TDD. The query sets used were obtained by sampling the *trembl* and *guten* databases in a similar way as for type 1 sequences.

The construction times for sequences with large alphabet size, *trembl* and *guten*, are shown in Figure 12, and are consistent with corresponding results for type 1 sequences. However, the inefficiency of TDD bitmap arrays was further illuminated when we tried to construct the ST for DNA sequence ($|\Sigma| = 4$) of size 440MB (chr1+chr2). While STTD64 completed the task in 30 minutes, we stopped execution of TDD after hour and a half, in which only a small part of TDD ST was completed.

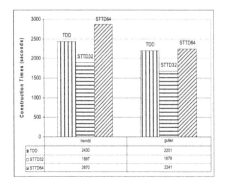

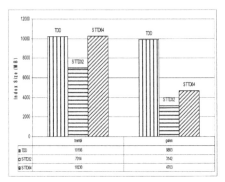

Fig. 12. Construction Times for *trembl* and *guten*   **Fig. 13.** Index Sizes for *trembl* and *guten*

The storage requirements for TDD, STTD32, and STTD64 are shown in Figure 13. While the index sizes for STTD32 and STTD64 are consistent with those obtained for short sequences, the size of the TDD index for *guten* sequence was found to be 25.5*n*, which was not expected.

The measured search times are reported in Figure 14. For *trembl*, the search using STTD64 is twice faster than STTD32. These two indexing techniques show similar search performance for *guten*. Because of the large alphabet size (95 characters) of this sequence, the height of the ST is relatively small. This leads to relatively small answer subtrees which would fully fit in the main memory, thus taking away the benefit of the *depth* values stored in STTD64. TDD search algorithm is memory based and is not applicable for type 2 sequences.

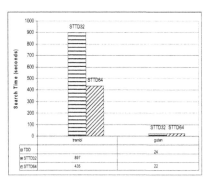

Sequence Size	Short < 250MB	Medium (250MB – 1GB)	Large > 1GB
ESA	√	NA	NA
TDD	√	√	√
STTD32	√	√	NA
STTD64	√	√	√
Index Size	STTD32	STTD32	?
Construction	ESA	STTD32	STTD64
Search	ESA (or) STTD64	STTD64	STTD64

Fig. 14. Search Times for *trembl* and *guten*   **Fig. 15.** Ranking of Indexing Techniques

6.3 Type 3 Sequences (above 1 GB)

For the largest input sequence we consider the entire human genome obtained by concatenating the 24 human chromosomes to obtain a sequence of size 2.8 GB. This sequence is compressed by encoding each character as a 4 bit binary value as done in

[29]. The only applicable techniques for type 3 are TDD and STTD64. The TDD code available to us does not support indexing sequences of such sizes. For comparison purposes, we use the TDD construction time reported in [29] which is 30 hours. We noted that running TDD on our machine for sequence types 1 and 2 results in faster construction time compared to those reported in [29], due to hardware differences. Using STTD64, we could construct the ST for the entire human genome in 16 hours. The improved construction time of STTD64 is the result of better utilization of the available main memory, obtained by using dynamic buffering and avoiding the leaf and rightmost bit arrays used in TDD. The size of the STTD64 index for the entire human genome is about 36GB (i.e., 12.8 bytes per character).

7 Conclusions and Future Work

We studied the representation and indexing techniques for sequence data, and evaluated their performance on construction time, space requirement, and search efficiency. As a first step, we considered the most space efficient ST representation, STTD32 [12] and developed a disk based construction and search algorithms. The comparison between STTD32 and the best known disk based ST technique TDD [29] show the superiority of STTD32. However, STTD32 representation has two limitations: (i) the size of the sequences it can handle is at most 2^{30} characters; (ii) the number of disk I/Os during search is high, due to excess traversals of the nodes in the ST. To overcome these problems, we proposed a new ST representation, STTD64, which indexes larger sequences and reduces disk I/Os by storing the *depth* values for the leaf nodes. While STTD64 has an increased construction time and storage space compared to STTD32 and TDD, it exhibits improved search efficiency and better scalability for larger sequences. We also compared the performance of STTD64 with a state-of-the-art suffix array based solution, ESA [1]. Our experimental results show that both approaches provide comparable search times and have similar storage requirements for short sequences. While ESA has an advantage over STTD64 in its construction time, it is a memory-based technique, thus limiting its application to sequences up to 250 million symbols on a typical desktop with 2GB RAM.

Figure 15 summarizes our findings in this study. For each sequence size type, the table indicates the applicable techniques, using a check mark ($\sqrt{}$). For each considered criteria, the table indicates the superior technique. We did not find any report on the TDD index size for the entire human genome and thus no comparison for index size is possible for type 3 sequences.

While we used exact match search as a measure for performance evaluation, the development of approximate search algorithms as additional measures to evaluate alternative indexing techniques is essential. The *depth* value stored in the leaf nodes of STTD64 representation is similar to the *lcp* value in SA. As the computation of the *lcp* values between all pairs of lexicographically subsequent suffixes is important for many approximate matching solutions [13], the presence of *depth* value might be of potential advantage. We believe it would result in more efficient approximate matching (k-mismatch search, k-difference search), which we are currently investigating.

Acknowledgments. This work was in part supported by NSERC Canada, Genome Quebec, and by Concordia University. We thank S. Kurtz for making executable of *vmatch* available. We also thank S. Tata for providing details on TDD.

References

1. Abouelhoda M., Kurtz S., and Ohlebusch E. Replacing Suffix Trees with Enhanced Suffix Arrays. In *Journal of Discrete Algorithms*, 2:53-86, 2004.
2. Altschul S.F., Gish W., Miller W., Myers E.W., and Lipman D.J. Basic Local Alignment Search Tool. In *J. Mol. Biol.*, 215:403-10, 1990.
3. Altschul S. F., Madden T. L., Schaffer A. A., Zhang J., Zhang Z., Miller W., and Lipman D. J. Gapped BLAST and PSI-BLAST: a new generation of protein database search programs. In *Nucleic Acids Research*, 25:3389-3402, 1997.
4. Andersson A. and Nilsson S. Efficient Implementation of Suffix Trees. In *Softw. Pract. Exp.*, 25(2):129-141, 1995.
5. Burkhardt S., Crauser A., Ferragina P., Lenhof H. P., Rivals E., and Vingron M. q-gram Based Database Searching Using a Suffix Array. In *RECOMB*, pp.77-83. ACM Press, 1999
6. Crauser A. and Ferragina P. A Theoretical and Experimental Study on the Construction of Suffix Arrays in External Memory. In *Algorithmica*, 32(1):1-35, 2002.
7. Delcher A. L., Kasif S., Fleischmann R.D., Peterson J., White O., and Salzberg S.L. Alignment of Whole Genomes. In *Nucleic Acids Research*, 27:2369-2376, 1999.
8. Dementiev R., Kärkkäinen J., Mehnert J., and Sanders P. Better External Memory Suffix Array Construction. In *Proc. 7th Workshop Algorithm Engineering and Experiments*, 2005.
9. ExPASy Server, http://us.expasy.org/
10. Ferragina P. and Grossi R. The String B-Tree: A New Data Structure for String Searching in External Memory and its Applications. In *Journal of the ACM* 46(2), pp. 236-280, 1999.
11. GenBank, http://www.ncbi.nlm.nih.gov/Genbank/index.html
12. Giegerich R., Kurtz S., and Stoye J. Efficient Implementation of Lazy Suffix Trees. In *Softw. Pract. Exper.* 33:1035-1049, 2003.
13. Gusfield D. Algorithms on Strings, Trees and Sequences: computer science and computational biology. *Cambridge University Press*, 1997.
14. Halachev M., Shiri N., and Thamildurai A. Exact Match Search in Sequence Data using Suffix Trees. In *Proc. of 14th ACM Conference on Information and Knowledge Management (CIKM)*, Bremen, Germany, 2005.
15. Hunt E., Atkinson M.P., and Irving R.W. A Database Index to Large Biological Sequences. In *VLDB J.*, 7(3):139-148, 2001.
16. Jagadish H. V., Koudas N., and Srivastava D. On Effective Multi-dimensional Indexing for Strings. In *ACM SIGMOD Conference on Management of Data*, pp. 403-414, 2000.
17. Kurtz S. Reducing the Space Requirement of Suffix Trees. In *Software-Practice and Experience*, 29(13): 49-1171, 1999.
18. Kurtz S. Vmatch: large scale sequence analysis software. http://www.vmatch.de/
19. Kurtz S., Schleiermacher C. REPuter: Fast Computation of Maximal Repeats in Complete Genomes. In *Bioinformatics*, pages 426-427, 1999.
20. Manber U. and Myers G. Suffix Arrays: a new method for on-line string searches. In *SIAM J. Comput.*, 22(5):935-948, 1993.
21. McCreight E.M. A Space-economical Suffix Tree Construction Algorithm. In *Journal of the ACM*, 23(2):262-272, 1976.

22. M. Morgante, A. Policriti, N. Vitacolonna, and A. Zuccolo. Structured Motifs Search. In *Proc. of RECOMB '04*, 2004.
23. Morrison D. R. PATRICIA - Practical Algorithm to Retrieve Information Coded in Alphanumeric. In *Journal of the ACM*, 15(4), pp514-534, 1968.
24. Navarro G. A Guided Tour to Approximate String Matching. In *ACM Computing Surveys*, 33:1:31-88, 2000.
25. Navarro G. and Baeza-Yates R. A Practical q-gram Index for Text Retrieval Allowing Errors. In *CLEI Electronic Journal*, 1(2), 1998.
26. Navarro G., Sutinen E., Tanninen J., and Tarhio J. Indexing Text with Approximate q-grams. In *CPM2000*, LNCS 1848, pp. 350-365, 2000.
27. NCBI: National Center for Biotechnology Information, http://www.ncbi.nlm.nih.gov/
28. Project Gutenberg, http://www.gutenberg.org
29. Tian, Y., Tata S., Hankins R.A., and Patel J. Practical Methods for Constructing Suffix Trees. In *VLDB Journal,* Vol. 14, Issue 3, pp. 281–299, Sept. 2005.
30. Ukkonen E. On-line Construction of Suffix trees. In *Algorithmica 14*(1995), 249-260, 1995
31. Weiner P. Linear Pattern Matching Algorithms. In *Proc. 14th Annual Symp. on Switching and Automata Theory*, 1973.
32. Witten I. H., Moffat A., and Bell T. C. Managing Gigabytes: Compressing and Indexing Documents and Images. *Morgan Kaufmann*, 2nd edition, 1999.

Author Index

Lecture Notes in Bioinformatics